Soyuz

A Universal Spacecraft

Springer
London
Berlin
Heidelberg
New York
Hong Kong
Milan
Paris
Tokyo

Rex D. Hall and David J. Shayler

Soyuz

A Universal Spacecraft

Springer

Published in association with
Praxis Publishing
Chichester, UK

Rex D. Hall
Education Consultant
Council Member of the BIS
London
UK

David J. Shayler
Astronautical Historian
Astro Info Service
Halesowen
West Midlands, UK

SPRINGER–PRAXIS BOOKS IN ASTRONOMY AND SPACE SCIENCES
SUBJECT *ADVISORY EDITOR*: John Mason B.Sc., Ph.D.

ISBN 1-85233-657-9 Springer-Verlag London Berlin Heidelberg New York Hong Kong Milan Paris Tokyo

British Library Cataloguing-in-Publication Data
Hall, Rex
Soyuz: a universal spacecraft. – (Springer-Praxis books in astronomy and space sciences)
1. Space vehicles – Russia (Federation) 2. Space vehicles – Soviet Union 3. Astronautics – Russia (Federation) 4. Astronautics – Soviet Union 5. Salyut-Soyuz Project
I. Title II. Shayler, David, 1955–
629.4′0947

ISBN 1-85233-657-9

A catalogue record for this book is available from the Library of Congress.

Printed by MPG Books Ltd, Bodmin, Cornwall, UK

Copy editing and graphics processing: R. A. Marriott
Cover design: Jim Wilkie
Typesetting: BookEns Ltd, Royston, Herts., UK

Printed on acid-free paper supplied by Precision Publishing Papers Ltd, UK

This book is dedicated to the inspiration of OKB-1 Chief Designer

Sergei P. Korolyov (1917–1966)

To the memory of pioneering Soyuz cosmonauts

Vladimir M. Komarov (1927–1967)
Georgi T. Dobrovolsky (1928–1971)
Vladislav N. Volkov (1935–1971)
Viktor I. Patsayev (1933–1971)

And to the unsung thousands of workers who throughout more than four decades have participated in the design, planning, construction, preparation and support for each flight of the remarkable Soyuz spacecraft

Other books by the same authors in this series

Rex D. Hall and David J. Shayler
The Rocket Men (2001). ISBN 1-85233-391-X

David J. Shayler
Disasters and Accidents in Manned Spaceflight (2000). ISBN 1-85233-225-5
Skylab: America's Space Station (2001). ISBN 1-85233-407-X
Gemini: Steps to the Moon (2001). ISBN 1-85233-405-3
Apollo: The Lost and Forgotten Missions (2002). ISBN 1-85233-575-0

Table of contents

Foreword

The Soyuz spacecraft is nearly forty years old. The first designs were produced in the Design Burean OKB-1 (later reorganised into NPO and then into RKK Energiya, named after Academician S.P. Korolyov in 1962). I did not take an entire course of training for the Voskhod 3 flight, as the mission was cancelled for a number of reasons. Later, all my activity was connccted with testing and carrying out successful Soyuz missions.

In 1969 I flew and commanded Soyuz 4 and performed the first docking with the manned Soyuz 5 spacecraft. The docking enabled us to carry out the transfer of two cosmonauts, as planned for domestic Moon programme; and the same was to be tested during the missions of Soyuz 6, 7 and 8. During my third flight, onboard Soyuz 10 in 1971, we tested the new docking unit designed for the orbital Salyut station.

From 1971 to 1991 I was responsible for the selection, training and crewing of the cosmonauts. I was in charge of the improvements in the teaching and training facilities, and all of the cosmonauts' activities. I was assistant to the Commander-in-Chief of the Air Force, and I later became Head of the Cosmonaut Training Centre named for Yuri Gagarin.

Soyuz is a wonderful spacecraft which has allowed us to achieve many remarkable firsts. It has proved very adaptable in meeting many different types of mission profile, and at present it is being used to support operations on the ISS.

I am very proud of the Soviet and Russian space achievements and of my role in the Soyuz programme, and I am very pleased that it is included in this book.

Vladimir Alexandrovich Shatalov
Twice Hero of the Soviet Union
Pilot Cosmonaut of the USSR
Lieutenant-General of Aviation

18 August 2002

Lieutenant-General of Aviation Vladimir A. Shatalov, Twice Hero of the Soviet Union, and Pilot Cosmonaut of the USSR.

Космическому кораблю Союз уже около 40 лет. Первые разработки Проводились в конструкторском бюро ОКБ-1 (позже преобразованном в НПО, а затем в РКК Энергия имени академика С.П. Королева), начиная с 1962 года. Мне не довелось пройти весь объем подготовки к полету на корабле Восход-3, который в силу ряда причин так и не состоялся. После закрытия этой Программы вся моя деятельность была связана с испытаниями и обеспечением дальнейшей эксплуатации кораблей Союз.

Так в 1969 г. в качестве командира корабля Союз-4 была выполнена первая стыковка в космосе с пилотируемым кораблем Союз-5, что позволило осуществить внешний переход двух космонавтов, как это предусматривалось нашей лунной программой. Эти же задачи предусматривались во втором полете трех кораблей Союз-6,-7,-8. В третьем полете на корабле Союз-10 в 1971 г. проводились испытания нового стыковочного узла, предназначенного для обеспечения внутреннего перехода из транспортного корабля на борт орбитальной станции Салют.

С 1971 года по 1991 год я отвечал за отбор космонавтов, их подготовку, формирование экипажей, постоянное совершенствование учебно-тренировочной базы и деятельности космонавтов вначале как помощник Главнокомандующего ВВС страны, а затем совместив эту должность и обязанности начальника ЦПК им. Ю.А. Гагарина.

Союз – прекрасный космический корабль, позволивший нам установить многочисленные выдающиеся приоритеты в космосе. Он доказал прекрасную возможность стыковки с различными типами космических объектов. В настоящее время он используется для обеспечения работ на МКС.

Я горжусь достижениями советской и российской космонавтики и своей ролью в этих успехах. Высоко ценю роль, которую сыграл космический корабль Союз, и рад быть действующим лицом в этой книге.

Владимир Александрович Шаталов
Дважды Герой Советского Союза
летчик – космонавт СССР
генерал-лейтенант авиации

18.09.2002

Authors' preface

The human exploration of space began in April 1961, with Yuri Gagarin's 108-minute flight onboard Vostok. Between 1961 and 1965, ten other cosmonauts flew onboard Vostok-type vehicles in Earth orbit, before a new family of manned spacecraft called Soyuz was introduced in 1967. The Soyuz spacecraft had a multiple role that included the potential to send cosmonauts around the Moon.

In America, the one-man Mercury spacecraft that flew between 1961 and 1963 was replaced by the two-man Gemini (1965–66), and subsequently by the three-man Apollo (1968–75). Since 1981, all but three American astronauts have entered space by means of the Space Shuttle – the fourth-generation US manned spacecraft. In contrast, the Russian shuttle – Buran – never flew manned, and so the majority of Russian cosmonauts have relied upon only their second-generation manned spacecraft – the versatile Soyuz. For almost forty years, Soyuz has been the mainstay of Russian manned access to space, with just a handful of cosmonauts experiencing spaceflight on the American Space Shuttle. For a spacecraft designed during the height of the Cold War and the dawn of the space age, Soyuz continues to be an integral part of human space exploration; and as an essential element of the International Space Station it could become the only spacecraft to celebrate forty years of manned flight operations.

This book relates the story of that remarkable spacecraft, evolved from design studies of Vostok and adapted to meet a wide range of roles – both unmanned and manned, and civilian and military in nature – for Earth orbital and lunar applications, in independent flight and as a resupply craft for space stations.

In a continuation of the story of Russian manned spaceflight from the pioneering era of Vostok and Voskhod described in *The Rocket Men* (Praxis–Springer, 2001), we describe the design and development of the next generation of Russian manned spacecraft, and how the original ground support and launch vehicle hardware was adapted for the new spacecraft. We also cover the selection and training of new groups of cosmonauts to crew these missions.

The story of Soyuz as a spacecraft is complex, but can be separated into distinct phases. It began with design studies during the early 1960s and the initial ill-fated test flights and military applications, before advancing to the equally frustrating manned lunar programme in the late 1960s. From 1971, Soyuz was primarily used as a space station ferry – initially to Salyut space stations, then from 1986 to the Mir complex, and from 2000 to the International Space Station.

During this varied career, Soyuz has undergone several upgrades to improve onboard systems and facilities, and each of these primary variants is discussed in separate sections of this books: the early test flights, the solo scientific research missions, the original two-man space station ferry, a return to a crew of three, unmanned space freighters, and the latest variants supporting ISS operations. As with Vostok, Soyuz has also supported unmanned applications, and these are also discussed, together with an appendix listing all flight data and crew experience as well as major contractors and leading designers of the Soyuz spacecraft.

The name Soyuz – Union – denotes the physical docking of two vehicles; but it is also a political statement relevant to the era in which it evolved. In a long and chequered career, and in all its variants, Soyuz has proved that it is, above all, a truly universal spacecraft.

Rex D. Hall
London, England

David J. Shayler
West Midlands, England

February 2003

www.astroinfoservice.co.uk

Acknowledgements

Initial Western reports of the first Soyuz flight operations were – like those concerning all early Soviet space missions – to say the least, sparse. Researchers had to rely on Novosti Press Agency News Releases or small booklets, articles in *Soviet Weekly* or *Soviet News*, and broadcasts by Radio Moscow, plus occasional access to publications such as *Pravda* and *Izvestia*, as well as the few books published in the Soviet Union. This contrasted sharply with the abundance of American information on the Apollo programme from both NASA and the primary contractors.

The authors' own research included the study of these early reports and releases, together with Western interpretations of the early years in *Aviation Week*, *Flight International* and the publications of the British Interplanetary Society. As the Soyuz programme continued, and from the 1970s expanded to include the Salyut space stations and cosmonauts from other nations, so more detailed information became available (the primary sources are encompassed below). This included information supplied by NASA during the Apollo–Soyuz Test Project, the former members of the Interkosmos group of nations, and commercial space ventures with European and Asian nations. With the demise of the Soviet Union in 1991, the space programme offered an openness that had been restricted under the Communist régime, and further information became available; and the Russian use of Soyuz in Mir operations, and more recently in ISS operations, has been particularly fruitful.

Communications with a world-wide group of Soviet/Russian space-watchers over many years has produced additional information on, and analysis of, Soyuz operations. Foremost of these 'space sleuths' are Colin Burgess, Mike Cassutt, Phil Clark, Brian Harvey, Bart Hendrickx, Gordon Hooper, Neville Kidger, James Oberg, Andy Salmon, Charles Vick and Bert Vis.

This book has benefitted greatly from Bart Hendrickx's knowledge and expertise concerning the complexity of the Russian space programme, translations of numerous documents – including the Kamanin diaries (1960–71) – and his unselfish help and guidance in authenticating many aspects of Soyuz operations. In many ways this is also his book.

Thanks are extended to Lieutenant-General Vladimir A. Shatalov for his Foreword, and to Elena Esina, curator of the museum in the House of Cosmonauts at Star City. We also acknowledge the staff and work of *Novosti Kosmonavtiki*, Russia's leading magazine on space exploration.

We express our thanks to the Council and staff of the British Interplanetary Society, London, for their cooperation. We also recognise the pioneering work of the Kettering Group – notably its founder, Geoff Perry – and that of Charles Sheldon and Marcia Smith of the US Congressional Research Service.

The works by William P. Barry (*The Missile Design Bureaux and Soviet Manned Space Policy, 1953–1970*), Phil Clark (*The Soviet Manned Space Programme*), Nicholas L. Johnson (*Handbook of Soviet Manned Space Flight*), David S.F. Portree (*Mir Hardware Heritage*), Asif Siddiqi (*Challenge to Apollo: the Soviet Union and the Space Race, 1945–1974*), and Tim Varfolomeyev (*Soviet Rocketry that Conquered Space*) were frequently consulted, and are highly recommended for further study.

We appreciate the assistance of the Public Affairs Office at NASA JSC, Houston, Texas, and the staff of the NASA Archive,University of Clear Lake, Houston, in providing access to the NASA archives on Soyuz in the ASTP, Skylab, Space Shuttle and Space Stastion collections.

We also acknowledge the artistic talents of Ralph Gibbons, Charles Vick and Dave Woods, and their kind permission to reproduce their work. Unless otherwise stated, all photographs are from the authors' collections or from NASA. Further information can be consulted on the web sites of Mark Wade (www.astronautix.-com) and Sven Grahn (www.svengrahn).

Special thanks go to Lynn Kelterborn for initial comments and observations on the draft text, and to Mike Shayler for countless hours of pre-editing and preparation of the submitted manuscript. Thanks also go to our Project Editor, Bob Marriott, for his extra efforts in preparing the text and illustrations for publication; and to Clive Horwood and his staff at Praxis, for their unlimited patience on a long road to completion of this book in 2003.

The lifetime of interest, support and encouragement of Derek J. Shayler (1927–2002), in the activities of all his family, is fondly remembered.

List of illustrations and tables

Progress, 1978–

Soyuz T, 1979–86

Soyuz TM, 1986–2002

Soyuz TMA, 2002

Conclusion

Tables

Front cover

A Soyuz TM spacecraft in Earth orbit. The three elements of the spacecraft have remained relatively unchanged for more than forty years: (*left to right*) Orbital Module, Descent Module and Propulsion Module. Soyuz has been the mainstay of Soviet and Russian access to space since 1967.

Back cover

(*Left*) Cosmonaut Anatoli Solovyov and Nikolai Budarin inside the Soyuz TM-21 Descent Module during a check of the systems whilst docked to Mir (the cramped confines of the spacecraft are evident). (*Right*) After a prolonged stay in orbit, a cosmonaut crew experiences the traditional end-of-mission ceremony as they are helped out of the Descent Module by members of the search and rescue teams at the landing site in Kazakhstan.

Prologue

A journey on a Soyuz vehicle into and from space is one of tradition, technology, and thrills, and is completely different from a flight on the US Space Shuttle. The launch-day begins at the cosmonaut hotel in Leninsk, Kazakhstan, with a wake-up call by the crew doctor. Before leaving their rooms for a final 10-minute medical check-up, each member of the prime crew leaves a letter to his/her family, and mentally resolves to perform well on the mission. Wearing lightweight training suits as they leave their hotel rooms, the cosmonauts sign the back of the door – the first tradition of a memorable day that, no matter how complex, remains leisurely and organised, with ample time to complete all the necessary procedures.

Soon they leave the cosmonaut hotel onboard the bus, and proceed towards the Cosmonaut Area, a small group of rooms in the MIK spacecraft preparation building near the launch complex. Here they receive an alcohol rub-down – the final sterilisation of the body before dressing for flight. Wearing flight undergarments, they are each fitted with a chest belt that includes an electrocardiograph, worn through ascent to orbit to monitor their heart-beat rate and breathing pattern. Then the Sokol pressure suits – worn for the ascent and re-entry phases of the flight – are donned, and with the gloves locked onto the arms and the integral helmet/visor closed, the crews are weight-tested and pressure-tested. These final tests are carried out in a seat that faces the curtained Press Area window. When completed, the curtains are pulled back, and the crew – now wearing their open-helmeted and unpressurised spacesuits, with their gloves stuffed into the leg pockets – conduct their final press conference.

They then walk through the MIK building to the outside, and to three adjacent white squares painted on the concrete apron. Each of these squares is reserved for one of the crew: KK (Komandir Korablya) for the Commander, BI (Bort Inzhener) for the Flight Engineer, and KI (Kosmonavt Issledovatel) for the Cosmonaut–Researcher. The crew face the Chairman of the State Commission, who stands on his own square with the initials PGK (Presdsedatel Gosudarstvennoy Kommissii). The spacecraft Commander then salutes, and reports that his crew-members have completed their training and are ready to carry out the assigned flight programme. The Chairman of the State Commission then returns the salute with permission to

Outside the MIK, Soyuz crew-members (standing on three white squares) report to the Chairman of the State Commission (standing on a fourth square, facing them) on their readiness to complete their mission prior to departing on the cosmonaut bus to the pad area.

fly, and wishes them a successful flight and (normally with a smile) a soft landing. This ceremony is attended by a large crowd of technicians, politicians, designers and well-wishers.

With their heads filled with messages of farewell and 'useful' tips (from veteran cosmonauts) for adjusting to weightlessness, they reboard the cosmonaut bus and set off towards the launch pad, stopping along the way to complete another tradition begun by Yuri Gagarin: all male crew-members 'wet' the bus tyres. Arriving at the base of the launch pad, they disembark from the bus and walk towards the base of the pad, wearing their Sokol suits but with helmets open and gloves off, and carrying the suit-cooling equipment in one hand. Here, more press, dignitaries and well-wishers greet them – although the families of the Russian crew-members are not allowed in this area. They then move up the gantry steps, pausing to wave and bid farewell to the onlookers. Apart from the prime crew, only two pad workers ascend the lift to the spacecraft access gantry. (At this point, another tradition is often enacted: the equivalent of the actor's 'break a leg'. Russians believe that if someone wishes you good luck, then the reply has to be 'Go to Hell', otherwise you will have *bad* luck.)

The Soyuz launch vehicle, which is based on the R-7 missile, always appears different on launch day, standing on its pad and pointing towards space. It is fuelled, and the main support gantry is tipped back to reveal the 'stack', with white condensation billowing down the sides of the stages. It creaks and cracks, and almost 'talks' to the crew, waiting for them to board. Layers of ice build up on the outer surfaces of the super-cold fuel tanks. From the base of the pad the rocket seems quite small, but at close quarters it becomes an enormous structure, and the crew feels the platform and sees the rocket swaying slightly in the wind.

Because of the design of the Soyuz, the Crew Module is in the centre of the three-module configuration, within the launch shroud. The only crew access hatch in the Crew Module is located in the nose of the spacecraft, which is normally connected to the Orbital Module (OM) until immediately prior to re-entry at the end of the mission. The OM therefore has three hatches. The forward hatch is for crew transfer in space, and at launch is blocked by the docking equipment, and so the side hatch – usable for EVA from Soyuz – is the sole entry path for the crew while the vehicle is on the pad. With the side panel of the launch shroud removed, and the side hatch of the OM opened, the two pad assistants crawl inside with a plank, which they suspend over the open rear internal hatch leading to the Crew Module beneath. One by one the cosmonauts enter the Soyuz. First the anti-scratch protective cover is removed from their helmets, and then, sitting on the lip of the OM hatch, the technicians remove the cosmonauts' outer boots, after which the cosmonauts slide inside, head first. They are then each lowered by hand into the Descent Module, helped by the two technicians in the OM. The first to enter the Crew Module is the Flight Engineer, whose job is to complete the final pre-boarding check of the crew compartment (including the correct installation of the crew seats) before taking up his position in the left-hand seat. Next is the Research Cosmonaut, who manoeuvres himself into the right-hand seat. Finally, the Commander is lowered down and straps himself in, assisted by his colleagues and by one of the pad workers leaning down from the overhead OM/DM hatchway. Inside the cramped DM there is not much room to move as the crew settle into the couches with their legs raised in the launch position. Elbows bump and the crew reach across each other as they settle in and attempt to alleviate pressure point in their suits. It is here that they at last feel comfortable in their suits, lying on their backs with their knees up. With everyone secure, the crew establishes communications (including one-way TV) with mission control, and the hatches are closed with a clang. The launch-shroud cover is then replaced and the pad area is cleared, leaving the flight crew alone, sealed inside the crew compartment perched on top of the R-7 rocket.

The crew normally enters the capsule two hours before launch, during which time a series of check-lists and tests have to be completed and the spacecraft has to be pressurised and its seals tested. Every ten minutes, a pressure reading is taken to check the integrity of the Descent Module. For most of this time, however, the cosmonauts sit and wait, thinking of their training and the mission to come. During the count-down for launch, each crew-member has a role to play, with check-lists to follow and measurements to record in the logbooks; and in such a cramped space, a hand-pump and valves have to be used to overcome the problem of condensation. Periods of intense activity and concentration are interspersed with others of sheer boredom and calm, and mementoes are hung and memories recalled as the crew perform their final checks during the last stages of the count-down. Music is also piped onboard to help the crew to relax.

Booster loading begins five or six hours before launch, and at T–30 min the service structures are moved away from the R-7. The launch escape tower is then

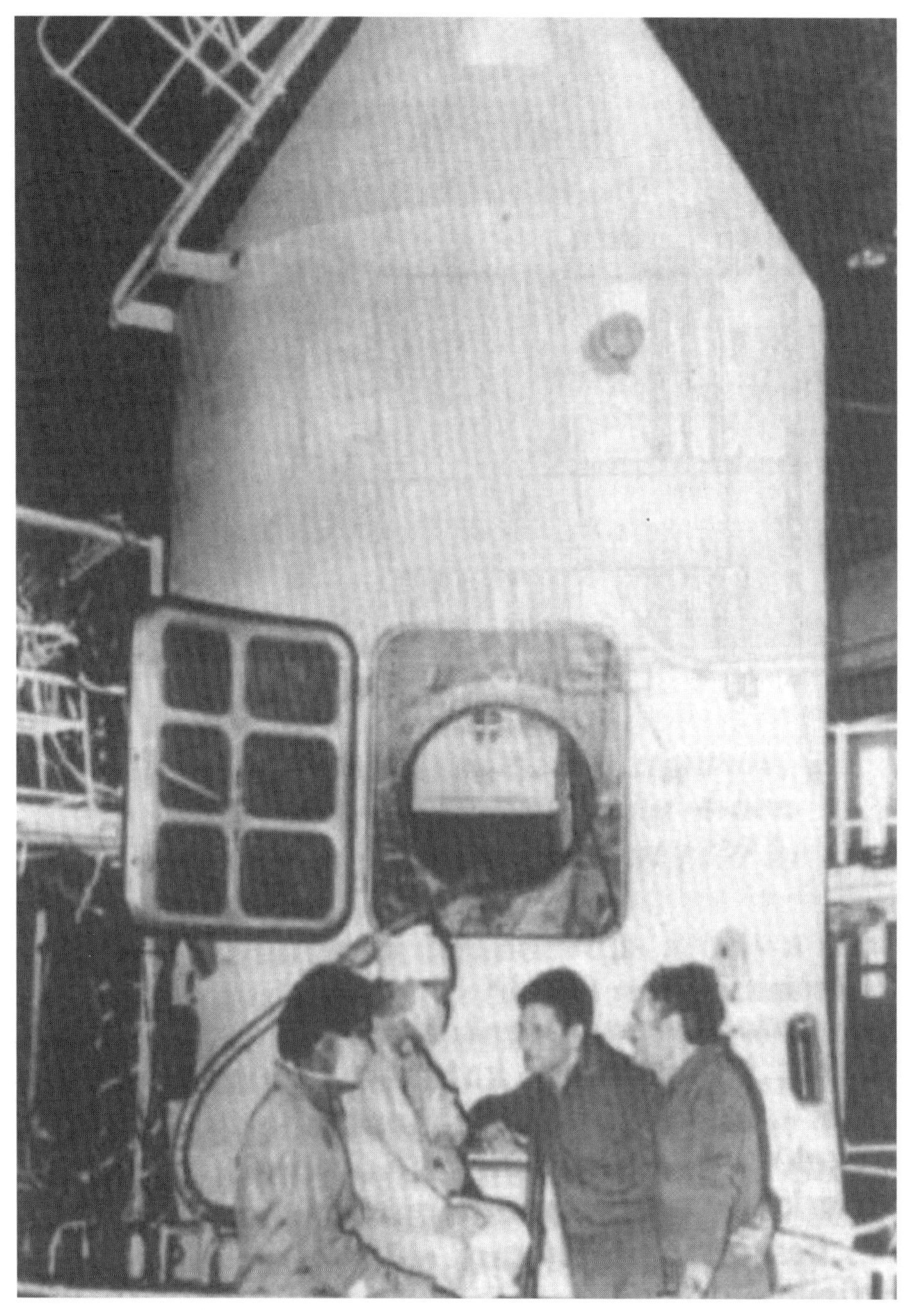

For many years the exact method of entry into the Soyuz on the launch pad remained unclear. In this photograph the launch shroud side hatch is clearly seen, and the Orbital Module side hatch is also open to allow the crew to enter the vehicle on the pad. In this photograph the Soyuz 21 crew perform inspection of their spacecraft in the MIK prior to its attachment to the launch vehicle and roll-out.

armed, and survival devices are placed in automatic mode – just in case. The crew start the onboard tape recorder at T–15 min, and ensure that their safety harnesses are tightened as the automated flight sequencer takes over. Combustion chamber covers on the first and second stages are blown off with nitrogen, while topping off of the liquid oxygen tanks is completed and the vent and safety valves are closed as the tanks pressurise and force the oxygen through the turbo-pumps.

At T–40 sec the launch vehicle is placed on internal power, and the ground umbilicals are disconnected from electrical power supplied from ground support equipment. At T–20 sec the generation of the 'Launch' command moves the upper umbilical mast away and starts the engine mechanism, allowing propellant into the combustion chambers of the first and second stages for ignition and the build-up of thrust. As optimum thrust is reached, the four supporting trusses holding the vehicle to the pad are released, and the Soyuz leaves the pad. There is no count-down as in the American programme; just a series of terse statements indicating the phases of the count-down and then powered flight.

At lift-off the crew-activated mission sequencer is initiated as the vehicle follows its allotted launch azimuth and nine-minute ride to orbit. From far below and behind them, the cosmonauts hear the 'roaring wave of thundering thrust coming from the blasting rocket engines'. From inside, as thrust builds, the rocket sways and appears to reach the point where it might fall over. Then the crew feel the loss of the supports, and, after a 2–3-second delay, the rocket takes off.

When Max Q is passed at 46 km, 115 seconds after lift-off, the LES is jettisoned. At 49 km the four first-stage strap-on boosters are burned out, and separate. At 85 km, 165 seconds into the flight, after the rocket has passed through the denser lower layers of the atmosphere, the nose shroud is jettisoned, uncovering the two crew compartment windows and revealing the layers of ice, which break off and stream past the windows.

At T+288 sec the second-stage central core separates and the third stage ignites for another 238 seconds. After the cut-off of the third stage the g-loads desist, and the crew experiences the onset of weightlessness. Four seconds later, Soyuz separates from the third stage into an initial orbit of about 240 km x 200 km, and control of the spacecraft switches from Baikonur to mission control in Moscow. The stringed mementos that hung down in gravity now dangle free, revealing to TV cameras that the spacecraft has entered 'weightless' flight. But the cosmonauts already know this, as, in orbit at last, they are no longer pressed into their couches or against their harnesses, and are suspended in free-fall, ready for their mission.

During the first orbits there is much to be done, including the deployment of solar arrays for electrical power and aerials for communications, a pressure integrity check of the Soyuz and its compartments, and further checks of the rendezvous and docking radar, radio and TV systems, and the flight control systems. With the Soyuz safely in orbit, the seat harnesses are released and the Sokol pressure suits removed – which, in the confined compartment, offers some relief. The hatches between the Crew Module and the Orbital Module are then opened, and the crew is ready to conduct the flight programme of anything from a few days onboard the Soyuz to many months docked to a space station.

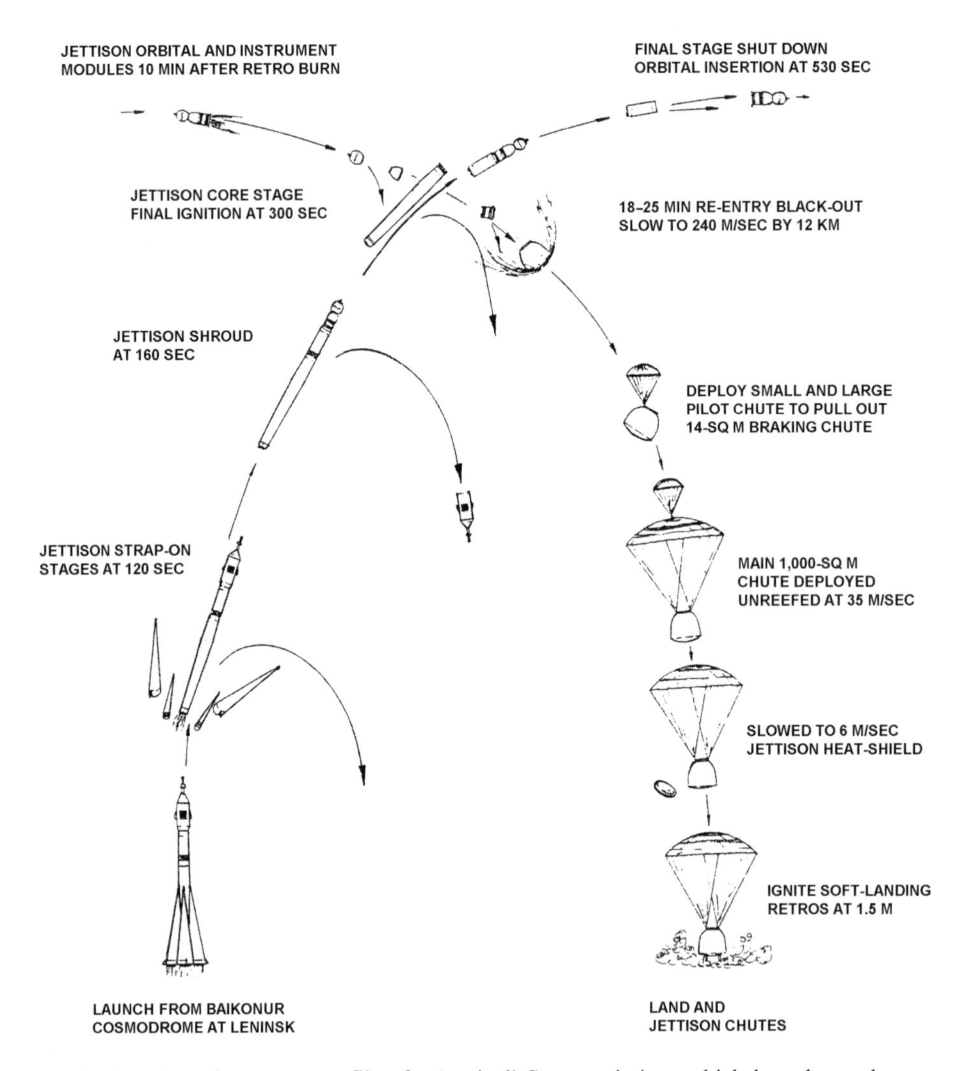

The launch and re-entry profile of a 'typical' Soyuz mission, which has changed very little since 1967. (Courtesy Ralph Gibbons.)

At the end of the prescribed mission, Soyuz returns home in a profile very different from that of the American Spce Shuttle. On a space station mission, undocking is carried out approximately three hours prior to landing, followed at a predetermined time by firing the spacecraft's retrograde engine against the direction of orbital flight at the beginning of the thirty-minute descent. The Soyuz normally follows an automated re-entry and descent profile, but for the entire descent the crew has to monitor the instrument panel and display readings, ready to intervene should something 'off normal' occur. Shortly after engine-firing, the spacecraft is turned 90° so that when the three compartments separate at 140 km altitude they follow separate descent flight paths into the atmosphere, to prevent collision with each other. The unprotected Orbital Module and Propulsion Module are burned to destruction, but the central Descent Module is coated with a heat-resistant ablative coating to protect the integrity of the structure (and the crew) from the searing heat of re-entry.

As the Descent Module descends, its shape provides a degree of lift sufficient to decrease loading to 3–4 g. If the angle were to be too steep, the much higher loads would become unbearable for the cosmonauts, and the vehicle would be destroyed in the atmosphere; and if it were too shallow, the module would skip out of the atmosphere (like a stone across water), and would enter an orbit from which the crew could never be rescued. The return profile ensures tolerable loads for the descent to Earth, but after a long mission even 1 g produces the feeling that the body weighs hundreds of pounds more than in reality. The centre of mass of Soyuz is offset in the base of the vehicle and centralises it as it slips through the atmosphere, dropping like a stone. Its rate of descent is controlled by a set of minute thrusters which, shortly after separation, orientate the vehicle heat shield first.

At 80 km, as the temperature rises across the surface of the heat shield, the plasma sheath that surrounds the vehicle cuts off radio communications for the next 40 km. Inside the module, the crew observe the dancing purple flares and sense the searing heat as they plummet earthwards. Gradually, the rate of descent slows through the thickening layers of the atmosphere, so that at a height of 10–12 km, the cosmonauts are travelling at 240–250 metres per sec, compares with the 7.8 km per sec at the beginning of entry a few minutes earlier. At 10 km, an ambient air pressure sensor sends a signal to eject the parachute cover, which is attached to a pair of small pilot parachutes that pull out a braking or drag parachute. These slow the craft to a velocity of 90 metres per sec, allowing the main parachute to be deployed after about 20 seconds, at 7.5 miles altitude. This parachute is initially deployed in a reefed configuration, and after a preset delay an explosive cutter cuts the reefing cord for full canopy deployment at about 5.5 km. This immediately follows the ejection of the base heat shield from the vehicle, to reveal the landing system. A back-up system is also carried in case the main parachute fails.

As the craft descends, a valve opens to equalise the internal pressure with the pressure of the air outside, and fuel is drained from the descent control system engines. Twenty seconds later, a gamma-ray altimeter is activated, the pressure suits are automatically inflated, and the seat shock absorber system is armed to help protect the crew from the impact of landing. The gamma-ray altimeter records the distance to the ground, and at just 1.5 metres the system triggers the soft-landing engines to slow the capsule to a 2–3-metre per sec impact, which is also absorbed by the seat impact system in the cosmonauts' couches. However, several cosmonauts have recalled that there is nothing soft about 'soft landing', despite the landing deceleration systems working perfectly. Crews returning on Soyuz train and expect to end their mission on dry land, and the activation of the retro-rockets – creating a plume of dust, soil or snow – has been termed 'dust-down' (rather than the American 'splash-down').

After landing, the vehicle often bounces, and does not always remain upright after stopping; and several returning Descent Modules – affected by winds and lateral velocity – have endured a landing that included more than one bounce and roll before coming to a stop, creating an uncomfortable ride for a few seconds. The landing area is predetermined, and a fleet of helicopters and search and rescue crews are on hand to retrieve the crew and the capsule.

Once the recovery team has arrived at the scene, the spacecraft is righted to stand on its base, and the top hatch is opened. A frame is erected and, helped by the recovery team, each cosmonaut is removed (in reverse order of entering before launch) onto recovery couches, for a brief welcoming ceremony with traditional gifts of bread, salt and flowers, followed by preliminary medical checks inside inflatable structures. The traditions continue, with the crew signing their descent capsule before being taken to a local airfield and then onto the cosmonaut training centre in Moscow for a series of mission debriefings. After the landing site has been secured, the Soyuz is transported, inside a specially-equipped all-terrain vehicle, to an airfield, and is then returned by aircraft to the Energiya design bureau near Korolyov, where it is examined and then stored in a large hangar. For the crew, the mission is consigned to history – the latest in a long line of variants of Soyuz spacecraft and missions which began in 1967.

Shortly after returning from a long mission in orbit, a cosmonaut is carried from the capsule to the makeshift medical tent.

Origins

The origins of the Soyuz spacecraft are entwined in the very beginnings of the Soviet manned spaceflight programme of the late 1950s. From there, the concept evolved into a proposal to send cosmonauts around the Moon and eventually to land on its surface. At the same time, variants of Soyuz would be developed to serve military objectives and, by the end of the 1960s, as a ferry craft to and from space stations. The role as a crew ferry and rescue spacecraft, as well as an unmanned resupply vessel, has been the mainstay of the Soviet/Russian programme since 1971, and continues to build upon the experience of the world's oldest operational manned spacecraft.

SOVIET MANNED SPACEFLIGHT AFTER VOSTOK

In deciding upon the design that would follow the initial Soviet manned spacecraft, Vostok, the designers re-examined the work of the father of Soviet cosmonautics Konstantin Tsiolkovsky during the early part of the twentieth century. Once human spaceflight had been achieved, the next obvious targets were the Moon and then Mars. Tsiolkovsky foresaw that to master cosmic flight, the meeting and joining (rendezvous and docking) of several components would be a key to the success of all future exploration in Earth orbit, to the Moon, and beyond. As work progressed on Vostok, so also did plans for what would follow. Vostok was incapable of achieving rendezvous and docking, but these would be major features of the next generation of manned spacecraft.

Design requirements

Work on the new vehicle began in Department 9 at Korolyov's OKB-1 design bureau, headed by M.K. Tikhonravov and Konstantin P. Feoktistov. Throughout 1958–59 they evaluated options for a spacecraft to follow Vostok, with the aim of its having a multipurpose role. The plan was to set up a piloted circumlunar mission and a vehicle capable of supporting the development of Earth-orbital space stations. Initial studies focused on Earth-orbital operations and resulted in two different designs; and these early designs had a profound influence on the final configuration of the spacecraft that became Soyuz.

The major consideration in all designs was in the ending of the mission by landing on Soviet soil. The option of a splash-down into water was originally not a consideration, and so the identification of the optimum re-entry profile and the selection of the shape of the re-entry module was crucial, and had a major influence on the final design. Two options were available: to draw on aviation experience to provide aerodynamic surfaces and a runway landing, or to adopt the missile principle of a ballistic re-entry under parachutes. By 1961 it was decided that winged designs were not suitable, mainly due to mass and the requirement for adequate heat protection. The designers evolved the concept of 'glancing re-entries', as opposed to direct ballistic re-entry which placed greater stress loads on the vehicle as it descended into the atmosphere. Further analytical studies focused on the even higher re-entry speeds of returning from the Moon. Tikhonravov's group devised a 'double dip' profile that not only reduced velocity in stages, but also eased the g-loads on the crew. It followed a re-entry corridor of some 3,000–7,000 km across the planet (south to north), enduring 3–4 g for a ground landing on Soviet territory, with an accuracy of ± 50 km.

Once the re-entry profile had been determined, the design of the re-entry vehicle could be established. OKB-1 engineers, in conjunction with scientists from NII-I (the Central Aerohydrodynamics Institute) and NII-88, evaluated three configurations which they designated the 'segmented sphere', a 'sphere with a needle', and a 'sliced sphere'. The guidelines for this research had to encompass the characteristics of the aerodynamics of each design, their effectiveness during descent trajectories, the selection of the structure's fabrication, and the selection of an adequate thermal protection system. The most promising design evolved from Vladimir Roshchin's group in Department 11 at OKB-1, which focused on a segmented sphere and a displaced centre of gravity. By 1962 this had evolved into an asymmetrical design that resembled an automobile headlight, with the base diameter the same as the overall length. The engineers found that this shape would increase lift at re-entry and avoid the problems of a high-temperature, high g-load ballistic entry.

Once the design had been chosen, the method of landing the vehicle had to be determined. Korolyov was interested in alternative proposals for landing on Soviet soil, and several organisations submitted ideas, including helicopter type rotors, fan-jet engines, liquid propellant rocket engines, controlled parachutes, ejection seats, and shock-absorbing inflatable balloons. Work on these recovery techniques began in 1961 and was not formally completed until 1966, and each method had an impact on the spacecraft designs being evaluated. In 1963 Korolyov finally approved the Department 11 proposal for a combination of braking parachutes and soft landing by solid propellant rockets.

With the method of landing determined, other areas could be finalised. The N1 launch vehicle required to send large payloads directly to the Moon was still on the drawing boards and years away from production; but in following Tsiolkovsky's pioneering theories, Earth-orbital assembly was within the grasp of the Soviets, because the R-7 was capable of placing payloads in orbit around the Earth. Several launches would be required, and once in orbit the different spacecraft components would be assembled. Thus Department 27 of OKB-1, headed by Boris Raushenbakh, approached the problems of rendezvous and docking in Earth orbit. It was apparent that sole dependence on launch power to bring the two spacecraft together would prove unreliable; and at that time, ground measurements were not sufficiently accurate for docking to be controlled from the flight control centre. In addition, in the early 1960s computer technology was in its infancy. Individual computer units were large and memories were small, and so total onboard measurements were also beyond Soviet technology at that time. Raushenbakh's department therefore developed what was termed the 'parallel approach'. In this method, the active spacecraft moves along a line of sight extending to the passive vehicle. To allow the two spacecraft to approach from great distances, onboard radar would be used, with automated docking for completion of the manoeuvre (although the cosmonauts and Air Force naturally preferred manual docking operations). From a short list of four methods, two were finally chosen. The Experimental Design Bureau of the Moscow Power Institute submitted the Kontakt (Contact) design, while NII-648 proposed the Igla (Needle) system. Following a detailed review of both systems, Kontakt was rejected in favour of Igla.

Sever and the 1L: the genesis of Soyuz

Sever (North) was a two-element spacecraft for Earth-orbital operations, incorporating a forward crew compartment and an aft unpressurised cylindrical equipment compartment. The design of the crew compartment featured the 'car headlight' shape, and would incorporate the crew provisions, life support and controls, and land recovery parachutes. The equipment compartment would house the guidance and control, propulsion, and power supply. (The Russians have never released detailed illustrations for this design.)

Sever was envisaged as a ferry vehicle for a space station. In his *Spaceflight* articles, Timothy Varfolomeyev says that Sever was the name of the station rather than the ferry, but this is not borne out by his sources. According to him, the ferry had the designator 5K, and the space station consisted of two elements: 5KA – the habitation module – and 5KB – the instrument module. (These designations have never been seen in open Russian literature. It is strange that Sever is not mentioned at all in the Energiya history, but this could reflect the military nature of the craft and its tasks.)

An alternative design was the 1L, which was evaluated for lunar distance missions (though not a lunar landing). As this was a more complicated mission profile, the comfort of the crew was a concern for the designers, who debated a range of configurations. In 1960, Department 11 proposed the addition of a third module – also pressurised – that allowed more habitable volume in the lunar spacecraft. This was called the Orbital Module (OM), and it was this three-module design that was finally selected – although exactly where the new OM was to be located was the subject of much debate. Initially, the crew compartment Descent Module (DM) was placed in the top of the configuration, with the OM and the Propulsion Module (PM) below. Access from the DM to the OM would be via a hatch in the base heat shield of the DM, in a configuration similar to that later evolved for the USAF Manned Orbiting Laboratory programme, with an entry hatch in the heat shield of the Gemini. Studies soon revealed that this was not the best location for the OM, as a hatch in the heat shield presented many structural and safety issues concerning the DM. The modules were therefore moved so that the OM was placed in the front of the spacecraft, followed by the DM and the PM. To gain access to the OM in this configuration, a hatch was placed in the nose of the DM and the base of the OM. This change altered the forward observation capabilities of the crew for rendezvous and docking, and so provision was made for the Commander (central seat) to use a periscope to look forward beyond the obstruction of the Orbital Module. The design approach of 1L was considerably different from that of Sever, in that the OM was more cylindrical (like the DM) instead of the headlight shape of Sever. In addition, the nose of the spacecraft featured a small propulsion system for attitude control during manoeuvres in Earth orbit. The PM was still cylindrical, but was skirt-shaped, with a boom extending from each side and carrying a circular solar array for electrical power. Elements from both Sever and 1L would evolve into the basic design of Soyuz.

The Vostok 7/1L Soyuz Complex

On 26 January 1962 – as both the Sever and 1L designs were undergoing final evolution – Korolyov himself presented an idea for a design incorporating elements of Sever, 1L, and the still active Vostok spacecraft – a design in which Tsiolkovsky's 'cosmic train' was revived. Korolyov planned no less than four modules, with three rocket blocks weighing 15–25 tonnes in total (the individual modules weighing between 4.8 tonnes and 6 tonnes). On 10 March 1962, Korolyov signed the scientific–technical prospectus for this design under the title 'Complex for the Assembly of Space Vehicles in Artificial Earth Satellite Orbit'. This whole project was assigned the identification name Soyuz (meaning Union, and pronounced 'sah-yooss', not 'Soy-uz').[1] The new programme was assigned three primary objectives: the creation of an 'orbital piloted station for military missions'; a spacecraft with the capability of circumlunar missions; and the creation of a satellite system for global communications.

Based on what has been published, it would seem that the plan (of 10 March 1962) involved two separate Earth-based assembly projects: one that required a 'rocket train', and one that did *not* require one. This is more or less confirmed in a document by Korolyov, in which he refers to the assembly of the space station and the rocket train as two different missions. On the one hand, there was a space station, which (according to a recently released Korolyov document dated 5 March 1962) would be used primarily for military reconnaissance of 'the countries of the imperialist blocks'. According to other sources, this space station was to consist of two modules, and crews would be flown to it onboard Sever spacecraft. The Vostok 7 assembly vehicle would probably have been required to dock the two space station modules (in the same fashion as the elements of the 'rocket train'); and the space station would probably not have required the rocket train, because it would have had to operate in a low orbit to perform its reconnaissance tasks.

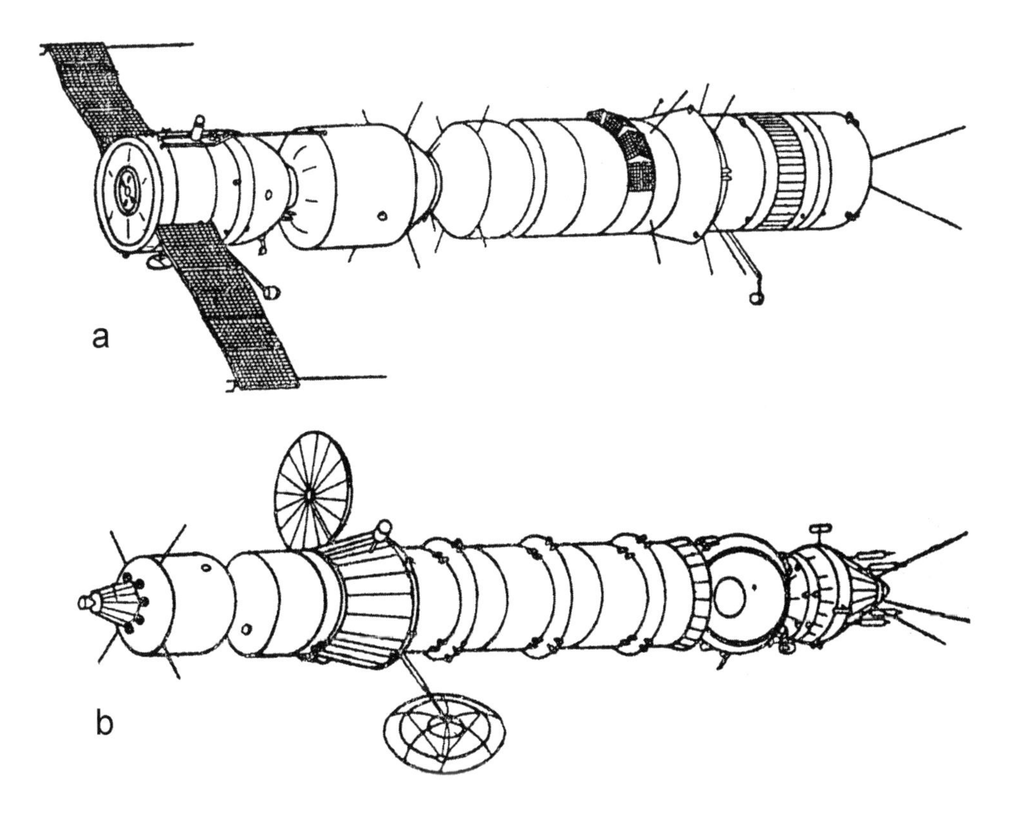

The 1L Soyuz Complex (a), and the Vostok 7K complex (b).

Alternatively, there was the 'rocket train' plan for any spacecraft that would have to go to higher orbits or to the Moon. The three possible 'payloads' for this rocket train (mentioned in the document of 5 March 1962) are communications satellites for geosynchronous orbit, a satellite interceptor (identified by Varfolomeyev as 5KM) capable of making significant manoeuvres and 'effectively fighting enemy satellites,' and a manned spacecraft for a lunar fly-around or other deep space missions.

In short, it would seem that there were two separate Earth orbit rendezvous programmes, with the Vostok 7 vehicle as a common element. Sever was specifically designed as a ferry vehicle for a low-orbiting space station, and 1L was specificially designed as a circumlunar vehicle. It is not clear how many features these spacecraft shared, but superficially, at least, there were some very basic differences. According to Varfolomeyev, Vostok 7 was also called 7K, and the rocket blocks 9K. These designators were then later inherited by the circumlunar spacecraft and the rocket blocks to be used in the second version of the Soyuz Complex.

For missions to the Moon, the addition of several rocket stages would provide the translunar injection burn from Earth orbit. From the front of this design, the 1L was the primary payload for the lunar distance flight. The aft docking port on the 1L was to link up only with the final rocket stage launched into orbit, and the Vostok 7 was to link up with the first of the rocket stages launched into orbit. Each rocket stage had a section capable of being jettisoned.

Vostok 7 was another derivative of the one-man Vostok spacecraft which had flown during 1961–63, and which had also been adapted into a multi-crew spacecraft called Voskhod during 1964–65. The basic design of the Vostok also evolved into the Zenit unmanned reconnaissance satellite programme.[2] Beginning with the basic Vostok spacecraft, the improvements would include rendezvous and docking equipment, a restartable orbital manoeuvring system, and attitude control engines, raising the mass from 1,100 kg to about 1,300 kg. Each of the rocket blocks weighed 4,800 kg, and was basically a cylindrical stage with a rocket engine to provide the additional boost in stages for the TLI. The launch vehicle for the lighter elements of the complex would be the 8K72K booster, with other elements launched by a shortened version of the 8K78 launcher, the fourth stage of which would be replaced by the orbital payload.

The mission sequence of the early Soyuz Complex

The assembly sequence would begin with the launch of a Vostok 7 carrying a sole 'pilot-assembler' cosmonaut. At an appropriate time, an unmanned rocket block would be launched into Earth orbit, and the pilot of the Vostok would rendezvous and dock with the rocket stage. The torroidal section would then be jettisoned, revealing a docking system for connection to a second rocket block. This sequence would continue until a four-section 'space train' – comprising Vostok 7 and three rocket blocks – had been assembled in orbit. At this point, the 1L would be launched with a crew of one, two or three cosmonauts. The 1L and the Vostok 7 would have been docked to opposite sides of the rocket train, and the Vostok 7 would undock after the arrival of the 1L. The rocket stage would then fire stage after stage, boosting the 1L towards the Moon on a circumlunar trajectory. The crew would use the DM for recovery at the end of the mission.

Several mission profiles were proposed, including deployment of geostationary communication satellites replacing the 1L lunar payload, and even a military space fighter configuration for military objectives.

In January 1962, Korolyov asked the Air Force to support the production of eight Vostok 7 spacecraft for flights in the mid-1960s. But the combination of several elements of Vostok 7/Soyuz was complex, and the Vostok could not be easily adapted to a rendezvous and docking spacecraft. However, although the Vostok 7/Soyuz Complex was promising, on 16 April 1962 the Soviet government issued a decree entitled 'On The Development of the Soyuz Complex for Piloted Flight to the Moon'. The question is whether this decree related to the first or the second version of the Soyuz Complex. Varfolomeyev's interpretation in *Spaceflight* (see Bibliography) is that it was approval for the second version (7K/9K/11K), but this is very improbable. Korolyov had finished his concept of the Vostok 7/1L plan only the month before, and so the decree was probably a stamp of approval for the *first* version of the Soyuz Complex. This is also supported by the fact that a preliminary draft plan for 7K/9K/11K was not finished until December 1962 and that the government did not approve the 7K/9K/11K until the decree of December 1963. Asif

Siddiqi's interpretation is that the decree of 16 April 1962 was mainly prompted by the possible military applications of the Vostok 7 plan, but this is not specifically stated in the Russian sources. The Vostok 7 plan was eventually abandoned in favour of 7K/9K/11K, but this does not seem to have happened until about mid-1962.

The turning point for Soviet cosmonautics was the decision that the concept of Vostok 7/1L would be redefined to address its weakest points, using Vostok as a major element in circumlunar planning. At this point, Sever was simply abandoned, to direct long-range planning towards circumlunar missions in response to American plans to send Apollo to the Moon. However, it is clear that several elements of the Sever and 1L designs had a lasting impact on the design of second generation Soviet/Russian manned spacecraft.

THE SOYUZ 7K COMPLEX

With a major redirection of future plans at OKB-1 in 1962, the design of the follow-on spacecraft to Vostok became more defined. Taking elements from the Sever and 1L proposals, the design engineers evolved the new design, designated Product 7K – or simply Soyuz. From Sever, the designers retained the headlight-shaped Descent Module and cylindrical Propulsion Module, and from the 1L they retained the three-module design. However, in order to decrease the overall mass to save launch weight, the diameter of the DM was reduced to 2 metres from the earlier 2.2 metres (without thermal protection) on Sever. When presented with this idea, an unconvinced Korolyov cleared part of his office and invited one of the authors of the new proposal to remain in the room for the duration of the meeting. The engineer continued to voice his support for the idea, and eventually Korolyov was convinced that the significantly more cramped internal arrangement for the DM was also workable, and thus signed off the design in late 1962. Ironically, this decision proved pointless six years later in 1968, when there was a requirement to decrease the mass of Soyuz by 200 kg, so that the reduction of diameter became irrelevant. However, by then the design had long been finalised, and it was far too late to incorporate a major design change. Since then, Soyuz has always been very cramped for a full three-man crew.

The new profile still required rendezvous and docking in Earth orbit as the primary route to circumlunar flight. On 24 December 1962, Korolyov signed a preliminary draft plan for the redesigned Soyuz. It featured the two-man Soyuz 7K manned spacecraft, a 9K translunar injection stage, and a propellant tanker designated 11K. Despite opposition to this idea from some of Korolyov's own deputies, Korolyov finally signed the draft plan for Soyuz/7K on 7 March 1963, committing OKB-1 to move ahead with the development of the Soyuz spacecraft. Over the next two months there were several discussions concerning the details and sequence of the proposed circumlunar missions. Many within the State Commission supported the plan, but there were objections from Chelomei and Glushko about the difficulty of realising the plan. Despite their caution, however, Korolyov had sufficient support from the other chief designers, and there was optimistic projection of the first test flight of Soyuz 7K by the summer of 1964. Despite the lack of governmental support or authorisation from the Communist Party to support the new project, Korolyov was confident that his new plan would be officially accepted. On 10 May 1963, OKB-1 issued the technical prospectus for the new project, called 'Assembly of Space Vehicles in Earth Satellite Orbit', detailing the 7K–9K–11K profile now identified as 'the Soyuz Complex', which, it was hoped, would take cosmonauts around the Moon before the Americans reached there.

Soyuz 7K (Soyuz A) design features

The main element of this design remained the manned space vehicle, Soyuz 7K. Measuring 7.7 metres long, the three modules retained the cylindrical 'instrument-aggregate compartment', the headlight-shaped crew compartment and the cylindrical living compartment.

The unpressurised instrument compartment at the rear of the vehicle featured four sub-divisions, resembling slices. These included a torus-shaped compartment at the base of the vehicle, which was jettisonable, and contained several sub-systems including rendezvous instrumentation, radio for orbital control, guidance command systems, thermo-regulation, automatic guidance and tracking. The next section included the aggregate compartment housing the approach correction engines for rendezvous and docking and the support structure for the twin solar arrays for electrical power. Above this was the instrument compartment containing all the sub-systems for prolonged spaceflight to and from the Moon, such as long-range communications, control and attitude rockets, thermo regulation systems, power sources for instrumentation, and guidance and control. The final section was the transfer compartment carrying the attitude control engines on the outer surface and their associated propellant tanks on the inside. This module provided the essential services and equipment to support the whole spacecraft on its flight in space. In the Apollo programme, the Service Module performed essentially the same function, and in the evolving Soyuz spacecraft this became known as the Propulsion Module (PM), or PAO (Priborno Agregatnyy Otsek – Instrument Aggregate Compartment).

Located above the PM was the pressurised crew compartment. This contained the life support system for entry through the atmosphere, the thermo-regulation system, optical (periscope) and TV systems for use in both observation and guidance crew control, display panels and associated relays and guidance equipment, an events timer, and parachute recovery systems including the base heat shield and soft landing system. In the Apollo, the equivalent vehicle was the Command Module, and on Soyuz this would become known as the Descent Module (DM), or SA (Spuskaemyy Apparat – Descent Apparatus).

Above this was located the cylindrical habitation module, which was also 2 metres in diameter. It contained life support systems for the duration of the mission to re-entry, thermal regulation, communications systems, TV and movie camera systems and support hardware, and mission-specific scientific experiments. The only detailed

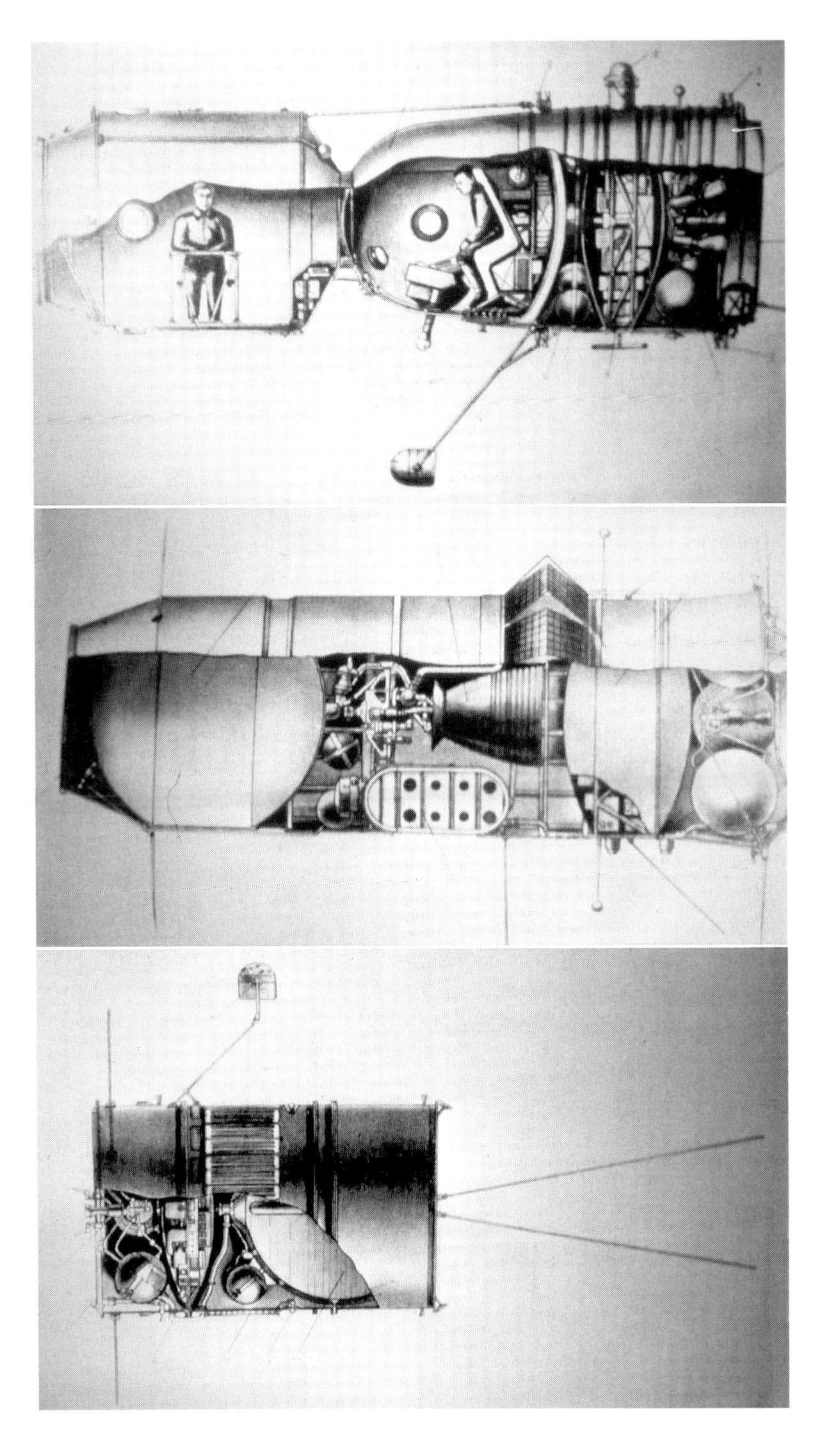

Cutaways of the Soyuz Complex.

Russian description of the 7K (as part of the Soyuz Complex) does not state that the OM could be used as an airlock – as was the case on the later 7K-OK version – and the only official drawing of the 7K does not show an exit hatch. However, there must have been a side-hatch for crew entry on the pad, although this does not necessarily imply that the OM could be used for EVA; after all, EVAs would not have been required on a circumlunar mission. In a document of September 1963, Korolyov describes L3 (lunar landing mission parent craft) as 'a modification of the L1 [circumlunar] ship which differs from it in that it has an airlock (the role of the latter can be performed by the [orbital module]).' In the plans circulating at the time. the L3 'Soyuz' itself would have landed on the Moon, and the cosmonauts would have had to depressurise it for their lunar EVAs, so that an airlock would have been necessary. The above excerpt from the Korolyov document would indicate that the circumlunar 7K of the Soyuz Complex (which he calls L1) did *not* have EVA capacity, and that an Orbital Module with EVA capacity was not introduced until the appearance of the manned lunar landing plans, and was afterwards inherited by 7K-OK. At the apex of this module was the docking apparatus. There was no equivalent module on Apollo, although early studies included a never-developed Experiment Mission Module for the Apollo Applications Program. On Soyuz this became the Orbital Module (OM), or BO (Bytovoy Otsek – Living Compartment).

The total mass of the 7K Soyuz was stated to be 5,500–5,800 kg. This spacecraft was also identified as Soyuz A when these 1963 designs of the multi-spacecraft complex were first disclosed. Soyuz A, B and V are cover designators created in 1980 when the Soyuz Complex was first described in Soviet literature. At that time the 7K, 9K and 11K designators were still classified, and so Soyuz A, B and V were not official designators in the 1960s.

The American General Electric concept

It has been observed that these early designs of Soyuz closely resembled one of the configurations considered by NASA for the three-man Apollo spacecraft.[3] The General Electric proposal progressed no further than paper studies or display models, but was widely reported in aerospace journals of the time; and the Soviets – in particular, Korolyov and his team – had full access to the Western technical press, and were aware of American ideas. At the time of this article, information on the earliest designs of Soyuz was emerging, thus allowing updated appraisals of the then 'hidden' history of Soyuz from earlier studies;[4] and as Clark and Gibbons pointed out, the early Soyuz – particularly the descent craft – *looked* like the GE proposal.

The GE proposal featured a three-module design with a total length of 10.2 metres, enclosed in a fairing (which remained attached until the end of the mission), and launched by a Saturn booster. The Propulsion Module (PM), located at the rear, was 4.6 metres long with a diameter of 3.0 metres and with a fairing of 5.4 metres diameter at the inter-stage of the Saturn. The Descent Module (DM), located above the PM, measured 2.4 metres long and 2.9 metres in diameter (though when scaled this changed to 2.5 metres long and 2.8 metres in diameter). Completing the design

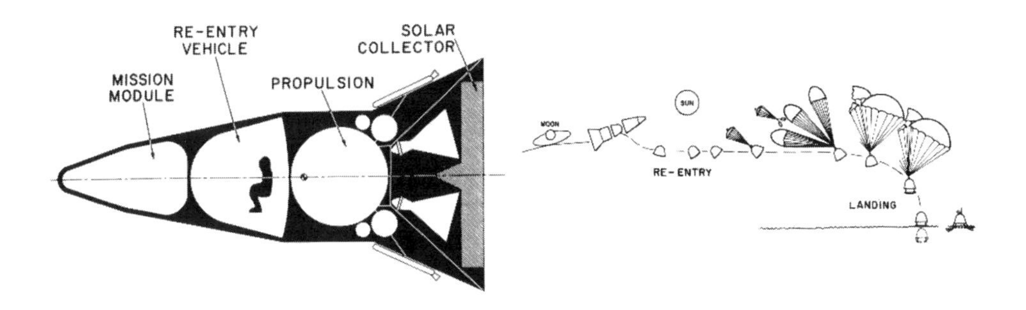

The General Electric proposal for the US Apollo lunar spacecraft, *c*.1960–61 and landing sequence. This design featured a three-module spacecraft design – an option which was also adopted for Soyuz. (Courtesy NASA.)

was a 'pear-shaped' Mission Module (MM), 3.1 metres long, with a maximum diameter of 1.7 metres.

The total spacecraft mass quoted in the proposal was 7,473 kg, with 2,184 kg for the DM. Using Apollo Service Module figures of a specific impulse of 314 sec for lunar orbit insertion, trans-Earth injection and mid-course corrections, Clark and Gibbons calculated a specific impulse of 315 sec for the GE concept. This produced 'a fuel mass of 4,035 kg and an empty payload of 3,435 kg' which would result in the MM and PM having 'a total mass of 1,350 kg, which is very light.' These figures therefore probably represent an early circumlunar concept, and not one intended for lunar orbit insertion.

As with Soyuz, the GE proposal envisaged the separation of the three modules for entry, with the PM and MM burning up on re-entry. Unlike Soyuz, however, the GE Descent Module would make a parachute splash-down in water, with a Mercury-type flotation bag as a cushion. Although the GE proposal failed to progress beyond conceptual studies, and lost the Apollo contract to North American Aviation on 28 November 1961, the design clearly reveals that the idea of a modular spacecraft capable of multi-mission objectives was in the minds of American designers as well as Soviet designers. It also defines a period during which the two nations reviewed similar concepts for their next generation spacecraft, but finally chose different configurations.[5]

Soyuz 9K and Soyuz 11K

Of the remaining two vehicles in this Soyuz Complex, the 7.8-metre long, 5,700-kg, unmanned single rocket stage, designated 9K, featured two main sections. The large rocket Blok (9KM) included the translunar rocket engine for the translunar injection burn with a thrust of 4.5 tonnes, and a smaller, jettisonable compartment (9KN) which served as the location for the orbital correction engine, associated control systems, rendezvous instrumentation, and a docking mode to receive the tanker vessels, with facilities for fuel transfer.

The remaining element – or rather, up to four elements – was the 11K unmanned cylindrical 'tanker' spacecraft, 4.2 metres long and with a total fuelled mass of 6,100 kg. This vehicle also had two compartments: the 11KA was the location for the oxidiser, and the 11KB carried the fuel. The remaining section carried the spacecraft attitude control engines, guidance electronics, Earth communications, and a docking unit for attachment to the Soyuz 9K with provision for automated fuel transfer to top off the tanks of 9K for the lunar flight. All three components of this combination would be launched by the 11A55 and 11A56 variants of the R-7 launch vehicle.

The Soyuz Complex mission profile

The piloted circumlunar mission was the primary objective of the Soyuz Complex. The first stage was the launch of the 9K element, and once it had attained its desired orbit the first 11K tanker would be launched to it to deliver 4,155 kg of extra propellant. The launch would occur as 9K passed near the cosmodrome, and the 11K would enter orbit some 20 km from the 9K. If the distance was greater, then 9K would be capable of closing that distance. The active vehicle (9K) would then dock with the passive (11K) vehicle. Propellant transfer would follow, using the fuel transfer lines routed through the docking connections (a system continued by the Progress resupply vessels from 1978 to the present day). After the fuel transfer, the tanker would be undocked and discarded, following a controlled re-entry using the 490 kg of fuel reserved for this manoeuvre. To fully supply the 9K with 25 tonnes of propellant, at least four 11K vessels would be required. Once fully loaded with propellant for the TLI, the spacecraft (7K), manned by either two or three cosmonauts, would be launched and, using its OM docking apparatus, would dock with the 9K stage at the forward end. Final orbital manoeuvres would be conducted by the 9K jettisonable stage, and once the spacecraft had been placed in the correct position this would be undocked and discarded. This action would deploy the main engine of the 9K rocket Blok, which would be fired for the TLI burn, boosting Soyuz towards the Moon with its PM facing the direction of flight. Mid-course corrections would use the 7K propulsion system. The lunar fly-around would be accomplished between 1,000 and 20,000 km, with the crew using cameras and scientific instruments to record surface data. After looping around the Moon and completing the return-to-Earth flight, the three sections of Soyuz 7K would separate, the DM would continue its descent into the atmosphere for a parachute recovery in the Soviet Union, and the other two sections would burn up in the atmosphere. The total flight time would be about 7–8 days. In an alternative assembly sequence, the 9K rocket stage would launch first as usual, but would then be followed by the manned 7K vehicle, with the cosmonauts manually docking their vehicle to the 9K. The 11K tankers could subsequently have been manually guided in by remote control (similar to the TORU (Teleoperatornyy Rezhim Upravlaniya – Television Remote Control) used onboard Mir and the ISS).

Contracts, funding and schedules

With this type of operation, mastering the techniques of rendezvous and docking was clearly the major objective; but it was also necessary, as there existed no rocket powerful enough (such as the N1 heavy lift vehicle) to render multiple rendezvous and docking unnecessary. (This was one of the reasons why the Americans opted for the Saturn V and LOR, instead of EOR with the smaller Saturn 1Bs or several Saturn Vs.) Although risky, it was still firmly supported by the leading figures in the design bureaux involved in the space programme (probably because they received valuable contracts to support the multiple launch/spacecraft approach). With several Soviet design bureaux competing for sizeable contracts, the Soviet lunar programme was not only competing with the American Apollo programme, but was also dealing with competition between its own design bureaux, thus diluting the focused effort of one agency (such as NASA) to drive the major objective of the programme: to send men to the Moon in a united national effort. At a time when Vostok was still flying, the Soyuz Complex was an achievable next step in the programme before the larger, more capable N1 was flight-proven.

By the end of 1963, Soyuz 7K planning included four vehicles at a cost of 80 million roubles, against an authorised budget of only 30 million roubles. Although the basic technology and goals of the programme had been approved, funding had not been sanctioned. A lack of any official government decision delayed full funding, and Korolyov constantly complained about the lack of funding for his Soyuz plan. On 3 December 1963, a joint decree issued by the Communist party and the USSR Council of Ministers committed the 7K–9K–11K space complex to manned circumlunar spaceflight. In this decree, the primary 'customer' of Soyuz was the Strategic Missile Force, with the Air Force and Air Defence Force only 'taking part' in technical and tactical test flights, but not the circumlunar programme. Under this plan, the first flight vehicle would be delivered in August 1964, and the second and third by September 1964, with test flights probably beginning in late 1964 or early 1965. Official sanction for Soyuz was a double-edged sword: there was authority to proceed to the Moon, but the Vostok spacecraft was still flying and required major support resources. Korolyov wanted to end the Vostok missions and proceed with Soyuz in order to compete against the growing American programme – especially the Gemini programme, which would mount rendezvous and docking exercises, accomplish long-duration flights of up to 14 days, and perform space-walk experiments from 1964 until 1966, paving the way for Apollo Earth-orbital test flights to take over perhaps as early as late 1966. Clearly Gemini would fly before Soyuz was ready, and so the Soviets required an answer to potential US gains in space. Korolyov had for some time, and unsuccessfully, attempted to gain support for extending the Vostok programme beyond the original six missions. Then, on 4 February 1964, after stopping work on Vostok and putting all efforts into Soyuz, he received a higher order to take the four remaining Vostok vehicles, redesign their interior to accommodate three cosmonauts (more than Gemini), and include capacity for EVA before the Americans flew Gemini in less than a year. This programme became Voskhod.[6]

Soyuz to the Moon

By 1963 some designers were also thinking of expanding their lunar plans to more than circumlunar flights – partly to recognise the dreams of Tsiolkovsky, and partly in response to the American Apollo lunar landing programme, even though there was much indifferent or non-committal public response to the American programme. The N1 heavy launch vehicle was beginning to look attractive as a potential carrier for a lunar landing mission, following a lunar orbital rendezvous approach and using a separate landing craft as with the American Apollo landing profile.[7]

By 23 September 1963, Korolyov and his colleagues at OKB-1 had produced a detailed technical document entitled 'Proposals for the Research and Familiarisation of the Moon', involving a programme of interwoven manned and unmanned missions and robotic lunar probes. To achieve a manned lunar landing by cosmonauts by 1967 or 1968, the plan envisaged five themes, each with a specific goal in mind but linked to the larger programme. The L1 theme was a circumlunar programme for the Soyuz Complex, using six launches to achieve the goal; the L2 objective was designed to place an automated rover (designated 13K) on the surface, using the 9K and 11K elements of the Soyuz Complex, again with six launches; the L3 was manned lunar landing in which the 7K Soyuz would be flown in a modified form along with a separate landing craft, and there would be one R-7 launch and three N1 launches; the L4 was a lunar orbital flight of a modified 7K vehicle, launched on a single flight by the N1; and the L5 carried an advanced lunar rover onboard a single N1.

The cost of the programme was not the only hurdle, as official support by the Kremlin was also required. In a meeting with Premier Khruschev on 13 June 1963, Korolyov indicated that with adequate financial and political support, the N1 programme could beat Apollo and place men on the Moon by 1969. Khruschev replied: 'I'll think about it. You prepare your proposals. We will discuss and decide this in the Presidium of the Central Committee.' It was not really the reply Korolyov wanted, as time and the Americans were against any delays. Korolyov tried to point out that the Americans were capable of flying around the Moon with Apollo, based on their experience gained with Gemini. Unfortunately, although OKB-1 had been working on a similar plan for years, it had never been officially supported or encouraged by the Kremlin and, as Korolyov pointed out: 'If additional urgent measures are not adopted on the Soyuz theme, then the Soviet Union will lag behind the US in this area too.'

A redirection for Soyuz

Early in 1964, Department 11 of OKB-1 redesigned the interior of the Soyuz 7K to support two or three cosmonauts, and in the spring the first boilerplate (mock-up) was produced at the Kaliningrad plant. When Korolyov saw the results he reportedly commented that this was indeed the spaceship of the future for the Soviets. By February 1964 the first full-size crew-trainer of the 7K had been installed at TsNII-30 in Noginsk, along with $^{1}/_{30}$ scale models of the 9K and 11K spacecraft, in order to allow cosmonauts to rehearse rendezvous and docking procedures. On 26

September 1964 a stripped-down mock-up of Soyuz was launched by OKB-1 engineers from an old proving ground at Kapustin Yar, on a sub-orbital demonstration flight to determine the aerodynamic qualities of the payload shroud and the DM. Unfortunately, due to excessive aerodynamic loads on the structure between T + 33 and T + 39 sec, it broke apart.

This occurred only a month after the Communist Party had awarded the Chelomei design bureau (OKB-52) the primary contract to conduct the circumlunar programme, which effectively stopped Soyuz dead in the water. The official Soviet Communist Party Central Committee decree number 655-268 was passed on 3 August 1964. Undeterred, however, Korolyov and OKB-1 began an even more dedicated pursuit for the goal of manned lunar landing under the N1/L3 programme. Why the OKB-52 design was chosen over OKB-1 remains unclear. Perhaps the multiple docking of 7K/9K and 11K elements appeared too complex compared with Chelomei's single-piloted LK-1 in just one launch onboard a modified UR-500 ICBM. Whatever the reason, L1 (the Korolyov vehicle) was officially abandonded in favour of LK-1 (the Chelomei vehicle), and OKB-1 changed its plan in favour of a more applicable goal against the American Apollo. On 25 December 1964, Korolyov presented his preliminary plan for landing cosmonauts on the surface of the Moon by 1967–68. However, the OKB-1 circumlunar programme continued, using a single launch of a four-stage Proton, and supporting the landing missions launched by N1.

The L1 programme used a stripped-down version of the Soyuz 7K launched by the Proton on circumlunar flights – initially unmanned, but later manned – which became known as Zond (Probe). The L2 missions would deliver a lunar rover (which later evolved into Lunokhod), and the landing would be achieved with the development of a separate landing craft called the L3. The flight profile for the landing mission would resemble the lunar orbital rendezvous mode being studied and prepared for by the American Apollo programme. OKB-1 also studied some simpler Earth-orbit rendezvous profiles for manned circumlunar missions using R-7 rockets with liquid hydrogen upper stages.[8]

The N1/L3 lunar landing mission profile

The mission would begin with the launch of the three-stage N1, which would place the two-man vehicle in Earth orbit. After checking the complex, the fourth stage (Blok G) would complete the translunar injection burn and reveal Blok D by jettisoning the fairings. This stage would be used for mid-course corrections and insertion into lunar orbit. Once in lunar orbit, the commander of the mission would perform an EVA from the LOK orbiter to the LK lander (a one-man vehicle). The lander would then be checked by the single cosmonaut, and separated from the mother spacecraft. The Blok D would be used for Powered Descent Insertion (PDI) to an altitude of 3 km, where it would be separated, allowing the LK engine to ignite to complete the rest of the landing approach. Once landing had been achieved, the Commander would perform a limited-duration surface EVA to collect samples and deploy experiments. After only 24 hours, the lunar take-off apparatus would initiate ascent from the surface for an approach, rendezvous and docking with the LOK. The Commander, carrying the collected samples, would need to perform an EVA from the lander back to the mother spacecraft, as there would be no internal transfer hatch between the vehicles (unlike the American Apollo), since an internal transfer system would have added too much mass. The LOK main engine would then perform the Trans-Earth Injection (TEI) burn, with any further required mid-course corrections performed by the engines of the LOK vehicle. Approaching Earth, the elements of the LOK would separate, allowing the DM to land in Soviet territory.

EXPLORING THE POTENTIAL OF SOYUZ

Towards the end of 1964, Boris Chertok headed a small team established by Korolyov to pursue some initial ideas of the previous year. Chertok's team would investigate exactly what missions the Soyuz 7K could achieve in conjunction with the changing Soviet programme after the end of the Vostok/Voskhod flights. The result was the idea that docking two 7K spacecraft in Earth orbit could form the new goal of the redirected Soyuz programme. Some of the crew could perform a Soyuz-to-Soyuz EVA transfer to simulate the transfer which the N1/L3 Commander would have to complete twice on the lunar landing mission. It would also have additional advantages in demonstrating the potential of crew rescue, and provide valuable experience before actual operations were completed in lunar orbit. In February 1965 this new profile was submitted to the Scientific Technical Council of the State Commission for Defence Technology. The Ministry approved this new objective for Soyuz, and all work on Soyuz thereafter began to focus on the manned docking and EVA transfer, and was relocated to Department 93 at OKB-1. Early in 1965 the 7K became Soyuz 7OK (Orbitalnyy Korabl – Orbital Ship), and would be more commonly known as the 'original Soyuz'. At the same time as these spacecraft were being prepared for Earth-orbital test flights, other variants of early Soyuz vehicles were proposed during the period 1964–70, although most of them did not progress to the flight stage.

Soyuz 7K-P: a piloted anti-satellite interceptor

As Korolyov developed his proposals for what became Soyuz in the early 1960s, he understood that funding for such an ambitious programme would have to be provided by the military. In his draft proposals, therefore, he also included a pair of military variants of the Soyuz 7K design. The military supported these variants at a time when the USAF was developing its own Manned Orbiting Llaboratory programme.

Soyuz 7K-P (Perekhvatchik – Interceptor) used the 7K-PPK (Pilotiruemyy Korabl-Perekhvatchik – Manned Interceptor Spacecraft) variant of the basic Soyuz design. Four decades after its proposal there are still relatively few details available about this spacecraft; but what *is* known is that a mission-dedicated launcher – the 11A514 R-7 – would have been developed for the project. It is thought that this vehicle would also have employed the Soyuz 9K rocket stage and Soyuz 11K tanker spacecraft in its missions to intercept 'enemy' satellites in higher orbits. During 1963 the programme was handed over to the Department 3 of OKB-1 in Kuibishev, under the leadership of Dmitri Kozlov, as Korolyov was overloaded with work on the Voskhod and N1 lunar programme.

A satellite interceptor was already part of Korolyov's Earth-orbit assembly plans involving the Vostok 7 assembly vehicle – as is clear from a recently released document written by Korolyov and dated 5 March 1962. It would be capable of making significant manoeuvres, which suggests that it would have required one or more of the rocket blocks. Writing on the evolution of the R-7, Timothy Varfolomeyev refers to the spacecraft as the 5KM, and states that it would weigh 5.6 tonnes.[9] It is, however, not clear how these plans changed after the switch to the second-generation Soyuz Complex, although G. Vetrov says that Soyuz P could have reached an altitude of up to 6,000 km.[10]

By 1964 the programme had been placed on hold, due to the increasing interest and support from the military in the unmanned automated anti-satellite system (IS – Satellite Destroyer), which had already been test-flown on two successful missions. Polet (Flight) was launched on 1 November 1963, and became the first satellite to manoeuvre in orbit. The second Polet was launched in 1964. In early 1965, Soyuz P was cancelled in favour of an unmanned automated military space vehicle being developed at that time.

Soyuz 7K-R: a piloted reconnaissance space station

The Soyuz R (Razvedchik – Spy) consisted of two separate spacecraft. The Soyuz 7K would be a ferry spacecraft (11F72 7K-TK) for crew transport to and from a small space station (11F71). Soyuz R would have retained the Soyuz PM, but its DM and OM would have been replaced by a single compartment housing photoreconnaissance and electronic intelligence surveying equipment (ELINT). It also featured an internal transfer hatch which allowed the cosmonauts to enter the 'station' without EVA or pressure suits. Once again, Kozlov's team at Department 3 handled the development of this vehicle.

On 18 June 1964 the USSR Ministry of Defence agreed a five-year plan (1964–69) for military space programmes, issued by Defence Ministry Marshall R. Malinovsky. The plan for a military space infrastructure included Soyuz R, the unmanned photoreconnaissance satellite programme based on the Vostok spacecraft bus (Zenit), the ocean reconnaissance satellite Morya 1 (also known as US1), and the military space-plane Spiral. Despite the subsequent approval of a pre-draft plan for the Soyuz R complex by the Interdepartmental Scientific–Technical Council for Space Research (the ministerial organisation that evaluated space proposals), there was no guarantee that Soyuz R – or indeed the Soyuz programme in general – would receive full support over other proposals from other design bureaux. On 12 October 1964, for example, the Chelomei bureau was granted permission to begin developing a military space station larger than Soyuz R. This station would have a mass of 20 tonnes, and would be launched on a single UR-500K Proton rocket with a crew of three cosmonauts. Clearly, by 1966 this station's prospects proved much more favourable in competing against the USAF MOL programme, and on 30 March 1966 Soyuz R was cancelled in favour of Chelomei's military space station, Almaz – which, ironically, was given the former Soyuz R designation, 11F71. When Korolyov unexpectedly died on the operating table on 14 January 1966, his design bureau OKB-1 and the Soviet space programme lost their main driving force, opening the opportunity for other design bureaux to secure larger slices of the space budget. With the cancellation of Soyuz R, Kozlov was instructed by the leadership to hand over all documentation on the Soyuz R complex to Chelomei's bureau, but to still continue developing the 7K-TK ferry craft, now in support of the Almaz station.

Soyuz VI: the military research spacecraft Zvezda

On 1 August 1965 a Military Industrial Commission (VPK) decree was issued proposing military variants of the Voskhod and Soyuz spacecraft. Part of this work centred on both Soyuz R and Almaz, but these were some years away from development – certainly after the planned orbiting of the USAF MOL programme around 1968. With OKB-1 attempting to reorganise after the loss of Korolyov, OKB-1 Department 3 head Kozlov was tasked with providing military cosmonauts with experience prior to the orbiting of the Almaz stations. Kozlov's team devised a completely redesigned Soyuz, designated 7K-VI, and also known as Zvezda (Star). But Korolyov's replacement, Mishin, began to have second thoughts about supporting only Almaz, and proposed a separate competing military space station programme, based on Soyuz and designated Soyuz VI (VI – Military Research). No-one has ever provided a clear explanation for the simultaneous existence of so many manned military projects (Soyuz R, Zvezda, Almaz, Soyuz VI), although the idea that Zvezda was seen as a stopgap measure until Almaz became available is just one possible interpretation, since Mishin also considered flying Zvezda at the same time as Almaz.

By January 1967 the Soviets had two different military manned space programmes: the OKB-52 Almaz space station, and the 7K-VI (Zvezda) developed by OKB-1 Department 3. (Shortly after Korolyov's death in January 1966, the Ministry of General Machine Building conducted a ministry-wide change in the naming of institutions, effectively replacing the OKB designations. On 6 March 1966 OKB-1 became the Central Design Bureau of Experimental Machine Building (TsKBEM), and OKB-52 became the Central Design Bureau of Machine Building (TsKBM).[11] For simplicity in this chapter the original OKB designations are retained.) On 15 July 1967 a meeting of the USSR Defence Council responded to a call by Brezhnev for an expansion of military space operations. Following approval by Air Force Commander-in-Chief Marshall Konstantin Vershinin, the proposal was sent to the General Staff for further study. By September 1967 there had been formulated an eight-year plan that would see the launch of no less than twenty Almaz space stations and fifty 7K-VI spacecraft between 1968 and 1975. If each crew were to exchange duty every fifteen days, then a corps of 400 cosmonauts would need to be selected, trained and flown by 1975. The cost to support such a programme stretched to billions of roubles, and (interestingly) the plan identified the development of a reusable space transportation system to offset some of this cost.

By the close of 1967, Soyuz 7K-VI was changing course. Due to the numerous failures on Soyuz, its design had been modified to have the DM at the top, the laboratory section in the centre and the PM at the rear, relying on two radioisotope generators to supply power. By 21 July 1967 the vehicle was scheduled to be operational in 1968, but by late August this had slipped to 1969. By October the programme was under severe criticism from Mishin. Clearly in competition to Chelomei's Almaz, during October 1967 Nikolai Kamanin recorded in his diary that Mishin had written to Military–Industrial Commission Chairman Smirnov and Minister of General Machine Building Afanasyev to request the termination of Kozlov's 7K-VI and the reallocation of resources to complete the fabrication of eight to ten Soyuz 7K-OK spacecraft during 1968. Kamanin also recorded that Mishin was not happy that Kozlov had adapted the Soyuz design and had proposed a competitor to the basic Soyuz, and would support the Almaz station design and create a programme in competition against other OKB proposals: 'Work on developing the [7K-VI] ship is in full swing, and it promises to be much better than the Soyuz.' Mishin was apparently unconcerned provided that he saw 7K-VI as an exact duplicate of the basic Soyuz, but when he found that Kozlov's design was different from and actually better than the Soyuz, he changed his mind, and worked to cancel the Kozlov vehicle. In February 1968, work on the 7K-VI was closed down by OKB-1, which then promoted the development of a smaller station supplied by Soyuz VI (11F730). This was the combination of the Orbital Research Station (OIK/11F731) and the transport ship (11F732).

There were two major flaws in the original 7K-VI design: the use of a radioisotope generator, and a hatch in the re-entry heat shield. In the new design, the orbital Blok (OB-VI) was a cylindrical vehicle, similar to Soyuz R, which could support a scientific/military payload of 700–1,000 kg, with solar arrays replacing the isotopes on Zvezda. One of the major advances in this design was the adaptation of a pin–cone docking system for internal transfer through a docking hatch and tunnels, avoiding the need for EVA transfer which was still the option on the 7K-OK 'basic Soyuz'. Again taking the development from Soyuz R, this was to prove a significant design milestone for future Soyuz operations.

Soyuz VI would operate at an altitude of 250 × 270 km in an orbit inclined at 51.6°, with missions of about 30 days in duration. In addition to the space station there were plans to fly *solo* Soyuz missions on both short-duration (the 7K S-I or 11F733) and long-duration (7K S-II or 11F734) missions, independent of the Soyuz VI complex programme. The 7K-S or 11F732 crew delivery vehicle (along with elements of the 7K-S-I and 7K-S-II) would form the basis of what would evolve into the Soyuz T transport vehicle, and a robotic cargo vessel, the 7K-G (or 11F735) that became known as Progress.

In adapting the basic Soyuz 7K-OK design, the engineers at OKB-1 reviewed all the weak points that had been identified while developing the spacecraft, and attempted to eliminate the faults by replacing many of the sub-systems. According to the official OKB-1 design bureau history, the upgrades had the goal of 'improving the tactical–technical, technological and operational characteristics in the design and onboard systems... Important changes were introduced, which affected the course of development and ultimately resulted in the creation of a new ship.'

In May 1968 the Soyuz VI project officially received further support with the issue of a new tactical–technical requirement document. During the summer, as intense activities in preparation for piloted lunar flights continued, there appeared theoretical drawings of the TK-S spacecraft, which were approved on 14 October 1968. Several cosmonauts were later reassigned from Zvezda to Soyuz VI, and many of them subsequently became involved with Almaz.

Although the pace of the programme had increased, it was clear that interest in Soyuz VI was waning – particularly with Kozlov himself. whose department was becoming more involved with the more important unmanned photoreconnaissance satellites. Mishin was promoting the development of the 7K-S as an alternative role for Soyuz, rather than the station itself. In 1968 the priority for the Soviet manned space programme was to reach the Moon, rather than develop military or civilian space station objectives. Soyuz VI was probably finally scrapped in February 1970.

ADAPTING SOYUZ FOR LUNAR MISSIONS

At a meeting of Chief Designers on 15 December 1965, Korolyov presented a preliminary design for the OKB-1 L1 lunar spacecraft. Drawing on experience from research conducted on the Soyuz and N1 programmes, and being familiar with Chelomei's OKB-52 work on the UR500K launch vehicle and LK-1 spacecraft, Korolyov proposed a fly-by spacecraft based on a lighter version of Soyuz spacecraft but launched by the UR500K vehicle. Using his authority, the experience from Vostok and the advanced work on Soyuz, he secured the authority to pursue a series of manned lunar fly-by missions. The OKB-52 design had fallen seriously behind its development plan, and on 25 October 1965 a government resolution ordered that the OKB-1 L1 spacecraft replace the LK-1 design. Unfortunately, Korolyov died in January 1966, and he would never see his efforts come to fruition on the Soyuz spacecraft in its various guises.

The 7K-L1 was essentially a Soyuz spacecraft without the Orbital Module, but with an additional upper stage (Blok D) on the Proton launch vehicle. On 31 December 1965 – two weeks after Korolyov's presentation – agreements were reached between OKB-1 and OKB-52 on the construction of fourteen L1 spacecraft. Spacecraft 1P would be used for ground tests; 2P and 3P would be used for tests in Earth orbit using the Blok D upper stage; 4–10 were assigned to unmanned circumlunar missions; and 11–14 were for manned circumlunar missions, although they could fly unmanned missions should the need arise.

The plan envisaged the construction of one spacecraft in the third quarter and one in the fourth quarter of 1966, while the rest would be spread over the first three quarters of 1967. The first launch was planned for around the end of 1966, with the first unmanned circumlunar flight by 15 April 1967, although it soon became clear that no missions to the Moon would be possible before 1968. Kamanin reacted to government statements of unsatisfactory progress in general in a typical way: 'As the fiftieth anniversary of the October Revolution [October 1967] is near, the big bosses will be requiring eye-catching space missions. But paperwork and shouting won't help the cause. Too much time has been wasted already.'

By the end of 1966, flight preparations were continuing apace, with two unmanned missions in Earth orbit and two unmanned circumlunar missions planned before attempting a manned launch on 26 June 1967. In order to gain additional experience with the Proton (by late 1966 it had been successfully launched only three times out of four launches) before placing a crew onboard, the first crew would be launched separately, rendezvous with the Blok D in Earth orbit, and transfer by EVA to the circumlunar vehicle. Flight planning was changed to accommodate this by flying five circumlunar missions using dual launches (flights 6–10), and then four additional missions on a single launch (flights 11–14). This adaptation allowed the Soviets to both man-rate Proton and advance the first manned circumlunar flights.

Once the decision had been made to put the crew and vehicle into orbit, there remained the question of how to recover the spacecraft at the end of the mission. Traditionally, Soviet manned spacecraft landed in Soviet territory, but returning from the Moon suggested that the double skip re-entry trajectory might also lead to an ocean splash-down in the event of a problem, or even a landing on foreign territory. Since the Air Force could not mount ocean-based recovery, on 21 December 1966 an agreement was reached between the Soviet Air Force and the Soviet Navy that the recovery would be shared between the two services, with an estimated 12,000–15,000 personnel to support landings outside the Soviet Union.

Spacecraft design changes

The restricted time-frame for developing the spacecraft for lunar flight, added to the restrictions in the launch capability of the Proton, meant that, like the Apollo Lunar Module, weight reduction was a major priority in selecting and preparing the Soyuz design for lunar flight. The overall mass of the circumlunar spacecraft could not be more than 5.1–5.2 tonnes, and so the most obvious reductions would be in the removal of the Orbital Module and the reduction of the crew to two – a Commander and a Flight Engineer. Some of the sub-systems in the re-entry vehicle – including the reserve parachute – were also removed, while others were improved and updated over the 'basic Soyuz'. Many of the instruments placed in the Soyuz for Earth-orbital operations (such as the horizon attitude indicator) were not required for lunar flight, and with no OM, the forward periscope was not required for docking. In the PM, one panel from each of the twin solar arrays was removed, and improvements were made to the attitude and control system and the radio communications, including the installation of a larger directional antenna for communication at lunar distances. Additional thermal protection was added to the DM because of the increased heat generated due to the return from lunar distances. As the vehicle was not planned to enter or leave lunar orbit, there was no requirement for a large propulsion system (as on the Apollo Service Modules), and so no back-up main engine was fitted, whereby the crew relyied on the Soyuz KTDU-35 (or S5.35) system with a 400-kg propellant mass.

The issue of the crew access hatch on the L1 remains a problem, and there are two Russian sources that refer to it. The RKK Energiya history says that the EVA transfer from Soyuz to L1 would be carried out via 'a side hatch' in the L1. Vetrov (in *Sergei Korolyov i ego delo*) states that one of the features which distinguished L1 from 7K-OK was 'the absence of a back-up parachute system, the hatch of which was used as an entry hatch'. Here, the Russian word used for 'entry' would suggest that it was a hatch used to enter the vehicle on the pad. Based on these two sources, therefore, it would seem that there *was* a side-hatch, and that it would be used both for entry on the pad and entry after EVA. A cut-away drawing of L1 in the Energiya history[12] does indeed show something that could be interpreted as a side-hatch. On the other hand, Siddiqi says that for entry after EVA the crew would use 'a curved tunnel in the support cone'.[13] At the moment (2003), therefore, this issue is still open for discussion.

Crew provisions in the spacecraft would be very basic, as the removal of the OM also eliminated the waste management system, food preparation area and additional storage lockers. The cosmonauts assigned to make the lunar flights commented on how much easier the new displays were to use and interpret, but they also complained about the cramped conditions they would have to endure for up to a week.

With no OM securing the L1 inside, the launch shroud had to be changed. The shroud was fitted with a launch escape system for the crew in the event of a launch abort, and was a more powerful system than that used on the Earth-orbiting Soyuz vehicles. The DM was attached to the fairing by a special support cone leading from the upper hatch of the DM, and the crew would enter the fairing through the side hatch and then via a passageway in the centre of the cone leading to the hatch of the DM.

The four-stage Proton featured three stages using nitrogen tetroxide and unsymmetrical dimethylhydrazine (UDMH) propellants. The fourth stage (Blok D) was planned for the N1 programme, and burned kerosene and LO. For the L1 missions it would be used for TLI, but on the lunar landing missions this stage would be employed for the lunar orbit injection burn and in assisting the lunar lander, the L3, with the powered descent insertion burn to descend to the lunar surface. The circumlunar missions provided an opportunity to gain valuable experience in operational use prior to committing to lunar landing missions.

Crewing for circumlunar missions

In 1966, twelve cosmonauts were assigned to a training group to prepare for possible lunar landing missions as well as a lunar orbital mission using a Zond spacecraft. They were drawn from the Air Force group based at TsPK, as well OKB-1 – the Korolyov bureau that had designed the Zond craft. They were Alexei A. Leonov (identified as the group commander), Valeri F. Bykovsky, Pavel R. Popovich, Pyotr I. Klimuk, Valeri A. Voloshin and Georgi T. Dobrovolsky; four civilian flight engineers from OKB-1 – Oleg G. Makarov, Nikolai N. Rukavishnikov, Vitali I. Sevastyanov and Georgy M. Grechko; and two Air Force engineers from TsPK – Yuri P. Artyukhin and Anatoly P. Voronov.

The members of the group were paired in possible crews, but were occasionally interchanged. Leonov and Makarov were selected to train for the first lunar landing mission, and Bykovsky and Sevastyanov were selected to fly a possible orbital mission in a Zond. Details concerning the crewing are a little confused however, as although the cosmonauts have talked about the training and possible missions, they have claimed that they were on several different crews.

The group worked on these missions until 1969–70, and when they disbanded they were assigned to Salyut and Soyuz training.

THE ZOND MISSIONS

During 1967–70, thirteen launches were attempted under the circumlunar Zond programme:

Cosmos 146 (designated 2P – 1P having been used in ground test programmes) was launched on the fifth Proton on 10 March 1967. Its primary objective was to test the Blok D upper stage. The L1P was a basic 'boilerplate' design not intended for return to Earth. On this mission the Blok D performed two burns as planned, and probably followed the scenario of the dual launch mission, with the second burn occurring 24 hours after launch, with the L1/Blok D waiting in its parking orbit for the launch of the manned Soyuz to rendezvous and dock with it. The spacecraft was launched towards a high elliptical orbit, and during its mission the thermal regulation system recorded an abnormal pressure reading and the radio beacon failed to turn off. At the time, dual launches were still a consideration, with a single cosmonaut performing EVA to complete the lunar mission. This mission may have reflected some of that planning. The DM was not recovered.

Cosmos 154 (3P) was launched on 8 April 1967. The first firing (an orbit insertion burn required to place the spacecraft in Earth orbit) went smoothly, but the ullage rockets designed to stabilise the onboard propellants (BOZ – Ignition Insurance Blok) and supposed to be jettisoned after the second Blok D ignition, were inadvertently cast off during orbital insertion, making it impossible for the Blok D to reignite 24 hours later. The automatic timer on L1 should have been turned off to prevent this error, but had not been programmed to do so. Eleven days after launch, the Blok D/L1 combination re-entered the atmosphere, having failed to complete its high elliptical orbital test mission.

4L launch abort. At a meeting of the L1 State Comission in June 1967 – in the wake of the Soyuz 1 accident – it was offically decided to abandon the Earth-orbit rendezvous for circumlunar cosmonauts and instead launch the crew on the UR-500K Proton booster; and the meeting also authorised the addition of two further automated circumlunar flights to the four scheduled before mounitng a manned attempt. The next unmanned mission (4L), to be launched on 28 September 1967, should have completed the first circumlunar profile. If the launch were to be delayed, circumnavigation of the Moon would not be possible, and so a high apogee trajectory mission was planned as an alternative mission. The circumlunar flight was considered to be a secondary objective, with re-entry from a lunar distance considered as the primary objective of the mission. Landing on 4 October (the tenth anniversary of the launch of Sputnik) would necessitate night-time recovery, which would be difficult, but there were back-up sites in the Indian Ocean. Unfortunately, vehicle 4L never reached anywhere near Earth orbit, let alone the Moon, as one of

Three views of the interior of the Zond crew module simulator, compared with Soyuz. Note the changes of the display layout and the apparent location of a navigation station on the left side (out of frame) of the simulator, with Kubasov and Yeliseyev.

the Proton's first-stage engines failed to ignite, and 60 seconds into the mission the vehicle began to deviate from the flight path. Detecting the change in trajectory, the launch escape system ignited, pulling the L1 capsule from the vehicle seconds before the launch vehicle exploded. Instead of travelling half a million miles to the Moon and back, the spacecraft DM travelled to just 65 km north of Baikonur. (Other reports indicated that the DM landed close to the burning wreckage of the Proton, and that the recovery teams faced hazardous conditions as they attempted to retrieve it). Within a week, the cause of the failure had been traced to a rubber plug accidentally left inside one of the fuel lines during construction at Perm in 1965. Although the pad (81L) was damaged, a second pad (81P) was available, clearing the way to launch another circumlunar flight in late November.

5L launch abort. Following the 4L launch abort it was decided that a least four to six successful launches of the Proton would be required before men could be placed on board. There was also discussion about the reintroduction of the reserve parachute, but this would have increased the mass of the vehicle by 200 kg, pushing it outside the capability of the Proton. The launch of 5L occurred as planned on 22 November 1967, and all appeared to be satisfactory as the first stage performed properly. However, just four seconds into the second-stage burn the Proton once again veered off course, activating the launch escape system and shutting down all the second-stage engines. Three months later the cause was still not identified, although it was thought to be due to the premature ignition of the fuel to over 200° C. The capsule was recovered and its automated destruction system deactivated before it could destroy the capsule. The parachute failed to separate, and dragged the DM 600 metres across the steppes. Despite the soft-landing rockets firing at 4.5 km instead of the planned 1.2 km, resulting in a heavy landing. The DM was in fairly good condition, with only scratches to its outer surfaces, and shattered portholes. Had cosmonauts been onboard, however, they would probably have sustained serious injuries, or worse.

Zond 4 (6L). The plan was to fly one L1 mission each month from March 1968. There were also plans to improve the landing conditions, but this contingency would not be ready for this mission, and so the capability to support a crew would need to be evaluated after the ensuing four unmanned missions. This spacecraft was planned not to fly towards the Moon, but directly opposite it, out to 330,000 km, returning at lunar velocity to test navigation and re-entry techniques before committing to a circumlunar trajectory. The launch window was therefore more flexible, and allowed more time for flight preparation. This time, three aircraft were launched to film the progress of the Proton at different altitudes should anything go wrong, but the launch on 2 March 1968 was flawless, placing the 6L spacecraft and its Blok D stage in the required parking orbit. A 459-second burn of the Blok D then sent the spacecraft to its high-apogee flight. A TASS press release indicated that the flight was designated Zond (Probe) 4, hiding the actual intention of the mission under the 'planetary probe' label of three earlier missions associated with robotic exploration of Venus and Mars.

Monitoring the mission from the Crimea tracking station were some of the cosmonauts assigned to the L1 training group. The first of three course corrections did not take place, as the 100K star sensor failed to lock on to Sirius. Later attempts could lock on to the star for only a few seconds, and an attempt by controllers to lock on to Venus also failed. After changing the filter automatically, the sensor managed to lock on to the star long enough for a 15-second course correction to take Zond 4 to its highest apogee on 6 March. This burn was so accurate that the third correction planned for 160,000 km from Earth was not required. After two trouble-free days of coasting, Zond 4 headed towards a landing on 9 March. Planned to skip into the atmosphere to an altitude of 45.8 km, it would then fly back to 145 km before completing its final re-entry and landing. However, monitoring communications ships reported that Zond 4 was experiencing 20 g, and had not skipped in the atmosphere but was making a ballistic entry. The self-destruction system was activated, and Zond 4 blew up at an altitude of 10–15 km over the Gulf of Guinea, some 150–200 km off the coast of West Africa. Kamanin was dismayed that the vessel was destroyed and that valuable data on the parachute recovery system and the state of the spacecraft had been lost, but overall the mission was judged a success. The post-flight analysis indicated that Zond 4 had deviated only 2 km from the planned re-entry corridor (well inside the allowed 10 km deviation), but that once again the 100K star sensor had failed shortly before the first entry, making it impossible for the onboard gyroscopes to maintain the proper orientation and double-skip re-entry. Apparently, the problem was due to contamination of the optical surfaces of the sensor, and for the next flight a cover would be installed.

7L launch abort. On 20 April 1968 the State Commission set the launch of 7L for 23 April. One of the discussions was the need to destroy the spacecraft in the event of an unscheduled re-entry; but Kamanin and the cosmonauts assigned to the programme naturally opposed this, because although the destruction system would be inert on the manned flights, they wanted to recover as many spacecraft as possible for post-flight evaluation to determine the reliability of the spacecraft and systems in support of manned flight. In the event of an ocean splash-down, ten naval vessels had been despatched to the Indian Ocean, and there were apparently recovery forces in the Pacific in the event of a final launch stage abort. The launch occurred on schedule on 23 April (MT), and the ascent proceeded as planned until T + 3 min 14 sec, when the escape system once again activated, initiating a normal recovery of the capsule 520 km down-range of the cosmodrome. A preliminary conclusion indicated that the problems were not associated with the launch vehicle, but with the spacecraft, which, due to a short circuit, had sent erroneous emergency signals to the rocket. It has also been determined that an altered course resulted in the shut-down of the second stage as the payload fairing was jettisoned.

8L launch accident. This mission was planned as a repeat of the Zond 4 trajectory for a 19 July launch, but during launch processing on the pad, on 14 July, the unfuelled Blok D oxidiser tank ruptured, causing the top of the rocket (including the spacecraft) to topple over and come to rest on one of the service arms of the launch tower. The fault was traced to the ground-based electrical system, which over-pressurised the tank. The accident claimed one life and a serious injury, but with 150 people on the pad at the time it could have been far worse had the stages been fully loaded with propellant. It took several days of 'heroic efforts' by launch crews to safe the vehicle and spacecraft and to roll it back to the assembly building for examination. The pad was not damaged, but although initial reports indicated that 8L could be reused, the spacecraft never flew under the 1L programme.

Zond 5 (9L). This launch, on 14 September 1968, occurred without incident, and the third stage burn-out placed Zond 5 *en route* to the Moon. On 16 September, problems with the 100K star sensor again affected a mid-course correction, but a successful burn took place the following day. On 18 November the craft looped behind the Moon at a distance of 1,950 km, and took pictures of the surface and far side before beginning the journey back to Earth. Because of problems with the 100K sensor, guided re-entry over the Soviet Union was impossible; and in addition, the guidance system was affecting the main engine, requiring the use of the two small orientation engines over a 20-hour period to keep the Zond on track for re-entry. The spacecraft had only a 13-km corridor to enter the atmosphere, but it detached its PM and antenna and entered the atmosphere as planned to complete a successful re-entry. The parachute deployed at an altitude of 7 km, and Zond 5 splashed down in the Indian Ocean 12 minutes later, at 32° 38′ S and 65° 33′ E, 105 km from the nearest recovery ship, which picked it up a few hours later.

Onboard the spacecraft – the first to go to the vicinity of the Moon and return to Earth – were a turtle (which returned alive and well), plants, flies and worms – all part of an experiment to investigate cosmic radiation on living organisms. A recorded voice was also transmitted back to Earth from 400,000 km.

Zond 6 (12L). Planning for the next flight included a guided re-entry over the Soviet Union, and it also confirmed that two further unmanned flights would be required for the system to be qualified for manned flight. During September and October 1968 there emerged details of the American plan to send men around the Moon during Christmas of that year. Kamanin felt that the Americans lacked experience in returning spacecraft from the Moon, and at such velocity, and that the Saturn V, like the Soviet N1, was not man-rated. He considered the American attempt to be 'sheer adventurism', and thought that they might fail, although he could not totally rule out the possibility that they might succeed. He considered the Soviets were better prepared for a manned circumlunar mission, but realised that they could not risk an attempt to beat the Americans with luck alone. If the Americans accepted the risk, Kamanin conceded that there was no way to stop 'their adventurous intentions to jump ahead in the Moon race.'

On 10 November the Proton worked perfectly to send Zond 6 towards the Moon – but it was soon apparent that the high-gain antenna had failed to deploy, rendering the 101K Earth sensor useless. Although communications were still possible using the low-gain system, it was obvious from previous experience with the 100K star sensor on Zond 4 and Zond 5 that expectations for a guided re-entry were not high. There was also the possibility that if the high-gain antenna could not separate from the DM, this might also affect the guided re-entry. The next day, cosmonaut Alexei Leonov completed a communications test, and successfully sent a spoken message, relayed via Zond 6, back to the control centre. Despite being transmitted via the low-gain antenna, the message was audible and clear. The 101K sensor failed, but the 100K sensor in the DM worked perfectly, allowing a mid-course correction and raising hopes that a guided re-entry was still a possibility.

On 14 November, Zond 6 looped behind the Moon, reaching an apogee in its high elliptical orbit over the far side at 2,420 km, and taking pictures. On Earth, the cosmonauts assigned to the manned circumlunar missions were discussing the chances of flying a manned L1 mission with Kamanin in response to Apollo 8. It was clear that they were unable to prepare a mission to beat Apollo 8; but if Apollo 8 failed, with a more careful approach it might be possible to send cosmonauts to the Moon in January 1969. If Apollo 8 succeeded, such a response would be unlikely to take place before April 1969. Onboard Zond 6, however, problems were being recorded in the hydrogen peroxide storage tanks (part of the attitude control

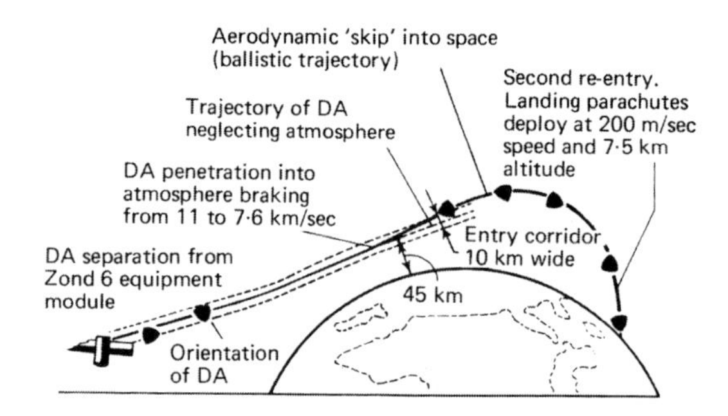

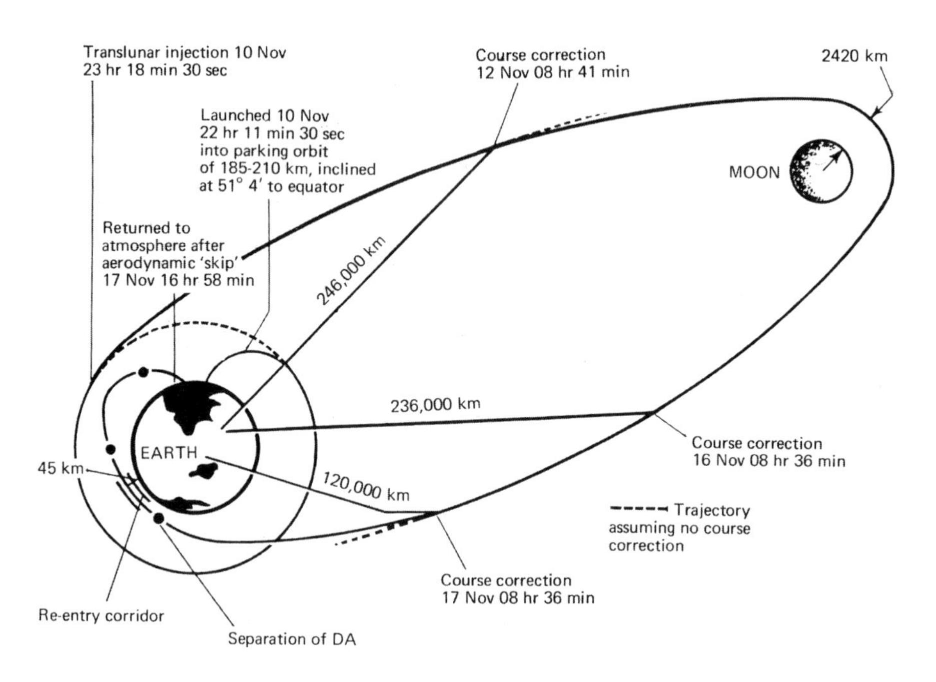

The Zond 6 flight profile. (Courtesy British Interplanetary Society.)

system), the temperature of which had fallen dramatically from +20° C to –5° C. If this trend continued, the thrusters used for guided entry would fail. On Zond 4 and Zond 5, a similar problem had been solved by turning on internal lights in the DM, but these lights were not present on Zond 6. The spacecraft could be turned to face the Sun to maintain the correct temperature, but doing so ran the risk of overheating the 100K star sensor. It was therefore decided to turn the spacecraft for the tanks to face the Sun for an hour at a time, and after about 24 hours the temperature returned to an acceptable +1° C; and the 100K sensor was working satisfactorily, allowing mid-course corrections. Data indicated that Zond 6 was slightly off course, with the low point of entry at only 25 km instead of 49 km. On 17 November the KTDU-53 was fired for 3.3 seconds to correct the re-entry into the 13-km wide corridor later

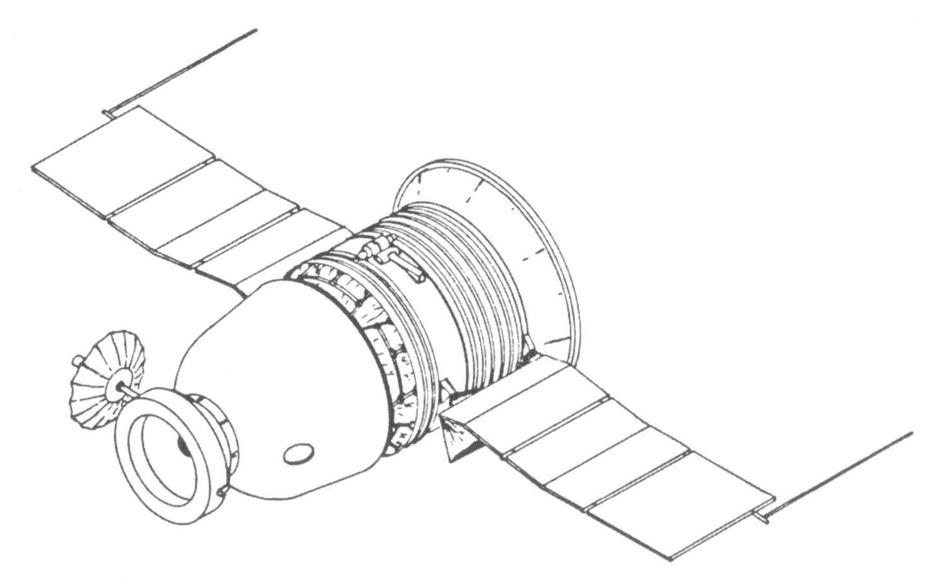

Zond unmanned circumlunar spacecraft (1967–70). (Courtesy Ralph Gibbons.)

that day. In the DM, the internal pressure showed signs of dropping from 718 mm to 380 mm, which posed a further risk to a successful guided re-entry as, in a depressurised DM, the automated system might easily malfunction.

Re-entry took place as planned, over the Soviet Union; but despite confirmation of DM separation, the expected signals were not received to confirm parachute deployment. Witnesses at Baikonur reported seeing the red-hot capsule streak across the sky, but no reports were being received from the expected recovery area. Telemetry data indicated that internal pressure had dropped from 380 mm to only 325 mm (which probably explained the lack of data from the transmitters) and also indicated that the parachutes had probably not deployed. What was certain was that somewhere in the 500-km radius of the landing zone, Zond 6 had returned to the Soviet Union – but its condition and exact position were not known. Search teams took 36 hours to find the parachutes and DM. The parachutes were found first, some 70 km from the planned landing site, on 19 November; and six hours later the shattered remains of the DM of Zond 6 were discovered 3 km from the parachutes. The final landing site was only 16 km from the pad at Baikonur. Initial inspection of the wreckage indicated that despite the damage, the self-destruct system's 10 kg of TNT had not ignited, but it was found and safely destroyed near the landing site. This was one of the most dangerous operations in the Soviet programme to date.

During the post-flight examination of the capsule it was determined that the high-gain antenna had indeed not separated during re-entry, considerably influencing the spacecraft's descent trajectory. In addition, the hydrogen peroxide thrusters had fired in only short bursts, causing the DM to deviate wildly during the descent. It also appeared that the shaping strip around the hatch had not been correctly tightened before the launch, and was the probable cause of the depressurisation of the DM. As a result, the capsule altimeter, which was not designed to work in a vacuum, sent incorrect signals to the recovery systems to fire the soft-landing rockets and deploy the parachutes while still at 5.3 km. If there had been crew onboard, their pressure garments would have protected them against the pressure leak – but they would have been unlikely to survive the landing impact. Despite all the problems, the exposed film taken during the flight had survived, and was recovered. In addition to the cameras, Zond 6 also carried biological payloads similar to those on Zond 5, and also a micrometeoroid detector. Despite the failed soft-landing, press reports from TASS indicated a successful re-entry and landing in the Soviet Union – a precursor for a manned circumlunar flight by cosmonauts.

13L launch abort. On the day that Apollo 8 landed back on Earth after a successful flight to the Moon, the L1 State Commission decided that the next launch in the programme would take place on 20/21 January 1969. There were four other spacecraft in various stages of production, and for the 13L mission the recovered vehicle from the 7L accident in April 1968 would be used. Launch occurred without incident, but just 8 minutes 21 seconds into the flight, the rocket deviated from the flight programme and headed back to Earth. Apparently, the no.4 second-stage engine had shut down 25 seconds early (T+5 min 13 sec) due to a turbo-pump bearing failure. Then, at T+8 min 20 sec, the third stage failed due to the loss of a gas generator fuel feed line, preventing the stage from compensating for the loss of the second-stage thrust to place the spacecraft in orbit. Responding to the deviation, the automatic escape system activated, sending the DM towards a narrow valley between 3,000-metre high mountains in Mongolia, about 350 km south-west of the Siberian city of Irkutsk.

N1 3L launch abort. The next Zond payload would have been sent to the Moon by means of the first N1 launch vehicle. The plan was to launch the first Ye-8 automated lunar rover on a Proton rocket, followed by a 7K-L1S spacecraft on the N1. The 7K-L1S resembled the L1, but featured systems intended for the LOK manned orbital spacecraft. At the nose of the spacecraft was the Orientation Engine Module (DOK), plus fuel tanks to place the vehicle into lunar orbit and for orbital manoeuvres. The unit would be jettisoned prior to leaving lunar orbit, and the spacecraft would use its main engine in the PM for the Trans-Earth Injection (TEI) burn. On 19 February the first part of the mission (the rover on the Proton) was launched; but just 51 seconds later the payload shroud and payload broke apart and the booster exploded, the debris falling over a 15-mile radius.

Post-flight analysis indicated that the specially-made payload shroud designed to support the rover had caused vibrations during maximum dynamic pressure, and that the shroud and payload simply broke apart. The debris struck the stages of the Proton, causing the destruction of the vehicle. The launch of the N1 proceeded, however, but due to bad weather was delayed for 24 hours from 20 February. On 21 February, the first N1 left Earth. Generating 4,590 tonnes of thrust from thirty first-stage engines, it took 13 seconds to build up enough thrust to lift the vehicle off the pad. After years of planning and delays, and thousands of hours of testing and simulation, the Moon rocket – Russia's reply to the highly successful Saturn V – was airborne... but not for long. First, two of the first stage engines shut down, but with the other twenty-eight engines working this was not a major problem. But then, 70 seconds into the flight, the onboard Engine Operations Control (KORD) suddenly shut down all twenty-eight remaining engines. The velocity carried the N1 to an altitude of 27 km before gravity took over and started pulling it back to Earth; by which time the onboard emergency detection system had triggered the 7K-L1S escape system, and once again a Moon-bound Zond was heading back to Earth minutes after leaving the pad. It landed without incident about 35 km from the pad.

Post-flight investigations determined that the KORD system had been inadequately tested before launch. During the seconds between engine ignition and lift-off it had incorrectly identified a problem on engine no.12 and its direct opposite no.24, and had shut them down. At 55 seconds into the flight, excessive vibrations caused a rupture in the fuel lines of engine no.12, and a fire had started, burning through the engine control system at 69 seconds into the flight, and shutting off the system and all the first-stage engines. Mishin was obviously disappointed in the result of the first launch, but confidently stated: 'This is normal for a first launch.' (The first two launches of the Saturn V were successful, despite small problems, allowing the Apollo 8 mission to proceed with only the third launch of the American vehicle.) It was clear that integrated systems testing was not as thorough as it could or should have been for the N1. Amendments to the design and testing of the systems would be required before the next flight, utilising a flight model further down the manifest production line (5L) in order to incorporate the required changes.

N1 5L launch abort. By early July 1969 the second N1 was ready for launch. America had successfully flown Apollo 9 in Earth orbit and Apollo 10 in lunar orbit, tested the complete Apollo CSM/LM/Saturn V system, and performed a test of the lunar EVA suit on Apollo 9. The Moon was within reach of the Americans, who the same month were preparing to launch Apollo 11 to attempt a first manned lunar landing. The Soviet launch was again to place a Zond 7K-L1S in lunar orbit. This time the thirty engines ignited and lifted the vehicle off the pad; but just after clearing the tower it fell back, and engulfed the whole pad in a huge fireball as it exploded. Witnesses remember feeling the ground shake and seeing the N1 bend over in silence before the sound wave washed over them.

In the early hours of 3 July, the N1 rocket had lifted 200 metres off the pad before loose metallic objects caught in the oxidiser pump on engine no.8 of the first stage. This initiated an explosion that took out several adjacent engines, depriving the N1 of thrust, and caused considerable damage at the pad. Fortunately, there were no fatalities – which was remarkable considering that the estimated force of the explosion was equivalent to 250 tonnes of TNT, and that chunks of hot molten metal were falling from the sky. Debris was found 10 km from the pad, and the shock-wave shattered windows 40 km away. A 400-kg spherical tank was blasted to a distance of 7 km, and landed on the roof of the installation and test wing building. Crews returned to the pad area 24 hours later to survey the damage. The only success of the mission was the flawless performance of the DM launch abort system at T + 14.5 sec – once again carrying the spacecraft clear of the explosion and landing it safely 2 km from the pad. Less than two weeks later, the Saturn V, carrying Apollo 11 to the Moon, was launched from the Kennedy Space Center in Florida on a mission into history. On 20 July 1969, while astronauts Armstrong and Aldrin collected samples of Moon rock from the Sea of Tranquillity, the Soviets were collecting the debris of their second Moon rocket accident in less than five months; and 21 July, their unmanned sample return craft Luna 15 crashed in the Mare Crisium.

Zond 7 (11L). Following the Apollo 11 mission, the momentum to reach the Moon waned. Having lost the Moon race, the Soviets had hardware to fly but no clear goal about what to do with it. Authority to continue with the circumlunar missions resulted in two more flights of Zond spacecraft. The piloted aspects of the L1 programme had been suspended in March 1969, but by August the penultimate Zond spacecraft was ready to fly. Launched on 7 August, the spacecraft was perhaps the most successful of the series, flying behind the Moon at a distance 3,200 km, and taking photographs in the process. The photographic programme had begun on 8 August, when its cameras recorded images of Earth from a distance of 70,000 km; then on 11 August, as the spacecraft swung around the Moon, two further sessions were completed. Zond 7 also carried a range of biological specimens, including four male Steppe tortoises. The only problem recorded was the failed deployment of the main parabolic antenna due to jammed securing cables, although this did not hinder any of the mission accomplishments. The re-entry on 14 August included the double-skip profile and a successful landing in the Soviet Union, with the parachutes opening at 7.5 km and the soft-landing rockets working as planned 1 m above the landing site south of Kustany in Kazakhstan, only 50 km from the planned site. At last, a 7K-L1 mission had been fully accomplished; but although it was two years later than planned, it supported the argument for sending cosmonauts on a future mission.

At the meeting of the L1 State Commission 19 September 1969, there was discussion about what to do with the remaining three spacecraft. An unmanned flight could launch as early as December 1969, followed by the first manned flight in April and a second manned flight on the last L1 later that year. However, there was little to be gained politically from such a programme; and with the Americans about to fly Apollo 12 in November 1969 and Apollo 13 in the spring, there was news from the US that about ten Apollo landings were planned by 1973 or 1974, and that hardware from Apollo was being prepared for orbiting a manned space station (which became Skylab in February 1970). Although Apollo was also under the budget axe, there seemed little to gain in sending cosmonauts to the Moon in 1970, and so the programme became more of a technological learning curve than an operational experience.

Zond 8 (12L). Launched on 20 October 1970, this mission completed the Zond circumlunar programme and took more photographs as it flew around the far side of the Moon. As well as black-and-white pictures, the spacecraft used colour film cameras and TV cameras. Despite a failure of the attitude control centre, it re-entered over the north pole and made a ballistic descent into the Indian Ocean, as planned. (This trajectory was in support of a possible similar scenario being discussed for the L3 missions.) Zond 8 was recovered from its splash-down site on 27 October, 730 km south-east of the Chagos archipelago, before being taken to Bombay docks, and from there, by plane to Moscow. The spacecraft was 7K-L1 no.14. In total, seven Zond 7K-L1 vehicles had accumulated more than 44 days' automated flight in space, including five deep-space tests, four of which circumnavigated the Moon. The last two L1 spacecraft were never flown.

The end of the Soviet lunar programme

Although the Zond circumlunar programme ended, the Soviets continued with plans to launch their N1 and resume their manned lunar programme into 1974. The next N1 was launched on 27 June 1971, but broke up at 12 miles altitude, due to excessive rates of roll. Also during 1971, the Soviets test-flew elements of the manned lunar lander in Earth orbit, but without a crew (similar to the American unmanned Apollo 5 test of the LM in 1968). A fourth attempt at launching the N1 took place on 23 November 1972, less than two weeks before the last Apollo mission (Apollo 17) left for the Moon. Lift-off was without incident, but at the end of the first-stage flight it was destroyed at 40 km altitude. As pieces fell from the sky across the Kazakhstan steppes, so did Russia's dreams of flying cosmonauts to the Moon. A fifth flight was planned for August 1974, but was cancelled during a major reorganisation and redirection of the Soviet space programme.

Thirteen N1s had been built, three of which had been used in ground tests and four destroyed in flight, while the other six never flew. The Soviets continued unmanned exploration of the Moon with the Luna programme that had begun in 1959, including sample return spacecraft, lunar orbital craft, and roving vehicles. This programme ended with Luna 24 in 1976, by which time the Soviets had turned to the creation of permanent manned space stations. Although the Zond variant of Soyuz had played a part in attempting to send cosmonauts to the Moon, the Earth-orbital ferry version of Soyuz was to become distinguished in the space station programme.

The lunar orbit module (7K-LOK)

Although no cosmonauts had ventured to the Moon, they had a spacecraft that was capable of taking them there. The L1 provided the opportunity to fly cosmonauts in a circumlunar trajectory without entering orbit, but the requirement for the manned landing programme was a vehicle capable of entering orbit, remaining there for the duration of the landing phase, and returning to Earth. This vehicle – designated 7K-LOK – had the capability to fly in space for up to thirteen days.

Two cosmonauts would ride the 7K-LOK to orbit atop the N1 launch vehicle (with the L3 lunar lander and Blok G and Blok D rocket stages – together called the Lunar Rocket System). There would have been a flight test version, designated T1K, but this was cancelled in favour of testing on the N1. The estimated mass of the LOK was 9.85 tonnes, and it had a length of 10 metres and a standard diameter of 2.2–2.3 metres, extending to 3.5 metres across the aft frustum. The estimated habitable volume was 9 cubic metres, and onboard systems were said to be more advanced than on the Earth-orbital Soyuz due to the complexities of the mission. The T1K would itself have been used for Earth orbital tests.

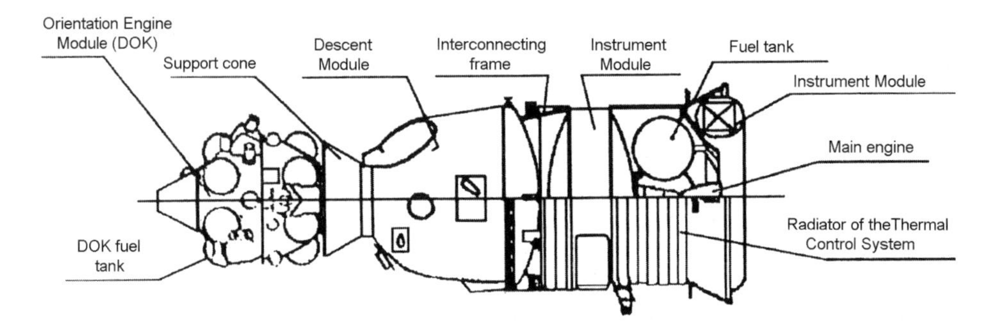

The 7K-L1S spacecraft that employed elements from both the circumlunar spacecraft and those intended for the LOK Orbital Module. Two of these craft were lost in the February and July 1969 launch failures of the N1 launch vehicle. The purpose of their missions included placing them in lunar orbit to take detailed photographs of proposed landing sites for the Soviet LK lunar lander. The DOK engine, designed for the LOK, orbiter would have placed the 7K-L1S into lunar orbit, and then separate before the vehicles main engines burned to take the spacecraft back toward Earth. (Courtesy Bart Hendrickx and the British Interplanetary Society.)

The PM had a diameter of 2.2 metres and a length of 2.82 metres, and was divided into three sections: the pressurised instrument compartment containing the necessary flight instrumentation for telemetry and radio communications, and a system of attitude control engines to assist in docking; an unpressurised transfer compartment; and the aggregate compartment. A large spherical propellant tank was located in the PM. This was the source of propellant for the main engine, which differed from the main Soyuz PM engine. This engine was not employed until after the lunar lander (L3) and Blok D had separated for the landing in lunar orbit. It would have been used for the TEI burn to begin the return to Earth after the landing and for mid-course corrections on the way home. The UDMH and nitrogen tetroxide were in the same 1.9-metre diameter tank divided by a membrane, part in the aggregate compartment and part in the skirt. The aft skirt also connected to the wider L3 lander and Blok stages carried at the rear of the Lunar Rocket System – a restartable unit with two chambers, capable of a thrust of 3.388 tonnes. In addition there was a separate Soyuz engine with 417 kg of thrust and a capacity for thirty-five restarts, allowing orbital changes during lunar orbital operations; and there were a further sixteen smaller reaction control engines for attitude control in the skirt, fuelled from the same tanks. Instead of solar arrays this variant carried Volna 20 (Wave) fuel cells – the first hydrogen–oxygen fuel cells incorporated on a Soviet manned spacecraft – with a mass of 70 kg and the capability of supplying 1.5 kW at 27 V for up to 500 hours.

The DM was built to the familiar 'headlight' shape of the Soyuz DM, but featured much thicker ablative thermal protection due to the increased loads and the speed of returning from the vicinity of the Moon. Measuring 2.19 metres long and 2.2 metres in diameter, it could support a two-man crew for launch and Earth recovery. Inside were the two cosmonaut support couches and the controls and displays to operate the spacecraft, plus the onboard computer and life support system. There was also a forward internal hatch for crew transfer into the OM section of the spacecraft. The heat shield and the parachute recovery system was similar to those on the 7K-OK Soyuz.

The foremost module consituted the living quarters – similar to the OM on the Earth-orbital Soyuz spacecraft – and featured storage compartments, water and food supplies, an area for the preparation of meals, the waste management system, and the location for the single Orlan (Bald Eagle) lunar EVA suit. In addition, there was a smaller control unit at the front of the sphere, where the Flight Engineer could control the spacecraft during the lunar orbital rendezvous and docking phase of the mission. At the front of the 2.26-metre long module was the 800-kg, 1.5-metre orientation engine complex, including six spherical tanks containing 300 kg of storable UDMH, as well as an additional four cylinders that could supercharge the tanks during use. This system serviced four sets of engines used for attitude control during critical operations in lunar orbit.

On the side of the OM was a hatch (larger than that on the Soyuz 7K-OK), but there was no internal docking transfer hatch, and so to transfer from the LOK to the L3 and back again, one of the cosmonauts would need to perform EVA. At the front of the vehicle was the active unit of the Kontakt (Contact) docking system. The active element of this system was a spring-loaded probe designed to penetrate and grip the passive unit on the top of the L3 lander. This unit was a honeycomb fixture, and was designed to be used just once on each mission. When the lander returned to orbit after leaving the lunar surface, the cosmonaut in the LOK would close in to bring the two elements of the Kontakt system together. Accuracy was not required, as the two spacecraft had only to be linked firmly enough to allow the returning Moon-walker to transfer back to the return spacecraft by EVA. A live LOK was carried on the fourth N1 launch.

A change of direction

By 1971 the 'original' Soyuz had flown nine missions, plus the five Zond missions. Although no cosmonaut had ventured to the Moon, the Soyuz was about to begin a new role as a space station ferry – a role which it would maintain, in four manned variants and an unmanned resupply vehicle, for more than thirty years.

REFERENCES

1 Oberg, James, 'There's no Soy Fuzz in Soyuz', *Spaceflight*, **37** (1995), 157.

2 Hall, Rex D. and Shayler, David J., *The Rocket Men*, Springer–Praxis (2001).

3 Clark, Phillip S. and Gibbons, Ralph F., 'The Evolution of the Soyuz Programme', *Journal of the British Interplanetary Society*, **36** (1983), 434–452.

4 Oberg, J., 'Russia Meant to Win the Moon Race', *Spaceflight*, **17** (1975), 163–171, 200; and 'The Hidden History of the Soyuz Project', *Spaceflight*, **17** (1975), 282–289.

5 Ertel, Ivan, and Morse, Mary, *The Apollo Spacecraft: A Chronology*, **1**, NASA SP-4009 (1969).

Lunar missions and hardware

Spacecraft (serial number) name	Design designation	Launch date	International designation	Docking date	Undocking date	Target spacecraft/port	Landing/ decay date	Flight duration dd:hh:mm:ss
Key: (U), unmanned								
11F91 (Zond lunar manned fly-by spacecraft)								
(01)	7K-L1M	*Mock-up (Maket); used for tests at Baikonur in 1967 January*						
(02) Cosmos 146	7K-L1P (U)	1967 Mar 10	1967-021A *Blok D high-altitude test; still in orbit(?)*					
(03) Cosmos 154	7K-L1P (U)	1967 Apr 8	1967-032A				1967 Apr 10	02:05:23:??
(11) Zond 7	7K-L1P (U)	1969 Aug 8	1969-067A				1969 Aug 14	06:18:25:??
(04)	7K-L1 (U)	1967 Sep 28	*First-stage failure of Proton LV; SA recovered by means of the SAS*				1967 Sep 28	00:00:00:37 +
(05)	7K-L1 (U)	1967 Nov 22	*Second-stage failure of Proton LV; SA recovered by means of the SAS*				1967 Nov 22	00:00:02:09 +
(06) Zond 4	7K-L1 (U)	1968 Mar 2	1968-013A				1968 Mar 9	07:00:20:??
(07)	7K-L1 (U)	1968 Apr 23	*Second-stage failure of Proton LV; SA recovered by means of the SAS; refurbished for vehicle no.13*				1968 Apr 23	00:00:03:14
(08)	7K-L1	*Planned 1968 Jul 21*						
(10)	7K-L1	*Possibly intended for manned spaceflight; constructed(?)*						
(09) Zond 5	7K-L1 (U)	1968 Sep 14	1968-076A				1968 Sep 21	06:18:26:??
(12) Zond 6	7K-L1 (U)	1968 Nov 10	1968-101A				1968 Nov 17	06:18:48:00
(13)	7K-L1 (U)	1969 Jan 20	*Failure of Proton LV; SA recovered by means of the SAS; used refurbished capsule from vehicle no. 7*				1969 Jan 20	00:00:08:20
(15?)	7K-L1	*Possibly intended for manned spaceflight; constructed(?)*						
(14) Zond 8	7K-L1 (U)	1970 Oct 20	1970-088A				1970 Oct 27	06:18:00:00

11F92 (Stripped-down L1 vehicle, to be placed in lunar orbit by two N1 test-flights)						
(03)		7K-L1S (1) (U)	1969 Feb 21	*First-stage failure of N1 LV; SA recovered by means of the SAS*	1969 Feb 21	00:00:01:10 +
(05)		7K-L1S (2) (U)	1969 Jul 3	*First-stage failure of N1 LV; SA recovered by means of the SAS*	1969 Jul 3	00:00:01:10 +
(?) (Stripped-down instrumented L1, in a study of weightless propellant behaviour in Blok D upper stage)						
(01)		7K-L1E (1) (U)	1969 Nov 28	*First-stage failure of N1 LV; SA recovered by means of the SAS; debris fell in China*	1969 Nov 28	00:00:09:16
(2K)	Cosmos 382	7K-L1E (3) (U)	1970 Dec 2	1970-103A *Still in orbit(?)*		
11F93 (LOK lunar manned orbital spacecraft of L3 lunar complex)						
			1971 Jun 26	*First-stage failure; spacecraft crashed*	1971 Jun 26	00:00:00:52
			1972 Nov 23	*First-stage failure; SA recovered by means of the SAS*	1972 Nov 23	00:00:01:47
11F94 (LK manned lunar cabin of L3 lunar complex)						
Not Soyuz-derived vehicles						

6 Hall, Rex D. and Shayler, David J., *The Rocket Men*, Springer–Praxis (2001), pp.214–262.
7 Siddiqi, Asif, *Challenge to Apollo*, NASA SP-4408 (2000).
8 Varfolomeyev, Timothy, 'Soviet Rocketry that Conquered Space', pt. 12, *Spaceflight*, **43** (2001), 28–31.
9 *Ibid.*, pt. 9, *Spaceflight*, **41** (1999), 207.
10 Vetrov, G., *Sergei Korolyov I ego delo.*
11 Siddiqi, Asif, *Challenge to Apollo*, NASA SP-4408 (2000), 517–521.
12 Hendrickx, Bart, 'The Kamanin Diaries', *Journal of the British Interplanetary Society*, **53** (2000), 409.
13 Siddiqi, Asif, *Challenge to Apollo*, NASA SP-4408 (2000), 559.

Mission hardware and support

The first chapter of this book discussed the origins of the vehicle that became known as Soyuz. During the 1960s, many studies were carried out on variants of this spacecraft, leading to a complicated genealogy; but one constant during the career of Soyuz has been the three-module design. The following is a discussion of the basic configuration and sub-systems of the vehicle that is common to all manned versions flown, and this is accompanied by a summary of the Soyuz launch vehicle and ground support infrastructure required to mount any Soyuz mission in Earth orbit. In researching this book, it became apparent to the authors that relatively little *detailed* information is available on the physical hardware and sub-systems of Soyuz, compared with data concerning American manned spacecraft.

HARDWARE AND SYSTEMS

The basic design and configuration of the manned spacecraft Soyuz has remained unchanged for almost forty years. Through numerous upgrades and modifications, the three-module design has remained the focal point of the spacecraft design since it was first introduced in 1966; and even the unmanned Progress vehicle features a three-module design inherited from its manned ancestor (see p.243). The design of the crew compartment has been amended and the sub-systems upgraded over the years, but the basic appearance of the spacecraft has remained unchanged. And throughout this period there has been a notable lack of large spacecraft identification marks or national emblems (unlike most US manned spacecraft).

For some time, details of the design features of the early vehicles were very limited, but with the creation of the American/Soviet Apollo–Soyuz Test Project in 1972, the West was for the first time provided with details of the spacecraft. Due to American concern over the safety of Soyuz (in light of the Soyuz 1 and Soyuz 11 accidents, the Soyuz 15 docking failure, and the Soyuz 18-1 launch abort), a series of NASA-generated Safety Assessment Reports (SAR) was produced from Russian data, and from meetings of the ASTP US/USSR working groups, to establish the safety of Soyuz for docking with Apollo, and that of the American astronauts when transferring into the Russian vehicle.[1]

With the demise of the Soviet Union, cooperative ventures with other countries, and the limited release of official documentation, the workings of Soyuz became more widely known. The following is intended as a guide to the rest of this book, and, based on information released at the height of the ASTP in 1975,[2] briefly describes the components and major systems of Soyuz, but is expanded to include the relevant details from more recent years.

Crew positions

Depending on the mission, a Soyuz can be manned by up to three cosmonauts – normally two or three. During the launch phase the Soyuz is usually operated automatically, and there is little for the crew to do during launch except sit back and hope that all goes well. Emergency procedures are also automatic, as, in principle, is docking, although the Commander will take over if problems arise. In such cases, the Flight Engineer assists by reading out velocity and range data. Landing is also usually automatic.

First crew-member (centre seat): Soyuz Commander (Komandir Korablya). The Commander is normally an Air Force cosmonaut, although Kubasov and Rukavishnikov flew as civilians. For the ISS this role has expanded to allow non-pilot cosmonauts from ESA member-states to fly in an emergency situation, and some RKK Energiya Flight Engineers have recently been qualified for ISS re-docking and emergency return operations, including Usachev on Expedition 2 and Budarin on Expedition 6.

Second crew-member (left seat): Flight Engineer (Bortinzhener, or BI). On Soyuz 12–40, this was the seat that was replaced by support equipment, and on all these flights the Flight Engineer or Interkosmos cosmonaut occupied the right seat. Beginning with Soyuz T, the flight engineer occupied the left seat.

Third crew-member (right seat): Researcher Crew-member (Kosmonavt Issledovatel). This was the seat occupied by the Soyuz Research Engineer on Soyuz 4/5, 10 and 11. When the crew was reduced to two, the Flight Engineer occupied this seat instead of the left seat. Once the crew size reverted to three on the Soyuz T/TM and TMA flights, this became the position occupied by commercial 'passengers'.

Call-signs: Following a practise begun on the Vostok missions, each Commander assigns himself a call-sign – usually derived from a geographical feature, a celestial body, a mineral, meteorological conditions, or from numerous other elements or phenomena. The Commander normally uses the same call-sign on each of his missions, while the remainder of the crew adapt the same call-sign, adding either '2' or '3' for personal identification. Several Flight Engineers have therefore, throughout their careers, used different call-signs when flying with different Commanders.

THE SPACECRAFT

The three main modules of Soyuz are, from the aft: the unpressurised Propulsion Module (PM), which acts as the vehicle's Service Module; the pressurised Descent Module (DM), which contains the crew and acts as the command centre of the spacecraft; and the Orbital Module (OM), which provides habitation facilities, storage and EVA access, and is a transfer compartment to and from a second vehicle. According to RKK Energiya, the materials used in the fabrication of these modules were chosen between 1961 and 1963, and titanium, beryllium and aluminium alloys were all studied before an aluminium alloy was finally chosen. The thermal insulation covering the modules consists of two layers – the upper layer consisting of an asbestos-cloth laminate, and the lower layer of 'a light, heat-insulating material'.

The Propulsion Module (PM)

This cylindrical compartment is full of instrumentation, and is divided into three sections: intermediate, instrument and assembly. The PM generally has a mass of 2,560 kg, an overall length is 2.3 metres, and a diameter of 2.2 metres flared out to a 2.72-metre skirt-shaped base flange that attaches to the upper stage of the R-7.

Intermediate. (Often translated as 'transfer section'.) This is the foremost section of the PM, and interfaces with the Descent Module (DM). It is the location for the fuel tanks for the approach and orientation engines, the majority of the approach and orientation engines themselves, and their sub-systems. On its exterior is mounted the smaller of the two thermal control radiators.

Instrument. This is a sealed and pressurised compartment that houses the thermal control system heat exchangers, associated equipment, radio and telemetry equipment, power supply systems, and guidance and rendezvous instrumentation.

Assembly. This unpressurised compartment contains the main propulsion system, additional approach and orientation engines, and on-orbit storage batteries. On the exterior is mounted the largest of the two thermal control radiators. The aft of this section is a base ring that mates to the upper stage assembly of the R-7 launch vehicle. In the early designs for a circumlunar Soyuz, a toroidal section, at the base of the assembly section, included electrical systems for rendezvous and docking in Earth orbit, which would then be jettisoned before TLI. On the 'original' Soyuz flights this unit was retained, but was not jettisoned in flight.

On Soyuz 1–11, each solo Soyuz (13, 16, 19 and 22, but not Soyuz 12) and all Soyuz T, TM and TMA craft, two sets of wing-like solar battery panels were mounted on opposite sides of the instrument–assembly module. The PM is separated prior to re-entry, and is destroyed in the denser layers of the atmosphere.

The Descent Module (DM)

This module – the only part of Soyuz that is recoverable – is mounted above the PM, and is occupied by the crew for launch, orbital manoeuvring, docking, undocking,

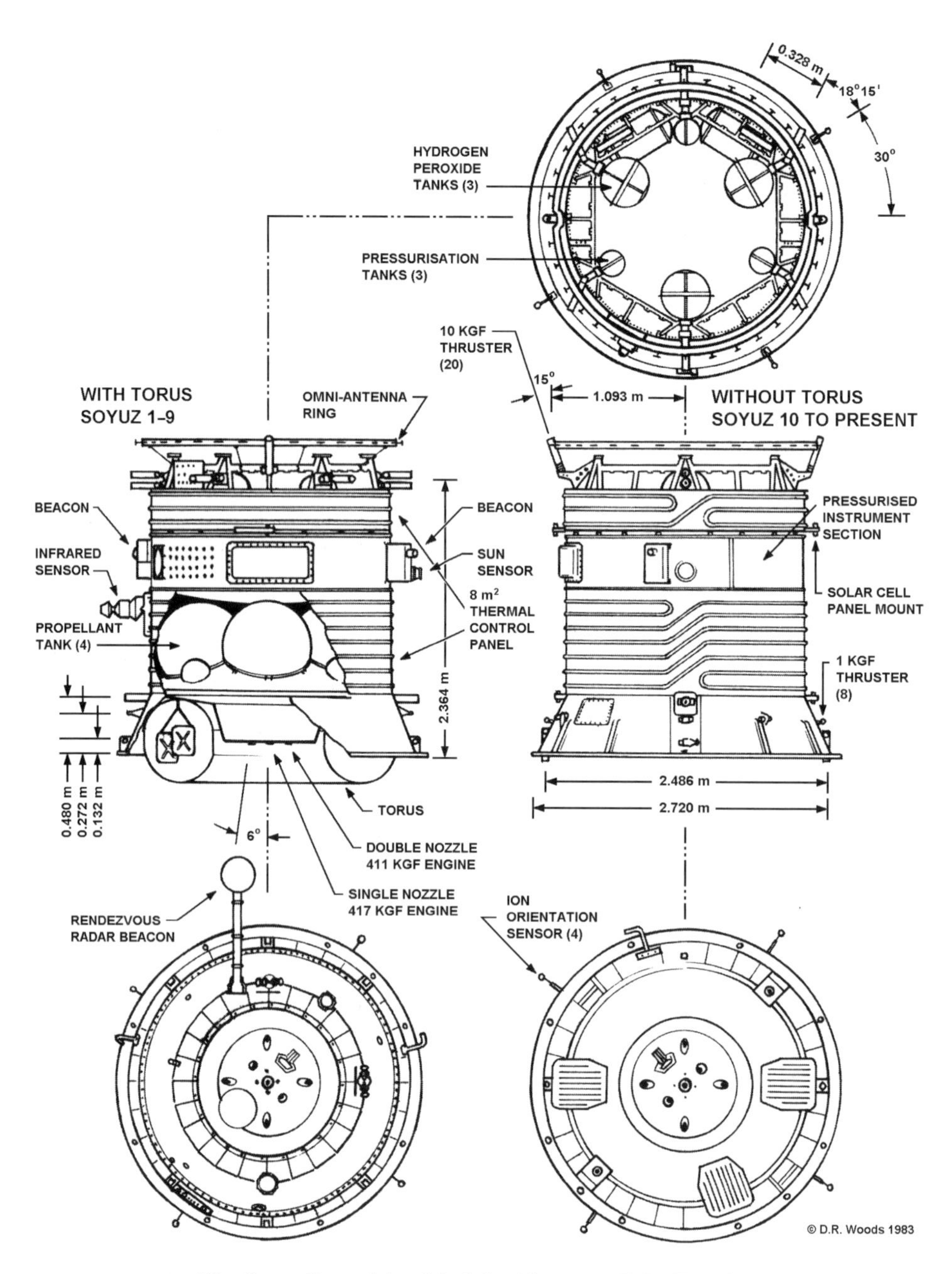

The Soyuz Propulsion Module. (Courtesy D.R. Woods.)

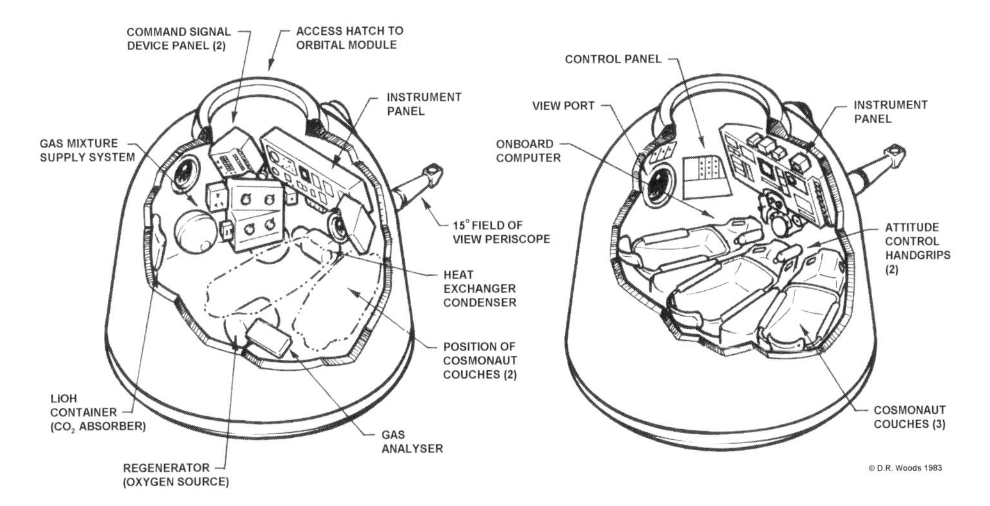

A cutaway of the Soyuz Descent Module. (Courtesy D.R. Woods.)

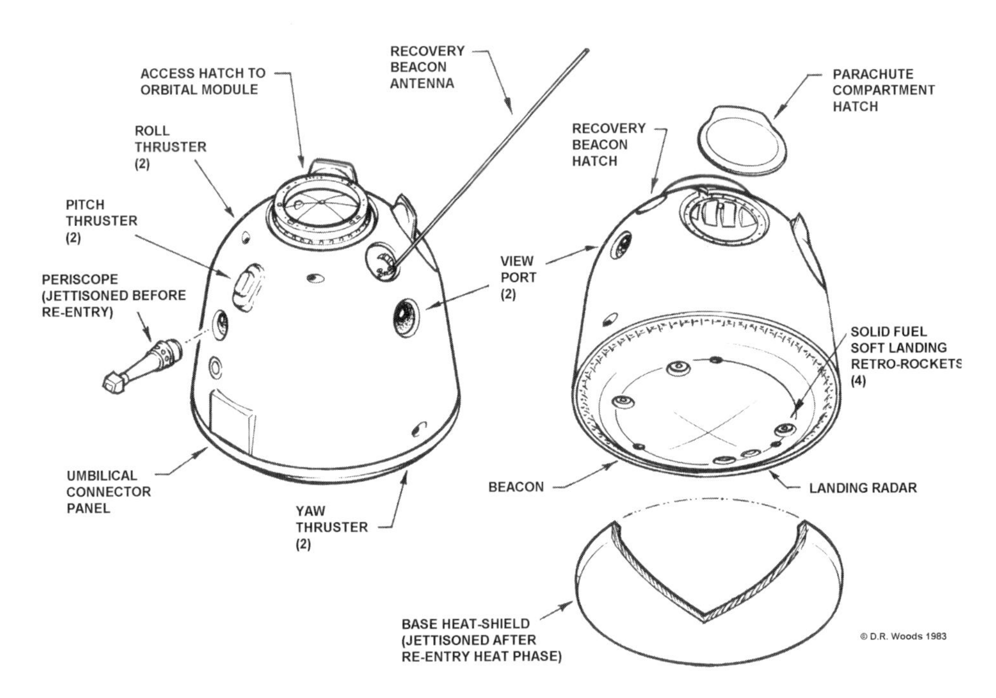

Details of the Soyuz Descent Module, showing the separated heat shield and soft-landing rocket system underneath. (Courtesy D.R. Woods.)

re-entry and landing. It has several hatches – one for the crew in the forward part (hatch no.1) for access to the OM and for entry and exit at the beginning and end of the mission, and a separate hatches for the primary and back-up parachutes. Its exterior is covered with an ablative coating for protection during re-entry, and a lower heat shield that, at the end of the mission, detaches to reveal the soft-landing retro-rockets used to cushion the landing. The details of the construction of the heat shield and the thickness of the ablative coating are not clear. Nick Johnson describes it as 'a high-temperature-resistant ablative material similar to that used on Vostok',[3] and a similar description is provided by Asif Siddiqi.[4] Details of the use of ablative materials and the design of the heat shield on Zond DM on lunar distance missions are also unclear. These heat shields had to withstand the higher speeds of a skip re-entry from the Moon (see p.28), but probably adapted research efforts already carried out for Soyuz, with slight changes in thickness and composition. The DM has a nominal mass of 2,800 kg, a length of 2.2 metres, and, to match the DM, a maximum diameter of 2.2 metres at its base. The module was designed with two main areas: the work area, containing the crew couches, controls and displays, and equipment bays; and the instrument area, incoporating the life support, vehicle attitude control and landing systems.

The Orbital Module (OM)

This is the foremost assembly, and is habitable during orbital operations. The forward end of this module has the facility for docking apparatus and/or an internal transfer hatch (no.2) and tunnel (no.2) (except Soyuz 6, 9, 13 and 22), or for locating scientific instrumentation (Soyuz 13 and 22). The module also contains a side hatch used for crew entry on the launch pad, and (on Soyuz 5/4 only) for egress and ingress during EVA. There is also an aft tunnel (no.1) with a hatch cover for entry into the DM (hatch no.1). Initially, two windows were provided – one forward of the side hatch for Earthward observations, and the second on the opposite side for celestial

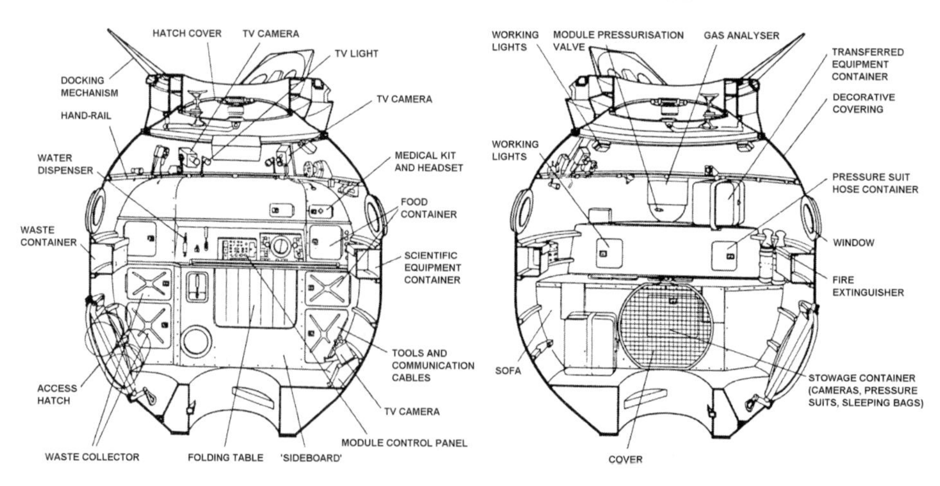

A cutaway of the Soyuz Orbital Module. (Courtesy NASA.)

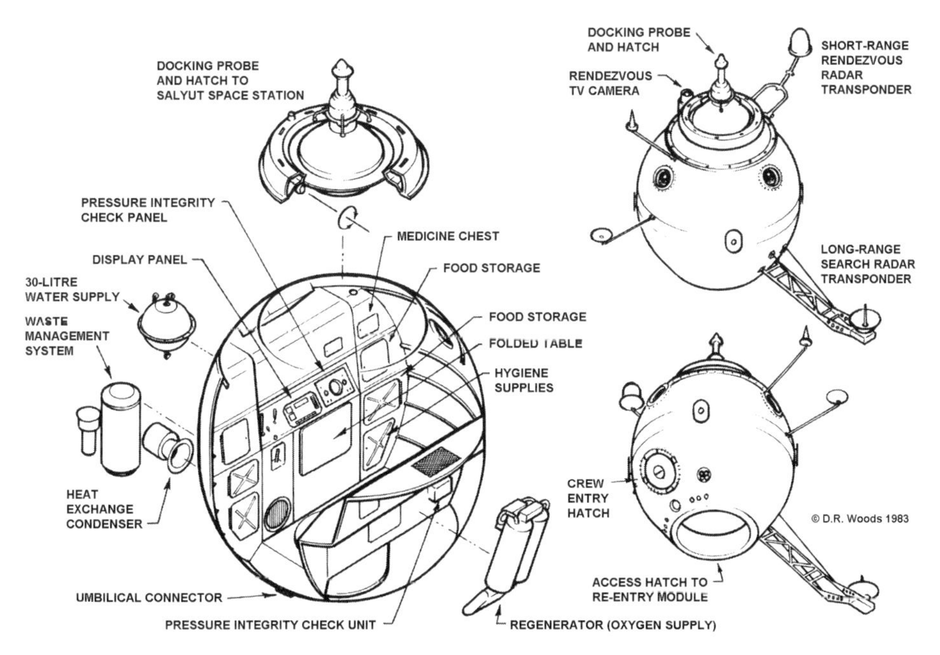

Details of the Orbital Module. (Courtesy D.R. Woods.)

observations. From Soyuz TM, a forward-looking window was installed for the Flight Engineer to use during docking with a space station. This module, like the PM, is separated before re-entry and is destroyed in the atmosphere. During early Soyuz T missions (1980–82), Western reports interpreted some Soviet information as suggesting that the OM could be left attached to a station as an extra small compartment; but this has never been officially confirmed by the Soviets, nor have they ever officially confirmed this or released details of such a capability. The OM has a nominal mass of 1,200 kg and a maximum diameter of 2.25 metres (see the accompanying table on p.404 for comparison of all Soyuz variants recorded here). The OM also features observation portholes, a bunk, cupboards for food and clothing, EVA equipment, experiments, navigational and photographic equipment, TV cameras, radio equipment, and other associated equipment and supplies. On space station ferry missions it can be used to transport additional supplies to the station and as a location for rubbish to be incinerated with the OM at the end of the mission.

Pyrotechnic devices

Soyuz includes a number of pyrotechnic devices (pyrocartidges and explosive bolts), which are used for the deployment and separation of spacecraft parts and hydro-pneumatic lines. The pyrotechnic devices are used to separate the modules of the spacecraft, prepare parachute systems by opening caps and latches, deploy antennae (four pyrocharges) and docking targets, and open and close valves in the hydro-pneumatic lines. They are not, however, used to deploy the solar arrays. Both pyrocharges and explosive bolts are used for module separation, and for all other

A cutaway Soyuz Orbital Module located on top of a Soyuz Descent Module at MAI.

A segment of an Orbital Module used to train young engineers at MAI.

operations only pyrocharges are used. The charges on Soyuz include six explosive bolts and six pyrocharges for the separation of the OM and DM, and six pyrocharges for the separation of the DM and PM. Eight pyrocharges are used to separate cables from the PM to the OM, eleven are used for separation of back-up control lines, two are used for separating the periscope from the DM, and four are used in controlling the cooling system lines.

During recovery, eleven pyrocharges are used to release the control unit of the Descent Control System (DCS) lines, twenty-four are used to jettison the cover of the parachute container; and eighteen are used to jettison the cover of the back-up parachute if required. Separation of the aft heat shield requires twelve pyrocharges, while four are used to arm the crew couch shock-absorbers, eight to fire the soft-landing engines, and four to unblock breathing vents.

SPACECRAFT SUB-SYSTEMS

The Soyuz spacecraft has a variety of similar sub-systems incorporated into its overall design, and over the years these have been upgraded as new variants of the spacecraft have been introduced. (A list of contractors for these systems and hardware can be found on p.443.)

Rendezvous, docking and transfer

Docking systems. Several variants of docking systems have been incorporated into Soyuz. Work on these systems began under the direction of OKB-1 engineers Viktor P. Legostayev and Vladimir S. Syromyatnikov in 1962. The initial design was a simple pin-and-cone device to allow the Soyuz 7K to dock with a series of 11K tankers, and so no internal transfer hatch was provided. When the Soyuz programme was redirected in 1965, Korolyov suggested the incorporation of such a hatch; but the design of the original pin-and-cone assembly was far advanced, and with the desire to place Soyuz in orbit as soon as possible, the idea was postponed (until 1971).

On the original Soyuz spacecraft (1967–69), therefore, a simple probe and drogue (or pin-and-cone) assembly was installed for evaluation and demonstration. The active spacecraft carried a probe and a set of latches, while the passive spacecraft carried a receiving cone, socket and capture latches. In order to accept the probe, the frustum of the passive unit was longer than the active unit. The probe was designed to be the shock absorber, with sensors in its tip to register contact with the passive cone and to disable the control system on the active spacecraft, while at the same time it would fire its thrusters to bring the two spacecraft together. At the apex of the passive cone, the probe encountered the socket in which a series of catches and a restraining ring locked into place to secure the two spacecraft together. On the rim of

The early Soyuz/Salyut pin-and-cone docking assembly.

both spacecraft docking units, plugs and sockets engaged in a male/female configuration to establish electrical and communications connections between the spacecraft. This type of arrangement was tested on the unmanned Cosmos 186/188 and Cosmos 212/213 missions in 1968 (see p.139), and was performed by the manned Soyuz 4/5 spacecraft in 1969 (see p.147). Similar arrangements were also planned and/or attempted for Soyuz 1/2 in 1967, Soyuz 2/3 in 1968, Soyuz 7/8 in 1969, and for the very first unmanned Soyuz, Cosmos 133. However, the launch of the second Soyuz (which should have been Cosmos 134) was cancelled when Cosmos 133 developed problems while in orbit.

From 1971, space station missions (both manned Soyuz and Progress resupply vessels) included an improved pin-and-cone docking system, and also provision for the removal and use of an internal transfer hatch. The active spacecraft inserts its probe into the drogue on the passive station. At the apex of this cone, latches in the socket engage, and motors draw them in to secure the two vehicles together. Gas, electrical, fluid and communications are connected in the rim of both collars. Once pressure and seal integrity checks are completed, the cosmonauts remove the probe and drogue assembly and open the hatches between the two vehicles, forming a passageway tunnel from the Soyuz/Progress into the space station. To undock, the probe/drogue unit is replaced and the hatches are closed, and four spring-loaded rods – aided by the remaining air between the exterior hatches – push the spacecraft apart. If for some reason the push-rods fail, the cosmonauts can fire pyrotechnic bolts to separate the spacecraft from the station or, as a last resort, detach the PM/DM from the OM for an immediate return to Earth.

The L3 manned lunar landing mission programme would have used a different system, but it never progressed to flight operations (see p.35). As with the 7K-OK system there was no provision for internal transfer, and a lone cosmonaut had to perform EVA from the LOK orbital parent spacecraft to reach and return from the landing craft (LK). The docking system utilised an active spring-loaded probe on the lunar orbital craft (LOK) to penetrate 108 honeycombed passive docking fixtures. This was a much simpler and lighter structure which required little accuracy or precision flying, as it would have been used only once on the mission, when the lunar lander returned to lunar orbit and rendezvoused with the command ship. The pilot of the LOK would align the pin or the probe with any location on the honeycomb structure on the roof of the LK, and allow the probe to penetrate and be captured by 'claws', within the passive system, that pulled the two spacecraft together. The whole system was mechanical, and had no electrical or power transfers between the two spacecraft.

For the joint mission with the Americans in 1975, a different docking system needed to be developed in order to link up with the American Apollo, which featured the probe/drogue system with twelve capture latches used on the manned lunar programme. To compensate for a change in atmospheres between the two spacecraft (see p.205), the Americans developed a Docking Module to include a jointly developed androgynous docking unit, designated the Androgynous Peripheral Assembly System 1975 (APAS-75), which fitted onto the Soviet end of the Docking Module to allow the Soyuz to dock with it. Compatibility between the Soyuz and the

The internal Soyuz transfer hatch from the Descent Module into the Orbital Module.

Docking Module was achieved by incorporating standard elements that connected or interacted. An erectable T-shaped docking target (of the type used on the Apollo LM) was installed on the Soyuz docking assembly, which was sighted through an Apollo CSM optical alignment sighting device mounted adjacent to a CM forward rendezvous window. In the event of a deployment failure of this target, a back-up fixed target was installed on the Soyuz. The Soyuz 19 spacecraft was also equipped with two flashing beacons for visual acquisition during final approach, and four orientation lights on the outboard corners of the solar arrays. The forward light on the left (+Z) side was red, the forward light on the right (–Z) side was green, and both aft lights were white. It was foreseen as a prototype of future international docking system, with the capability of rescuing stranded crew-members.

On the Russian segment of the ISS there is now also a hybrid of the pin-and-cone and androgynous systems. This has the linking mechanism of the pin-and-cone, but the circumference of the APAS-89 (used on Soyuz TM-16). The hybrid system is situated on the multiple docking adapter of Zvezda (nadir, zenith and front ports) and on the aft port of the FGB. It was originally planned that the first three Soyuz vehicles for the ISS would be equipped with the hybrid docking port (presumably to link up with the Zvezda nadir port), but these plans have since been scrapped, and it appears that all Soyuz vehicles to the ISS will incorporate the pin-and-cone system. All the current docking ports for Soyuz (Zvezda aft, Pirs nadir and FGB nadir) incorporate the pin-and-cone system.

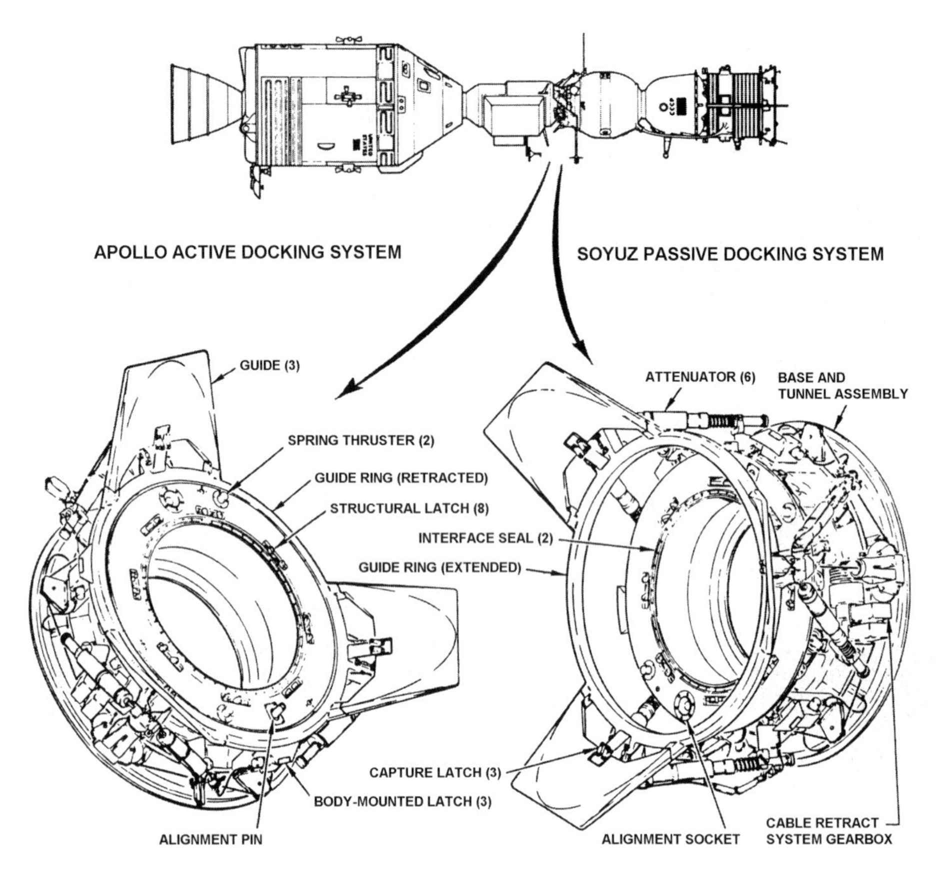

Details of the EPAS docking system and its installation on the American Apollo and Soviet Soyuz spacecraft.

Rendezvous systems. A number of rendezvous systems have also been incorporated in Soyuz. From 1960 to 1963, the techniques of orbital rendezvous and docking in space were studied at OKB-1's Department 27, headed by Boris V. Raushenbakh's, and the supervision of this work was directed by OKB-1 Deputy Chief Designer Boris Ye. Chertok. The research focused on two separate spacecraft orbiting the Earth, and was further divided into long-range rendezvous and close-range rendezvous. One spacecraft would be passive in the operation and have facilities to respond to and receive the second active spacecraft that would perform the rendezvous and docking operation. The accuracy of long-range rendezvous focused on the ability to place the active spacecraft in the most optimum orbital trajectory, which would be based on ground measurements, computer models, the timing of the launch, and the performance of the launch vehicle. Any errors could be compensated by ground stations up-linking commands to the spacecraft for a series of orbital corrections, to refine the trajectory to the desired parameters. The results of theoretical calculations by the rendezvous engineer indicated that Soyuz could bring

Ground tests of the active and passive units of the ASTP docking hardware.

the two spacecraft together, at a relative velocity of ± 40 metres per sec, to within a position in space measuring $25 \times 15 \times 15$ km.

From this position it was clear the two spacecraft would be unable to complete a physical docking, and so the Rashenbakh group focused on a 'parallel approach' (line of sight from the active spacecraft to the passive spacecraft) to bring the two spacecraft close enough together to physically dock. It was clear that ground-based measurements could not be used to identify two distant and separate vehicles in such close proximity, and onboard measurements were also impossible due to the limits of Soviet technology in sufficiently miniaturising computer hardware to be accommodated within the confined spacecraft. Onboard radar was instead to be employed to assume the control of the spacecraft to accomplish the final stages of the rendezvous approach. Four institutions submitted proposals for supplying the Soyuz rendezvous systems, and the two final designs chosen were Kontakt (Contact), from the Moscow Power Institute's Experimental Design Bureau, and Igla (Needle), proposed by NII-648. After detailed evaluations the Igla system was chosen over the more advanced Kontakt system, while the latter was later evaluated for the manned lunar landing programme.

The Igla system – used on all original versions of the Soyuz spacecraft, Soyuz T and Progress vehicles through 1986 – manoeuvred the passive spacecraft so that the

Space writer Andy Salmon operates the TORU remote docking controls at the 1994 ILA Air Show in Berlin. TORU is used for training cosmonauts to dock Progress transport craft automatically from onboard Mir. (Courtesy Andy Salmon.)

The TORU training facility at TsPK.

chosen docking port faced the approaching spacecraft; while the Kurs system – used on Soyuz TM, Soyuz TMA and Progress M and M1 – allowed the active spacecraft to approach and fly around to the chosen port. (In 1971 an alternative rendezvous system, called Lira, was briefly considered and discussed. It was planned to use it to fly 7K-S to DOS 4, but it was abandoned in favour of Igla. Nothing further is known of this system.)

On Mir, TORU was also installed. This allowed a cosmonaut inside Mir's core module to remotely control the approach and docking of a Progress automated freighter. For rotational and transitional control of the approaching freighter, the cosmonauts used a pair of hand-controls. These were installed on the main control panel, and functioned in the same way as the Soyuz hand-controls. A camera, located in the docking unit of Progress, transmitted live TV pictures (with numerical data overlay) of its approach to a monitor screen in front of the cosmonaut, and this allowed the operator to fly Progress from within Mir as if he was onboard the approaching spacecraft. This system was tested on Progress M-15 in 1992. In June 1997, the operation of TORU was partly responsible for the collision of Progress M-34 with Mir (see pp. 264 and 355).

Electrical power

On the majority of Soyuz spacecraft (except the Soyuz 12–40 ferry craft), the electrical system provides dc power from solar arrays, with storage batteries providing a buffer mode, to produce an onboard voltage of 23–34 volts. As a back-up in case of main system failure, the storage batteries provide power for short-term operations on the spacecraft, sufficient for an emergency return to Earth. Small batteries are also provided in the DM for operation during re-entry and descent, and for a few hours immediately after landing (see also Soyuz 23 entry on p.196). For Soyuz 1–11, the spacecraft carried solar arrays, of up to four panels, with a total area of 14 square metres, a length of 3.6 metres, and a width of 1.9 metres. However, on Soyuz 16, 19 and 22, and on Progress, Soyuz T, TM and TMA, these consisted of three segmented panels, 3 metres long. These were folded flat against the PM during launch, and were deployed on entering orbit. The four panels had a bend in the segment joints, producing a bird-wing appearance, and the three-segment panels were deployed flat. On Soyuz T, the four panel wings were reintroduced, and were also used on Soyuz TM and TMA. These wings are also deployed flat. Whip-like antennae for communications and telemetry systems were deployed from the ends of the arrays. During the independent orbital mission, the active (upper) side of the panels was orientated towards the Sun by manoeuvring the vehicle with the solar orientation system and attitude control engines. The Soyuz was placed in an orientation to rotate the panels towards the Sun to constantly feed solar power into the storage batteries during the sunlit side of each orbit, using the storage batteries for power during the night pass. The operation was controlled by the cosmonaut, using the attitude control system to roll Soyuz and to allow the solar disc to appear in the cross-hairs on the Vzor optical orientation device. He then commanded the vehicle to rotate around a Sun/spacecraft axis for constant illumination of the solar arrays, to ensure that the temperature was equal across the spacecraft. On Soyuz 9 in

1970, the crew completed the longest solo flight in a Soyuz spacecraft (18 days), and reported experiencing unpleasant sensations due to the spin stabilisation of their spacecraft (see p.168).

For the Salyut/Almaz space station missions between 1973 and 1981, the independent flight duration was planned at only two days from launch to docking and recovery on the same day as undocking. Instead of solar arrays, chemical batteries were provided, which limited Soyuz to a 2–3 day independent flight and, if docking did not occur, an immediate return to Earth. In an emergency situation (such as on Soyuz 23 in 1976 and Soyuz 33 in 1979; see p.229), the spacecraft could conserve power by shutting down non-essential equipment, but this could extend the mission by a only few hours to the next available landing opportunity. This severely limited the flexibility of these missions, but once docked to a Salyut, the Soyuz could be mothballed and sustained directly from the station resources.

Had it not been for the Soyuz 11 disaster, the Russians would probably have continued using solar panels on subsequent Soyuz vehicles. The removal of the panels saved mass required elsewhere for post-Soyuz 11 modifications.

Thermal control

Soyuz employs both active and passive thermal control methods. The active system features two loops – one used to maintain the correct temperature and humidity of the habitable modules (the DM and OM), and the second to control the temperature in the Propulsion Module. Both loops are thermally interconnected by means of a liquid-to-liquid exchanger. For passive thermal control in Soyuz, a combination of thermal control coating, multi-layer insulation and thermal bridges is incorporated in the system to minimise unregulated heat exchange between the spacecraft and space, while the primary method of heat exchange is by conduction and radiation in the unpressurised compartments and by forced air convection in the pressurised compartments. Equipment cooling is, for the most part, controlled by the direct transfer of heat from the equipment to the atmosphere, and to the liquid stored within the heat exchanger units by means of forced air convection. In some cases, items of equipment are cooled by passing a heat-carrying liquid through a network of channels on the spacecraft structure. Air temperature is controlled with three automatic settings at 288, 293 and 298 K, while liquid temperature is controlled with four settings at 276, 278, 280 and 282 K.

Life support

The design of the system for supporting the crew inside the Soyuz was influenced by experience gained on the Vostok series of spacecraft (1960–66), and is more complex due to the increased size of the crew compartments (and the addition of the OM), the number of crew, provision for atmosphere replenishment after EVA or crew transfers, and longer mission durations. The life support system consists, of course, not only of breathable air, but also includes food and water supplies, waste collection, medical support by data collection, and the provision of personal rescue kits with survival equipment and clothing, rations and signal aids in the event of emergency landing.

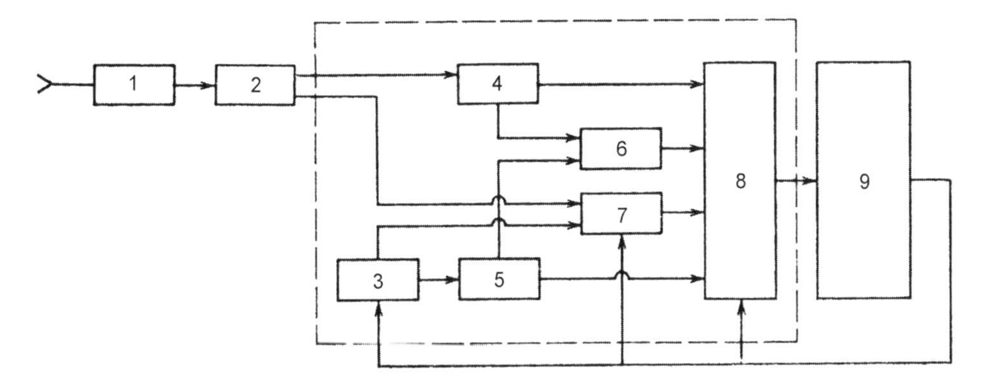

A schematic of the Control System of Onboard Systems (SUBS): 1, receiver; 2, decoder; 3, control and display panel; 4/5, command matrixes; 6, timer; 7, digital information unit; 8, unit to generate commands; 9, onboard systems.

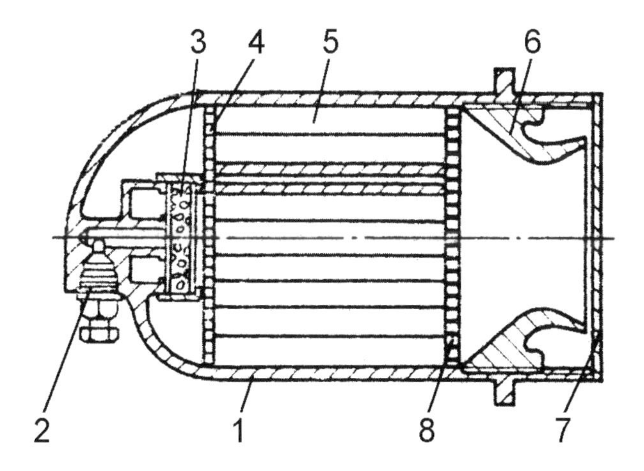

The soft-landing engine: 1, body; 2, explosive charge; 3, ignitor; 4, grid; 5, solid propellant; 6, nozzle unit; 7, blow-out plug; 8, diaphragm.

Using similar modes of operation as on Vostok,[5] the internal temperature of Soyuz is maintained at $20 \pm 3°$ C, with a cabin pressure of 710–850 mm (140–200 mm of which is due to the content of oxygen) and 40–55% relative humidity. The temperature and humidity is controlled by a series of single loop heat exchangers. Oxygen and carbon dioxide levels are maintained to within safe limits with a superoxide that releases oxygen and lithium hydroxide to absorb carbon dioxide. Part of the system is a special sensor that constantly measures the partial pressure of oxygen in the atmosphere, and controls the rate of air-flow through oxygen regulators and the carbon dioxide absorbers to within the required atmospheric levels. Dust and odours are removed from the air by a series of filters within the system. The operation and status of the atmospheric system is displayed on the control panel so that it can be monitored by the cosmonauts throughout their flight.

Pressure equalisation and relief is provided by three valve assemblies, located in the DM, the OM and the forward docking tunnel (tunnel no.2). If necessary, the third of these can be used to pressurise/vent the tunnel prior to the opening of the hatch or undocking, while the DM/OM pressure equalisation valve, located on hatch no.1, is used to equalise the pressure in the DM and OM prior to opening that hatch. The pressure valve in the OM cabin is used to vent the compartment and for checking the hatch no.1 seal in the DM prior to separating the OM towards the end of the mission. A pressure integrity check system monitors the integrity of the DM and OM, tunnel no.2, docking interfaces and the hatches; and this same system also vents the OM prior to docking, pressurises the interfaces between docking seals, and equalises the OM/tunnel no.2 pressure levels. All valves and gauges are mounted in the OM, together with check unit that warns of excessive pressure loss.

The pressurisation system consists of an external spherical steel tank containing 4.5 kg of air. This is used for repressurising the OM and DM after undocking, and can also be used for emergency repressurisation in case of accidental pressure loss (part of the Soyuz 11 recommendations). The air supply for the Sokol pressure garments, located in the DM, provides a mix of 40% oxygen and 60% nitrogen, and the system incorporates a contingency pressure relief valve, descent and landing pressure equalisation valves, vents and fans.

Onboard the Soyuz, food supplies are determined by the duration of the mission. The daily nutritional value of a cosmonaut's diet on non-space station missions was about 3,000 kilocalories rationed around four meals (breakfast, lunch, dinner and supper), multiplied by the length of the mission, and with the addition of a built-in contingency percentage. The largest amount of food stored on a Soyuz was for the 18-day Soyuz 9 flight of two cosmonauts, during which a supply of up to 36 man-days of meals ($36 \times 4 = 144$ meals) was required, with probably a two- or three-day extension contingency. The Soyuz food is packaged in cans, tubes, and plastic containers, and the crew is provided with a folding work surface (table) with restraint devices, and an electric heater for heating food tubes, both of which are located in the OM. Emergency rations are available in the DM, but there is no means of heating food, so if any has to be consumed in orbit (with the OM separated, as on Soyuz TM-5 in September 1988), it is eaten cold. The food in the OM includes hot meals, snacks, beverages such as soup, meats, vegetables, fruits and desserts. Hot coffee is also available, as are soft drinks and a certain amount of fresh provisions.

Water is supplied from a 0.03-square-metre storage tank in the DM. To obtain water, the cosmonaut operates a hand pump to pressurise an air chamber in the tank, which pushes a diaphragm against the water supply and forces the water through a flexible hose into the DM. Each cosmonaut uses an individual mouthpiece, and operates a valve assembly to drink the water. A safety unit maintains sufficient pressure for water delivery without over-pressurising the tank.

The waste management system, located in the OM, consists of a fold-away toilet for the collection, isolation and storage of body wastes. To use the device, the cosmonaut has to place an insert (bag) in the collection receptacle to contain the expelled faeces. After use this receptacle is closed, removed from the unit, placed in a container, and sealed. In the same toilet device, liquid waste is collected in an individual soft urine receptacle, and is transported to a liquid waste receptacle by a flow of air created by two centrifugal fans. The air passed through the liquid waste receptacle is then fed to a purifying filter, after which the purified air is vented into the cabin. In case of system failure, a back-up receptacle is provided to collect urine in bags for storage. Similar rubber containers are located in the DM. The early urine receptacle was reportedly criticised by cosmonauts as being 'inconvenient' (that is, it probably leaked, or was uncomfortable to use).

All Soyuz crews are fitted with biomedical monitoring devices, via sensors attached to the body, to gather data on ECG rates, breathing rate and temperature – the data being transmitted to ground control during ascent, orbital flight and re-entry. TV cameras located in the DM also monitor crew actions during ascent.

In the Soviet and American programme, the value of crew post-flight survival equipment was recognised as necessary due to the experiences of downed military aircraft crews. The first such equipment in the Soviet programme was used during the Voskhod 2 recovery in March 1965,[6] when the cosmonauts had to endure a 48-hour delay in being recovered from the snow-covered forests of Siberia. Over the years, the accrued experience of missions such as Soyuz 5, 15, 23, 24 and T-5 has led to the improvement of crew rescue and survival aids, including emergency rations, signalling equipment, life rafts in case of a water landing, desalting kits, water survival suits, winter and wilderness clothing, and the learning of skills to live off the land until rescue (see p.75).

The water survival suit – the Forel (Trout) – is a one-piece orange nylon flotation suit with attached rubber-soled feet and integrated hood, flotation collar, emergency beacon, and signalling devices on the shoulders. The thermal protection suit – the TZK (termozashchitnyy kostyum) – is made of quilted blue nylon, and has front closure zippers and waist draw cords, trousers, thermal boots, gloves and headgear.

Habitability

In Glushko's encyclopaedia,[7] figures for the OM are 6.6 cubic metres total volume, but only 4 cubic metres habitable volume; and for the DM, 3.85 and 2.5 cubic metres respectively, producing a total habital volume of 6.5 cubic metres for the crew. On a Soyuz (which follows Moscow Time) a working day lasts about 16 hours, including time for meals, exercise and personal hygiene. Once orbital flight has been achieved, the crew remove their pressure suits and don light flight coveralls until the docking phase or re-entry, when pressure suits are again worn. No pressure suits were worn by the crews of Soyuz 1–11 (the Soyuz 4 EVA crew did not don their EVA suits until just before their spacewalk), and they instead wore three-piece polyester woollen flight suits (trousers, jacket and hood).

Since 1971, only five Soyuz flights – Soyuz 13, 16, 19, 22 and TM-5 – have lasted more than three days in independent flight, and so orbital routine has been limited to the two days prior to docking with a space station. As a result, the flight plan is usually concerned with checks of the spacecraft and its systems, the rendezvous and docking profile, and adaptation to weightlessness. Very few in-flight experiments and research objectives have been included on space station ferry missions since the full evaluation of Soyuz as a 'long-duration vehicle' performed by the Soyuz 9 crew on their 18-day mission in June 1970.

Sleep can be taken either in the couches in the DM, or in the OM using the sleep restraints (bags) or the couch (with suitable restraints) provided. Shaving kits include either an electric razor or, if desired, foam and safety razors. Washing is performed by using damp and dry towels, and at least one change of underwear and socks (to keep the feet warm) is supplied. A vacuum cleaner is also provided for cleaning the contaminants out of the cabin air.

Up to three shock-absorbing Kazbek-U seats are located in the DM for the crew to use during launch and entry and during dynamic manoeuvres of the spacecraft (such as docking) in orbit. (Kazbek is a mountain in the Caucasus, near the Russian–Georgian border.) The Kazbek-UM seat is the modified version on Soyuz TMA, and is increased in size to accommodate larger ISS crew-members from America or Europe as required (see p.384). The seats are angled at 80° to the horizontal, the bent knees being supported in a harness to withstand landing impacts of up to 7.5 metres per sec on the DM body. Each seat incorporates a structural framework with harness

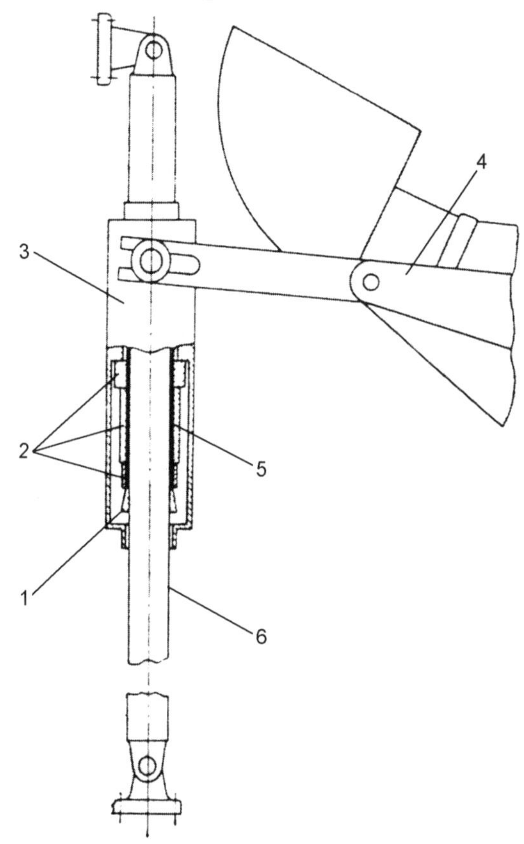

The shock absorber of a Soyuz seat: 1, conical blade; 2, stretchable rings; 3, support housing; 4, frame with seat liner; 5, immobile sleeve/plug; 6, rod/stem.

and squib-actuated ram-jack shock absorbers to withstand landing loads. This supports an individually moulded shock-absorbing couch liner of fibreglass and plastic foam construction, with provision for harness webbing attachment to the permanent seat supports. They can be removed and exchanged in-flight on Soyuz crew/spacecraft exchange operations on long-duration station missions. A back-up couch liner is also available as a spare or for use during a post-flight emergency, to immobilise an injured or ill cosmonaut.

Propulsion and attitude control

The development of the primary engine system for Soyuz (7K) began at the OKB-2 plant in Kaliningrad during 1962, and for the past four decades the spacecraft has retained a four-engine system: the primary (and for Soyuz 1–40 only a back-up) manoeuvring and retro engines; the attitude control system engines; the vernier translation engines; and the re-entry attitude control engines.

Primary (KTDU) and back-up engines

KTDU denotes Correction and Braking Engine Installation – which is what was used for 'public consumption' during the Soviet era. Other designators being used by the engineers at the Isayev bureau (mainly the 'S') were SKD (primary) and DKD (back-up), although these are not widely used. Located in the assembly compartment of the Propulsion Module, the main single combustion chamber hypergolic propulsion system for Soyuz was the S5.35 engine (used on Soyuz 1–40), producing a thrust of 417 kg, and designated KTDU. Also available is an almost identical twin combustion chamber back-up engine, with additional nozzles on either side of the main engine nozzle, and with a thrust of 411 kg. Both engines use unsymmetrical dimethyl hydrazine (UDMH) as a fuel, and nitric acid as an oxidiser. The propellants are stored in four spherical tanks (two for fuel and two for oxidiser) mounted at the base of the PM assembly compartment, and are turbine-fed to the engines. The engines are used for major orbital manoeuvres to a quoted (but never attained) maximum 1,300-km apogee, and for the de-orbit burn at the end of the mission. The back-up engine was not installed on the Soyuz T, TM or TMA versions.

Soyuz 1–40. KDTU-35/11D62/S5.35. The main engines could be fired up to twenty-five times, with burn times lasting from fractions of a second to several hundred seconds, and a quoted overall burn time of more than 500 seconds. These were also used on the first-generation Progress, and modified versions were used on Zond lunar missions (without the back-up engine) and DOS 1–4.

Soyuz T. 11D426 (other designators are not known). The thrust is 315 kg, burn times are not known. It used nitrogen tetroxide/UDMH, and was part of a Combined Engine Installation (KDU) that also incorporated the DPO and DO thrusters, and which all drew from the same propellant supply. As with Soyuz 1–40, there was no back-up, as this capability was provided by four of the DPO thrusters. It was also pressure-fed, and as with the KTDU-35, turbo-pumps were not required.

Soyuz TM. KTDU-80/S5.80. An improved version of the Soyuz T engine, it was also used on Progress M/M1. Burn times are not known, but the thrust is 316 kg, and it can be fired forty times. The improvements over the Soyuz T engine included the combustion chamber – made of special ablative materials without any requirement for cooling – and the use of a metallic rather than an elastic membrane between the pressurisation gas and the propellants

Attitude control

The Russians refer to these as Approach and Orientation Engines (DPO). They are used for fine-tuning the orbit during the final approach, and also for attitude control. On Soyuz T/TM they are also used as back-ups for the main engine, and are fuelled by hydrogen peroxide (which is also used to power the turbines in the main propulsion system).

Vernier translation

The Russians refer to these as Orientation Engines (DO). They are used only for attitude control, and have a much lower thrust than that of the DPO thrusters. For Soyuz TM, the designator DPO-B (High Thrust Approach and Orientation Engines) has been used for what is usually called DPO, and DPO-M (Low Thrust Approach and Orientation Engines) for what is usually called DO.

On Soyuz 1–40, the DPO (attitude control) and DO (vernier) systems featured ten DPO thrusters (10 kg) on the PM intermediate compartment, and four DPO thrusters (10 kg) and eight DO thrusters (1 kg) on the PM assembly compartment.

On Soyuz T there were fourteen DPO thrusters (13.5 kg) and twelve DO thrusters (2.45 kg). The four DPO thrusters also acted as back-up for the main engine. DPO and DO thrusters were part of the Combined Engine Installation. The DPO thrusters were assigned the index 11D428A-10 and 11D428A-12 – the only difference between them being the fuel lines. Specific impulse was 251 sec, and burns could last from 0.03 sec to 2,000 sec. Theoretically, they could be restarted 500,000 times with a total burn time of 50,000 sec.

Installed on Soyuz TM were fourteen DPO thrusters (also called DPO-B – 13.5 kg) and twelve DO thrusters (also called DPO-M – 2.5 kg). Early Soyuz TM and Progress M vehicles used the same DPO thrusters as Soyuz T (the A-10 and A-12). An improved type, introduced in 1997, had the designators 11D428A-14 and 11D428A-16 (the only difference again being the fuel lines). They had the same thrust as their predecessors, but an improved specific impulse of 290 sec, resulting in less fuel consumption during the final approach manoeuvres. Possible restarts and total burn time were the same. A few of these were initially installed on Progress M-36, M-37 and M-38, and a full set was flown for the first time on Progress M-39. The first Soyuz to use the new thrusters was TM-28 in 1998.

The A-14 and A-16 were also used on the Zvezda (ISS) Service Module. The 11D428A-10, 11D428A-12, 11D428A-14, and 11D428A-16 thrusters were built by the Scientific Research Institute of Machine Building in Nizhnyaya Selda in the Sverdlovsk region. No additional information is known concerning the DO 'verniers' for Soyuz TM or any improvements incorporated on Soyuz TMA.

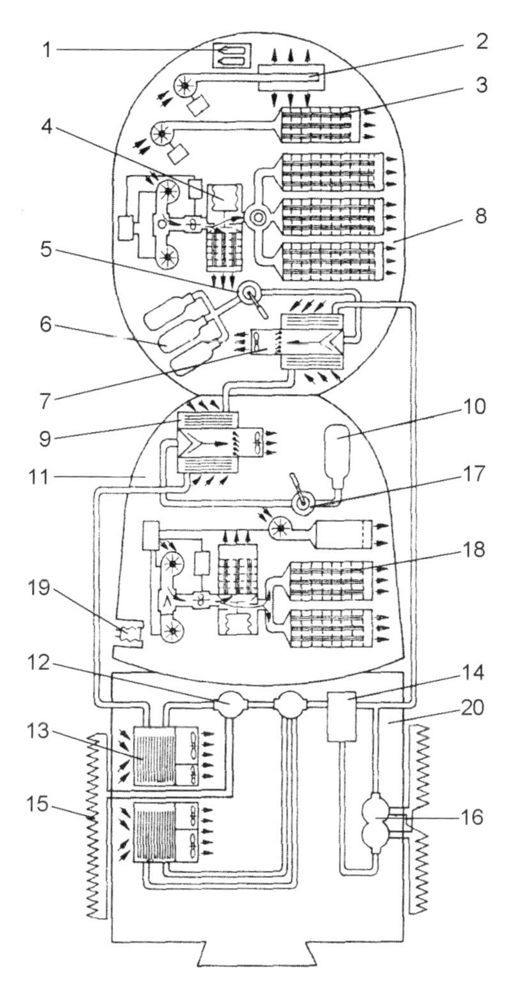

Soyuz life support systems. 1, heater for food packets; 2, autonomous carbon dioxide scrubber; 3, 4, 18, regenerative fixtures (oxygen); 5, 17, hand-pump; 6, 10, apparatus for condensation; 7, 9, freeze-drying equipment (dehumidifier); 8, Orbital Module; 11, Descent Module; 12, liquid–liquid heat exchanger; 13, gas–liquid heat exchanger; 14, regulator for liquid expenditure; 15, radiative heat-exchanger; 16, sectional liquid–liquid heat exchanger; 19, pressure regulator block; 20, equipment compartment.

Re-entry attitude control

Six 10-kg hydrogen peroxide thrusters are located in the DM for use during re-entry, to control the attitude of the descending capsule. and for lift during the re-entry profile.

According to RKK Energiya, the original plan was to install the tanks of these thrusters inside the DM, but in 1964 it was decided to move them to a small pressurised compartment in the back of the DM behind the crew couches. Almost four decades later, hydrogen peroxide is still used in this system – the official reason for its use being that it is environmentally clean (since the engines are used in the atmosphere). However, there are disadvantages. It is not stable; and the presence of the hydrogen peroxide tanks in the DM is one of the primary factors in limiting the on-orbit lifetime of Soyuz, because much of the oxygen evaporates and causes an increase in pressure inside the tanks. Burn times are not known.

Vehicle control

The left-hand controller in the DM allows the cosmonaut (usually the Commander) to fire the attitude control rockets in rates of 0°.5 or 3°.0 per sec in the lateral movement axes (X, Y or Z) in six directions: forwards or backwards, left or right, up or down. The right-hand controller governs the attitude of the vehicle in pitch (nose up or down), yaw (turn left or right) and roll (rotate left or right). Translation controllers allow the cosmonaut to make fine adjustments to the speed of these movements. By using a combination of both controllers, the Soyuz can move in any orientation to a second target spacecraft and remain in close proximity to it, move to it and station-keep, dock or undock, and perform separation manoeuvres.

This system is part of the System of Orientation and Motion Control (SOUD), which also uses the attitude control sensors and Vzor optical sighting device, the gyros and an onboard computer, and the Igla (or Kurs) system, as well as the attitude control engines. The cosmonaut rolls his spacecraft until the Earth appears in Vzor, and then activates the three step gyros. Two of these have 2° of freedom for inertial orientation, and the third has 3° of freedom for sensing angular velocity. In addition, Soyuz carries a second orientation system, using IR sensors relative to Earth's vertical, and stellar, solar and ion sensors for velocity vector measurements, all of which are located on the exterior of the PM. Reading from the sextant, the crew can also set an optical sensor so that the angle between it and a solar sensor is the same as that between the Sun and a known star.

The location of the attitude control system sensors on the Soyuz spacecraft has been studied by Sven Grahn.[8] The infrared vertical sensor uses an opto-electronic instrument that measures angular misalignment on the –Y axis and the local vertical using radiation from the Earth and its atmosphere converted into a command signal for pitch and roll. Three electronic ionic sensors (two on the +X axis and one on the –X axis) are used to measure misalignment of the Soyuz in the +X axis from the orbital velocity vector. This system uses incident ion flow on the Soyuz, and converts this into pitch and yaw command signals. The opto-electronic 45K Sun sensor produces a signal that corresponds to the solar attitude misalignment of the Soyuz in the +Y axis to within ±6° of the sensor's line of sight. The crew monitors the attitude by means of a shade gauge, and if the sensor fails, solar orientation has to be performed manually. The gyro system is composed of two units (KI-38), each of which incorporates a gyro, a caging mechanism, angle sensors, and a programming device. These are used simultaneously without gimbals aligned to the roll and yaw axis of Soyuz. This system allows 0–360-degree manoeuvres about the roll and yaw axis and for selected inertial attitude hold. The allowable variance of angular deviation about any axis is only ±8°, beyond which an emergency (Avaria) signal is activated. Another system – the rate gyros – contains three gyros which emit signals proportional to spacecraft angular rate vectors in the C, Y and Z axes. The operator can select an integrated (time) mode by which the unit can provide inertial or attitude hold over a period of time, allowing programmed attitude manoeuvres. Deviations of more than $\pm 6^\circ$ also initiate an Avaria signal.

Communications

In the original design of Soyuz, the communications sub-system – developed at NII-885 – included radio and TV links, a telemetry channel, and voice communications. It was developed to support long-range deep-space circumlunar flights, and in 1964 was modified to operate in Earth orbit, with the work to develop the new system being divided among several institutions.

To maintain voice communications during the flight (including re-entry) and after landing, an ultra-short-wave and short-wave system (Zarya) was developed by NII-695, which had also worked on the Vostok communication system. The work of providing Soyuz with a TV system fell to NII-380, which drew on the experience of providing the first images of the lunar far side from Luna 3 in 1959. This system – Krechet – has developed into a system designed to monitor rendezvous and docking, provide interior views of the crew compartments, exterior views, and, during the ASTP in 1975, televise joint activities with the Americans. The TV pictures can be displayed to the crew by means of onboard monitors, as well as being transmitted to the ground via the communications links. One 625-line per frame, 25 frames/sec TV camera was located in the main display panel, two outside the spacecraft, and one in the OM.

A two-way radio telephone and telegraph system provides direct contact between the crew and mission control or, via ground stations, with other spacecraft (Soyuz, Salyut, Mir, ISS and, for the ASTP mission, the American Apollo and NASA MSC in Houston via US communication links). The system also transmits direction-finding signals for location of the DM during parachute descent and landing, as well as providing communications (using headphones and microphones in a soft headgear) with other members of the same Soyuz crew.

The radio-telemetry system was developed at NII-885, which had provided the Vostok system. The system included the installation of forty small T-shaped antennae around the aft end of the DM. Automatic data recorders onboard the Soyuz (Mir 3), supplied by NII-88, capture telemetry data from various systems onboard the spacecraft, and this is used for post-flight analysis. Some have described Mir 3 as the 'black box' of Soyuz. It was destroyed on Soyuz 1, thus complicating the accident investigation, and was strengthened in the wake of that accident. The up-link facility to the spacecraft is used to transmit commands and digital display data, various reception signals, and trajectory measurements. The down-link is used for transmitting spacecraft system status data, and medical data on the crew.

Antennae on the DM are used to communicate with the ground during re-entry after PM separation, a small VHF antenna is embedded in the DM hatch, and there are also small VHF and short-wave antennae in the parachute lines.

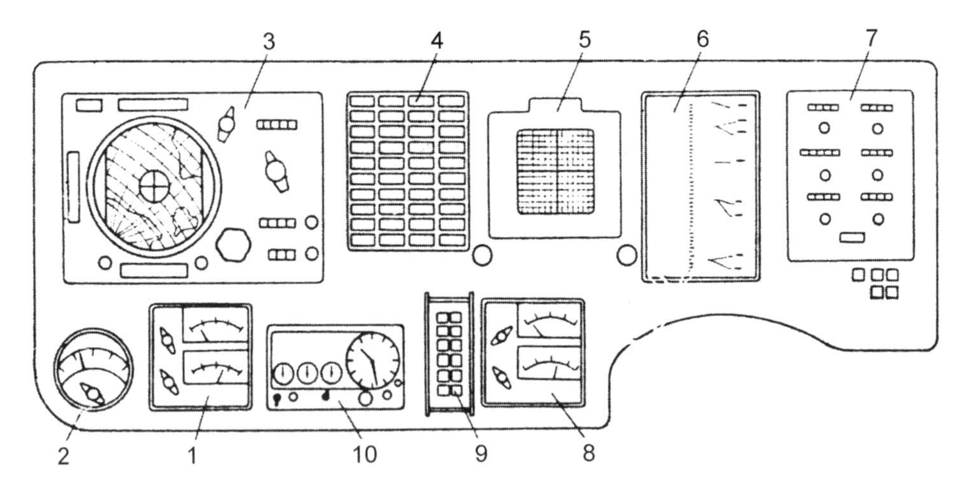

The instrument panel: 1, pressure and temperature displays; 2, voltage and current display; 3, navigation display; 4, panel with electroluminescent indicators; 5, combined electron-ray display; 6, programme control display; 7, device to enter digital information; 8, range and approach speed displays; 9, keys for entering very important commands; 10, clock.

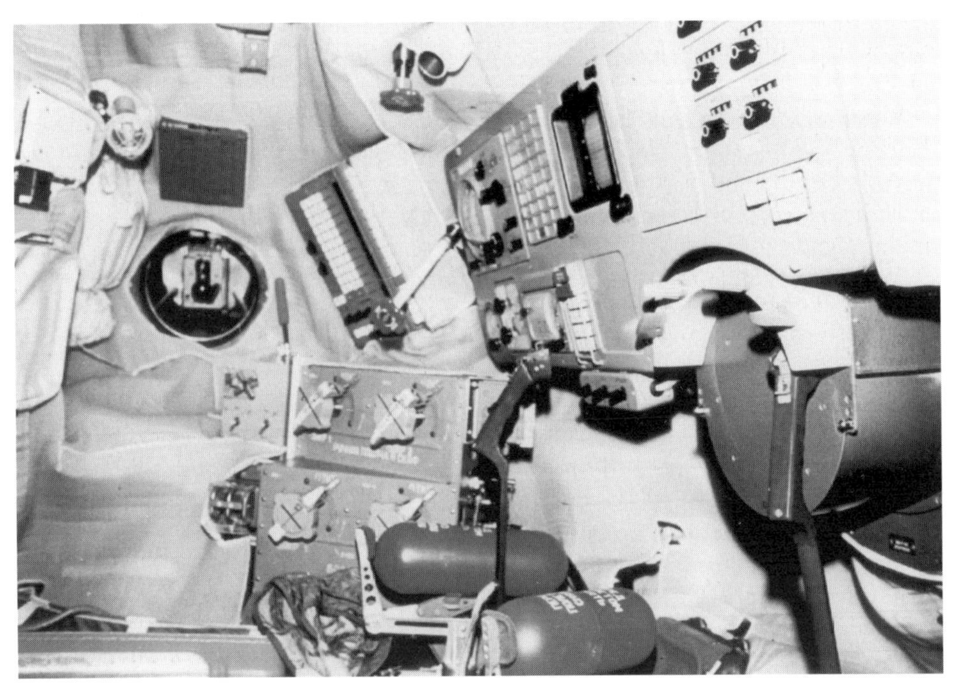

The crew display panels in the Descent Module.

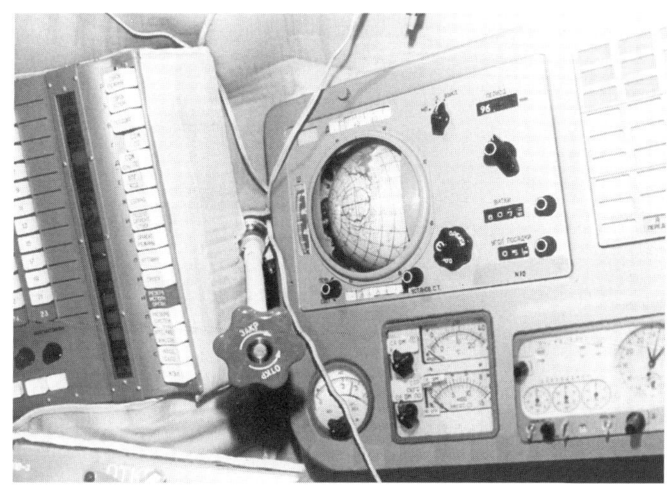

The central display console located in front of the central couch (Commander).

Display and controls

Over the years, the layout and composition of the controls and displays of Soyuz has changed many times. In early Soyuz craft, these consisted mainly of read-outs and visual displays, reflecting the lack of direct input from the cosmonauts in controlling the Soyuz. Although control of the spacecraft was possible, many of the operations in the early missions were pre-programmed into the onboard computers, and were controlled from the ground. As a new variant of Soyuz emerged, so upgrades to these controls and displays were introduced.

The Globus instrument – located in the central control panel (KSU), most common to all variants of Soyuz, and also used on Vostok – is a revolving Earth globe used to determine the spacecraft's position as it passes over Earth. The Vzor porthole system includes a periscope protruding out of the DM for the centrally-seated Commander to view the docking approach from the DM and to orientate the spacecraft relative to the Earth's horizon for de-orbit and re-entry. To assist in these operations, the optical viewer is etched with a grid.

When the crew-members are strapped to the seats in their pressure garments, they use extension sticks to reach the control buttons. At the Commander's station are the two joysticks for control of his spacecraft, the left of which handles the velocity of the spacecraft, and the right its attitude (pitch, yaw and roll).

The left and right panels constitute the command control system displays and controls to activate or shut down a variety of primary and back-up systems and to monitor medical read-outs. These include automatic programme monitors, voltage and current indicators, pressure and temperature indicators, a clock with current (Moscow) time and mission elapsed time, radio and TV controls, and a digital display unit to allow input and monitoring of control data and to display levels of propellant and changes in velocity resulting from engine burns. On the right is the spacecraft control panel for the pressure suits worn from Soyuz 12 onwards, and in front of the Flight Engineer's station is the radio communications equipment.

At either side of the cabin is a circular observation window used, with a sextant, for star sightings as part of the stellar navigation system. Most of the area around the sides and rear of the crew compartment in the DM is void of instrumentation (apart from lights and loudspeakers), and the fabric-covered walls is used to attach storage bags. Above the heads of the crew, and out of reach, are the main and reserve parachute compartments, while the majority of the electronic sub-systems for spacecraft control are located beneath the couches.

On the original Soyuz there were, in total, more than 200 push-buttons and 250 warning lights and indicators (green or blue for correct operation, and red for failure), and many of these (70 buttons and 96 indicator lights) were used for the control of the spacecraft in flight (attitude, rendezvous, docking and re-entry).

The OM also contains smaller control and display panels for radio and TV communications, lighting, atmospheric control, and pressure integrity checks, and a junction box connects radio and TV equipment to the DM.

The Argon 16 computer (80,000 Kbps) – used on the Soyuz as well as Progress, Salyut, Almaz and Mir during the mid-1970s – was designed by the Research Institute of Computer Engineering (NICEVT) of the USSR Ministry of Engineering (NII Argon since 1986), and was built by the Kalmykov SAM factory in Moscow. The development of this computer was completed in 1973, production commenced in 1974, and by 2002 more 380 Argon 16 units had been manufactured in a continuing production programme.[9] The unit is installed in the instrument compartment of the spacecraft, and is destroyed on re-entry.[10]

One notable departure in the layout of 'Soyuz' panels was revealed in photographs released during the 1970s, showing cosmonauts V.N. Kubasov and A. Yeliseyev in what was initially identified in the West as a Soyuz simulator. The control panel layout did not resemble any Soyuz variant known at that time, and it was suggested that it might have been the crew compartment of the Zond (7K-L1 simulator). Soviet space analyst Sven Grahn has conducted a study of these photographs,[11] and has noted that the apparent differences were the removal of the rotating Earth globe (which was not required for a lunar mission), the replacement of the programmer/sequence and CRT equipment with changed designs, the removal of the ΔV monitor from the main panel, the removal of the time-line indicator, the relocation of the spacecraft clock to the centre of the main panel, the relocation of cabin atmosphere meters and volt/amp meters to a new location above the main panel, the removal of critical command buttons, an expansion of the caution and warning displays, and the addition of a still or TV camera to the top of the panel.

Grahn also noted that the two cosmonauts appeared to be positioned in a similar way in two different photographs, and suggested that they may have been working at the Zond navigation work-station to the side of the capsule, similar to the Apollo navigation station which was located separate from the main control panel. To support this assertion, he enlarged the photograph of Kubasov to reveal the cosmonaut apparently holding a celestial sphere, which could have been a navigational device for the lunar crews.

On Zond spacecraft, the Argon 11c computer (data rate not stated) was used,[12] with development being completed in 1968. During the pilot production programme only twenty-one units were completed, and full-scale production was not initiated. This development was headed by G.M. Prokudaev and N.N. Solovyov at NIEM.

Recovery

Emergency recovery of the cosmonauts in the DM can be accomplished from the launch pad or during the first 2.5 minutes of the powered ascent by means of the launch escape system mounted at the apex of the launch vehicle. The Earth-landing parachute system and the soft-landing rockets in the base of the DM are employed to cushion the landing. A nominal recovery from orbit can be achieved using the main propulsion system in the PM and the parachute/soft landing rockets in the DM. Throughout the forty-year history of Soyuz, the emergency systems have been employed only twice to rescue cosmonauts. In September 1983 a Soyuz launch pad abort was initiated (see p.303), and in April 1975 a high-altitude abort was necessary to recover the crew (see p.188). All but one manned Soyuz mission has been recovered on land – the exception being Soyuz 23, which, in October 1976, inadvertently fell into a frozen lake after an emergency (see p.195). Due to this and similar occurrences (for example, the Soyuz 1, Soyuz 11, Soyuz 18-1 Soyuz T-10-1 accidents), the design of these systems has, over the years, been improved and upgraded.

The launch escape system, placed on top of the launch shroud, employs solid propellant rockets, and features a single 676-tonne solid rocket engine with several exhaust nozzles. It can be used from T–20 min to T + 160 sec, after which it separates from the ascending vehicle. For use on TM spacecraft it underwent a major revision, and it is that system, designated the SAS (Emergency Rescue System) which is described here. From the pad, the twin combustion chamber main engine is used to pull the OM/DM from the PM and to separate the payload shroud at the emergency joint. Four deployable canards control the trajectory to an altitude at which the separation of the DM from the OM/escape system allows for parachute deployment (see below).

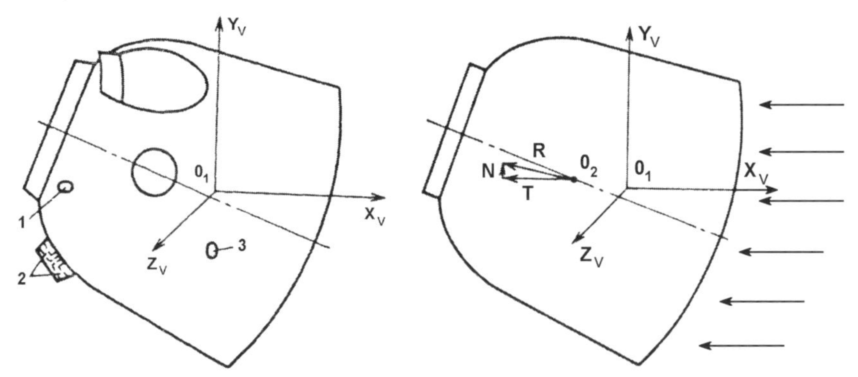

(*Left*) The location of the thrusters of the Landing Control System (SUS) on the Descent Module: 1, yaw thruster; 2, pitch thruster; 3, roll thruster; (*right*) aerodynamic forces acting on the Descent Module in a body-axis coordinate system: O_1, centre of mass, O_2, centre of pressure.

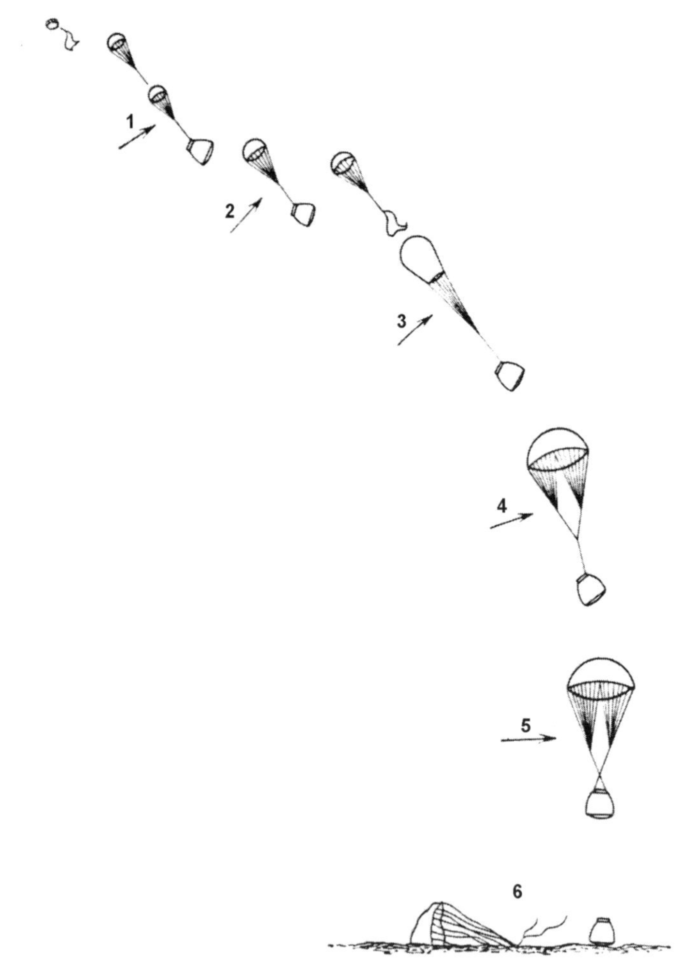

Parachute deployment sequence: 1, deployment of pilot chutes; 2, deployment of drag chute; 3, deployment of main chute; 4, complete deployment of main chute; 5, repositioning of main chute; 6, landing, chute jettisoned.

Around five different versions of the SAS have been used on Soyuz. Initially, the tower was jettisoned at about T + 160 sec, to be followed 10 seconds later by payload shroud separation. During those 10 seconds, the Soyuz was essentially trapped inside the shroud, and the crew had no means of escape.

During the early design of 7K-S (later Soyuz T) in 1968–72, an improved SAS was designed. This had more powerful separation motors, enabling the vehicle to be fired to higher altitudes and farther away from the pad in case of a pad abort, and also to use the more reliable main parachute rather than the back-up parachute. Another modification involved the installation of four extra separation engines on the shroud to pull the Soyuz OM/DM and the upper part of the shroud away from the rocket

Details of the parachute recovery sequence.

after the jettisoning of the escape tower. It was then possible to jettison the escape tower at T + 123 sec, and to retain a means of escape before the shroud was separated. This earlier jettisoning of the tower also compensated for the higher mass of the entire escape system.

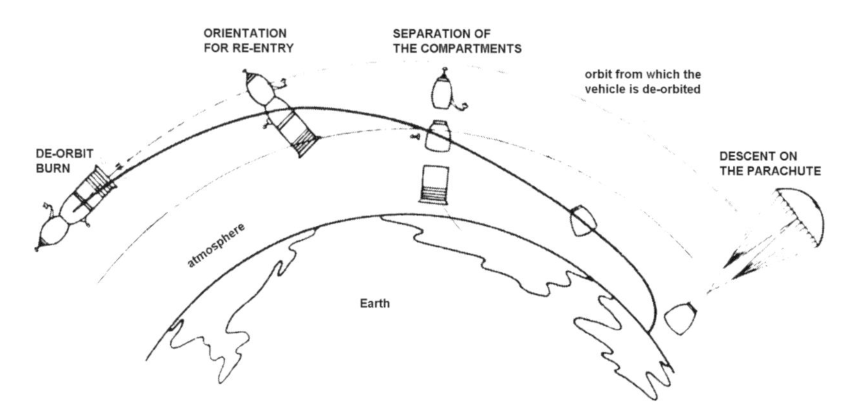

The orientation during the separation of the three modules of Soyuz prior to re-entry. (Courtesy Ralph Gibbons.)

Some of the modifications planned for 7K-S were incorporated into the SAS for the ASTP version of Soyuz, and here it was necessary to take into account the higher mass of the ASTP vehicle and the shift of the CG due to the heavier docking adapter. The ASTP system had a more powerful escape tower, and was also equipped with the additional four separation engines on the shroud, although it is not quite clear if the tower was jettisoned earlier as on Soyuz T. On Soyuz TM, separation of the tower moved even further forward to T + 115 sec.

A drawing in the book by RKK Energiya shows possible emergency procedures during a (first-generation) Soyuz launch after shroud separation. It shows two modes after separation of the payload shroud: 1) separation of the OM/DM from the rocket and a controlled decent; and 2) separation of the entire Soyuz vehicle and a ballistic descent. The latter mode would apparently be activated after T + 522 sec, just before third stage burn-out. However, the account of the 1975 Soyuz 18A abort would seem to indicate that even though the abort took place well before T + 522 sec, the entire Soyuz vehicle was separated.

Unlike the crew of the American Space Shuttle, the Soyuz crew does not have the option of an abort-to-orbit in order to buy time to assess the situation and select a more suitable emergency landing site. On the Shuttle orbiter, if one of the engines cuts off, the crew can fire the others a little longer to achieve orbit, and use the Orbital Manoeuvring System to help compensate for the lost engine. On Soyuz launches, however, if the single engine on the third stage of the R-7 rocket fails, the Soyuz KTDU does not have the reserves of fuel and additional thrust to compensate for the lost engine and still complete its orbital mission safely.[13]

Soyuz capsule descent relies on two parachute systems operating on a cold reserve basis, and ensures a total operational reliability of 0.96, providing a confidence probability of 0.95. The main chute container (which is stored throughout most of the mission) has a volume of 0.27 cubic metres, and the back-up parachute container has a volume of 0.17 cubic metres. The use of either of the two parachute systems without the other provides a design experimental reliability of 0.93. The main parachute sequence begins with the pilot and drag chutes at about 10 km, and the main chute is fully deployed at about 5 km, at a dynamic pressure of 10–1,700 kgf per square metre for an 8 metre per sec vertical velocity. To support the mass of the payload (the Soyuz DM), the mass of the parachute system needs to be 4% of the mass of the descending capsule, with the volume of the occupied space of the deployed canopy of 250 l. The second or reserve parachute (surface area 570 square metres) is designed to deploy at altitudes of 3–6 km and dynamic pressures of 40–1,600 kgf per square metre after failure and separation of the main chute. It supports a spacecraft descent rate of 10 metres per sec. Here, the parachute mass needs to be 3% of the mass of the descending spacecraft, with the occupied space in the deployed canopy of 180 l{?}. Maximum shock loads on the system do not exceed 17,000 kgf.[14]

The thermal shield is jettisoned at 3 km (although one source states 5.5 km – immediately prior to full main chute deployment) and at 1.5 metres from the ground, and a gamma-ray altimeter commands the firing of four solid-propellant rocket motors in the base of the DM to reduce the final landing speed to 2–3 metres per sec. In a worst-case scenario, use of the back-up parachute and failure of the solid rocket motors would result in a landing at 4–9 metres per sec.

Soyuz was designed to return by parachute towards a 'soft landing' on the steppes of Kazakhstan, although the descent has not always gone according to plan.

The first parachutes were evaluated in 1961, in a cooperative programme between OKB-1, the M.M. Gromov Flight Research Institute, Plant 918, Plant 81, and the Parachute Landing Service's research and experimentation institute. This resulted in a deployment system similar to that on the Voskhod spacecraft, with a solid-propellant braking engine at the base of the prime canopy to reduce the landing velocity to 8.5 metres per sec.[15] If the braking engine failed, the landing velocity would increase to a much harder 10 metres per sec. The reserve system did not use a braking engine, and relied solely on the canopies. In this Voskhod-type parachute design there was probably insufficient mass reserve to include a separate set of soft-landing engines in the back-up chute. To test this parachute system, the first drop

At the end of the operational phase of a Soyuz mission, a Descent Module lands with its parachute in the agricultural landscape of Kazakhstan.

tests of mass models of the 7K Soyuz took place at the Flight Research Institute. During 1963–64 the system was modified, revealing problems with the reliability of the back-up system when used in conjunction with the primary system. By early 1965 a decision was made to redesign the parachute system to produce a landing speed of no more than 6.5 metres per sec, by increasing the size of the parachute dome from 574 to 1,000 square metres, and by moving the braking engine from the parachute risers to the base of the DM, where four small solid-propellant rockets would be installed underneath the detachable base heat shield. With the switch to the soft-landing engines beneath the heat shield, these could used in all eventualities, irrespective of whether the prime or back-up chute were to be used. However, these soft-landing engines had to be able to fire first time after prolonged exposure to the vacuum of space.

Throughout 1965–66, tests of the new parachute system were conducted in a programme of seven drops initiated from the Air Force test station at Feodosiya. In various conditions and over different terrains, capsules were dropped from An-12 aircraft at an altitude of 10 km. During the tests, a problem arose when hydrogen peroxide leaked a from small attitude control thruster and contaminated the deployed parachutes – and this had to be remedied before the system could be man-rated. In the final design and tests, the soft-landing engines worked successfully if the back-up parachute was used.

In case of an emergency landing in remote areas or on foreign soil, and to assist possibly injured crew-members, the side interior wall of the empty parachute compartment of the DM and the base of the main structure (under the detachable heat shield) are marked in Russian, with brief instructions in English. The Cyrillic text translates as follows. In the 11–12 o'clock position: 'Take Key, Open Hatch' – with an arrow pointing to the top hatch, so this indeed seems to be a key to open the top hatch. (There are similar markings in the 3 o'clock and 8 o'clock positions). In the 1 o'clock position: 'Attention! Step Aside! The cover may be jettisoned!' (This is also in the 5 o'clock and 9 o'clock positions). Just underneath the USSR/CCCP markings: 'People Inside! Help!'. There are also various internationally recognised hazardous material warning decals displayed near the exhausts of the soft-landing engines (four sets in approximately the 2–4–8–10 o'clock positions) and the backscatter altimeters, showing the areas to avoid around the base of the DM.

The American Shuttle has, around the world, more than forty-five contingency landing sites where it can perform an emergency landing. It appears that Soyuz also has contingency plans for emergency landing points outside of the normal Kazakhstan or Russian territories on each mission. The late Geoff Perry's world-wide Kettering Group of space observers (originally a group of schoolboys at Kettering Grammar School) 'stumbled' across the fact that emergency landing sites were considered for Soyuz 33 in 1979.[16] On that mission, the docking with Salyut 6 was cancelled due to an in-flight failure of the main engine (see p.229), and an immediate return to Earth was required. Analysis of air-to-ground transmission by Mark Severance (living in Fort Worth), of the Kettering Group, revealed discussions concerning 'ugol pasadki' (landing angles, for retro-fire), which would result in a landing in the Indianapolis, Indiana, area of the United States, and not the Soviet Union. This, however, seemed to be a misinterpretation of the commentary, and it was not until 8 October 1985 – during an interview with Vladimir Shatalov at the

The base of an unflown Descent Module, underneath the detachable entry heat shield, and showing the descent soft-landing rocket motors and international hazard warning signs.

IAF Congress in Stockholm – that Sven Grahn received this explanation from the cosmonaut: 'Each night before the crew goes to bed, we send them an orbit number and a table which tells them where to fire the retro-engine in order to reach the designated landing spot. It is the angle measured along the orbit, subtended at the Earth's centre.' It thus became clear that the intercepted communication from Soyuz 33 had concerned retro-fire and landing coordinates.

Further research revealed similar transmissions during ASTP in 1975 and from Mir in 1988. By plotting these data on a world map, it appears that there are three emergency landing zones: the Sea of Okhotsk (for reduced landing loads?); North American prairies, which are similar in terrain to the third (primary and emergency site); and the steppes of Kazakhstan near the Baikonur cosmodrome.

SUPPORT INFRASTRUCTURE

The journey of a Soyuz spacecraft and its crew into space depends, as with any manned spaceflight, upon a ground team of several thousand to plan the flight, prepare the hardware, train the crew, launch the vehicle, and control the flight operations. This can take place over many months prior to the mission, depending on the objectives assigned to that flight. Most of the pre-flight activity occurs at several locations. At the Yuri Gagarin Cosmonaut Training Centre (also known as Star City) near Moscow, the crews are prepared for their mission; while at the facilities of Energiya in Korolyov, the spacecraft is prepared for shipment to the Baikonur Cosmodrome in Kazakhstan. Here, the vehicle and the launcher – delivered from its factory in Samara – are brought together and prepared for flight. During the mission the flight control centre at Kaliningrad (TsUP), near Moscow, directs the mission's progress, supported by a network of tracking stations and satellites around the globe. At the end of each mission a huge force is dispatched to recover the crew and spacecraft from the primary landing sites in Kazakhstan. This infrastructure of ground support is a vast improvement over the original facilities used in the early Vostok and Voskhod programme (detailed in *The Rocket Men*, Springer–Praxis, 2001).

FLIGHT PLANNING

During the era of the Soviet Union, new projects were formally proposed by the design bureaux and the Council of Chief Designers, in response to required objectives (resolutions) or due to military requirements. They were then reviewed by the Academy of Sciences (AN-SSR) and the State Committee for Defence Technology (GKOT), which would amend or modify them to suit the current space programme. The proposals would then be set before the Military Industrial Commission (VPK) for budget approval, and then so on up the levels of government, leading to final approval from the Central Committee and the Council of Ministers. These decrees date the official beginning of a particular project, although much preliminary work had usually been completed beforehand. These decrees, prepared by the VPK, included significant input from the Chief Designers, set the preliminary flight schedule, assigned key events and milestones (such as flight tests) over a specific period, and led to the creation of flight schedules at the design bureaux.

The early flight schedule proposals were based on the production schedule of the spacecraft. These were then passed to the hierarchy of either the Military Industrial Commission (VPK) or the State Commission in charge of the particular project. In the 1960s and early 1970s there was additional pressure from higher levels to time certain missions to coincide with anniversaries or political events, or to 'beat the Americans' to a certain objective or milestone. State Commissions were responsible for setting launch dates for each mission, and used information supplied by the design bureaux concerning the state of the spacecraft, the launch vehicle, ground support and the readiness of the crew. In October 1966 the Soviet government established the first Soyuz State Commission for Flight Testing. The State Commission would conduct formal evaluations of the proposed programme, and would report its findings to the Soviet leadership for approval. As the Soyuz programme evolved, so a new State Commission was established to authorise a new programme of flights within the next phase.

Overall plans for the missions are developed within NPO Energiya, based on confirmation from the State Commissions. Within Energiya there is a Flight Documentation department which compiles the initial documentation that the crew uses as the basis for their mission. Crews in orbit base their work programme on a daily schedule issued by mission control in Moscow, the specialists in Energiya, and other design bureaux. The schedule is always fluid, as the repair and maintenance programme needs to be balanced against science and research objectives.

SOYUZ COSMONAUT TRAINING

While detailed flight plans are formed to fly Soyuz spacecraft, cosmonaut training groups are selected from the active cosmonaut team to form crews to fly the missions, who then have to complete an arduous training schedule to prepare for their assigned mission. Over the years, Soyuz cosmonaut training has changed and developed as the specialists at the Cosmonaut Training Centre have developed their techniques; and among other factors influencing change has been the economics of taking cosmonauts to remote parts of the Russian Federation for specialist training. Some elements, however, have remained very similar to those developed during the 1960s. Cosmonauts have to be very fit, and so Star City is provided with a specialist and well-equipped gymnasium; and during the 1980s they were expected to be able to run a kilometre in 3½ minutes and a cross-country course in 11 min 40 sec. Many of the training classes are administered by example, with constant repetition until the task is understood by the cosmonauts, who undergo numerous examinations on the systems of the craft and flight experiments.

A Russian/American ISS crew undergo wilderness survival training as part of the post-landing exercises completed by every Soyuz crew.

American astronaut Frank Culbertson reaches into the Descent Module upper hatch during winter wilderness survival training in Russia. Note the additional lower protective covering on the Descent Module.

Winter training

In the 1980s and 1990s, cosmonauts were regularly taken to Vorkuta, in the Arctic, for winter survival training. A Soyuz capsule would be set up in the snow, and the cosmonauts would be expected to stay there for two days. They had already undergone some training to prepare them for the prospect of a Soyuz capsule being stranded in deep snow on their return to Earth, and had learned how to build an igloo to provide shelter, as conditions in the capsule can be extreme. The specialists were looking to evaluate the psychological qualities and the physical condition of the crew, who were medically examined every evening. (This, of course, mirrored a real situation in March 1965 in which the Voskhod 2 crew was stranded in Siberia for more than 24 hours before being rescued by ski teams, as helicopters could not fly due to the weather.) The parachute could be cut up to act as a shelter, and each capsule had a survival pack, including a hunting knife and a Makarov pistol to shoot game. Currently, crews destined for the ISS are left in a remote part of the grounds of Star City as part of their training. Crews are trained to stay near the capsule in unfamiliar terrain for up to three days.

Desert training

In the early days of the programme, cosmonauts were exposed to extended testing in a heat chamber based at Star City. This chamber – measuring about 5×3 metres – simulated desert conditions to test the cosmonauts' ability to work in humid and high-temperature conditions, and to evaluate their mental facilities and condition. The tests were monitored with instruments and by observation via a window from the control room. This type of testing was gradually phased out of the training programme, and in July 1978 the first training sessions under real conditions were held with a 24-hour summer survival test using a Soyuz DM capsule set up in the Kyzylkum desert in Uzbekhistan. The cosmonauts had to spend time in the capsule before exiting, after which they had to use the parachute to make a tent and hammock. They also learned how to start a fire, and their supplies included a radio, weapons, a flashlight, food and water.

Mountain training

Cosmonauts also simulated landing in mountainous areas. The training was similar to that undertaken in the snow and in desert conditions. This also proved useful when Lazarev and Makarov landed in the Altai Mountains in April 1975. During a break in survival training in a mountainous region near the Black Sea in July 1993, cosmonaut candidate Vozovikov was drowned when he became entangled in a net while diving for fish.

Recently, cosmonauts have trained with American and European astronauts in the highlands of Utah, scaling a 4,500-metre mountain and walking 60 km.

Sea recovery

Although the Soyuz is designed to return on land, the Russians simulate the potential and dangers of an ocean splash-down landing by using a capsule dropped from a ship into the Black Sea. The crew learns how to exit the craft, get into the inflatable dinghy, and inflate their Trout suits (sea survival outfits). During this test the crew is monitored by divers and specialists, and they are recovered by helicopters. A blue Soyuz capsule called Ocean is used, and there is also an Orange version with a dolphin painted on the side. The crew also also train in a water tank on the deck of the ship to learn how to stay together while awaiting rescue. (Some of

Candidate cosmonauts undergo mountain training.

the Americans carried out this test in the hydro tank at Star City before simulating it in the ocean.) The necessity of this type of training was justifed when Soyuz 23 splashed down in a lake at night, in high winds and during a snow-storm. The cosmonauts were in danger of sinking with their capsule, but were rescued by very brave helicopter crews. The recovery from the Pacific Ocean of the lunar version of Soyuz – Zond – was monitored by an Australian Air Force aircraft – much to the annoyance of the ship's crew.

Currently, a specially equipped rescue ship – *Antarktida* – is located in Vladivostok in case a Soyuz has to ditch in the sea after a launch abort.

Swamp training

This again tested the survival skills of the crew, and used the capsule Ocean – the same as used for sea training.

Parachute jumping

All Soviet and Russian cosmonauts undergo extensive courses in specialist parachute training, which teaches them to control their emotions, and psychologically prepares them for encounters with danger. The training includes free-fall jumps as well as jumps from helicopters. Many cosmonauts have completed more than five hundred jumps, and many have received the Instructor Parachutist award.

Zero-gravity training

To simulate zero-gravity conditions, a converted IL-76 MDK flying laboratory is flown by the Seryogin regiment based at Chkalov Air Force Base. The passenger compartment is converted into a weightlessness training facility, with the floor and

In the Black Sea, cosmonauts conduct water egress training with a dedicated Descent Module and support craft, as well as a team of divers.

ceiling covered by mats to prevent injury. The aircraft goes into a dive, and accelerates to a speed of 620 kmh before ascending in a curve. At an altitude of 7,200 metres the controls are set to idle, and the aircraft continues to follow a path resembling a Keplerian parabola. Weightlessness occurs for 25–30 seconds as the aircraft's speed drops to 420 kmh on the descending branch of the trajectory and the aircraft moves along the curve in horizontal flight. The cycle is repeated every 3–4 minutes, and ten cycles are usually flown, although more can be flown depending on mission requirements. During these short bursts of weightlessness, cosmonauts test equipment, don Sokol suits, and exit from the Soyuz hatch. They also practice at moving large objects, and medical tests are conducted to assess their performance.

Earth-observation

A Tu-154 MDK has been modified to train cosmonauts in space navigation and Earth-observation of ground-based and sea-based objects. Space equipment is also tested on the aircraft, which is operated by the Seryogin regiment based at Chkalov.

The simulator

The integrated simulator of the Soyuz TM enables the crew to practise all spaceflight stages on the ground: launching, placing a spacecraft in orbit, orientation and manoeuvre, orbital complex search and detection, docking and undocking, re-entry from orbit, and other operations. The simulator proviedes specialists with the opportunity to test the cosmonauts in emergency operations.

A number of Soyuz simulators have been updated to take account of the Soyuz variants developed over the years. Soyuz TDK-7ST was used for Soyuz T training, but has recently been converted for TMA training and is now called TDK-7ST3. Soyuz TDK-7ST2 completed its 4,000th simulation session during 2002, after more than sixteen years of operation. There were plans to convert it into TDK-7ST4 for

Soyuz TMM training, but TMM development has been suspended. In the 1960s and the early 1970s there existed a Soyuz system called Volga that enabled the cosmonauts to practise docking. Shatalov has said that he used it more than eight hundred times. Volga was replaced by Don-732 – a specialised simulator for manual dockings, with the Commander seated in the Descent Module and the Flight Engineer in the Orbital Module. In recent years this simulator has operated with *two* descent capsules: one for Soyuz TM training, and the other for Soyuz TMA training.

In spring 1998, two Soyuz simulators were delivered to the Johnson Space Center, Houston, and are used by US astronauts for preliminary Soyuz training prior to their main Soyuz training at Star City. One of these simulators is not a Soyuz vehicle, and is instead a set of four computers to acquaint astronauts with undocking and landing procedures. The other is a Soyuz TMA mock-up consisting of an Orbital Module and the refurbished Descent Module of Soyuz TM-22. This is used to acquaint astronauts with the general internal layout of Soyuz (with some of the life support systems), and for training with Sokol suits. In addition to these, in December 1997 JSC borrowed a Soyuz mock-up from the Kansas Cosmosphere and Space Center for a two-year period. Since this was a museum piece and not an exact replica of a Soyuz TM, it has been used only to familiarise astronauts with the internal layout.

Simulator ballistics

This is used to train pilots in the skills required for manual approach and docking. It includes a special facility module for simulating the docking of unmanned craft to stations, using the TORU system.

Simulators for space station operations

The Russians have built full-scale mock-ups of the various core modules and scientific modules for all space station operations. The mock-up for Mir and the modules for the Russian segment of the ISS are at Star City. Similar mock-ups were made for all the civilian and military Salyuts, and the mock-up for Salyut 4 is in the Energiya museum. An Almaz is at the Chelomei OKB premises at Reutov, and there may well be others at Chkalov Air Force Base near Star City, although there has never been a military Almaz simulator at Star City. Crews spend many hours here, learning how to operate and repair systems in as real an environment as possible. During the 1980s, crews simulated entire missions in such craft.

The neutral buoyancy laboratory

Crews practise EVA operations in a very large water tank measuring 5,000 cubic metres, with a diameter of 23 metres and a depth of 12 metres. This incorporates mock-ups of the modules with which cosmonauts practise in an environment that is very similar to the conditions in space. It can currently be used by Russian cosmonauts using their Orlan suits, and by American astronauts using their space suits. Located near the Mir simulator hall there is also a facility called Vykhod (Exit), which simulates EVA training by hanging cosmonauts on a weight system.

Simulation controllers monitor the progress of the latest Soyuz TM docking simulation being undertaken by a future crew as part of the training programme.

The layout of the Soyuz TM simulators and control panels. The control panel at top right is shown in the photograph above.

Rear view of the Soyuz Descent Module/Orbital Module configuration used in the training of Soyuz TM crews at TsPK.

The centrifuge

The TsF-18 centrifuge has a rotation radius of 18 metres and a maximum load of 30 g. It enables specialists to test how cosmonauts react in controlling spacecraft under similar flight conditions, and is also used for practising operations during launch and re-entry. Cosmonauts have been tested up to 20 g, although 3 g is currently normal during launch simulations. There is also a smaller centrifuge, the TsF-7, in the medical facility at Star City.

Vacuum and pressure chambers

During flight preparation of the spacecraft it is necessary to place it in vacuum and

A view though the side hatch of the Soyuz ferry simulator, *c*.1970s. The hatch appears only in the training models and not in the flight vehicles, in which access is only via the central top hatchway.

This Orbital Module, outside the water emersion facility at Star City, clearly displays the effects of prolonged underwater storage during EVA simulations, followed by the ravages of the Russian winter.

pressure chambers to simulate, as near as possible, space conditions. This has been carried out not only for Soyuz, but also for Mir, its modules, and Buran. In certain circumstances, cosmonauts and testers are also put into vacuum chambers to help with training and equipment testing. These are not located at TsPK, but are at Energiya and Zvezda, as well as other research facilities. At Zvezda, such chambers are used to test spacewalk techniques and experiments, which supplements work in the water tanks.

Foreign language training

In 1973 the first Americans were selected to train for the Apollo–Soyuz Test Project – and they had to learn Russian (as have the Americans assigned to Mir and ISS missions), as the periods of joint training and flight operations demanded an understanding and command of each other's languages. All the Interkosmos cosmonauts already spoke Russian, as many of them had undertaken pilot training in the Soviet Union, and all subsequent foreign cosmonauts had to learn Russian to work with their fellow crew-members.

With the ISS imminent in the mid-1990s, English became an important part of cosmonaut training. It was very extensive in their weekly training schedule, and many of them are now very conversant. They have to read NASA technical manuals and conduct some of their training in English, and the language is considered so important that one cosmonaut was recently removed from his assignment due to his inability to learn it properly.

Examinations

Examinations constitute a major part of the life of a cosmonaut. They are conducted on a regular basis, and are based on a top mark of 5 for knowledge of an individual system, experiment or technique. (One cosmonaut had a picture of the Soyuz control panel on his study wall so that he could spend every possible moment familiarising himself with its layout.) When they are assigned to a crew, the cosmonauts are tested as a unit in all the key areas of their mission, and are also assessed individually – and the results are posted on the wall of the Soyuz simulator hall for all to see. When a crew is confirmed by the State Commission – usually about three weeks before a

Ivanov, head of the State Commission, announces the names of the prime and back-up crews for the mission.

mission – they sign a certificate at the training centre, and also visit Gagarin's office (which has been located at the House Museum in Star City) to sign a crew book. During the era of the Soviet Union they also visited the Kremlin Wall and Lenin's Office.

Departing for Baikonur

Training for the flight continues at TsPK up to the time when the crew transfers to the Baikonur Cosmodrome, usually about a week before the launch. The crew will have been in isolation for about a week beforehand, except for visits from close family members in a facility in Star City. On arrival at Baikonur they test the integrity of their suits and inspect the Soyuz craft before its integration with the launch vehicle. There is a formal ceremony with the State Commission, which confirms the crew and gives them permission to fly. This is followed by a press conference, after which the crew stays in semi-isolation at the Cosmonaut Hotel until the beginning of their mission.

THE SOYUZ LAUNCH VEHICLE

As the crew arrives at Baikonur, so preparations are being completed on the spacecraft and launch vehicle assigned to their mission. The Soyuz launch vehicle is a variant of the original Soviet R-7 ICBM designed in the early 1950s and first flown in May 1957. On 4 October 1957, a variant of the R-7 ballistic missile was used to orbit the first space satellite, Sputnik. Three-and-a-half years later, on 12 April 1961, Yuri Gagarin rode another variant of the R-7 into orbit onboard Vostok. Five-and-a-half years after that, on 28 November 1966, the first unmanned Soyuz was launched by another uprated version of the same R-7 booster. For the last thirty-five years, almost all the Soyuz and Progress variants have been launched into Earth orbit onboard improved variants of the original R-7, the only exception being the Zond circumlunar spacecraft launched by Proton during the period 1968–70. The development of the R-7 and its launching facilities used for the early Sputniks and the Vostok spacecraft is discussed elsewhere (*The Rocket Men*, Springer–Praxis, 2001), while the launcher's development is summarised below.

Raketa 7 ICBM 8K71

The concept of grouping similar rocket stages together (clustering) originated in the Soviet Union in 1947, with the work of M.K. Tikhonravov at the NII-4 (Nauchno Issledovatelsky – Scientific Research) Institute for the Soviet Defence Department. Tikhonravov's ideas interested S.P. Korolyov, the Chief Designer at the OKB-1 design bureau, who was working on design configurations for the first Soviet ICBM programme. From 1949 to 1953, several design configurations were evaluated, and these eventually featured propulsion units developed by V.P Glushko's OKB-456 (also known as GDL-OKB) from early 1953. The configuration of what became the R-7 ICBM was approved early in 1954,[17] and received the manufacturer's designation for development of Izdelie 8K71 (Product 8K71) in June 1954. (Soviet

ICBM designs were usually assigned the development number 8A or 8K, followed by the design bureau's two-digit designation, which for OKB-1 was in the 50s or 70s. Thus the ICBM design from OKB-1 became 8K71).

The design of the R-7 featured a central sustainer stage (Blok A), to which four boosters (Blok B, V, G and D) were attached to form Stage 1. Each stage featured four chamber engines and four vernier engines on the central sustainer, and two each on the strap-on Bloks. For the first stage this totalled twenty main chambers and twelve verniers, all firing at the same time to lift the vehicle off the pad. When the strap-ons burned out they would separate, leaving the central core to continue as Stage 2, and any upper stages added to this configuration would be termed Stage 3. It is this three-stage configuration, with upgrades, that has been employed for all Soyuz and Progress launches since 1966.

Power for the R-7

Early design studies on the R-7 focused on single-chamber engines, fuelled by a combination of liquid oxygen (LOX) and kerosene, developed at OKB-456, headed by Valentin Glushko. The sustainer core would have an RD-106 engine producing 53 tonnes thrust (all tonnes are approximate mass values) at launch, and 65.8 tonnes in vacuum. The four strap-ons would employ the RD-105, and would each produce 55 tonne of thrust at launch. However, during development these engines displayed problems in stable burning inside the single chambers, and by 1953 it became evident that the vehicle could no longer lift the increasing weight of the thermonuclear warhead. Before 1953 the idea was to use the ICBM to launch atomic bombs, but this shifted to (heavier) hydrogen (or thermonuclear) bombs. The move from atomic to thermonuclear bombs was the main reason why the payload capacity had to be increased. The specifications had grown to produce a vehicle capable of sending a 5.4-tonne warhead to a distance of 8,500 km. Fully laden, therefore, the ICBM would weigh approximately 283 tonnes, and would require a lift-off thrust of almost 400 tonnes.

To meet this requirement, Glushko's bureau developed the RD-107 for the strap-ons and the RD-108 for the central core. The RD-108 provided 75 tonnes of thrust at lift-off (96 tonnes in vacuum) and a burn time of 304 seconds, while the RD-107 supplied 83 tonnes of thrust and a burn time of 122 seconds. Both engines continued to use LOX/kerosene. They would remain the core propulsion units for Soyuz first- and second-stage engines (with upgrades), with extra propulsion deriving from a third or upper stage. The RD-107 and RD-108 were not the original choice for the R-7. Early designs of the engines for the launch vehicle had featured single-chamber LOX/kerosene engines (the RD-105 and RD-106) with a thrust of 50–60 tonnes; but it soon became clear that these would be unable to lift a 5.5-tonne payload, and that they were already exhibiting poor performance in ground tests as a result of combustion chamber instability that led to severe vibrations. The problem was solved by a design tested by the Chief Designer at NII-88, Aleksei Isayev, who had tested a multi-chamber engine, developed from single-chamber 40-tonne thrust engine, that revealed a much higher cumulative thrust than its original single component. This evolved into a four-chamber engine fed from a single turbo-pump, thus reducing the effects of unstable burning as well as saving engine mass; and by using the same chambers, all phases of construction and testing were considerably simplified, leading to the creation of the RD-107 and RD-108 engines that were to power the R-7.[18]

Basic design features of the 8K71

The central core of the 8K71 measured 26 metres, with a diameter of 2.95 metres tapering to 2.15 metres to accommodate the strap-on stages (see the illustration on p.88). Each strap-on was 19 metres long, with a base diameter of 2.68 metres tapering to a point where it was attached the central core by a ball-and-socket joint at the apex and a tension band at the base. Inside the core stage was 92.65 tonnes of LOX/kerosene, and each of the strap-on Bloks was filled with 39.64 tonnes of propellant. With the payload shroud attached, the basic R-7 was 33.74 metres long and, combined with the strap-ons, had a maximum base diameter of 10.3 metres. Standing empty, but with four strap-ons and the payload, it weighed 27.7 tonnes; and with 246.6 tonnes of total propellant loaded this became 274.3 tonnes. With a launch thrust of 398 tonnes, it could carry a 5,400-kg payload.

Adapting the R-7 for Soyuz

Between 1957 and 1966, improvements to the R-7 engines, structure and upper stages resulted in the creation of a reliable and versatile family of launch vehicles that has sustained the Soviet/Russian space programme for forty-five years, and the following briefly summarises the improvements made for launching Soyuz/Progress vehicles from 1966 to 2002. (The improved R-7 required man-rating, and its career with the Vostok and Voskhod spacecraft is discussed in *The Rocket Men*, Springer–Praxis, 2001).

The launch vehicle used to lift the Vostok (8K72K) featured a new upper stage, designated Blok E (Ye), which featured a single RO-7 engine burning LOX/kerosene. It was developed by Semyon Kosberg's OKB-154 and gave a total thrust of 5.6 tonnes and a 430 second burn time to place the spacecraft in orbit. Blok E measured 3.1 m long, with a 2.6 m diameter.

From as early as 1956 – a full year before the launch of Sputnik – Korolyov had been developing plans for the most 'urgent' goals in exploring space. These included future research using 'satellite stations' in space, and the development of the means for controlling two spacecraft during a rendezvous. Over the next four years, these plans developed to include manned flights around the Moon and the creation of an orbiting space station. These craft were much larger than the R-7 could launch at one time, and Korolyov and his team – not wishing to wait for the large heavy-lift launch vehicles still on the drawing board – evolved plans for following the Vostok series with a programme of missions for spacecraft with larger mass, focused on using the Uprated R-7 and Earth-Orbital Rendezvous, for which a more powerful upper stage was required. However, these new spacecraft never progressed further than paper studies or model mock-ups, and in 1961 the development of an orbiting complex was cancelled in favour of a replanned circumlunar programme which in April 1962 became known as Soyuz.

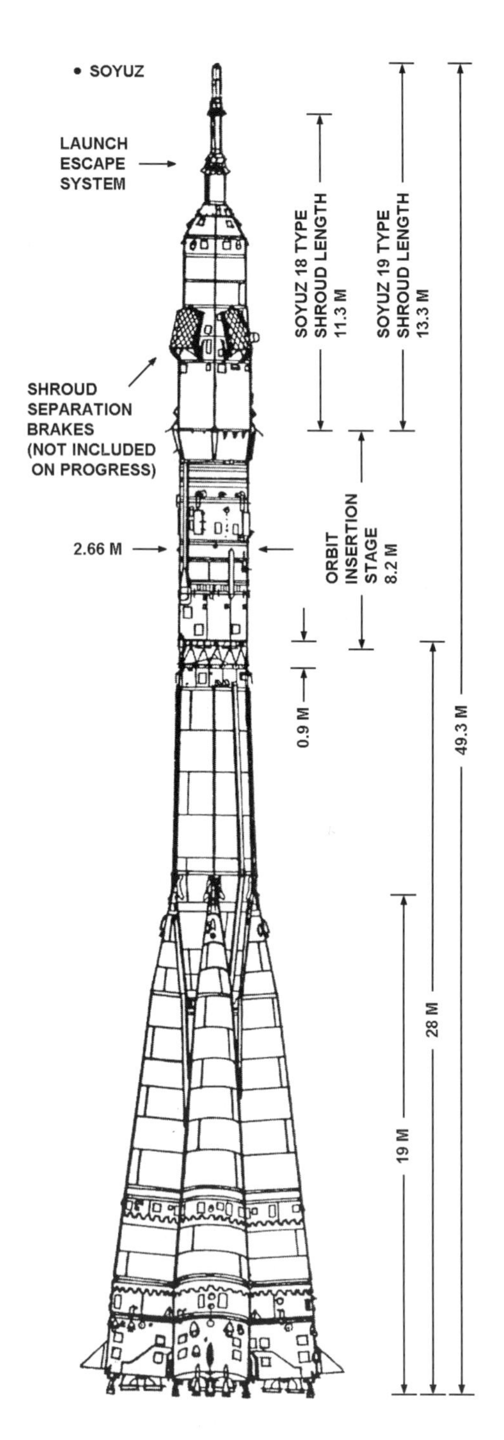

A Soyuz launch vehicle based on the R-7 rocket.

11A55 and 11A56 early Soyuz launch vehicles

Improvements in early studies for sending Soyuz around the Moon included the uprating of the central sustainer by 5%, and this increase in specific impulse would allow for an orbital mass of 5.8–6.0 tonnes. The 8K711 R-7 variant could deliver only 5.6 tonnes using the modified RD-107 and RD-108 engines, but it was conceived as a universal launcher for elements of Koloryov's Earth-orbit rendezvous programme (which included not only circumlunar missions but also missions in low Earth-orbit and to geostationary orbits). Basically, the 8K711 was the 8K78 Molniya vehicle without the fourth stage. Two versions of the R-7 were therefore developed for early Soyuz circumlunar operations. The 11A56 would launch the unmanned elements of the complex, while the 11A55 would launch the crew in Object 7K (Soyuz). The manned launcher would be developed first, with uprated first- and second-stage engines. Concurrently, work on an upper stage (RD-0107) for unmanned missions had been conducted by the Kosberg design bureau, which was asked to upgrade this for use on the new manned launcher, the 11A55. By 1963 this engine had been designated RD-0108. It soon became evident, however, that the payload capacity of the 7K complex would soon exceed that of even the new vehicles, and the 11A55/11A56 project was consequently terminated in favour of a much improved launch vehicle, designated the 11A511.

11A511 standard launch vehicle

In 1963 this launch vehicle was designed specifically to launch the Soyuz, and to provide launch support for early variants of the proposed Soyuz Complex. In this variant, telemetry systems were reduced in weight to 150 kg, and engines were tested to ensure improved performance. The upper stage (Blok I) had a length 6.7 metres, a diameter of 2.66 metres and a dry mass of 2.4 tonnes, and was 100 kg lighter than the 11A57. The RD-0110 engine used LOX/kerosene, producing a 30.4-tonne thrust and 246 seconds of burn time. The vehicle (with payload and LES) stood 49.91 metres tall, had a mass of 308 tonnes and a lift-off thrust of 411.7 tonnes, and could lift 6,450 kg to a 200-km orbit at 56°.6. It was a considerable improvement over the 11A57 vehicle used for Voskhod, as it could launch an additional 550 kg of payload. The first launch of this variant took place on 28 November 1966, when it carried Cosmos 133 into orbit. The Cosmos spacecraft was the first Soyuz spacecraft (7K-OK) to enter orbit, and when, from April 1967, it was identified for the first Soyuz mission, the Western media began designating this version as the 'Soyuz launch vehicle', to distinguish it from the 'Vostok launch vehicle'. The 11A511 was finally used in the Soyuz programme for Soyuz 23 on 14 October 1976. Other improved variants were proposed for the Soyuz R and P programmes and in support of the manned lunar programme (see p.16).

11A511U launch vehicle

The next launch vehicle variant was the Soyuz U – 11A511U. Work on this unified vehicle began in 1969, with a government resolution on the development of the 11A511U issued on 5 January 1973. The vehicle was to replace the 11A57 (Voskhod), 11A511 and 11A511M, and featured a number of improvements to both the launcher and the ground support equipment to test and prepare the vehicle at Baikonur. The engines were improved, and employed chilled higher-density fuel in the central core stage to improve performance and to allow for an increase in the overall payload-to-orbit weight to 6,220 kg up to a higher orbit at 450 km. The first launch, with an unmanned photoreconnaissance satellite, took place at the Plesetsk cosmodrome on 18 May 1973, and the first man-related launch took place on 3 April 1974 with Cosmos 638, a test version of the Soyuz spacecraft designed for the ASTP. The first crew was carried on Soyuz 16 (the ASTP test flight) on 2 December 1974, and over the next twenty-seven years the launcher was used for Soyuz, Soyuz T, Soyuz TM, Progress and Progress M spacecraft. In addition, it was the launch vehicle that took the Pirs facility to the ISS in September 2001.

When Soyuz 18-1 (using a 11A511 vehicle) was aborted during the launch phase, NASA expressed concern over the reliability of the 'Soyuz' rocket to support the ASTP launch two months later. However, the Soviets indicated that this was the last of the older design and that the ASTP would use a newer, improved Soyuz launcher – the 11A511U.

11A511U2 launch vehicle

On 5 June 1975 an order was issued by the Ministry of General Machine Building for the development of a new version of the R-7/Soyuz using synthetic kerosene (sintin) for the first stage of launch, thereby offering an improvement in payload delivery over that of the standard 11A511. This vehicle could deliver 7,050 kg to a 200-km orbit at 52°.6. The first payload, launched on 23 December 1982, was the Yantar-4KS1 photosurvey satellite (the Soviet Union's first digital imaging reconnaissance satellite), and the first manned launch, with the T-12, took place in July 1984. However, in 1996 production of sintin ceased, forcing the continued use of 11A511 for manned and Progress missions to Mir and early ISS missions. The problem with reverting from the Soyuz U-2 to the Soyuz U was that the Soyuz U was powerful enough to launch only a two-man Soyuz TM to Mir, although this could be overcome by slightly improving the characteristics of the Soyuz U and slightly lowering the mass of the Soyuz TM. The last manned mission launched by this vehicle was Soyuz TM-22. There were seventy-two 11A511U2 launches – all of them were successful.

Soyuz FG

This is the most recent uprating of the basic R-7 design begun almost fifty years ago. This vehicle includes improvements to the main engines, incorporates modern avionics packages, and includes mostly Russian-originated equipment, reducing the reliance on non-Russian sources and thus hopefully improving reliability and supply flow in support of Russian commitments to ISS operations.

The central core now features an RD-108A engine (upgraded RD-108), using LOX/kerosene for a vacuum thrust of 101,931 kg and burn time of 286 seconds. The four strap-ons use the RD-107A (upgraded RD-107) with LOX/kerosene, producing a thrust of 104.1 tonnes and a burn time of 120 seconds. The main improvement in both engines is that they have modified fuel injectors, and the upper still carries the

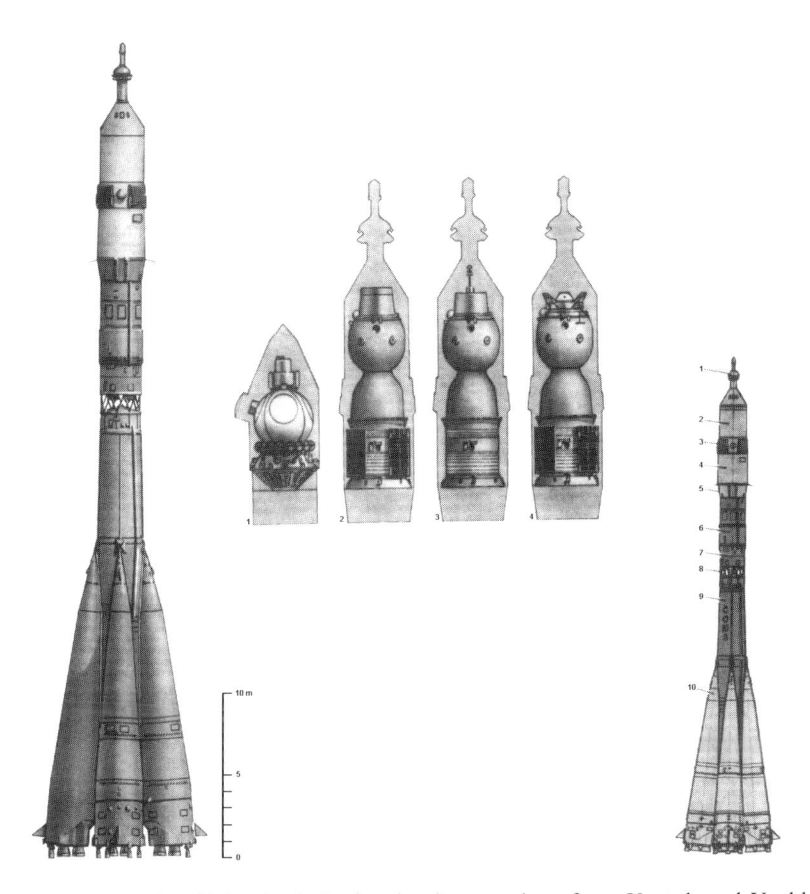

The Soyuz launch vehicle, the R-7, showing its genealogy from Vostok and Voskhod, and the location of Soyuz inside the launch shroud. The smaller diagram illustrates the stages and tankage of the launcher.

RD-0110 engine from the Soyuz U. Overall, the Soyuz FG (Forsunochnaya Golovka – Fuel Injector) can lift 7,420 kg into a 193-km orbit at 51°.8. It has a total lift-off thrust of 422.5 tonnes and a launch mass (including payload) of 305,000 kg, offering a 5% improved performance rate. It was first used for the launch of Progress M1-6 to the ISS on 20 May 2001.

Soyuz-2

Soyuz-2 is the next step in a Soyuz modification programme known as Rus. The first step was Soyuz FG, which uses uprated core and strap-on engines with improved fuel injectors. Soyuz-2 will have the same modifications as Soyuz FG in the core and strap-on stages, and in addition will have a completely new third-stage engine: the RD-0124 (which will replace the RD-0110 used on Soyuz launch vehicles to date, including FG). This will have a higher specific impulse than the RD-0110 (359 sec compared with 326 sec), and a longer burn time (300 sec compared with 250 sec).

The launch of Progress atop an R-7 carrier rocket, beginning another mission in Earth orbit.

This will lead to a payload increase of about 950 kg. The RKK Energiya history states that Soyuz-2 can orbit a Progress weighing 8,350 kg – compared with 7,420 kg when launched by the Soyuz FG – and a specialised module weighing 8,100 kg.

SOYUZ FACILITIES AT BAIKONUR

Work to create a test launch complex for Soviet ICBMs began in the Kazakhstan region of Tyuratam in May 1955, with the creation of a small support town – later called Leninsk – that grew over the years into a large city. The main facilities at this cosmodrome in support of Sputnik, Vostok and Voskhod were retained in the early Soyuz programme, and were gradually replaced or expanded. Construction of the first R-7 launch pad (Site 1) began in July 1955, and in April 1957 was completed with systems tests of a structural test model of the R-7 to simulate launch preparations and fuel loading. The site supported all the early launches of the R-7, including the first launch on 15 May 1957 (8K71), the launch of Sputnik on 4 October 1957, and the flight of Gagarin on 12 April 1961. It continued to support many Soyuz launches; and in order to confuse Western intelligence agencies, the Tyuratam facility was for many years called Baikonur, which is some 370 km from the actual facilities. It was not until the fall of the Soviet Union in 1991 that the name Tyuratam was assigned to the cosmodrome, but it is still generally referred to as the Baikonur Cosmodrome. In December 1995 it was ordered that the town of Leninsk be renamed Baikonur (Baikoyr in Kazakh); and it therefore appears that there are now two places called Baikonur in Kazakhstan: the original town (that never had anything to do with the cosmodrome) and the former Leninsk. (For a more detailed description of the creation of Baikonur and its early facilities and range capabilities, see *The Rocket Men*, Springer–Praxis (2001).) Over fifty years the complex has been greatly expanded to include launch facilities to support the (cancelled) manned lunar and Buran shuttle programmes, as well as several commercial launcher programmes, Soyuz manned missions, Progress resupply missions, and launches of the Proton vehicle carrying modules of space stations as well as commercial payloads.

Site 1: launch complex (PU) 5

Following the flight of Gagarin this pad was renamed Gagarinskiy Start (Gagarin's Pad), and it has since been the site of more than four hundred launches in support of both manned and unmanned programmes. The official designation of the complex is 17P32-5 (or Object 135), and an extension to this area (1A) was completed in the

In the MIK, the stages of the R-7, and the Soyuz in its launch shroud, are assembled horizontally onto a rail carriage for transportation to the launch pad a few days before launch.

mid-1970s. Extensions to the support arms and access platforms were required to support the larger Soyuz, but the pad afterwards remained essentially the same, without major modifications for the change from Vostok and Voskhod to Soyuz. (For a description of the development of this pad, see *The Rocket Men*, Springer–Praxis, 2001.)

Site 2: the launcher processing area and the MIK facility

In 1956–57, work was completed on a processing area (MIK-1) for the R-7s and their payloads; and in 1975 a new extension called 1B was completed, and was used for the ASTP. In 1997 this area was abandoned, and Soyuz and Progress processing was moved to Site 254. In 1958 the MIK 2A processing area was completed, and from 1960 this would prepare warheads for the R-7 ICBM. Shortly afterwards the site was abandoned, and the R-7 was decommissioned as an ICBM. A second MIK (MIK-2B) was used for processing military (Zenit) satellites, and is still used for processing and integrating Soyuz launch vehicles

Site 2B: the launcher processing area and the MIK 2A facility

This is the replacement processing area for Soyuz, its launcher and Progress. Also designated 135R, work here was planned to be relocated to Site 112 from 2002.

The Soyuz inside the launch shroud atop a Soyuz launch vehicle on the pad, with the access arms retracted ready for launch.

Site 31: launch complex (PU) 6, or 17P32-6

Work on this site – a second R-7 launch pad – began in 1960, and was intended as an 'R-7 battle-station' When the R-7 was taken out of ICBM service, this site was modified for the Soyuz launch vehicle with Fregat upper stages for deploying Molniya communication satellites. This is also the second Soyuz launch pad for manned and Progress launches although it has not been used for a Soyuz launch since 1984. A number of Progress ships were also launched from here. It also continued to be used throughout the 1970s and 1980s for some launches of the older 'Vostok'-type launch vehicle, which may explain why it is also known as the 'Vostok' pad. Yet another designation is Object 353.

Site 31 is situated on the so-called 'right flank' of the cosmodrome, some 50 km from Leninsk. The first R-7 ICBM launch from this site took place on 14 January 1961, and the first payload sent into space from here was Cosmos 28 (a Zenit 2 spy satellite) launched, using the 8A92 version of the R-7, on 4 April 1964. The site was also used for the first Soyuz launch (Cosmos 133) in November 1966, but on 14 December 1966 the pad was completely destroyed in an accident involving the second Soyuz vehicle. Soyuz launches from Pad 31 resumed with Cosmos 186 in October 1967. The pad is also used to launch unmanned payloads using variants of the R-7 with a Freget upper stage, with the initial test launch taking place in February 2000, although the launch of Molniya comsats using this configuration are launched solely from the Plesetsk cosmodrome. As of 1 January 2000, Site 31 had been used for 339 launches.

Site 31 also has at least one Soyuz assembly building (MIK 40) to support launches, and also a Soyuz spacecraft fuelling station called 11G12. This fuelling station was originally built in the mid-1960s for Soyuz, but was later also used for fuelling other spacecraft and satellites. Even if Soyuz is launched from Site 1, it first has to be transported to Site 31 for fuelling before being returned to Site 2 for mating with its rocket. In 1972 another fuelling station (11G141) was completed at Site 91A (in the Proton area) to fuel Almaz stations, and was later also used for fuelling Blok D upper stages. It remained operational until the early 1990s, but is to be refurbished in various stages until 2005. When it is finished it will take over from 11G12, which will afterwards be refurbished.

Site 32: R-7 residential housing

Begun in 1960, this facility is a residential area for people working at Site 31. It includes the assembly buildings MIK-32 and MIK-32GCh and was once used to store R-7 components.

Site 112: launch vehicle assembly building

Work to convert the huge N1/Ernergiya assembly building at Site 112 (designated 11P591) was scheduled for April 2002. Once completed, all assembly and preparation of Soyuz launch vehicles will be focused on this site. In addition to the recycling of the older building, the rail line connecting Site 112 site to Pad 1 was renovated to support the transfer of flight hardware from the assembly building to the pad. It is also planned to connect Site 112 with Site 31, from which Soyuz-2 will be launched, by reusing the rails connecting Site 112 with the Energiya launch pads. The site will support Soyuz U preparation and the forthcoming Soyuz-2 next-generation R-7 vehicle, which will be manned. The hope was to have Site 112 ready for Soyuz launch vehicle processing in 2003, but this may be delayed because of a recent collapse of the high bay roof (where the flown Buran and some Energiya hardware were located). Fortunately, processing of the Soyuz rocket is to take place in the (unaffected) low bay area, which also has clean rooms for payloads from the Starsem organisation.

Site 254: spacecraft assembly building

This facility (designated 11P592) housed the Buran shuttle during launch preparation, was afterwards used for the fabrication of Russian space station modules, and is now the location for the processing of Soyuz and Progress spacecraft.

The bulk of the Soyuz vehicles have been launched from Site 1, and the following vehicles used Site 31: Cosmos 133, Cosmos 186, Cosmos 213, Cosmos 238, Soyuz 3, Soyuz 4, Soyuz 6, Soyuz 8, Soyuz 9, Cosmos 638, Cosmos 672, Cosmos 1074, Soyuz 32, Soyuz 33, Soyuz 34, Soyuz T(1), Soyuz 35, Soyuz 36, Soyuz T-10, Soyuz T-11, and Soyuz T-12 – the last three because Site 1 had been destroyed in the Soyuz T-10A pad explosion.[19]

SOYUZ MANUFACTURING

Construction of the Soyuz spacecraft was planned to begin at the same location as the Vostok series of spacecraft. However, the building of such a pioneering vehicle required a dedicated facility for the production of flight hardware, and the most promising site was an old field-gun artillery plant – located on the opposite side of Kaliningrad from OKB-1 – which during the 1950s had been designated Central Scientific Research Institute 58 (TsNII-58). With the beginning of the missile age, artillery field-guns became obsolete, and so the production of fast nuclear reactors was the primary activity at the site until it was agreed to assign it to OKB-1 for work on Vostok and, up to 1962, solid propellant missiles. On 3 July 1959 the ‘second territory’ officially became part of the OKB-1 organisation, and within a month, 5,000 employees, under the direction of Deputy Chief Designer Konstantin Bushayev, began work on building the Vostok shariks.[20]

In May 1963, Deputy Chief Designer Sergei Kryukov assumed the role of Soyuz programme Director, and by the spring of 1964 the first Soyuz boilerplate vehicle was rolling off the assembly line. As with the Vostok and most Soviet spacecraft, their long-term success lay in the assembly-line principle; and this has certainly been true of the Soyuz design, as it is a process that has continued for almost forty years.

Construction of the Descent Module is divided into the base area and the upper bell-shaped structure. First, the lower supporting frame is assembled in the base, instrumentation is located on this supporting frame, and on top of this are placed the supports for the crew seats. Brackets for supporting instrumentation are separately

The interior of a Soyuz Orbital Module during fabrication.

fitted inside the bell-shaped structure, followed by the attachment of each instrument. (Each instrument or component for inclusion inside Soyuz is separately tested for strength, vibration, high pressure, and vacuum). Only when all instrumentation is fitted into the two parts are they joined and sealed together. In a separate assembly line, the Orbital Module is fabricated in two half-spheres and outfitted with the equipment and provisions to support the crew in orbital flight. A third assembly process is followed for the assembly of the cylindrical structure of the Propulsion Module. Each component is tested individually and in sequence with the rest of the module in which it is located. Following component assembly, all three sections are brought together for a series of what the Americans term 'fit and function tests' to ensure their compatibility with the rest of the spacecraft subsystems.

Ground testing

The components and spacecraft are subjected to a series of individual and combined tests in ground facilities and airborne. Scale-models and engineering test articles are initially used in the various stages of the tests, but full-size boilerplate mock-ups are also included in the test programme, as are flight articles prior to shipment to Baikonur. In some cases, the cosmonauts conduct part of their training in these mock-ups and simulations.

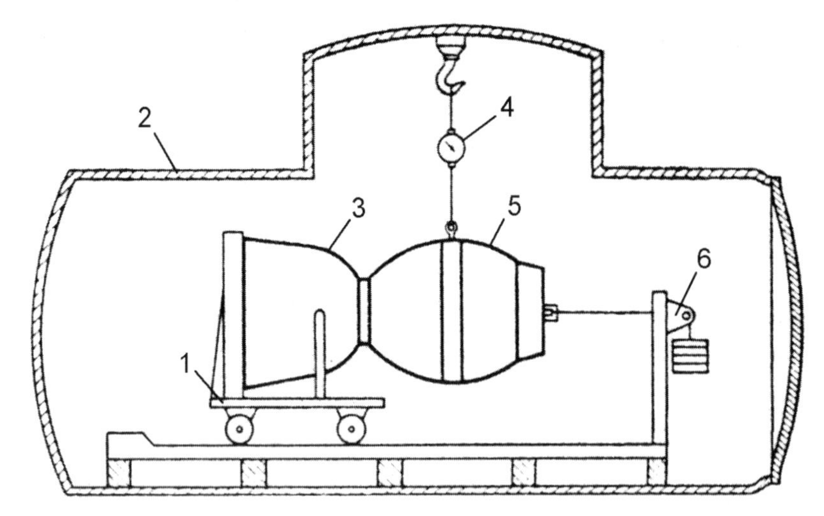

The stand for testing the separation of the Descent Module from the Orbital Module: 1, trolley; 2, pressure chamber; 3, Descent Module; 4, dynamometer/load gauge; 5, orbital module; 6, device to keep the Orbital Module in place after separation.

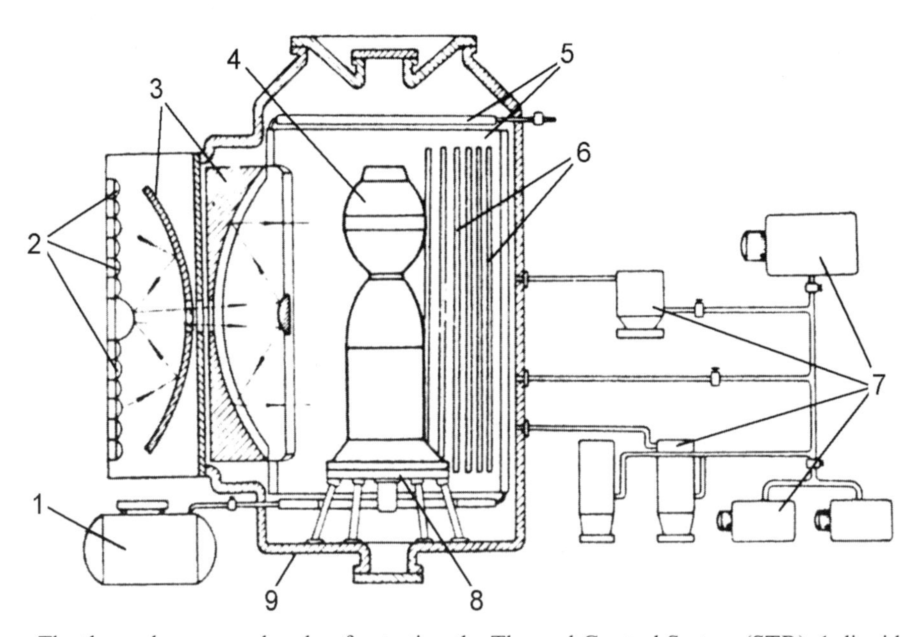

The thermal pressure chamber for testing the Thermal Control System (STR): 1, liquid nitrogen tank; 2, xenon gaseous discharge lamps; 3, optical system; 4, spacecraft mock-up; 5, cryogenic shields; 6, device to simulate Earth radiation; 7, pumps; 8, rotatable support structure; 9, chamber.

Vacuum testing. Thermal models of the spacecraft are placed in vacuum pressure chambers, in which their pressure integrity and the operation of mechanical systems (such as the deployment of antenna and solar arrays) are checked.

Thermal testing. The heat shield of each spacecraft is subjected to ultra-high temperatures in special 'space ovens'.

Bench tests. A series of bench tests is conducted to determine the capability of the spacecraft to withstand the rigours of launch, in-flight procedures and landing conditions. The three components and the combined spacecraft are shaken in a series of tests which simulate the vibrations encountered during the launch and landing phases, and this is followed by a programme of dynamic strength tests which determine the reliability and efficiency of the structure of the Soyuz under simulated high g-forces. The Descent Module is then subjected to a series of impact tests in which it is dropped from different heights, and with different combinations of full, partial and minimal retro-rocket firing, onto rocks, soil, snow, ice, swamp and water.

Systems testing. Once the spacecraft has been fabricated and tested, all individual flight systems and sub-systems are tested and evaluated to ensure that, during the flight, performance is as designed. A simulation of nose shroud separation is conducted with an engineering test model of the spacecraft inserted into a test rig to simulate the location of the Soyuz inside the shroud and on top of the upper stage of the launch vehicle. High-speed cameras document the repeated separation of the nose cone, while telemetry sensors gather data from across the vehicle and shroud. The speed and angle of shroud separation in flight can thus be determined, and the size and force of separation devices calculated. From here, a spacecraft is subjected to tests of the deployment of the various appendages, including docking equipment, antennae and solar arrays, after separation of the shroud and insertion into orbit; and physical, mechanical and electrical systems are tested on test vehicles and flight spacecraft in vacuum chambers. To simulate the reduction in gravity encountered on orbit, working model spacecraft are suspended on a test rig. Wires several metres in length are attached to the antennae, and tension in the wires neutralises the effects of gravity on the antennae as they are deployed in the test. The solar arrays are supported on rollers, which move freely along a level surface as the arrays are commanded to deploy from their folded position at the side of the PM. Explosive bolts are activated and the arrays and antennae deployed in sequence, and the deployment forces on the appendages are recorded by attached sensors. The DM is placed in a test rig which simulates the separation from the OM and PM, in normal end-of-mission scenarios, launch abort and other situations, and film and telemetry data are again gathered to analyse the sequence of events and the level of capsule separation charges. Similar programmes of dynamics tests are conducted on the docking mechanism (either separately in test rigs or on the test and flight spacecraft) and on the operation of hatches, for crew transfer through the forward hatch and for EVA from the side access hatch.

The Soyuz Orbital Module during pre-flight tests.

Flight testing

Experimental flight testing of Soyuz systems was conducted during the development of the spacecraft, and continues in accordance with the implementation of new flight systems and upgrades and to expand the database of test results. The flight testing focuses on the atmospheric portions of the flight, during launch and landing under 'normal' and contingency scenarios.

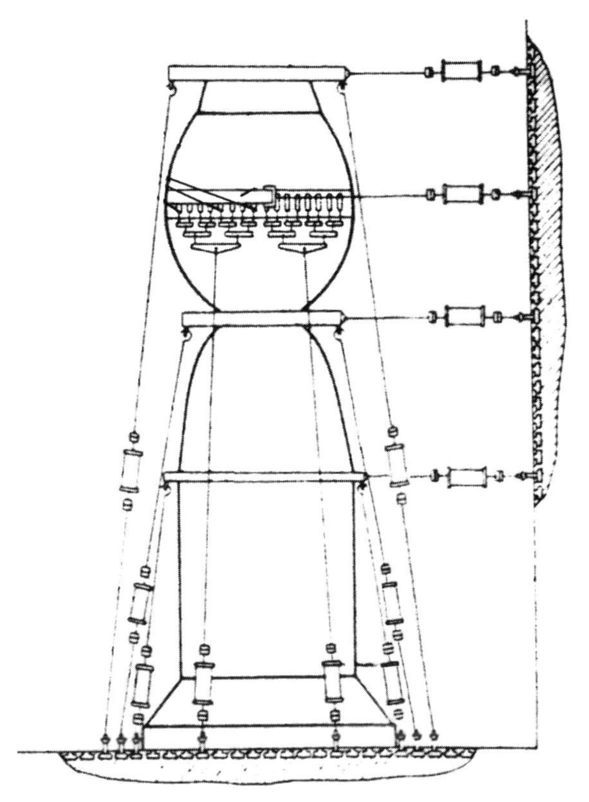

The system for testing the static strength of the spacecraft.

Early test models of the heat shield and spacecraft were mounted on geophysical rockets and launched into the upper atmosphere, where the capsule was separated to make a controlled descent. Inside the 'spacecraft' were telemetric sensors that recorded the temperature and ablation of the shield material, while structural, thermal and ballistic data were recorded inside and on the surface of the DM as it followed realistic descent profiles.

Aircraft were used in tests of the parachute system, during which accurate parameters of an real descent were simulated by using model spacecraft and full-size mock-ups, as well as flight-certified hardware. These tests also determined the maximum angles at which the parachute could be deployed and still perform within the safety limits, which parachute material and harness configuration worked best, and how the reserve parachute activated upon failure of the prime system. During the descent, various tests were conducted on the structural elements of the spacecraft and parachute system, as well as analyses of the hatch firing systems, cover separation of the heat shield, retro-rocket firing, tracking and communication systems, and the capability of the structure of the DM to withstand the impact of a landing from a great height.

The test objectives determined at which altitude and speed the transport aircraft

Testing the system for (*top*) separating the Descent Module from the Propulsion Module, and (*bottom*) for deploying the solar arrays from the sides of the Propulsion Module.

The complete spacecraft undergoes integrated systems checks on a test rig.

The flight vehicle, with solar arrays deployed, undergoes pre-flight systems tests.

would fly, and in what weather conditions. At the predetermined point, the test capsule was ejected from the rear of the aircraft cargo hold to fall towards Earth, stabilise, and for the parachute system deploy (or fail) at 10 km above the ground. Deployment from the aircraft and system operations was filmed from above and from the ground for later evaluation. As with the early drop tests, the airborne test programme included simulated landings in various environments.

Sea trials were carried out on the DM to determine its buoyancy in lakes and in the open ocean (usually the Black Sea). These sea trials were initially unmanned or were staffed by test engineers representing a flight crew, but (as previously mentioned) flight crews continue to conduct a programme of sea trials as part of their mission training.

Another element of aerodynamic testing carried out on Soyuz hardware was the test involving the launch escape system after verification of the abort and separation sequence on test rigs. Models of the launch escape tower and spacecraft were mounted on a small test vehicle (similar to the Little Joe tests conducted on Apollo CM escape towers), fitted with sensors, and either (unmanned) on-pad or in-flight launch aborts were filmed. This type of test was designed to simulate deviations of the launch vehicle in flight (as happened on Soyuz 18-1 in 1975), engine failures, excessive g-force during ascent, fires, and explosions (as experienced on Soyuz T-10-1 in 1983).

The range of Soyuz system mock-ups

An early opportunity to determine the range and depth of system mock-ups and tests conducted during Soyuz manufacture and flight preparation came with the Apollo–Soyuz Test Project of 1972–75. NASA engineers were allowed access to various Soviet check-out flow data to ensure safety and integrity in flying Apollo to, and docking with, a Soyuz in orbit. In part, these were:[21]

1 Orbital Module structural mock-up for static tests.
2 Orbital Module structural mock-up for dynamic tests.
3 Mock-up for the final development layout of compatible equipment in the living compartments and of exterior elements (such as targets).
4 Mock-up for the final development of the life support system, incorporating new and modified equipment.
5 Mock-up for Orbital Module thermal conditions development.
6 Spacecraft antenna mock-up.
7 Mock-ups for docking dynamics development.
8 Mock-ups for docking system development and interface pressures integrity control check in the thermal/pressure chamber.
9 Bench set for development and verification of docking system structural components.
10 Docking system fit check.

After completion of the test programme, the flight spacecraft is prepared for shipment to the Baikonur Cosmodrome, where a further programme of tests is conducted prior to and during launch integration. At this point, test and design engineers and quality inspectors responsible for manufacture verify all test procedures in sequence:[21]

1 Module and hydro-pneumatic line pressure integrity check after installation onboard the spacecraft.
2 Pyros circuit check.
3 Autonomous check of docking system operations.
4 Check of RF communication links and antennae.
5 Check of external device deployment mechanism and accuracy of installation, including alignments.
6 Sensor and measurement systems equipment checks.

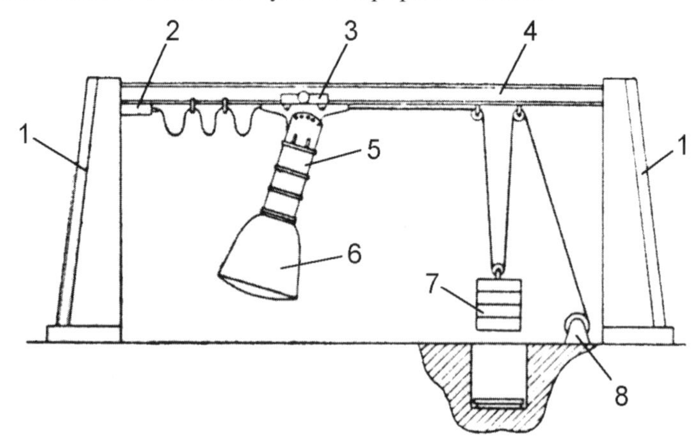

The stand for testing Descent Module landing loads: 1, support structure; 2, braking device; 3, slide; 4, beam; 5, suspension device; 6, Descent Module mock-up; 7, weight; 8, winch.

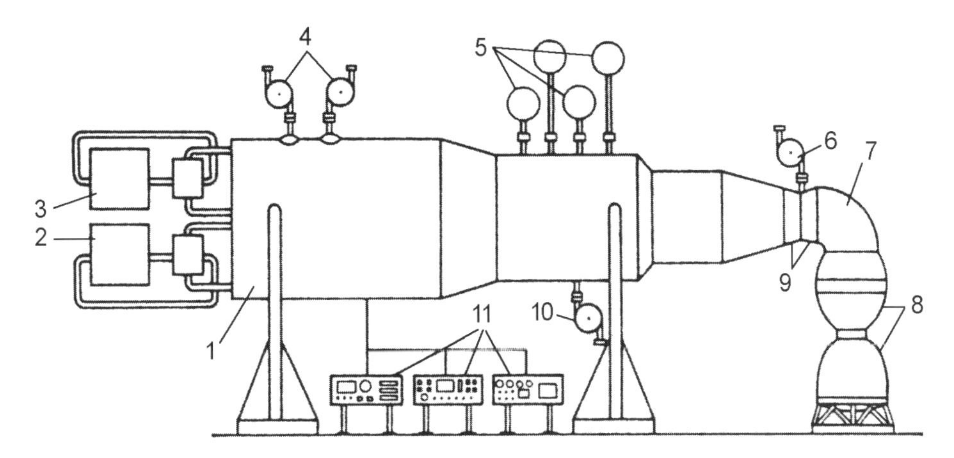

An experimental installation to test the life support system of the Salyut–Soyuz complex: 1, mock-up of Salyut station; 2, simulated external heating loop; 3, simulated external cooling loop; 4, pumps to remove gas from the airlocks; 5, tanks to pressurise the compartments with gas; 6, pump to remove gas from the area between the docking mechanisms; 7, transfer tunnel (only intended for use on the ground); 8, mock-ups of the Soyuz Orbital and Descent Modules; 9, docking mechanisms; 10, pump to simulate leaks; 11, control and display panels.

7 Cable connections to ground equipment.
8 Autonomous electrical verification tests of spacecraft systems.
9 Test activation of radio systems.
10 Test activation of life support system, everyday usage equipment, photographic and movie equipment, orientation lights, and flashing beacons.
11 Elimination of discrepancies and preparations for integrated tests.
12 Integrated tests (electrical).
13 Test result analysis.
14 Elimination of discrepancies and repeated tests (as required).
15 Disconnection of test cables.
16 Preparations for transport.
17 Transportation to technical site of the launch complex.

SOYUZ AT BAIKONUR

The three elements of the mission (launch vehicle, spacecraft and crew) merge at the Baikonur cosmodrome but arrive at different times, depending on mission planning. It is here that the final training and tests are performed on both the hardware and the crew, and the final items are cleared for flight.

Launch preparations

The central core and four strap-on elements of the Soyuz launch vehicle arrive by rail

A Soyuz spacecraft with a flown Descent Module on top of an upper stage of a launch vehicle at MAI.

from the Progress factory in Samara and are delivered to Site 32 for launch integration, which is conducted in the MIK-2B at Site 2.

Soyuz is processed at MIK OK Area 254 in the Baikonur maintenance area, where it is suspended in a variety of rigs for pre-mating testing of independent and integrated systems. Final tests on the radio engineering systems, including the important rendezvous and docking radar, are conducted in an anechoic chamber, where radio waves emitted from the spacecraft's equipment are prevented from false reflection by absorbent materials on the chamber walls. Onboard radio equipment is checked for electromagnetic compatibility and to ensure there is no interference with other equipment. This is followed by the charging of the temperature control system with the heat transfer fluids.

Approximately two weeks prior to launch, the crew visits Baikonur to tour the MIK facilities and to view the launch processing of their spacecraft. During this brief visit, the crew-members slide inside their spacecraft to perform final checks of where their equipment will be located and to once more familiarise themselves with the internal arrangement of their flight vehicle over the trainers and simulators. (In the US, this more detailed crew involvement exercise is called the Crew Integration Test, and includes fit and function activities in the crew module of the Shuttle orbiter). The

Soyuz in various stages of mission processing in the MIK at Baikonur.

flight crew enters the spacecraft for the last time before it is enclosed inside its shroud. (In 1982, the Soyuz T-5 crew, while opening the hatch from the OM into the DM during the inspection of their spacecraft, dislodged a loose nut, which fell into the Descent Module – and it took some time before an assembly worker found it. The crew were apprehensive, as they felt it reflected on their performance – and it is considered bad luck to damage the spacecraft, if only accidentally, so close to a launch.) After visiting the facility, the crew returns to TsPK for the final stages of the training programme, and does not return to the launch site until approximately a week to ten days before the planned launch.

The Soyuz is then loaded with its fuel and compressed gas supplies before being placed in the launch shroud. The combination is then moved across to MIK-2B, where the Soyuz is mated to the Blok I upper stage and again tested for integrity.

The current Soyuz processing cycle at Baikonur is as follows. After arrival at the launch site, the Soyuz spacecraft and the Soyuz rocket's upper stage are placed in the Site 254 assembly building, and the lower stages of the Soyuz rocket are placed in the assembly building at Site 2. Just over a week before launch, the Soyuz spacecraft is sent to the 11G12 fuelling station at Site 31, and is afterwards returned to the Site 254 MIK for mating with the Soyuz rocket's upper stage and payload fairing. Subsequently, the spacecraft and upper stage are sent to Site 2 for mating with the core and strap-on stages. The whole system (called the 'package') then completes a further systems and integrity check.

The roll-out from MIK-2B to Pad 5 Site 1 traditionally takes place at 7.00 am. This operation is different from that at the second Soyuz facility at Site 31, where MIK 40 is used to process the payload and upper stage, and the sustainer and strap-ons of the launch vehicle. As the integration facility is quite close to the LP6, it all takes place under the auspices of Site 31 where, following another tradition, vehicle roll-out is always scheduled for 7.30 am. This site also includes vehicle fuelling facilities for Soyuz, ISS modules and other Russian satellites, and is located adjacent to MIK 40.

With the R-7 on top of a flat-bed rail carriage, which also incorporates an erecting arm, it is slowly rolled tail-first from the MIK towards the launch pad, two days prior to launch. The speed of this movement has been described as 'walking pace' and the slowly moving train is photographed by the media and accompanied by officials and workers at the complex. As the train rolls across the rails, coins are placed in its path so that the weight of the train and its R-7 payload flattens them, to produce souvenirs for the pad workers. It is traditional for the cosmonauts to see the roll-out of their vehicle to the pad.

At the pad, as the vehicle is held firmly, the erector arm hoists the vehicle vertically towards the pad retention devices; and once secured on the pad, the arm is lowered. After the launcher carriage is backed away from the pad area, service platforms and umbilical masts are raised to surround the vehicle, and the azimuth guidance is then set. A network of ground support equipment – both mobile and

The Soyuz and its carrier rocket are transported horizontally from the processing area at Baikonur to the launch pad by means of a rail network, and are then hoisted vertically onto the pad.

The Soyuz/R-7 combination is hoisted from the rail carriage to its vertical position on the launch pad.

stationary, and upgraded from those first used in 1957 (see *The Rocket Men*, Springer–Praxis, 2001) – is installed around the rocket and the launcher facilities. The time of launch depends on mission requirements and landing conditions.

Air temperature inside the shroud is regulated by forced air, operating continuously until lift-off to maintain the temperature required for the spacecraft and its internal equipment, and a series of mechanical, electrical, telecommunications and pyrotechnics pre-launch tests of the launcher and spacecraft are conducted. When the area is cleared, the launcher is filled with fuel and compressed gases. The fuelling status checks are completed automatically and remotely – a lesson learned the hard way after a number of launches were delayed and ground support personnel were killed. This affected only unmanned flights, but the dangers were obvious with the Soyuz T-10 explosion. Propellant loading begins 5–6 hours prior to launch, and was completed by further checks and purges of the fuel lines, followed by their detachment from the spacecraft.

Launch site test and verification sequence

At Baikonur, the programme verification procedures conducted during the integration phase and at the pad were also made available to NASA representatives during the ASTP. The launch processing test sequence was as follows:

1 Technical inspection of individual modules in preparation for assembly.
2 Module power supply circuits check.
3 Pressure integrity checks of the modules and hydro-pneumatic lines of the spacecraft.
4 Autonomous verification tests of the docking assembly.
5 Spacecraft assembly.
6 Spacecraft, mass adjustment and weight determination; instrument module propulsion system installation.
7 Antenna and docking system deployment check.
8 Ground test cable connection.
9 Test activation of radio/telemetry system.
10 Test activation of movie and photographic system, equipment for everyday usage, and life support system.
11 Test activation of orientation lights and flashing beacons.
12 Spacecraft integrated tests (electrical).
13 Assembled spacecraft pressure integrity checks in the altitude chamber.
14 Pressure integrity checks of the assembled spacecraft thermal control system.
15 Spacecraft provisions.
16 Life support and power systems verification of activation.
17 Manned tests of system operation (crew fit-and-function tests).
18 Spacecraft pyro circuit check.
19 Spacecraft propulsion system loading with fuel and compressed gases.
20 Solar cell installation and check-out on the spacecraft.
21 Hatch pressure integrity check.
22 Mating with the launch shroud.
23 Mating with the launch vehicle and transportation to the launch pad.

The pad check-out followed the sequence:

24 Radio system activation verification.
25 Pyro circuit check.
26 Hatches integrity check.
27 Emergency escape system pyros connection integrity check.
28 OM and DM environmental and thermal control system preparation.
29 Crew ingress.
30 Hatch pressure integrity check.
31 Onboard system initial monitoring from the cosmonauts' panels.
32 Air-to-ground radio communication verification.
33 Medical monitoring.
34 Launch readiness confirmation, and launch sequence.

The sequence of early Soyuz launch preparations was completely different from the launch procedures at Cape Canaveral. The first Westerners to view the launch processing area were accustomed to ultra-clean white-room facilities, and were amazed to see a layer of sand deposits – apparently blown in from the steppes – on the surface of the R-7 as it underwent processing for launch. Querying this as a potential problem during launch, when hundreds of instruments and pieces of equipment had to operate in sequence and with split-second timing to ensure a smooth launch, they were told by their Soviet hosts that this never seemed to be a problem at Baikonur. 'The rocket is rolled out, it is hoisted, it is launched, and then the dust just falls off!'

Launch phase

The launch of a Soyuz is governed by several factors, including daylight hours, mission duration (timing where and when the vehicle lands), the abort situation, and whether the launch is to a second vehicle (such as a space station). On a mission to a space station in orbit, launch is governed by the rotation of Earth beneath that station, and the lift-off can take place only when the orbital plane of the station passes over the Cosmodrome, allowing the spacecraft to enter the same orbital plane, behind and below the target station. As the orbital plane shifts each day, it creates a repeating cycle of about 60 days, and so the launch window for the optimum times to launch from Baikonur to a space station occurs approximately every two months. (The sequence of events during the count-down, the launch and the ascent are detailed in the Prologue.)

Riding the 'package'

In his *Diary of a Cosmonaut* (published in 1983), Valentin Lebedev recalled his second launch onboard Soyuz T-5 in 1982: 'From somewhere below I hear the roaring wave of thundering thrust coming from the blasting rocket engines. The rocket begins to sway to the left and to the right as if it was losing balance. Then we rise above the pad, we feel the loss of support. There was a 2–3-second delay, and then all at once the rocket takes off as if it's suddenly unleashed, and we yell 'Go-o-o-u'.'

A cosmonaut's recollection of the separation of stages depends on each individual rocket. Some are smooth – others are not. The upper-stage flight generally seems rougher than the early part of the ascent, and during launch the crews experience 2–3.5 g. Four seconds after the shut-down of the upper-stage engine, the Soyuz separates from the upper stage into its initial orbit, and control of the spacecraft is switched from Baikonur to MCC Moscow, with the crew ready to carry out their assigned flight programme.

THE FLIGHT CONTROL CENTRE

With Soyuz safely in orbit, the main support network centres on the mission control facility located near Moscow. The current facility replaced the original command facility located in the Crimea, and has been in operation since the early 1970s.

During the late 1950s, a network of fifteen stations was established within the USSR, as a tracking and telemetry network known as the Scientific Measuring Points (NIP), which was used for communications for both unmanned and manned spaceflights. During the 1960s the Soviets gradually introduced a fleet of ocean-going ships fitted with space tracking and communications devices that could communicate with spacecraft outside the range of Soviet territory. It was a far cry from the more sophisticated American Manned Spaceflight Network of ground stations across the globe, although during 1969–70 the Soviets managed to site some stations in Chad, Cuba, Guinea, Mali and the United Arab Emirates.[22] Control of the ground stations in the USSR fell under the auspices of the Strategic Missile Forces NII-4 just outside Moscow, directly controlled by Military Unit 32103. Each mission had its own dedicated 'flight control team' consisting of about ten specialists from the military forces, design bureaux, production factories and the Academy of Sciences. The team (GOGU) was located in NII-4's Moscow facilities, or in the Ministry of Defence Control Centre, which was also in the capital. The State Commission never left Tyuratam during the Vostok or Voskhod missions, as they only lasted a few days, but they maintained contact with the NII-4 team in Moscow for the duration of the short missions.

Yevpatoria: the original Soyuz mission control

With the introduction of the Soyuz missions – intended to complete far more complicated and much longer flights to the Moon and to space stations – a new central dedicated control facility was required. The site chosen was the 16th NIP located at Yevpatoria in the Crimea, and this remained the primary control centre for almost all Soyuz spaceflights from 1966 to 1975. At the new flight control centre, a team of about five hundred worked in three eight-hour shifts for the duration of the relatively short missions, in much the same way as the American FCC in Houston. The site also featured the deep-space tracking facility built in the early 1960s to support the Venera and Mars planetary probes. During the early Soyuz missions, the Head of GOGU flight control was Major General Pavel A. Agadzhanov, and in the late 1960s Colonel Nikolay G. Fadeyev headed flight operations during the development of hardware in support of the manned lunar programme.

The Chief Operations and Control Group (GOGU) was headed by a military officer (the Flight Director), while the 'technical leader' (or Deputy Flight Director) was a leading Korolyov bureau representative (Chertok, 1966–68; and Tregub, 1969–73). The ultimate decisions concerning flight operations were made by the Soyuz State Commission, and GOGU had only an advisory role. It seems that the dual civilian/military leadership of GOGU somehow played a role in the loss of a DOS station in 1973 (Cosmos 557), after which it was decided to have a sole civilian flight director. Those who have held this post have all been cosmonauts: Alexei Yeliseyev (1973–82), Valeri Ryumin (1982–89), and (from 1989) Vladimir Solovyov.

Kaliningrad: mission control Moscow

During final preparations for the ASTP in July 1975, the Soviets launched Soyuz 18 to the Salyut 4 space station. The mission was originally to last only 28 days, but was extended to 62 days while in orbit, so that the crew would remain in space during the joint mission. This surprised and worried some Americans, who doubted the Soviets' capability to support a space station mission and the international docking mission

Since the mid-1970s, all Soyuz flights have been controlled from the flight control centre in Kaliningrad, Moscow, serving a role similar to NASA's Mission Control in Houston.

at the same time. CIA Deputy Director for Science and Technology, Carl Ducket, in a hearing before the Independent Agency Subcommittee on 2 June, voiced this concern, stating: 'I do not think [the Russians] are in good shape to handle two missions at once from the command point of view... The added complexity of having two missions going at once should be avoided at all costs. Soviet communications capabilities and central management facilities are greatly inferior to those of the US.' The NASA enquiry revealed that the Russians were using a separate flight control centre for station operations, and planned to use a new facility for ASTP. Indeed, they had operationally tested it with Cosmos 638 in 1974.[23]

The new flight control centre was located at Kaliningrad (renamed Korolyov in 1996), 10 km north-east of central Moscow and 10 km north-west of TsPK. This Tsentr Upavleniye Polyotom (Centre for the Control of Flight) is known by its abbreviation, TsUP (pronounced 'tsoop'), although from orbit the cosmonauts refer to the call-sign Moskva (Moscow). Construction of the building that became TsUP began in autumn 1959, and on 3 October 1960 it was officially established as the Computing Centre of NII-88 (the latter being renamed TsNIIMash in 1967). This is considered the official 'birth date' of TsUP, and the fortieth anniversary was celebrated on 3 October 2000. Later in the 1960s it was renamed the Coordinating Computing Centre. However, during the 1960s the its role was almost entirely passive, as it was responsible only for processing telemetry, and it was not until the

1970s (when it was renamed TsUP) that it was used for controlling spacecraft. It remains part of TsNIIMash. The Americans were impressed when they were taken on a tour of the facility in mid-May 1975, prior to the launch of Soyuz 18; but they did not consider that it had the capacity of their own MCC facility in Houston. TsUP took over all flight control of Soyuz (and space station) missions from 1975, and is the upgraded FCC of Moscow for Mir and the ISS. By the time Mir was in orbit in 1986, this facility had grown to include a computer support complex with seven machines, each with the capacity to handle 12.5 million instructions per second (MIPS), and with developments underway to incorporate computers with 40–50 MIPS.

The floor layout consisted of five rows of consoles, with twenty-four flight controller positions covering all aspects of the flight, including flight dynamics, orbital trajectory, spacecraft systems, crew health, communications, and electrical power. At the front wall of the room is a large map of the world. The current position of the Soyuz is displayed in its orbital path, either alongside or docked to a station, as it passes over the network of stations or ships around the world. Each side of the main screen are large data display panels that record each spacecraft's Mission Elapsed Time, the orbital revolution, the parameters of the orbit and other telemetric data, docking approach distances, and speed during docking. TV screens project images from orbit, either from inside (crew activities) or outside the spacecraft (such as Progress dockings). Outside the main control hall are a dozen support rooms for specialists in aspects of flight, such as tracking, orbital flight paths, and rendezvous and docking. Physicians man one room to monitor the work and health of the cosmonauts, while another room is reserved for regular private family conversations, and there is a similar control room at Institute of Medical and Biological Problems (IMBP). Around the upper gallery at the rear of the room are observation seats for officials, the media and guests.

By 1990 – twenty years after the establishment of the facility – it was estimated that the FCC employed a staff of five thousand, with fifty in the main flight control room on any given shift, and five hundred in support rooms. The controllers work in four shifts, for 24 hours on and three days off. During the flight of the Soyuz to and from the station, separate support rooms handle communication and control with the crew, and then hand over to the main room once Soyuz docks to station. During the 1980s a second control room – intended for the subsequently cancelled Buran shuttle operations – was added to the facility, and it is now used for controlling the Russian segment of the ISS. A number of cosmonauts have worked at the facility, and the role of TsUP role has recently been extended to the control of civilian satellites such as Okean and Meteor.

As with the American MCC in Houston, a shift Flight Director heads up the flight controller team, and cosmonauts usually handle the role of CapCom. In 1984, Vladimir Lobachev became the Director of the FCC in Moscow, and since the building was completed, the Flight Director has always been an experienced Energiya cosmonaut.

RECOVERY FORCES

The Federal Aerospace Search and Rescue Administration is responsible for locating returning spacecraft or capsules and evacuating cosmonauts from the landing site. The main recovery site is in Kazakhstan, between Tselinograd and Dzhezkazgan. As the Soviet and Russian space authorities have discovered, re-entry is always a dangerous procedure in manned spaceflight, and in February 2003 this was starkly demonstrated with the tragic loss of the Space Shuttle *Columbia* and her crew of seven astronauts.

Soviet documentation has revealed that recovery of manned spacecraft – at least from Earth orbit – is governed by two main factors: that the landing is in daylight, and that retro-fire in orbit is conducted in sunlight, allowing for manual orientation of the spacecraft. Just as launch to a target spacecraft occurs when the target spacecraft passes over the launch site, so undocking from a station for landing can take place only at certain time each 24 hours as the spacecraft passes over the landing zone. Due to the orbital plane of the spacecraft moving slowly over the Earth, the landing time moves to earlier in the day as the mission proceeds. Normally, optimum landing times fall between three hours before sunset, and sunset.

Re-entry usually takes place with retro-fire over the Gulf of Guinea. The craft enters the atmosphere over Egypt, and experiences maximum heating and communications black-out over the Caspian Sea. The normal landing area is north east of Leninsk, in Kazakhstan.

The recovery force is equipped with a variety of aircraft, helicopters and all-terrain vehicles, and is directed to the landing site by radio equipment, operating at 19.995 MHz, on the aircraft. As the Soyuz descends, suspended by its parachute, communications with the crew are established at 121.75 MHz.

Tractor recovery of the unoccupied Soyuz TM-34 spacecraft.

Once the capsule lands (normally recorded on film, from helicopters) the parachute is separated, and when the capsule comes to rest, the recovery forces land nearby. Recovery personnel and specialists from the Cosmonaut Training Centre rush to the capsule to open the upper hatch, to check that the crew inside are uninjured; but great care is required, as the outer shell of the capsule is normally still hot from the re-entry heating, and the signalling devices continue to emit a small radiation signature.

The specialists right the capsule and then, around the craft, place a metal structure which includes a slide for the cosmonauts to exit the vehicle. Once the cosmonauts are out, they are usually placed in seats and taken to a tent for an initial medical examination. Many recoveries are carried out in winter, and the cosmonauts are wrapped in furs. After brief welcoming ceremonies, and after signing the spacecraft and greetings from friends, the crew is transferred to a local air base and then flown back to Baikonur and on to Moscow. Once back at Star City, the crew undergo debriefing, medical tests and post-mission celebrations.

The craft is loaded onto one of the all-terrain vehicles (although sometimes it is slung under a helicopter), and is taken back to the main base. Spare fuel and pyrotechnical devices are removed, and the craft is then returned to Energiya at Korolyov, where most Soyuz craft are stored.

REFERENCES

1 *Design Characteristics for Soyuz*, ASTP 40001, 30 April 1974; *Safety Assessment Report for Soyuz Pyrotechnic Devices*, ASTP 20204, 10 February 1975; *SAR Soyuz Propulsion and Control Systems*, ASTP 20202.1, 1 May 1975; *SAR for Soyuz Fire and Fire Safety*, ASTP 20203.1, 1 May 1975; *SAR on the Soyuz Habitable Modules Overpressurization and Depressurization*, ASTP 20205.1, 1 May 1975. NASA JSC History Archive, ASTP Collection, Houston, Texas.
2 Apollo–Soyuz Mission Evaluation Report, NASA JSC-10607, December 1975, pp.A-14–A-32.
3 Johnson, Nicholas, *Handbook of Soviet Manned Spaceflight*, AAS Science and Technology Series, **48**, p.102.
4 Siddiqi, Asif, *Challenge to Apollo*, NASA SP-4408 (2000), p.467.
5 Hall, Rex D. and Shayler, David J., *The Rocket Men*, Springer–Praxis (2001), pp.94–102.
6 *Ibid.*, pp.248–251.
7 Glushko, Valentin P., *Kosmonavtika Entsiklopedia*, Moscow Research 'Soviet Encyclopaedia', 1985.
8 Grahn, Sven, 'Characteristics of the Soyuz Attitude Control System', web site, http//:www.svengrahn.pp.se/histind/Soyuz1Land/soyAtti.htm and 'The Igla radio system for rendezvous and docking', web site, http://www.svengrahn.pp.se/histind/RvDRadar/IGLA.htm.
9 Web site, http://www.computer-museum.ru/english.argon.16.htm.
10 Interview with Viktor Przhiyalkovsky, former General Director of NICVET,

Moscow, 24 May 2002, 'Computing in the Soviet Space Programme', http://hrst.mit.edu/hrs/apollo/soviet/interview/interview-przhiyalkovsky.htm.
11 Grahn, Sven, 'Zond (7K-L1) cockpit layout', web site. http://www.svengrahn.pp.se/histind/Zondmiss/Zondcock.htm
12 Web site, http://www.computer-museum.ru/english.argon.11c.htm.
13 'Propulsion unit of emergency rescue system of Soyuz TM spacecraft', ISKRA Machine Building Design Office press release, 1990.
14 *Parachute Systems for Descent of Manned Spacecraft*, Aviaexport, USSR, *c*.1990.
15 Hall, Rex D. and Shayler, David J., *The Rocket Men*, p.226.
16 Grahn, Sven, 'Soyuz emergency landing zones: the 'ugol pasadki' story', web site, http://www.svengrahn.pp.se/histind/Ugol/Ugol.htm.
17 Varfolomeyev, Timothy, 'Soviet Rocketry that Conquered Space: Evolution of Korolyov's R-7 launchers', *Spaceflight*, **7**, August 1995, pp.260–263.
18 Siddiqi, Asif, *Challenge to Apollo*, NASA SP-4408 (2000), pp.128–144.
19 *Novosti Kosmonavtiki*, No.4, 2002, pp.71–72.
20 Siddiqi, Asif, *Challenge to Apollo*, NASA SP-4408 (2000), pp.195, 462.
21 *Safety Assessment Report for Soyuz Manufacturing and Test Check-out*, ASTP 20206, 1 May 1974. NASA ASTP Collection, Box 1276, File 11, Houston, Texas.
22 Siddiqi, Asif, *Challenge to Apollo*, NASA SP-4408 (2000), pp.543–538.
23 Ezell, Edward, and Ezell, Linda, *The Partnership: a History of the Apollo–Soyuz Test Project*, NASA SP-4209, 1978, pp.307–310.

Docking missions, 1966–70

After more than five years of design studies, the first orbital flight tests of the Soyuz spacecraft were completed between 1966 and 1970. These missions were flown using a Soyuz (7K-OK) variant to test and evaluate methods of rendezvous and docking directly related to the manned lunar programme, and later, additional objectives related to the space station programme. This type of Soyuz vehicle evolved from a combination of the previous designs and variants evaluated during the early 1960s, and was the first operational manned variant from which all subsequent Soyuz spacecraft originated. It has often been termed the 'original Soyuz'.

THE 'ORIGINAL SOYUZ'

When the first cosmonauts flew the early Soyuz missions, the spacecraft was described as a three-part vehicle with a forward working module (Orbital Module), a central crew compartment (Descent Module), and an aft service module (Propulsion Module). (This is the basis of the systems detailed on pp.39–74.)

Initial design studies for this spacecraft began in 1964, from recommendations derived from earlier studies of the Soyuz Complex. The manned spacecraft continued its development, but work on the 9K tanker and 11K lunar transfer stage was terminated. On 18 August 1965, VPK approved the 7K-OK, and on 23 October 1965 the draft plan was signed. The space station ferry vehicle and the Soyuz T, TM and TMA versions all evolved from this spacecraft (designated 11F615) called 7K-OK (Orbital Ship). The original Soyuz first flew unmanned in November 1966 (Cosmos 133). In this section, we discuss the early unmanned and manned missions flown between 1966 and 1970.

The launch vehicle for these initial Soyuz missions was the 11A511 variant of the R-7, which in the upper stage featured an uprated RD-0110 producing 30.4 tonnes of thrust. This vehicle was developed from the variant used for Voskhod, but without the RD-0108 upper-stage engine. The launch vehicle was also (confusingly) designated 'Soyuz', and increased the overall length of the combination with a Soyuz spacecraft and its shroud to 49.91 m, with a sea-level launch thrust of 411.1

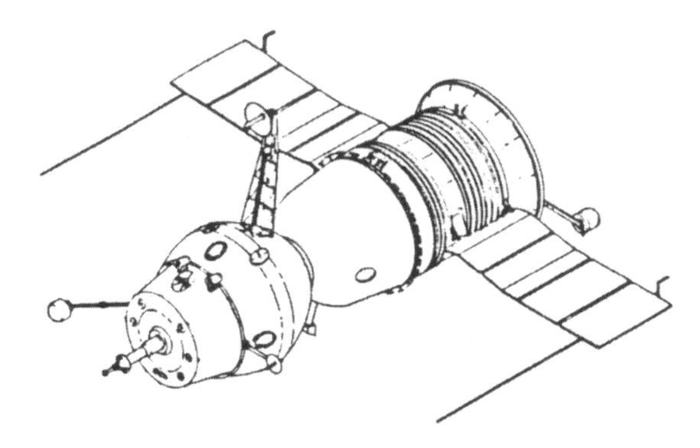

The 'original' Soyuz, flown from 1967 to 1981. (Courtesy Ralph Gibbons.)

tonnes. This vehicle was capable of delivering a 6,900-kg payload into a 200 × 450-km orbit.

PREPARING FOR THE FIRST FLIGHTS

The delivery schedule for the first 7K-OK spacecraft for an ambitious series of test flights was issued by the Military Industrial Commission on 18 August 1965. In this plan, two spacecraft would be delivered in the fourth quarter of 1965, followed by an additional two during the first quarter of 1966 and vehicles in the second quarter of 1966. The plan also detailed that the very first flight would include a link-up of two unmanned spacecraft in the first quarter of 1966. Less than six months later, in early 1966, plans were amended to follow the unmanned docking of spacecraft 1 and 2 (thereby planned for July/August 1966), with a manned docking mission between spacecraft 3 and 4 in September/October 1966, and a second manned docking mission using spacecraft 5 and 6 during November 1966. In this plan, there would be no solo test flights of Soyuz before the ambitious docking missions had been attempted.

As preparations continued, the launch schedules slipped further, so that by October 1966 it was clear that the first unmanned launch would not take place before November of that year, and that no cosmonaut would fly Soyuz in Earth orbit before January at the earliest. The production of training hardware for the crews, and flight hardware for the mission, was falling behind schedule. No Soviet cosmonaut had flown in space since March 1965, while between March 1965 and November 1966, ten American Gemini crews had been launched to perfect aspects of rendezvous and docking, EVA, tethered operations, extended-duration missions of 4, 8 and 14 days, and a variety of additional onboard experiments. In addition, the first manned Apollo was manifested for early 1967, and it was clear that the Soviets were falling behind the Americans in the race to the Moon. On 3 November, while the crews were preparing for the mission and the hardware was being prepared for the first manned Soyuz mission, the qualification of the parachute descent system suffered a setback when, during drop tests held in Feodosiya, the back-up parachute system failed. The test featured the deliberate separation of the main parachute and a successful development of the reserve canopy, but at 460 metres the hydrogen peroxide, used in the small stabilising thrusters of Soyuz during re-entry, was vented as planned and eroded the parachute lines, causing them to break. The tests also caused the descent capsule, under the parachute, to turn at 1 revolution per second, and produced pulsations in both the main and reserve parachute systems. While revisions to the parachute hardware continued, the crews for the first missions were being finalised and trained – all of which still indicated no manned mission before February at the earliest.

Early Soyuz assignments for cosmonauts

The initial training group, assigned to Soyuz training in September 1965, consisted of A.G. Nikolayev, V.F. Bykovsky, V.M. Komarov, P.I. Kolodin, Y.P. Artyukhin and A.N. Matinchenko, all of whom were members of the Air Force team. Feoktistov was also assigned from OKB-1, and Korolyov reserved the right to add more of his own bureau's engineers at a later date. After the cancellation of the Voskhod 3 mission, the Air Force also added Y.V. Khrunov, V.V. Gorbatko, G.S. Shonin and V.A. Shatalov to the group; and after much lobbying of senior Air Force officials and political leaders, Yuri Gagarin was restored to flight status and joined the Soyuz group. In September 1966, Korolyov's OKB-1 added A.S. Yeliseyev, V.N. Kubasov, V.N. Volkov and G.M. Grechko to the team.

The make-up of the group was fluid; for example, Artyukhin joined the lunar group during this period. In autumn 1966 the first crews were finalised for the ambitious plan to fly two Soyuz craft and conduct an exchange of crews:

Soyuz 1 prime crew	V.M. Komarov
Back-up crew	Y.A. Gagarin
Soyuz 2 prime crew	V.F. Bykovsky, A. Yeliseyev and Y.V. Khrunov
Back-up crew	A.G. Nikolayev, V.N. Kubasov and V.V. Gorbatko

The first Soyuz was launched in April 1967, and when the in-flight problems occurred, the Soyuz 2 flight was cancelled. When Komarov died on re-entry, all crews were stood down for a while as mission planners and design engineers addressed the problems. In further changes, Bykovsky was transferred to the lunar group, and in March 1968 Gagarin was killed in an air crash.

It was eventually decided that the docking and EVA mission profile of Soyuz 1 and Soyuz 2 would be repeated, but it was also decided to conduct a test of the Soyuz, with a solo flight to dock with an unmanned Soyuz. In early 1968, cosmonauts Beregovoi, Shatalov and Volynov were named to train for a Soyuz test flight, and crews were also formed for the impending repeat of the Soyuz 1/2 mission. All the Commanders – Shatalov, Volynov, Filipchenko and Shonin – were new on both the prime and back-up crews, but the rest of the crews were the same as were assigned to the 1967 mission.

When Soyuz 4 and Soyuz 5 succeeded, it was decided to repeat the mission with Soyuz 7 and Soyuz 8. This mission would be preceded by a solo Soyuz with an

The Soyuz 1 and Soyuz 2 crews at Baikonur on the morning of 23 April 1967. (*Left–right, foreground*) Komarov (Soyuz 1), Bykovsky, Khrunov and Yeliseyev (Soyuz 2). They report their readiness to carry out the joining flight programme.

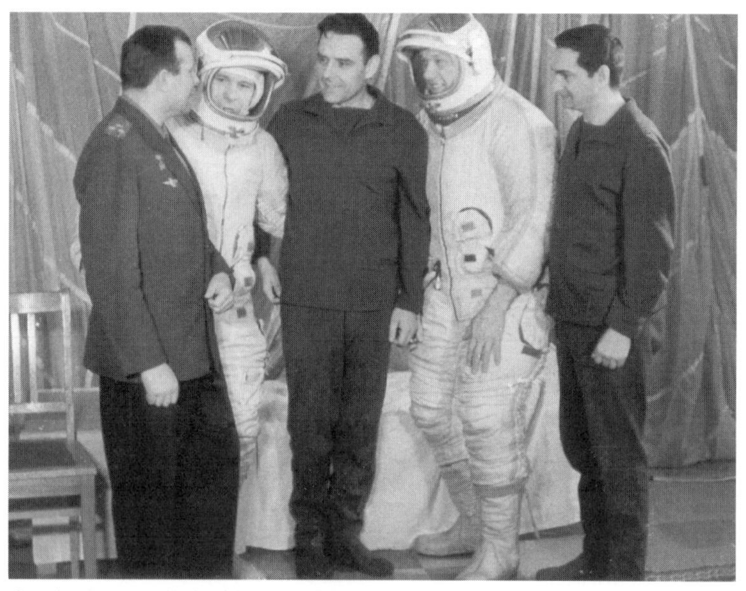

Crews for the Soyuz 1/2 docking and EVA transfer mission discuss their objectives with Yuri Gagarin. (*Left–right*) Gagarin (back-up Commander, Soyuz 1), Khrunov (Flight Engineer, Soyuz 2, in EVA suit), Komarov (Commander, Soyuz 1), Yeliseyev (Flight Engineer, Soyuz 2, in EVA suit) and Bykovsky (Commander, Soyuz 2).

experiment sponsored by the Paton Institute in the Ukraine, and a small group of cosmonauts was selected to train for it. The group consisted of Shonin and Kuklin from the Air Force group, and Grechko from TsKBEM (the old Korolyov design bureau). These three were joined by V.G. Fartushny (a cosmonaut specially selected from the Paton Institute), who would operate the welding experiment (Vulcan). Due to planning and political indecision, however, this mission was delayed for several months, and during this period Fartushny failed his medical and was stood down, to be replaced by a Korolyov design bureau engineer. Eventually it was decided to fly the three missions, launching over three days, and a number of crews was finalised; but it was also decided to have only one back-up crew for all three missions, chiefly due to the shortage of simulation time. Thus only four crews – rather than six – needed to be trained. The selected crews were:

Soyuz 6 prime crew	G.S. Shonin and V.N. Kubasov
Soyuz 7 prime crew	A.V. Filipchenko, V.N. Kubasov and V.V. Gorbatko
Soyuz 8 prime crew	A.G. Nikolayev and V.I. Sevastyanov

Shatalov and Yeliseyev were the back-up crew on all three missions, and Kolodin was back-up for Gorbatko. The Soviets had to change the crew of Soyuz 8 when they failed their examination, and the prime crew was replaced by Shatalov and Yeliseyev. Nikolayev and Sevastyanov acted as back-ups to the 'troika' missions.

After these flights were accomplished, the Soviets decided to fly an 18-day Soyuz mission. This long-duration mission would be very demanding, and so a special training group of three crews was created:

Prime crew	A.G. Nikolayev and V.I. Sevastyanov
Back-up crew	A.V. Filipchenko and G.M. Grechko
Support crew	V.G. Lazarev and V.I. Yazdovsky

There was much debate about whether to fly Lazarev, as he was an Air Force doctor and this was a medical mission; but the crews remained unchanged.

Cosmos 133: the first Soyuz in orbit

On 25 November 1966, the rendezvous and docking mission of the first two unmanned Soyuz spacecraft was authorised by the State Commission. On 28 November the first spacecraft – 7K-OK no.2 – would be launched. This would be the active participant in the docking, while the passive spacecraft – 7K-OK no.1 – would be launched 24 hours later. In Soviet flight planning, the active spacecraft always received even-number designations and the passive usually odd-number designations. The flight plan allowed for a docking on the first or second orbit of the passive spacecraft, if it could be inserted into orbit within 20 km of the active spacecraft. Should this not be possible, the docking would occur 24 hours later. The plan was for them to remain docked for three days, and to land on the fourth day of their respective missions. It was believed by Soyuz engineers that the cosmonauts should be able to mount a similar mission by the end of December, if everything went according to plan. This included delaying the launch of an unmanned military Zenit satellite to allow the Soyuz to launch from two pads. In his diary entry for 28 November 1966, Kamanin recorded this important step, which was crucial to the next phase of Soviet manned space exploration: 'We've been waiting for this to happen for more than four years. Today and tomorrow we see launches on which the immediate future of our space programme will hinge... All the Moon spacecraft are based on Soyuz.'

On 28 November 1966, spacecraft no. 2 was launched, and some nine minutes later reached a 171 × 223-km orbit at 51°.82 and with a period of 88.9 minutes. It was an almost flawless launch, marred only by a slight under-performance of the R-7. The launch was the culmination of years of dedicated work in design bureaux, factories and at the launch site. Not wishing to disclose the true nature of the vehicle, the TASS news release indicated that a new Soviet spacecraft – designated Cosmos 133 – had been successfully placed in orbit.

However, it was soon realised that all was not well with Cosmos 133. As the spacecraft separated from the upper stage, the internal pressure within the approach and orientation system (DPO) tanks suddenly dropped from 340 atm to 38 atm in only 120 seconds. The engines then continued to burn fuel for the next 15 minutes – almost completely depleting the supply – and created a slow roll of 2 rpm.

With the depletion of the fuel necessary for the attitude control required for the docking manoeuvres by the second spacecraft, it was clear that such an operation would be very difficult, and the planned roll-out and launch of spacecraft no.1 was cancelled. All effort then focused on returning the spacecraft from orbit. But the DPO thrusters were used not only for docking, but were also critical for stabilising the Soyuz during the de-orbit burn by the main SKDU engine. It was suggested that the back-up de-orbit engine (designated DKD) should be used, as it had its own set of stabilising thrusters; but tests of these thrusters revealed that they turned the ship in the direction opposite to the direction commanded, essentially marooning Cosmos 133 in orbit until, after 39 orbits, natural decay brought it down, when it could be destroyed by the automatic destruct system.

A solution to this problem was to use the SKD main engine in a short burst of 10–15 seconds, and another set of thrusters – the Orientation Engines (DO) – for stabilisation during the burn. This was far less accurate than the primary method, as although it would enable re-entry, the landing point would not be precise. Engineers employed this technique until the thirty-fourth orbit of Cosmos 133, and following the alignment of the spacecraft by using the solar sensors, landings were attempted on the seventeenth orbit, but were abandoned when it became unclear whether the correct commands had been used. It was therefore decided to use the ion orientation sensors on subsequent attempts.

On the morning of 30 November, the spacecraft again passed into Soviet tracking range and was commanded to de-orbit on revolution 33. The engine burn was too short, but on the next orbit the spacecraft was slowed enough for it to begin its re-entry profile, However, Air Defence Force radars lost track of the spacecraft over the region of the city of Orsk, and despite several days of aerial surveys across the region of Aktyubinsk to Semipalatinsk, nothing could be found. Cosmos 133 had apparently overshot the Soviet Union, and as it headed for China, the 23 kg of TNT on board was ignited and the spacecraft was destroyed, to finally descend east of the Mariana Islands in the Pacific Ocean.

Kamanin had long been opposed to fully automated systems on Soviet spacecraft, and stated, together with other members of the State Commission, that the mission might have been saved had there been a crew instead of a mannequin on board. However, the mission could be called a qualified success, as it provided ample opportunities for an evaluation of systems and procedures in flight. Four commissions were established to investigate the failures: the depletion of the onboard propellant in the DPO, the lack of stabilisation during the firing of the de-orbit engine, and the failure of the telemetry instrument (Tral) on the fifteenth orbit. On the positive side, the ion orientation system proved stable, the SKDU was able to fire repeatedly in a vacuum, and, despite a fault in stabilisation, the Soyuz re-entered. In addition, the launch processing and launch phase proceeded correctly, indicating a milestone in Soyuz mission preparations.

The investigation determined that the design of the system was sound, but that the assembly and test procedures were at fault. The service lines were tangled in the jet vane control sub-system of the orientation engine, and a faulty system was installed on Cosmos 133. The faults in the orientation system prevented the correct orientation of the vehicle, which in turn prevented full firing of the engine burns. It was also recommended that the next Soyuz (no.1 spacecraft) should be flown on a solo unmanned test flight to qualify the improvements and to man-rate the spacecraft for an initial manned docking mission in early 1967. Any desire for a docking mission was overruled, as it became clear that a solo unmanned test flight would have been a better option than the ambitious unmanned docking mission.

A fire at Pad 31

The difficulties that plagued the no.2 spacecraft were easily rectified, and on 8 December 1966 it was decided to launch the no.1 spacecraft on an unmanned solo test flight around the middle of that month. If successful, this would clear the way for the manned docking flight of vehicles no.3 and no.4, to be launched around 29 January – indicating that a second unmanned link-up was not required to allow the manned docking to proceed. A manned docking and Soyuz-to-Soyuz EVA transfer was planned for spacecraft no.5 and no.6, and then a solo Soyuz – no.7 – would fly a research mission that would include welding experiments.

On 14 December, after a flawless countdown spacecraft no.1 was set to launch from Pad 31 at 16.00 hours local time. Seconds after the command to ignite the main engines of the R-7, they shut down, and thousands of gallons of water drenched the pad to quench any fires. After determining that the pad was safe from fire, a team of specialists, including Mishin, departed for the pad to make the vehicle safe. With several workers at the pad area, the gantries began to be moved back around the vehicle to prevent it swaying in the wind – when there was suddenly a loud bang and 'whoosh' as the Soyuz launch escape system ignited, initiating separation of the OM and DM from the top of the booster. The escape rocket then ignited the third stage of the R-7, which was consequently about to explode. In a race for their lives, the workers scrambled towards the nearby control bunker. Luckily, no-one was near the upper stage, and within two minutes, most had escaped safely as the rocket blew up and the pad became engulfed in flames. The sole fatality was Major Korostylev, who had unsuccesfully taken cover behind a concrete bunker, although many others were seriously injured.

As tragic as this event was, it was far less disastrous than the 'Nedelin catastrophe' of October 1960, which claimed the lives of more than a hundred people as an R-16 ICBM exploded on the pad.[1] At the precise time at which the SAS fired, Kamanin arrived in a building only 700 metres from Pad 31. After hearing a muffled explosion he descended quickly from the second floor to the outside area, where he saw the Soyuz capsule, under the recovery parachutes, slowly descending to Earth. On returning to the third floor windows, he could clearly see the third stage of the rocket on fire, with flames running down the side of the R-7. Realising that the whole rocket was about to explode, he ordered everyone out of the room and into the corridor. Kamanin later recorded that he was the last person to leave the room, '... and as I was about to close the door I saw a big flash on the pad. Two or three seconds later, a series of massive explosions followed. The walls and ceilings [came to life], the [plaster] came raining down [from the walls], and glass flew out of the windows. Going to the shattered windows, we saw the burning skeleton of the rocket and enormous clouds of black smoke. All the rooms were strewn with glass and plaster. Big pieces of glass had penetrated the walls opposite the windows like bullets. Had we stayed in the rooms for just a few seconds longer, we would have been riddled with pieces of glass.'

Twenty minutes after the explosion, a meeting of State Commission members was held in the Soyuz MIK assembly building, with concern for those missing. Luckily, Mishin, Kerimov and Major General Kirillov (Chief of the First Directorate at the Cosmodrome) had found safety in another bunker, and had escaped injury. The explosion had wiped out all telephone communications, and TV cameras had been turned off after the shut-down, and so no footage was available for analysis of the events. By the next day it was clear that the pad was seriously damaged, and rocket debris lay strewn across a wide area.

A commission was subsequently created to investigate the accident, and on 16 December the findings were presented. The problem lay in one of the strap-on boosters of the first stage. According to Chertok an ignitor in one of the strap-ons was faulty, which caused the abort of the entire ignition sequence. The problem with the shut-down of the R-7 engines was found not to be serious; and the fault in one of the first-stage boosters could have been easily rectified, allowing for launch a few days later had nothing else occurred. The more serious question was why the SAS had activated. During a flight, the escape system could be activated either by command from the ground or by onboard systems sensing either a deviation in the flight path, premature engine cut-off, or early booster separation and changes in the internal pressure in the combustion chambers of the rocket engines. Clearly, none of these took place during this incident. The post-flight investigation concluded that the gantries had perhaps shifted the position of the R-7 when moved back into place, and that this had triggered the escape system. Two days after the accident, Kamanin theorised about the SAS having been activated by the gantries moving back into place. However, the official history of RKK Energiya, and Chertok, blame the gyroscopes in the core stage of the rocket as having accidentally activated the SAS after being powered down. It seems that this conclusion was reached after more thorough analysis.

The other obvious puzzle was why the activation of the escape system resulted in the R-7 catching fire – which it certainly was not supposed to do! It was found that at the separation of the Descent Module and Service Module, the lines of the thermal control system in the SM were inadvertently severed, causing the iso-octane coolant to leak – and being highly flammable, it was set on fire by the exhaust flames from the ascending escape system. It was not long before the fire spread to the Service Module, and from there onto the R-7 stages. Ironically, the SAS system had been tested at Vladimirovka, near Kapustin Yar, on 11 December, only three days before the incident at the cosmodrome. On that occasion the system created an identical fire, but as the SM was not fuelled there was no explosion, and the incident was ignored. In his diary, Kamanin states that he was dismayed to learn of this incident only after the explosion at Baikonur that could have cost his own life. On a positive note, at least the activation of the SAS proved that the system would have worked to rescue a crew in the event of a pad abort, with the capsule landing safely. And seventeen years later, in 1983, this system would indeed be used to save a crew from their exploding launch vehicle on the pad.

It was also obvious now that no manned flight would be launched by January 1967. Hidden under the Cosmos label, the difficulty of the initial Soyuz mission (Cosmos 133) was masking some of the difficulties experienced by the Soviets in launching cosmonauts to orbit. Against the huge success of Gemini, no cosmonauts had flown since Voskhod 2, but rumours were rife of a pending Soviet return to space with a new manned spacecraft. Reports in the US media suggested that Cosmos 133 was no ordinary scientific satellite, but an unmanned precursor for such a new manned vehicle. On 20 December, Kamanin recorded in his diary: 'The Americans know that we have begun preparations for a new series of manned flights, but they have no idea in what kind of trouble we really are.' What Kamanin or anyone else could not have realised was that matters were about to become much worse.

The mission of Cosmos 140

On 16 December, the State Commission decided that a further unmanned mission should be mounted before launching a crew on Soyuz, and that the maiden manned launch would not take place before March 1967. If the 14 December launch had been successful, manned flights would have proceeded immediately, and several other delays pushed the unmanned flight into February. Pad 31 – the primary pad for Soyuz launches – had been seriously damaged, and was expected to be out of commission for at least six months, and so Pad 1 had to be quickly modified to accept the larger Soyuz vehicle. However, all the fuelling systems and hardware for Soyuz could only be handled in Area 31, and the vehicle therefore had to be loaded with propellant there and then transported 30 km by rail to Area 2, which housed support processing facilities for Pad 1 but did not yet have the dedicated Soyuz spacecraft checking facilities installed.

This already slow process was additionally hindered by the derailment of a small shunting locomotive and the need to stop several times to check the fixings and strapping on the vehicle mounted on the flatbed railway transporters. There was also discussion about the procedures devised to prevent a repetition of the December pad abort, and arguments for and against total automated systems as opposed to the traditional inspections by pad teams, who left the pad just five minutes before lift-off. However, in the early morning of 3 February, in strong winds and a temperature of –30° C, the third Soyuz was mounted on the pad.

But there were further setbacks. With just four hours left in the countdown on 6 February, the launch was scrubbed due to technical problems which could not be immediately identified, leading to fears of a 2–3-week delay resulting from a roll-back to Area 31 for more detailed inspections. Luckily, the fault was traced to a short circuit in the spacecraft's OM, which could be fixed in time for the new launch attempt the next day. Working at Baikonur is never easy, as conditions can be harsh in any month of the year, but on 7 February 1967, in temperatures of –22° C, and wind gusting at 7–8 km per second, the launch of the first Soyuz from the Gagarin Pad finally took place – albeit 20 minutes late. Because of a low trajectory, the vehicle went out of range of Baikonur earlier than planned, but the signal was picked up by the Sary Shagan tracking station, which confirmed that orbit had been achieved. The official name of the spacecraft was Cosmos 140; but in reality, Soyuz spacecraft no.3 was in space.

The insertion into a lower trajectory was probably caused by an under-performing booster, and resulted in an orbit lower than planned and the loss of signal at Baikonur. A main engine burn on the fifth orbit then failed to extend the orbital lifetime beyond forty-eight orbits; and as if this was not enough, the Soyuz apparently could not turn its solar arrays towards the Sun due to a problem with the 45K solar/stellar sensor. Tests of the astro-orientation system had drained more than 50% (far more than planned) of the fuel assigned to that system.

After several hours out of range of Soviet tracking stations, communications were re-established early on 8 February, on the thirteenth orbit. On the twenty-second orbit, the spacecraft's SDKU main engine was burned for 58 seconds, changing the orbital parameters to allow the vehicle to (theoretically) remain in orbit for more than 200 days. However, two further attempts to orientate the spacecraft to lock the solar arrays on to the Sun and recharge the onboard batteries had failed, and by the twenty-month, early on 9 February, the batteries were close to depletion. The decision was then made to attempt a re-entry on the thirty-third – at least a full day earlier then planned. With only the ion orientation system (already successfully proven on Cosmos 133) available for positioning the spacecraft, the concern was that exhaust gases from the retro-fire might confuse the sensors and set the vehicle off course. But all proceeded according to plan – until after deployment of the parachutes, when the VHF transmitters failed and the short-wave systems ceased to function properly.

It was determined that Cosmos 140 had landed in the vicinity of the Aral Sea, but it was impossible to pinpoint the site. However, after a four-hour aerial search, the DM was finally located in the ice covering the lake, about 510 km short of its intended landing point, which revealed that it had completed a ballistic re-entry instead of the planned controlled profile. Recovery of the spacecraft had been a long process, which was also hampered because the capsule had broken through the ice and sunk in ten metres of water by the time the recovery crews arrived on the scene. It had struck an iceberg which had punched a hole in it, and although it at first floated, the water which seeped into it caused it to sink. This alarmed the engineers, who had designed and tested the DM to survive descent onto a variety of surfaces, including water, and had incorporated a flotation capability – which in this case clearly failed. This was the first time that a Soviet capsule had descended onto water.

Recovery proved difficult, and required the assistance of the Air Force and a team of naval divers from the Black Sea Fleet. Being waterlogged, the capsule could not be lifted by helicopters (a situation reminiscent of the recovery of Gus Grissom's Liberty Bell 7 in July 1961), but an Mi-6 helicopter was able to drag the capsule 3 km through the ice and water to dry land some 48 hours after landing. The post-flight inspection revealed that the maintenance hole plug, specially fixed with glue for a temperature sensor, was incorrectly mated in place, thus, during re-entry, causing a sequence of events that bored a 30×10-mm hole into the spacecraft. Several kilometres above the ground, the heat shield had been ejected as planned and had smashed into the ice; but some of the recovered shards were from the areas where the plug had been, revealing that the burn-through into the DM had originated there. This resulted in a drop in internal pressure to 200 mm, but the internal temperature rose to no more than 20° C. Had a crew been on board they would have certainly died, as at this time there were no plans for Soyuz crews to wear space pressure suits other than for EVA. To prevent this happening again, the heat shield plug was eliminated and the unit was fabricated in one-piece sections. And as an additional back-up facility, all suspect areas were reinforced with a secondary material.

On 16 February, a meeting was held at TsKBEM to discuss the investigation and the results of the flight. This meeting was attended by the cosmonauts assigned to the prime and back-up crews for the impending manned docking mission. It was clear that although a crew could have overcome the problem with astro-orientation, they would in all probability have died during re-entry or drowned after landing. With all of these setbacks, it was clear that Soyuz was not ready for a manned crew, and that a further unmanned test flight should have been authorised. However, most of the programme officials seemed to have decided that the problems with Cosmos 140 were not serious enough to stop the planned space spectacular of docking and EVA transfer of two manned Soyuz spacecraft. The modifications to the heat shield were expected to prevent a depressurisation burn-through in further missions, and Kamanin reported that his crews could be ready in 40–45 days, providing that the flight programme was not further burdened with additional tasks or experiments. Kamanin recorded: 'Based on the results [summarised by Mishin] of the Soyuz flights and the ground tests, we can aim at preparing for manned flights in early April. Everyone agreed with this timeline, provided that all the tests and modifications are carried out in time.' What was not recorded was whether any of the cosmonauts present voiced any opinions concerning safety or lack of confidence in the Soyuz which they were about to fly.

In April 2002, an interview with spacecraft designer and former cosmonaut

Konstantin Feoktistov was published in the newspaper *Trud*. In this interview, he commented on the flight of the first manned Soyuz and how it still plagues the conscience of those involved: 'When the question of the manned spaceflight was discussed [at the 16 February meeting], only Prudnikov, one of our designers, appeared against it. He considered that one additional pilotless launching was necessary. But we, as always, hurried. The launch was planned in the third ten-day period of April – probably they [the Soviet leadership] wanted to make a 'gift to the country' on 1 May. To me, up to now, it's heavy to recall that tragedy.'

THE TRAGEDY OF SOYUZ 1

Crew preparations for the first manned Soyuz mission began in September 1965 with the assignment of a group of Air Force cosmonauts, followed in May 1966 by the first of several TsKBEM engineers. The mission was planned to include the launch of a single cosmonaut in Soyuz 1, followed the next day by three cosmonauts in Soyuz 2, which would rendezvous and dock with Soyuz 1 in orbit. Following this, two of the Soyuz 2 crew would perform an EVA transfer over to Soyuz 1, and would return in that vehicle. (In the West, this scenario has often been interpreted as a potential dress-rehearsal for lunar orbital transfers by lunar crews from the LOK mother craft to the N1/L3 lunar landing module and back again after the landing on the lunar surface.) For the following eighteen months, the Soyuz training group prepared for their mission.

Chief Designer Valeri Mishin (*left*) with the Soyuz 1 and Soyuz 2 crews at Baikonur.

Concerns for flight

As Soyuz 1 launch preparations continued, during February–March 1967 there was a series of detailed briefings to assess each individual system of the Soyuz and the results from the three unmanned attempts. Despite what might have been thought of as a logical call for further unmanned testing, most agreed that the system was ready for launch and manned testing, and that the faults and mishaps encountered during the three unmanned attempts had been understood and could be developed to ensure that they would not occur again. The puncturing of the Descent Module from the heat shield burn-through on the Cosmos 140 flight raised some concern among the cosmonauts, and was voiced by Kamanin; but work on the rectification of this problem seemed to have satisfied most of the doubters.

It is worth noting that these discussions also took place a few weeks after the Americans lost the three Apollo 1 astronauts in a pad fire on 27 January 1967. Kamanin referred to this as 'shocking news', but went on to claim that the complacency of the Americans after the success of Gemini had caused them to race ahead with Apollo before it was ready to fly. The American loss also brought back memories of the 1961 Bondarenko incident, in which the young cosmonaut trainee lost his life in a test chamber filled with pure oxygen.[2] It can be argued that had the events of 1961 been known publicly, then perhaps the Americans might have thought twice about incorporating a pure-oxygen atmosphere in Apollo.

With Soyuz having a mixed nitrogen/oxygen atmosphere, the implications of the Apollo fire did not have an immediate effects on the Soyuz programme. An earlier proposal to place a crew inside a mock-up Soyuz to determine whether there was sufficient air to support a full three-day mission was again suggested, but it was not conducted as it would have delayed the launch by two or three weeks. The Ministry of Public Heath's Institute of Medical and Biological Problems (IMBP) proposed the experiment, and the designers of the life support system suggested that consumption would be 22 litres per hour – sufficient for the full three-day mission. However, the Air Force presented figures for a 25-litre per hour consumption rate (taking into account the EVA), suggesting that the mission should last no longer than 66 hours. It was also pointed out that since there was no onboard system to advise the crew about the amount of air remaining, it seemed prudent to not push the limits too far on this first flight.

Meanwhile, as the crews continued their preparations, production of the hardware for the missions continued to slip behind schedule. Launches had been expected to occur on or around 12 April (Cosmonautics Day); but when the cosmonauts were sent to Baikonur two weeks ahead of schedule to perform tests on the flight spacecraft, they had only 7K-OK no.5 ('Soyuz 2') on which to work, and had a wasted trip due to problems in delivering 'Soyuz 1' (7K-0K 4).

On 25 March, the pre-flight readiness of Soyuz 1 and Soyuz 2 was discussed at a meeting of the VPK in the Kremlin. This meeting was chiefly ceremonial, and merely confirmed that party officials were ready to support the flight upon the recommendations of the VPK. It was not the time or place to voice any concerns about the mission. Kamanin dutifuly reported, 'We are completely confident that the mission will be carried out successfully', and at the meeting of the State Commission on 3 April, the launch dates were set for 22 and 23 April. It had long been accepted that automated docking could be the biggest hurdle, and it had taken some time to convince Korolyov that manual docking was easier in order to achieve success. Korolyov had changed his mind only a month before his death in January 1966. To achieve the docking, Soyuz 2 had to be within a 12-km radius of Soyuz 1 (the active spacecraft) at orbital insertion, and Komarov would take manual control from 350 metres.

The crews left for Baikonur on 8 April, and despite some fears that technical problems would delay the launches into May, they continued their training for the launches at the end of April. In Kamanin's diary entry for 15 April, he expressed concern for the impending mission, and that Korolyov's absence from manned flight preparations for the first time was being felt as they approached launch. 'I am personally not completely convinced that the whole flight programme will be carried out successfully, but having said that, there are not enough reasons weighty enough to object to the flight. On all previous flights we were confident enough about the successful outcome, but now there is no such confidence.' He noted that despite the extensive training of the cosmonauts (some reports noted that they were able to complete less than 50% of their training programme) and even though everything seemed to have been carried out for the flight, there was still a lack of confidence.

Launch preparations

Soyuz 1 was transferred to the fuelling station in Area 31 on 15 April, and the second spacecraft was fuelled two days later. On 20 April the State Commission cleared both spacecraft for launch, with the launch of Soyuz 1 on 23 April and Soyuz 2 on 24 April. There had also been frank discussions between the designers and cosmonauts on the mode of docking. Mishin preferred a fully automated approach, but the cosmonauts quite naturally opposed this and wanted to perform a manual docking; and after several minutes of silence, Komarov suggested a semi-automatic approach. He would allow the automatic system to close to within 50 to 70 metres, and would then take over and manually dock the two spacecraft. These arguments seem to have continued up to the final days of the launch.

In his diary, Kamanin reflected on the importance and the concerns for the impending mission: 'We are heading for a very important and crucial mission, which should once again give our country the lead and pave the way for the conquest of the Moon. Now two days before launch... preparations for the flight have dragged on excessively, there were many mistakes, shortcomings and even disasters. The condition of the ships and their level of testing do not give 100% confidence in the complete success of the whole flight programme [the docking and EVA], but there is every reason to believe and hope that the crews will get to spend some time in space and will safely return to Earth.'

On 22 April, after the traditional pre-launch ceremonies (conducted on the launch pad at Area 1), Mishin visited the cosmonauts later in the day. He outlined the situations which might lead to the cancellation of the launch of Soyuz 2 – due either to the failure of Soyuz 1's Igla system, or to a lack of electrical power from the solar arrays. Kamanin told Komarov that the major priorities were a safe launch and landing, and that the docking and EVA transfer were secondary objectives and were not to be risked unnecessarily. There had been 203 faults recorded on Soyuz 1 alone, and Komarov was unusually sombre and quiet, and was evidently concerned. He had discussed his misgivings with his colleagues, but was singing and joking with them on the way to the launch pad on the morning of 23 April.

The mission of Soyuz 1

Soyuz 1 was launched at 03.35 MT on 23 April 1967, and from that moment Komarov's fate was sealed.[3] Nine minutes later, Soyuz 1 was placed in a 196.2 × 225-km orbit inclined at 51°.72, and twenty-five minutes after launch, cosmonaut Pavel Popovich informed Komarov's wife that her husband was in orbit for a second time. She also told reporters that Komarov never told her when he went on a 'business trip'.

Almost as soon as Soyuz was in orbit, two ground tracking stations at NIP-4 and NIP-15, in the Soviet Far East, revealed data which showed that the left solar array had not deployed, and this also prevented the back-up telemetry antenna from deploying. In addition, there were problems with the 45K stellar sensor. It was originally thought that it was still stuck behind the protective panel blocked by the stuck solar array, but it was most probably fogged due to moisture settling on its internal surface. Komarov entered the OM, and commented on what he could see out of the windows during the second orbit: 'I feel fine... but the left solar panel hasn't opened... the charging current is only 13–14 amps, and short-wave communications are not working. An attempt to turn the ship to the Sun did not succeed. I tried to turn the ship manually using the DO-1 [attitude control thrusters]... now the pressure in the DO-1 [tanks] has dropped to 180.'

Komarov tried to orientate the spacecraft during the third orbit, but without success. Some unconfirmed reports also mentioned that Komarov had tried to knock the hull of his spacecraft in an attempt to dislodge the stuck solar array, but that this was also unsuccessful. The ground controllers determined that, in the current attitude, it would be impossible to lock on to the Sun to charge the batteries, and there was a strong possibility that the storage batteries might be drained before the end of the first flight day. Despite recommendations to cancel the launch of Soyuz 2, however, the State Commission proceeded with the preparations and tried to correct the orbit of Soyuz 1. Kamanin recorded the situation as being serious, but also that they were confident they could overcome the problem, launch Soyuz 2, and perform the docking and EVA transfer – perhaps even allowing Yeliseyev and Khrunov to manually unfurl the jammed array. What was clear was that if Soyuz 2 were to be launched with three men on board, it would be impossible for it to return with four cosmonauts on board. Komarov would somehow still have to attempt to re-enter in his own spacecraft if he was ever to return to Earth. During the fifth orbit, the lone cosmonaut again failed to manually orientate his ship, while attempts to use the ion orientation system also failed, being very difficult to accomplish during the night-side pass.

Between the seventh and thirteenth orbits the Soyuz would be out of range of Soviet tracking, and Komarov was told to sleep during this period – although he probably did not do so. The launch or cancellation of Soyuz 2 would be decided when contact was re-established with Komarov on the thirteenth orbit. One of the principal problems facing the cosmonaut and the controllers was the loss of the secondary arrays that provided electrical power to the spacecraft, and another was the orientation system. Soyuz has three different systems, and if all three were to be rendered inoperable it would be almost impossible for Komarov to affect a safe re-entry. Attempting re-entry with incorrect attitude could result in a too steep an approach which would incinerate the spacecraft and the cosmonaut, or a too shallow approach that would cause the spacecraft to skip out of the atmosphere and into an orbit from which Komarov could never be recovered. With the ionic system already recording with two failures, there was no confidence in the system, because an early morning re-entry would result in ion pockets in the atmosphere disturbing the sensor. The 45K stellar sensor was totally inoperable, and therefore only manual orientation was available. Komarov would normally have been in a position to cross the Earth's terminator into the daylight area and easily locate the horizon, but it would be difficult now that he had already begun manual orientation in the Earth's shadow. Landing at the earliest opportunity was the preferred option, and he would have to orientate his spacecraft for re-entry while still in the dark.

On the thirteenth orbit, when communications had been re-established with Komarov, he explained that he had tried to orientate the ship but had not succeeded due to the difficulty of holding the sighting target on the moving Earth. The solar array was still folded against the side of the Soyuz, and so the situation had not improved. During the period when Komarov was out of communications range, officials in Moscow decided on the next stage of the flight. Reviewing the status of the vehicle in orbit, it was decided to cancel the launch of Soyuz 2 and to try to land Soyuz 1 between the seventeenth and nineteenth orbits. The State Commission confirmed the cancellation of Soyuz 2, and the three bitterly disappointed cosmonauts were stood down while all efforts were made to bring Komarov home safely. None of the flight controllers had had any sleep for 24 hours, and they were now presented with the challenge of preparing for the landing, which everyone knew would be difficult. It had been decided to de-orbit Soyuz on the seventeenth orbit (the best opportunity, presenting two further options on the eighteenth and nineteenth orbits), using the ion sensors to orientate the spacecraft. During the fifteenth and sixteenth orbits, the data for Komarov was radioed to him, from the Yevpatoriya control centre, by Gagarin, who had flown there after the launch.

During the seventeenth orbit, as the spacecraft again came into range of the ground stations, it was obvious that Soyuz was still in orbit. Komarov reported that although the ion system had worked well close to the equator, the vehicle had deviated in pitch from its correct course and had automatically blocked the retro-burn. There was little time to instruct Komarov on an alternative method of orientation before the Soyuz passed out of range, and so the landing had to take place on the nineteenth orbit – the last opportunity to rely on storage batteries. The back-up battery could support three additional orbits, but the Soyuz would then have to land outside Soviet territory – which had to be avoided at all costs.

During the period out of communication, an alternative plan was devised – a plan for which Komarov had not trained. During orbital sunlight, Komarov would have to rotate the ship 180° so that the engines in the SM would face forward. Prior to passing into orbital darkness he would have to put his trust in the KI-38 gyros, and once he came out of the darkness he would have to manually correct any deviation that might have been introduced during the night-time pass. During the eighteenth orbit, Gagarin radioed up the final set of parameters to his friend and colleague. The plan was to ignite the SKDU engine at 5 hrs 57 min 15 sec, and to shut it down after 150 seconds. Should it not shut down automatically, Komarov would stand by to terminate the burn manually. When Komarov came back into radio contact, he reported that he had had a 146-second burn, with cut-off at 5 hrs 59 min 38 sec. Fifteen minutes later, data revealed that the Soyuz would follow a ballistic rather than a guided re-entry – and so it was at least heading for a safe landing on Soviet soil. From information recently revealed, it would seem that just before retro-fire, Komarov moved from the centre seat to one of the side seats – probably because of the shift of the centre of gravity of the Soyuz with only one man, rather than the planned three, on board. (This move probably compensated for the change, and may well have had something to do with the undeployed solar panel.) After the burn, he moved back to the centre seat. Komarov had demonstrated remarkable courage and skill in placing the troubled spacecraft in the correct orientation and position for re-entry, and shortly after he reported the de-orbit burn the Soyuz entered radio black-out as it entered the dense layers of the atmosphere and headed for Earth.

The members of the State Commission remained at the cosmodrome in the administrative building of Site 2 for the duration of the mission, and monitored the progress of the spacecraft as it hurtled towards Earth. Some five minutes after retro-fire, Soyuz 1 entered the ionisation layer, which cut off all communications for several minutes. Shortly after exiting the radio black-out, Komarov's voice was reportedly heard to be 'calm, unhurried, without any nervousness'.[4] After separation of the Service Module (with its antennae), communications were possible only via short-wave antennae in the main parachute lines (which had not been deployed), or via a VHF antenna near the hatch (although because of the off-target landing, the recovery teams were not near the landing site at the moment of impact, and probably did not pick up these signals). The tape recorder that would have recorded Komarov's voice was destroyed on impact, and anything which he might have said was lost. Ballistic data recorded that Soyuz 1 would land about 65 km east of Orsk at about 06.24 MT. The landing took place to the far west of the planned landing zone for guided re-entries, in the reserve landing zone.

It was a bright sunny morning at the landing zone as the rescue helicopters closed into the landing area to find the parachute on the ground. Witnesses saw what they thought were the soft-landing engines igniting– which worried specialists on the helicopters, who knew that this usually took place above the ground to cushion the touch-down. Landing 100 metres from the capsule, they realised that the DM was engulfed in a cloud of thick black smoke. There was an intense fire inside the spacecraft, and the base of the spacecraft had been completely burned through. Molten metal dripped from the wreckage as the recovery teams tried to extinguish the fire with foam extinguishers, and even dirt from the field. Soyuz 1 was completely destroyed, and a mound of earth and molten twisted metal was the only sign of the recently landed spacecraft.

Initial reports from the landing site were transmitted on open channels. One pilot cryptically stated that the 'object' had been found, and that 'the cosmonaut needs urgent medical attention out in the field.' Soon afterwards, all news from the landing site was suppressed – probably to stop the spread of rumours, although it also prevented members of the State Commission from learning the details. Once re-entry had been confirmed, Kamanin boarded an aircraft to go to the landing strip nearest to the landing site. Two hours after landing, he was expecting to see Komarov at the airfield, but was instead met by Air Force officials, who informed him that the spacecraft was still burning and that Komarov had not been found. 'My hopes of seeing Komarov alive faded,' he later wrote. 'To me it was clear that the cosmonaut had perished, but somewhere deep in my heart there was still a glimmer of hope.'

Several rumours that Komarov had been found alive and was in hospital added to the confusion, and Kamanin decided to go to the landing site by helicopter. However, navigational errors further delayed his arrival for another ninety minutes, and when he finally arrived he came to realise the true horror of the closing moments of Soyuz 1's mission. 'When we landed, the ship was still burning... Local inhabitants said that the ship had come down at high speed... the parachute had been turning around and did not inflate. At the moment of landing there had been several explosions and a fire had begun. Nobody had seen the cosmonaut. The local inhabitants had tried to quench the fire by throwing a thick layer of dirt onto the ship. A quick inspection of the ship convinced me that Komarov had died and was somewhere in the wreckage of the ship, which was still burning. I gave an order to remove the dirt from the ship and look for the body of the cosmonaut. After digging for about an hour, we discovered Komarov's remains amid the ship's wreckage. At first it was difficult to see where the head, the arms and the legs were. Apparently, Komarov had died at the moment of impact, and the fire had turned his body into a small burnt and blackened lump measuring 30 × 80 cm.' Komarov's remains were found in the central couch liner after the top hatch and the remains of the instrument panel had been carefully removed. At the scene, the cause of death was pronounced to be due to multiple injuries to the head, spinal cord and bones.

Kamanin flew back to Orsk, and telephoned Dmitri Ustinov with the terrible news that the spacecraft had been destroyed and that Komarov had died. It was apparent that the primary parachute had not opened and the reserve did not fill with air. Soyuz 1 struck the ground at 35–50 metres per sec, and after the impact there was an explosion of the braking engines followed by a fire. Ustinov then called Communist Party General Secretary Leonid Brezhnev, who was attending an international conference of communist parties in Czechoslovakia. Ustinov also edited the official TASS news release that appeared twelve hours after the landing, which stated that the parachute was deployed at 7 km, the cords became entangled, and Soyuz 1 crashed at great speed, resulting in the tragic death of cosmonaut Komarov.

Shortly after midnight on 25 April, the remains of the cosmonaut, accompanied by the prime and back-up crews of Soyuz 2, were returned to Moscow, to be met by

Komarov's widow at the airport. After an official autopsy, Komarov was cremated, and later that day his ashes, in an urn, were placed in the Central House of the Soviet Army, where scores of people afterwards filed past the urn to pay their last respects. The next day, 26 April – the day on which he would have completed his mission – Komarov was buried, with full military honours, in the Kremlin wall.

The accident investigation

The Soyuz 1 accident resulted in the indefinite postponement of any further manned flights. On 27 April a meeting of the leading industrial and military officials involved with the manned spaceflight programme was convened by Ustinov, who headed the Commission that was split into seven sub-commissions designed to investigate specific aspects of the flight. Each sub-committee would complete their investigations by 5 May, and the final report on the Soyuz 1 accident would be published by 25 May. The commission determined that the parachute failure was the result of the design of the elliptical parachute container, which resulted in the drag chute failing to pull out the main chute. The container was not only too small, but was also compressed by the difference in pressure that arose between the cabin and the parachute container. Thus more force was required from the drag chute in order to pull out the main canopy. This had not been seen on Cosmos 140 due to the leak in the heat shield of that spacecraft, which had significantly reduced the internal pressure. There were two failures: first, the failure of the drag chute to pull out the main chute; and second, the failure of the back-up chute to fully deploy because it was hampered by the drag chute. No-one had foreseen that the back-up chute would have to be deployed with the drag chute still deployed. The back-up chute had been designed to be deployed after jettisoning the drag and main chutes. Wind tunnel and drop tests were conducted to simulate the situation in which the back-up chute would have to be deployed with the drag chute still there. These tests confirmed that the back-up chute could not properly deploy under such conditions.

The commission recommended changing the shape of the canister from cylindrical to conical, polishing its interior to ensure smooth ejection of the chutes, and enlarging the inner volume. Further recommendations included the removal of the back-up chute and replacing it with a double canopy main chute, which would ensure a safe landing even if one canopy failed; a system for casting off the drag chute; the introduction of manual control of the parachutes by the crew; and the revaluation of the automatic deployment systems, although not all of these recommendations were implemented. There was no conclusive reason why the parachute had failed, and while tests with the parachute planned for Soyuz 2 repeated the non-deployment of the parachute, this time there was no deformation of the container. The reasons for this have never been clarified; but more recently, an alternative explanation has been suggested. Here, the entire capsule was covered by thermal insulation and subsequently placed in a high-temperature chamber to polymerise the insulation's synthetic resin. It was a standard operation to install the parachute containers before this operation, because the lids had to be covered with insulating material. However, for each of the unmanned vehicles that had flown before Soyuz 1, this had not been carried out, because of delays in the manufacturing of the parachute compartments. This would explain why these vehicles had not suffered the problem of the drag chute's failure to extract the main chute. For both Soyuz 1 and Soyuz 2, the containers were ready in time but the lids were not, and it therefore seems possible that the spacecraft went into the chamber with either poor covering or no covering at all to protect the inside walls of the parachute compartments (not the parachutes themselves). As a result, tiny particles of thermal insulation, which came loose during polymerisation, could make the inside walls of the parachute compartments rough and gluey, so that it would be much more difficult for the drag parachute to extract the main parachute. This theory explains why the unmanned systems worked, and why drop tests of mock-ups did not result in the same problem. If the Soyuz 1 and Soyuz 2 parachute containers had been incorrectly treated, then this was a major error that at the time may have been intentionally covered up. It is also worth mentioning that had the Soyuz 1 solar array been deployed as planned, and Soyuz 2 been launched, then both spacecraft could well have suffered parachute failure, resulting in the loss of four cosmonauts, and a tragedy from which it would have been far more difficult to recover.

The other problems of the solar array and the sensor could be resolved before the next flights, but it seems clear that Komarov was doomed the moment he left the pad in a vehicle clearly unfit to complete its mission. His early difficulties forced the cancellation of Soyuz 2, and probably saved the lives of Bykovsky, Yeliseyev and Khrunov. Komarov's problems in orbit had been overcome by his skill, and were completely unrelated to the parachute problem. He did not know that he would pay such a price.

It is interesting to reflect that since the Soyuz 1 accident there has not been a single parachute failure during more than ninety missions in thirty-five years. The flight of Soyuz 1 was influenced by political pressure to return cosmonauts to space at a time when the Americans were planning to fly the first manned Apollo after ten successful Gemini missions, during which time no Soviet cosmonaut left the launch pad. It was a difficult time for the Soviet programme, and like the American Apollo accident, the Soyuz 1 accident would remain a painful memory to all those concerned. The legacy of Soyuz and each crew that flies it follows in the footsteps of Komarov, whose death provided the impetus to ensure that such an incident would never again happen.

THE SOYUZ/COSMOS UNMANNED DOCKING MISSIONS

Following the inquiry into the Soyuz 1 accident, the Council of Chief Designers held a meeting, on 29 May 1967, to discuss the future flight manifest. The decision was made to launch two unmanned spacecraft on a joint mission in August of that year, to be followed in October (the fiftieth anniversary of the October Revolution) or November by the manned docking and EVA transfer mission planned for Soyuz 1 and Soyuz 2. This operation was still seen as critical to future planning, but only if the unmanned flight was completed successfully would the manned flight be mounted. The crewing for the two manned flights had been agreed a few days earlier, and was discussed at the meeting.

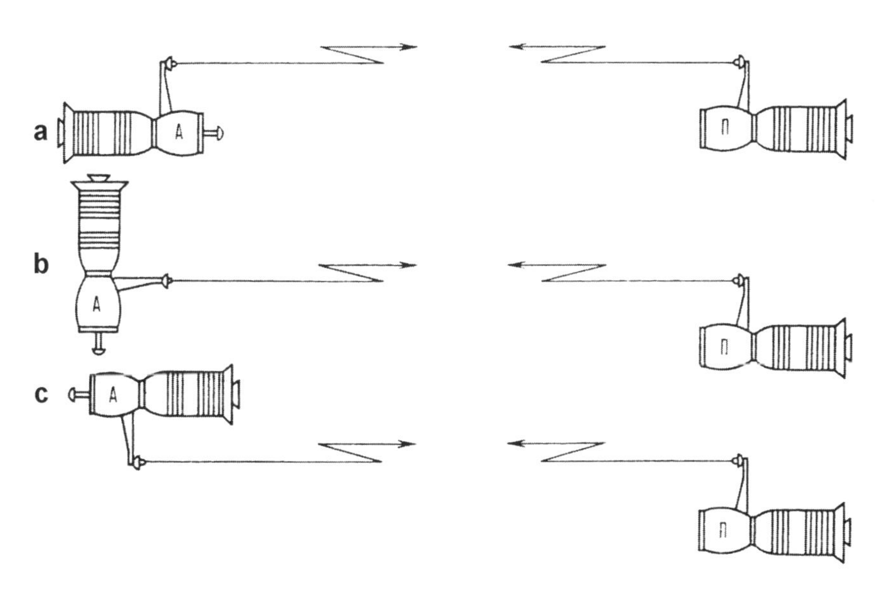

The Igla rendezvous system, showing the mutual position of the active and passive spacecraft in the distant approach phase: a) while increasing approach velocity; b) while reducing the angular velocity of the line of sight; c) while reducing the approach velocity.

The Cosmos 186/188 mission

During preparations for the unmanned launch, several delays had been necessary during the tests of the modified parachute system, which pushed the launch beyond August. The first test of the new parachute system finally took place on 23 August, and the same day, the Soyuz State Commission held a meeting to discuss the status of the Soyuz programme. The Commission reviewed the two hundred recommendations from the seven sub-committees of the Soyuz 1 investigation board (a third of these recommendations had been rejected by the Council of Chief Designers), while the others were under implementation or were about to begin. Before the launch of the unmanned mission it was decided to conduct at least twenty drop tests of mass models of the DM using inert FAB-300 aviation bombs, prior to two test drops of a mock-up DM – a programme which would be completed by 20 September. The two flight units (spacecraft no.5 and no.6) would be shipped to the cosmodrome no later than 5 September, and the launch could therefore take place between 15 and 20 October.

As the test programme continued, it became obvious that a manned launch in 1967 was highly improbable. Serious damage was inflicted on some of the electrical circuitry of one of the flight spacecraft's solar arrays, resulting in a processing delay while the damaged parts were replaced and re-checked. In Feodosiya, at least half of the FAB-300 drop tests had recorded problems; and further disappointment prevailed with the first drop tests of the mock-up DM (witnessed by Kamanin and several cosmonauts) on 6 October when, after normal parachute operation, the soft-landing engines ignited at 2 km instead of 1.2 km, resulting in a landing speed considerably higher than normal, with the capsule landing on its side. Had a crew been on board, they would certainly have been seriously injured. The next day, the consensus of the top officials of the programme was that no manned flights would occur before 1968. The second DM drop test, conducted on 12 October, was considered a success, although it too was not without problems. The State Commission decided to clear the parachute system for unmanned flight on 16 October, but it was not yet man-rated. It was therefore decided to launch the unmanned Soyuz mission between 25 and 27 October, and to conduct two more drop tests of a mock-up DM and another two to four unmanned Soyuz launches before the whole system was considered man-rated. It was also revealed that docking was not a primary objective of the two unmanned flights, although approach manoeuvres would be included. The primary objective was further qualification of the Soyuz spacecraft and, more importantly, the modified parachute system. Apparently, Mishin now favoured docking by the cosmonauts instead of automated systems – a turnaround from his earlier objections prior to Soyuz 1 – and also decided that the two unmanned Soyuz craft would approach to within only 50–70 metres before they returned to Earth, although his own engineers were not aware of this plan.

At Baikonur, the launch had to be delayed for three to five days due to damage to a membrane during fuelling operations for the orientation and stabilisation engines on spacecraft no.6. On 24 October, the flight profile for the mission was agreed by the State Commission at Baikonur. This time, the launches would occur two days apart. The active ship (7K-OK, no.6) would be launched first, on 27 October, to be followed by the passive craft (7K-OK, no. 5), which would then approach to an undisclosed distance before the return to Earth. Apparently, the decision to forego a docking had been influenced by the Air Force – the idea being to perform the first docking manually, with a totally automatic docking being attempted only after a manned docking had succeeded.

On 27 October, spacecraft no.6 was launched from the refurbished pad at Area 31. The TASS news release identified the launch as Cosmos 186 (and it was also assigned the codename Amur, although this was not indicated in the press releases). Initial operations were performed without incident, and this time both solar arrays unfurled as planned. However, during the second flight-day, several problems with the 45K solar sensor and the ion orientation system were tracked. It was therefore decided to delay the launch of the second spacecraft by 24 hours. It has long been speculated by Western observers of the Soviet space programme that this spacecraft was also used to simulate a lunar orbiter in the manned lunar mission, to which the next spacecraft would act as the returning lunar craft from the surface, but this has never been confirmed by Russian sources. The Americans had practised similar techniques during the Gemini programme.

On 30 October, spacecraft no.5 lifted off from the Gagarin Pad in Area 1. This was originally the Soyuz 2 manned spacecraft, but in orbit it was designated Cosmos 188 – codename Baikal. (The choice of codenames was appropriate when considering the proposed docking operation and the major civil construction operations in

Russia to link the Amur river region with the Lake Baikal region (some 4,300 km distant) by railway.) After orbital insertion, Cosmos 188 was only 24 km from Cosmos 186 (then in its forty-ninth orbit), and so the chance to attempt a docking of the two spacecraft was seen as too great an opportunity to miss. Apparently, the GOGU flight control team at Yevpatoriya had secretly prepared up-link commands to accomplish the docking, but required State Commission approval to initiate them. They expected a rebuke for their efforts, but were surprised to receive this approval. Whatever the accuracy of this statement, it is clear that the decision to dock Cosmos 186/188 came very late, after the launch of the first spacecraft.

The controllers knew that they could dock the spacecraft, but there was little confidence in being able to do so on this mission, as the operation would be concluded outside the zone of radio visibility. However, fifteen minutes before the spacecraft were in UHF range, a short-wave signal indicated that they had docked; and this was confirmed when the TV pictures received from Cosmos 186 showed Cosmos 188 attached to its sister ship. However, the success was somewhat dampened by the data, which showed that they had not hard docked and were still 85 mm apart, thus preventing electrical connection. In addition, Cosmos 188 had consumed far more fuel than expected (its SKDU had fired twenty-seven times, and the Approach and Orientation engines seventeen times), and there was concern that there might not be enough fuel left for landing. Later calculations revealed that there was enough fuel, but only for one attempt.

The two spacecraft undocked after 3.5 hours, and the procedure was witnessed by the controllers via live TV down-link. Cosmos 186 would land first, and since this was the first operational landing attempt from orbit by a Soyuz since Soyuz 1, there was great anticipation and apprehension as events unfolded. The de-orbit burn took place on 31 October, during the sixty-fifth orbit; and once again the 45K sensor encountered problems, leading to a ballistic re-entry instead of the desired guided re-entry. On this occasion, however, the parachute recovery system functioned as planned, although the short-wave antenna in the shroud line was broken off, and trajectory determination was more difficult. It was also determined that the landing rockets had fired as planned. The descent was tracked by radar from the Air Defence Forces, and visually from recovery helicopters. It was an important step on the road to requalifying the vehicle for manned flight. The next step was to determine whether Cosmos 188 would provide a second set of positive data.

Onboard the still-orbiting Cosmos, there was enough fuel in reserve for several days of tests, but it was decided not to risk the mission and to instead return the craft on 2 November. Once again, the 45K sensor encountered problems, and this time the craft entered an ion pocket over Brazil, which resulted in a re-entry path that was not steep enough and which would take it beyond the normal landing zone to land 300–400 km east of Ulan-Ude (still within Soviet territory). Unfortunately, as the onboard sensor detected a deviation in the craft's descent parameters, the automatic destruct system (used only on unmanned vehicles in case the spacecraft headed towards foreign territory) activated, blowing the vehicle apart as it flew 60–70 km over Irkutsk, and spreading pieces of spacecraft near the Soviet–Mongolian border.

In the Soviet press, the apparent success of the Cosmos 186/188 mission was reported as a major step in cosmonautics, although the American Gemini 8 astronauts had achieved the first-ever docking (with an unmanned Agena target in March 1966). The failures were not revealed, and the docking was hailed in celebration of the fiftieth anniversary of the October Revolution. As successful as the mission had been, it was clear that it was not yet prudent to launch a manned mission, and at the meeting of the Soyuz State Commission on 15 November it was agreed that after a period of modification, a second pair of unmanned Soyuz craft would be launched in March or April 1968. The first crew would not be launched before May or June 1968.

The Cosmos 212/213 mission

At a meeting of Chief Designers on 21 February 1968, it was reported that the tests of the Soyuz parachute system to qualify for manned flights would probably not be completed before the end of May 1968, and there was also a programme of life support system tests and sea trials to be completed. Because of these delays, the first crew could not possibly be launched before the second half of 1968. At a subsequent meeting of the State Commission on 26 March, it was concluded that the main parachute was reliable, but that further work was required on the back-up parachute before qualifying it for manned flight. However, the next two unmanned missions could proceed between 9 and 14 April, as planned. The new spacecraft were to test a new orientation system, using the 76K infrared vertical sensor in the roll and pitch modes, while the ion system (together with others) would be used for the roll axis. This system had been used on the Zenit spy satellites, and was employed on Soyuz when early tests revealed that the ion system alone was insufficient. The mission would also repeat the docking experiment, and further qualify the parachute landing system. On this flight, a primary objective was to achieve a guided rather than a ballistic re-entry profile, designed to reduce the deceleration forces from 7–8 g to 3–4 g.

But as the Soyuz programme began to look forward, yet another blow was dealt on 27 March 1967, with the death of Yuri Gagarin, the first man in space, during a training flight in a MiG 15.[5]

Just over two weeks later, on 14 April, spacecraft no. 8 (the active vehicle) was launched. On entering orbit it was designated Cosmos 212, and it performed almost flawlessly. Twenty-four hours later, on 15 April, spacecraft no.7 (the passive craft) reached orbit, and was designated Cosmos 213. As during the previous mission, the docking would again be out of range of tracking stations, and as the spacecraft slipped out of range, the data indicated that they were only 335 metres apart, with a relative closing velocity of 2 metres per sec. When communication was re-established, the data indicated not only a docking, but also that electrical connection had been achieved. The celebration of the success was shared by the cosmonauts in training for the next manned docking mission at the control centre – and they were even happier when it was revealed that the active spacecraft had not used any excess fuel.

The total docked time was 3 hrs 50 min, after which the two spacecraft were undocked to enable several days of independent tests and evaluations. The only concern arose during a test firing of the SKDU on Cosmos 212 during its fifty-first orbit, which was aborted due to a diagnostics problem during orientation of the spacecraft for the de-orbit burn. Flight controllers devised a back-up plan in which images of the Earth from the onboard TV camera were used to orientate the spacecraft (replicating the cosmonauts' procedure in viewing the Earth to orientate their craft). There was concern, however, that yet another Soyuz might be heading for self-destruction during re-entry.

Cosmos 212 consequently became the first Soyuz to perform the much-delayed guided re-entry, and on 19 April landed near the city of Karaganda, just 55 km short of the pre-planned point. However, a new problem emerged at landing, when winds of 22–23 metres per sec dragged the DM across the steppes for several kilometres, resulting in serious damage to the exterior surfaces. On manned Soyuz missions, the plan was to jettison the parachutes to prevent such an incident, but this procedure was not adopted for the Cosmos missions. Post-flight evaluation determined that static electricity from the parachute had eventually built up enough charge to separate the chutes from the DM. The following day, Cosmos 213 landed 157 km from its planned site near Tselkinograd; but it suffered the same fate, and was dragged across the steppes by winds of 25 metres per sec – creating such a dust cloud that it delayed the landing of the helicopters. Although causing a problem at the time, this system would not be employed on the manned spacecraft, and so there was no indication of any further delay in the launches.

SOYUZ 2 AND SOYUZ 3: SUCCESS AND FRUSTRATION

During the weeks following the second Cosmos docking mission, there were disagreements in the management structure of the Soyuz programme as to exactly what the next stage should be: either further unmanned flights, or the resumption of the manned flights. As a result of the loss of Komarov and Gagarin, the crewing for the next missions was also under review, with several options for the docking and transfer mission. These included the 1 + 3, 2 + 3, 2 + 2 and 1 + 2 manned scenarios; and later, the 0 + 1 scenario was proposed to precede the docking and transfer mission.

By 10 June the State Commission had decided on a final plan that would be suitable for all, with an unmanned flight to be flown in July, followed in September by a further docking mission involving one unmanned and one manned Soyuz (0 + 1). Should all these missions be successful, then the long-awaited docking and EVA transfer, with a single crew-man in one ship and three in the second (1 + 3), and featuring a two-man EVA, would be flown in November 1968. It was also decided at this meeting to begin construction of an additional six Soyuz 7K-OK spacecraft.

The Cosmos 238 unmanned shake-down mission

As preparations for the next unmanned flight proceeded, another setback was encountered during a series of five drop tests of a DM mock-up near Feodosiya, when, during the second test of the series, the parachute cover failed to operate, causing the DM to smash into the ground. As a result, the unmanned orbital mission was delayed by two weeks, and finally lifted off on 28 August. The spacecraft – designated Cosmos 238 – was the ninth 7K-OK vehicle, and it remained in space until 1 September, completing a textbook flight. The vehicle was of the passive type, but in the West, very little detail is known of the flight other than that one important manoeuvre was completed. This was a major success for the Soviets, as the Soyuz 7K-OK was planned for Earth orbital and lunar operations, and the military Soyuz VI was subject to its own development problems. On 23 September the parachute system was cleared for manned flights, although it was agreed that more work was required on water recovery efforts. The decision came three days after the latest drop test failed when the parachute container lid did not separate. On this occasion, however, it was due to a worker incorrectly fitting a plug, and therefore had no influence in qualifying the system for manned flight. These developments slipped the next manned flight into October – the same month in which the Americans resumed their manned flights with the three-man 11-day flight of the first Apollo Command and Service Module in Earth orbit. The priority for that mission, flown during 11–22 October, was to qualify the CSM for manned flights, and to test the large Service Propulsion System to be used to place the craft into and out of lunar orbit. However, Apollo 7 is remembered chiefly from the television broadcasts from orbit, which may have influenced the Soviets to add a TV camera to the Soyuz.

Manned flights resume

On 23 October the State Commission met to set the launch dates at 25 October (7K-OK no.11 – the unmanned passive vehicle) and 26 October (7K-OK no.10 – the active manned spacecraft). This was the first time that a passive vehicle would be launched ahead of the active vehicle, emphasising that only the one vehicle was manned. It also provided a further opportunity to ensure that the target vehicle was safely in orbit before committing to the launch of the manned vehicle.

Three days before launch, Beregovoi met with Kamanin to agree on the code words to be used to identify the status of the mission: 'excellent' indicated that there was not a problem; 'good' indicated a need to consult with flight controllers and systems experts; and 'satisfactory' indicated the requirement to return to Earth at the earliest opportunity. This was a continuation from the earlier Vostok programme, but was amended after Popovich, on Vostok 4, mistakenly uttered the code for a problem when he actually meant something else.[6] Beregovoi and his back-ups were also the first cosmonauts to break with the tradition of sleeping in the cottage (know as Gagarin Cottage) at the cosmodrome during the night before launch. A new and more comfortable hotel – Cosmonaut – was located in the neighbouring city of Leninsk, and it was here that all future crews would spend the night before launch.

The launch of the passive vehicle designated as 'Soyuz 2' took place on 25 October. The following day the Soyuz 3 vehicle, with Beregovoi on board, lifted off from Pad 31 – the first manned spacecraft to do so, ending a hiatus of eighteen months since the Soyuz 1 accident. Entering orbit only 11 km from the unmanned Soyuz 2, the Igla rendezvous system locked on to the unmanned target vessel and brought Soyuz 3 to within 200 metres of Soyuz 2. Here, Beregovoi took over manual control. While the two vehicles were out of radio contact during the dark-side pass of the orbit, they were only 40 metres apart; and an hour later, when contact was established, initial telemetry indicated that instead of docking, Soyuz 3 had deviated from its intended flight path and had used a considerable amount of fuel.

The problem was confirmed when each spacecraft was tracked as a separate signal and not as docked unit. Beregovoi reported that docking had not taken place because the ships were misaligned; and in addition, the approach and orientation engines had consumed 80 kg of fuel, which prevented a second attempt at docking. All that remained of the fuel was the 8–10 kg required to position Soyuz for de-orbit. Over the air, Beregovoi provided no explanation for the failure, and controllers attempted to cheer the obviously disappointed cosmonaut. Later in the day he managed to complete some orientation tasks, but the primary purpose of his mission had been defeted. Initial releases of the activities, and Western interpretation of the mission, indicated that a docking was never planned, and that Beregovoi was only to have rendezvoused with Soyuz 2, because after Soyuz 1 the Soviets did not wish to add any undue risks to the cosmonaut. However, the reason why no docking occurred, when it was clearly possible, continued to puzzle Western observers for many years; but details recently revealed make it clear that docking *was* planned, and that Beregovoi had failed to achieve it. During an interview in 2002,[7] Konstantin Feoktistov stated that Beregovoi 'committed the grossest error... He did not turn his attention to the fact that the ship to which he was meant to dock [Soyuz 2] was overturned [upside-down in relation to his own Soyuz 3]. The flashing lights of the unmanned ship proved to be on top, and should have been below. Therefore, the approach of Soyuz 3 [caused] the pilotless object to turn away. In these erroneous manoeuvres, Beregovoi consumed all the fuel intended for the ship docking... and when the cosmonaut left the [range of] communication with TsUP, the decision was taken about the landing.'

With the conclusion that blame for the docking failure rested solely with Beregovoi, there were a number of mitigating circumstances that also had to be addressed. (These are highlighted by Bart Hendrickx in his translation of Kamanin's diaries.) The first obvious fact was that this was the first Soviet attempt to manually approach and dock with a target, by a lone cosmonaut, without an onboard Flight Engineer to assist him. The early docking on the first orbit was a challenging operation that had been linked not only to lunar missions (the American Gemini crews also first practised orbital rendezvous and docking with Agena targets), but also, according to Boris Chertok, a 'demonstration of a killer satellite that could approach to destroy an enemy satellite in a very short period of time.' It was later emphasised that although Beregovoi was a test pilot with considerable experience, he was an inexperienced cosmonaut on his first mission, and had suffered from a mild case of space adaptation sickness.

The timing of the approach and docking also seemed to influence the result. The docking was planned to take place out of radio contact, and it would therefore be impossible for the cosmonaut to consult with the ground if he encountered a problem. In addition, he was attempting to dock in the dark side of the orbit – a constraint based on the need to ensure that the recovery would not take place during the windiest time of the day in the landing zone. It was clearly a safety priority that the landing should take place at this time, and to achieve this, the timing of the launch also necessitated docking during the dark-side pass. Previous unmanned Soyuz missions had also encountered ion pockets, and the resulting confusing in the sensors also influenced the decision to bypass this phenomenon.

Some changes were introduced on the Soyuz 4/5 mission to prevent a repeat of the Soyuz 2/3 failure. Shatalov carried out the docking 48 hours after launch, in orbital daylight, and while within range of the Soviet ground stations.

The planners clearly considered that night-time docking would not cause a problem, and argued that daylight dockings were hindered by sunlight entering into the cosmonaut's periscope. The Gemini astronauts had also experienced the problems of trying to dock to a target in daylight, and from above, where the target can be lost against the Earth passing below. In his interpretation, Hendrickx also questions Beregovoi's training proficiency to achieve the docking. The attempt at night-time docking was apparently decided only two weeks before his mission, which left him precious little time to familiarise himself with the procedures. It also has to be remembered that at this time (1968), space simulators were not very sophisticated, and the Volga docking simulator did not accurately reproduce conditions in the space environment. Nor were there accurate models of the spacecraft with which the cosmonauts were planning to dock. For example, the Igla rendezvous antenna was not indicated on the models used in Volga, and so there was no clear indication of the correct position or orientation of the antenna on the 'top' of the Orbital Module. No-one seems to have bothered to brief Beregovoi on exactly what he should expect to see on Soyuz 2, and he had trained only in aligning the solar arrays of his spacecraft with that of the target. In addition, the complacency about the success of the unmanned docking during the Cosmos missions, and possibly the relief that the problems with the parachute system had finally been resolved, probably led to the docking phase being considered to be lower priority. Kamanin wrote that the blame had to be placed on a combination of training, simulators, designers, and members of the State Commission, as well as on Beregovoi.

On the positive side, however, the flight finally man-rated the Soyuz system from launch to landing. Following the rendezvous attempt with Soyuz 2, and the failure of the 45K star-sensor, Beregovoi succeeded in manually orientating Soyuz 3 in preparation for re-entry during flight day 4 (although he used a great deal of fuel), and maintained the orientation using the gyros for about one orbit (90 minutes), after which he ignited the de-orbit engine for 3 seconds. He also broadcast TV pictures of his activities inside the Soyuz and views out of the portholes (following the recent TV 'shows' broadcast), conducted observations of the Earth, weather patterns and the stellar background, photographed clouds and snow, and was able to track typhoons, cyclones and forest fires. This was path-finding information for extending the use of Soyuz as an observation and research platform in addition to its role in lunar flights, and possibly for its involvement in manned space station operations. One of Beregovoi's final experiments included the photography of photometrically marked black-and-white images through an orange filter.

Soyuz 2 was brought back to Earth during its forty-eighth orbit on 28 October, and touched down uneventfully, 42 km short of its planned landing point near the city of Karaganda. Beregovoi, in Soyuz 3, remained aloft for two more days, and accomplished a safe landing after the eighty-first orbit. Luckily, a blizzard had passed through the landing area earlier that day, and – witnessed by a confused local boy and a donkey – he descended into a soft snow drift near the city of Karaganda. Although there was disappointment at not having achieved the docking, the flight had finally provided a successful manned flight test of the 7K-OK Soyuz vehicle that was so important for the immediate future of Soviet manned spaceflight operations.

During his post-flight debriefing, Beregovoi produced valuable data on crew habitability on Soyuz. He stated that most of the onboard systems on his spacecraft worked perfectly, but that the view-ports tended to fog up on entering orbit (which had implications for future docking and lunar missions), and also reported that his control systems were too sensitive (implying that there was more room for improvement in the system). When asked whether his age at the time of flight was a limiting factor (at 47, he was at that time the oldest man to fly in space), he replied that more than anything else, his *height* (1.80 metres) was a problem, as he had had difficulty in fitting into the couch.

Overall, the flight provided a valuable boost in confidence in the Soyuz, in readiness for the next team of cosmonauts and the long-awaited manned docking mission.

A DOCKING AND A TRANSFER

After Beregovoi's flight, it seemed improbable that a further mission would take place before the end of the year. The composition of the crews for the two manned missions was finalised at a meeting of the VPK on 23 December, with launches planned for 12 January 1969 for Soyuz 4 (7K-OK no.12 – the active craft) and 13 January for Soyuz 5 (7K-OK no.13 – the passive craft). The back-up launch dates were also confirmed as 14 and 15 January respectively. This was the first time that a manned space mission would be launched in the harshness of a Baikonur winter.

Amending the flight plans

Some thought was given to changing the de-orbit opportunity to the second of three daily windows instead of the first, in order to eradicate the slight possibility of descending into the non-frozen part of the Aral Sea. However, this would necessitate a landing in diminishing daylight during a short day of the Russian winter, and once the cosmonauts voiced their opposition to the plan, the landing attempt remained with the first window. This decision was supported by inspection flights over the area, which revealed that it would be safe to land there and that it could support helicopters of the search and rescue fleet.

A second discussion centred on the launch sequence of the two missions. In 1968 the cosmonauts suggested changing the sequence so that the passive ship launched ahead of the active one. This decision may have been connected to the manned lunar programme, as the 'active–passive' sequence (employed by Cosmos 186/188, Cosmos

212/213 and eventually also Soyuz 4/5) may have been a simulation of the LK/LOK rendezvous. The active vehicle, already in orbit, simulated the LOK, and the passive vehicle, launched subsequently, simulated the LK. The 'passive–active' sequence that the cosmonauts suggested in (late) 1968 was actually the *opposite* of the LK/LOK scenario. However, the Russians have never confirmed that the active–passive sequence was a simulation of an LK/LOK rendezvous. The 'passive–active' scenario had taken place only once before – during the previous Soyuz 2/3 mission – because the Soviets did not wish to launch Beregovoi first in the manned (active) craft, and then be faced with a failure in the launch of the passive vehicle. The Soyuz 4/5 cosmonauts argued that by first placing the passive (Soyuz 5 crew) in orbit, they have time to acclimatise to zero g before attempting their EVA; and in the event of a Soyuz 4 launch cancellation, the true nature of the mission could be disguised as a three-man flight including an EVA. Kamanin, however, considered that it would be too complicated to change the plans so close to launch.

The third matter for debate was the role that the crews would play in the approach and docking – either fully manual or totally automatic – an argument that had continued since before the 1967 Soyuz 1 and 2 planned docking mission. With the automatic docking of the two pairs of Cosmos craft, and the difficulties Beregovoi had had in manually docking on Soyuz 3, Afanasyev and Ustinov preferred fully automated docking, which seemed more favourable in assuring the success of the other major objective of the mission: the EVA. Kamanin supported his Air Force cosmonauts, who naturally preferred manual docking to demonstrate their piloting skills. Although normally in favour of the automatic systems, the designers (notably Mishin), and the chairman of the Soyuz State Commission, Kerim Kerimov, this time supported the manual approach. The discussions continued until the launch, but the mission plans had been changed to allow the cosmonauts more time to prepare for the docking. On Soyuz 3, Beregovoi would have rendezvoused and docked with Soyuz 2 on the first orbit of his mission. (This again was possibly connected with the lunar programme, although some reports indicate that it was connected with a demonstration of hunter-killer satellites).

On his flight, Beregovoi had encountered difficulties in his attempt at a first-orbit rendezvous. He had had to adapt to weightlessness, and, alone in his spacecraft, he had had to configure it for orbital flight, while preparing to dock in darkness and while out of range of ground tracking stations – all within 90 minutes. With this experience behind them, the planners decided to launch Soyuz 4 a full 24 hours before Soyuz 5, as normal; but the docking would not take place until the second flight day of Soyuz 5, with both spacecraft in daylight and in radio contact with mission control. This allowed Shatalov two days of orbital adjustment and preparation to dock with Soyuz 5, whose crew also had 24 hours to acclimatise to spaceflight. This change to the 24-hour orbital chase was later adapted to the Soyuz ferry flights to Salyut space stations from 1971 (Soyuz 10, with Shatalov performing the first attempt) until 1985 (Soyuz T-14).

The first docking of two manned spacecraft

The launch dates of Soyuz 4 (13 January) and Soyuz 5 (14 January) – a day later

The Volga docking simulators used in the early phases of Soyuz rendezvous and docking operations training.

than earlier planned – were confirmed at a meeting of the Soyuz State commission on 10 January 1969, and the crews were officially confirmed the following day. The day before launch, Kamanin advised Shatalov that he was to immediately revert to automatic docking should he have any doubts about, or problems with, manual docking. On the day of the planned launch, several problems plagued the countdown, including communications difficulties and a failure in the launch vehicle's gyroscope just nine minutes before launch. With a small launch window dictated by the constraints of the daylight landing window, there was little time to affect a repair in freezing temperatures; and so, for the first time in the Soviet programme, a launch scrub was called for 24 hours – with a cosmonaut already in the spacecraft. He had already been inside for more than two hours, and as he exited he joked that he could claim the shortest time in a spacecraft and the most precise landing; while others pointed out that the intended thirteenth cosmonaut was not destined to enter space on 13 January (a numerical sequence that would affect a certain Apollo lunar mission fifteen months later!). The gyro problem was found to be related to high humidity and extremely low temperatures (some Western observers had previously doubted the Soviets' capability to launch in such conditions), and as the gyros were replaced, the reliability of the SAS system was also reviewed, although it was determined that it was designed to operate in such conditions if necessary.

The following day – 14 January – Soyuz 4 was launched from the pad at Area 31. The launch had been threatened with delay when a fault was discovered in a system of the launch vehicle, which normally could only be repaired with the R-7 horizontal, rather than on the pad with the cosmonauts inside. A young launch-pad worker volunteered to strip off most of his clothes and, in freezing temperatures, squeezed into the R-7 via a narrow hatch to fix the problem as the count continued. He succeeded with thirty minutes to spare before launch, and thus prevented a further lengthy delay. Upon entering orbit, Shatalov exclaimed 'I like it up here!'

The following morning the Soyuz State Commission cleared the launch of Soyuz 5 carrying Volynov, Yeliseyev and Khrunov. It lifted off on schedule and entered orbit nine minutes later, to set the scene for the long-awaited docking. Shatalov's call-sign was Amur, and Volynov's call-sign was 'Baikal', reactivating those used on the Cosmos 186/188 docking mission. Mission control was code-named Zarya (Dawn).

On 16 January, Soyuz 4 was on its thirty-fourth orbit and Soyuz 5 was on its eighteenth orbit, at an altitude of 211×253 km; and after a long programme of orbital corrections, they closed in for the docking. During the flight to Soyuz 4, Khrunov had handled the navigation by using a sextant. Shatalov was surprised that he could see the other spacecraft ten minutes earlier than he expecting, but the spacecraft passed each other before Shatalov could orientate Soyuz 4. He then turned on the Igla system on Soyuz 4 to initiate the automatic approach up to 100 metres, at which point both he and Volynov switched to manual mode and successfuly steered the two vehicles into a smooth docking, during orbital daylight and within communication range of Soviet tracking stations:

Shatalov (Soyuz 4): Everything is normal. Everything is going normally.
Zarya: Roger, I am watching you.
Volynov (Soyuz 5): Roger Amur, Roger Zarya. This is Baikal reading you loud and clear... Range is 130, the spacecraft is responding perfectly.
Shatalov: Request permission to begin docking
Zarya: Permission granted... This is Zarya. If possible, give us a brief account of what you are doing.

Shatalov later explained: 'The automatic controls had brought [the two spacecraft] to within 300 feet of each other. At this point, I went over to manual control, and Boris Volynov did the same. The problem was to make sure that the docking units of both spacecraft were properly orientated towards each other. Throughout this time I was manually controlling the appropriate thrusters. With the control stick on the left- hand side I regulated the spacecraft's linear velocity – slowing it down or speeding it up – and damped out the lateral velocity. When we were over the shores of Africa – some seven or eight thousand kilometres from the borders of the Soviet Union – we approached to within 130 feet of each other and started to hover. At this range, Boris Volynov and I performed several manoeuvres.'

Shatalov: Roger. I now have Baikal on my screen. Velocity 0.82 feet per second. We'll proceed. [The solar] panels [are] clearly visible.
Khrunov (Soyuz 5): Beautiful, very beautiful! Just magnificent! Amur is flying like a bird from a fairy tale! She's coming up like an aircraft – approaching like an aircraft!
Volynov: Everything is fine... Waiting for contact.

Shatalov: I'm moving in. Everything is normal. Contact light. Link-up firm. Docking.

On the ground, controllers at Yevpatoriya were able to follow the approach and docking via a live TV link.

Zarya: We can see it clearly. Every detail is visible right on the nose!
Shatalov: Welcome Baikal! The docking went off splendidly. The spacecraft are matched. Fastening is being continued. No relative motion between the spaceships.

On each spacecraft's display panels, the indicators 'Mechanical Contact' and 'Coupling' flashed on, and after the throwing of a series of switches and the automatic connection of the electrical plugs in the docking ring, the 'Docking Completed' light illuminated. Having at last succeeded, both crews and controllers rejoiced – but only for a short while, for now that they were linked, the next task was to perform the EVA. Meanwhile, TASS was describing the achievement as the creation of the 'world's first experimental space station' of four compartments with a reported joint mass of 12,924 kg and a total internal volume of 18 cubic metres. It was, of course, impossible to transfer internally between the compartments, as the docking system did not incorporate an internal transfer hatch, and the docking mechanisms on the 7K-OK vehicles were different from those planned for LOK/LK. Soyuz had a drogue–probe system, whereas LOK/LK used a system in which 'claws' would penetrate a honeycomb structure, and so the docking apparatus on 7K-OK was not a test and evaluation device for the lunar programme. The 7K-OK and LOK/LK even used different rendezvous systems (Igla *versus* Kontakt), and although the Kontakt system was originally planned for use during the Soyuz programme, it was cancelled.

Stepping out of Soyuz

The EVA – the only one ever to be accomplished from the OM of a Soyuz spacecraft in more than thirty-five years – was to take place during the next pass over the Soviet Union. The two EVA cosmonauts – Yeliseyev and Khrunov, onboard Soyuz 5 – had been training for this event for almost three years. To conduct the EVA, they would move from the DM into the OM of their spacecraft, which would act as an airlock for the exit into space, and Volynov would help them don their EVA space suits before retreating to the DM and closing the inner hatch. The two cosmonauts would then exit via the side hatch in the OM, and translate hand over hand from Soyuz 5 to the OM of Soyuz 4. When they had sealed the hatch and repressurised the compartment, Shatalov would join them to help them take off their suits. Yeliseyev and Khrunov would then return in Soyuz 4, while Volynov would return alone in Soyuz 5.

When the early Soyuz missions were planned by Korolyov, several years earlier, they were designed as developmental missions for learning the techniques of rendezvous and docking, EVA and long-duration spaceflight, with applications for future spaceflight operations both at the Moon and elsewhere. The comparison in

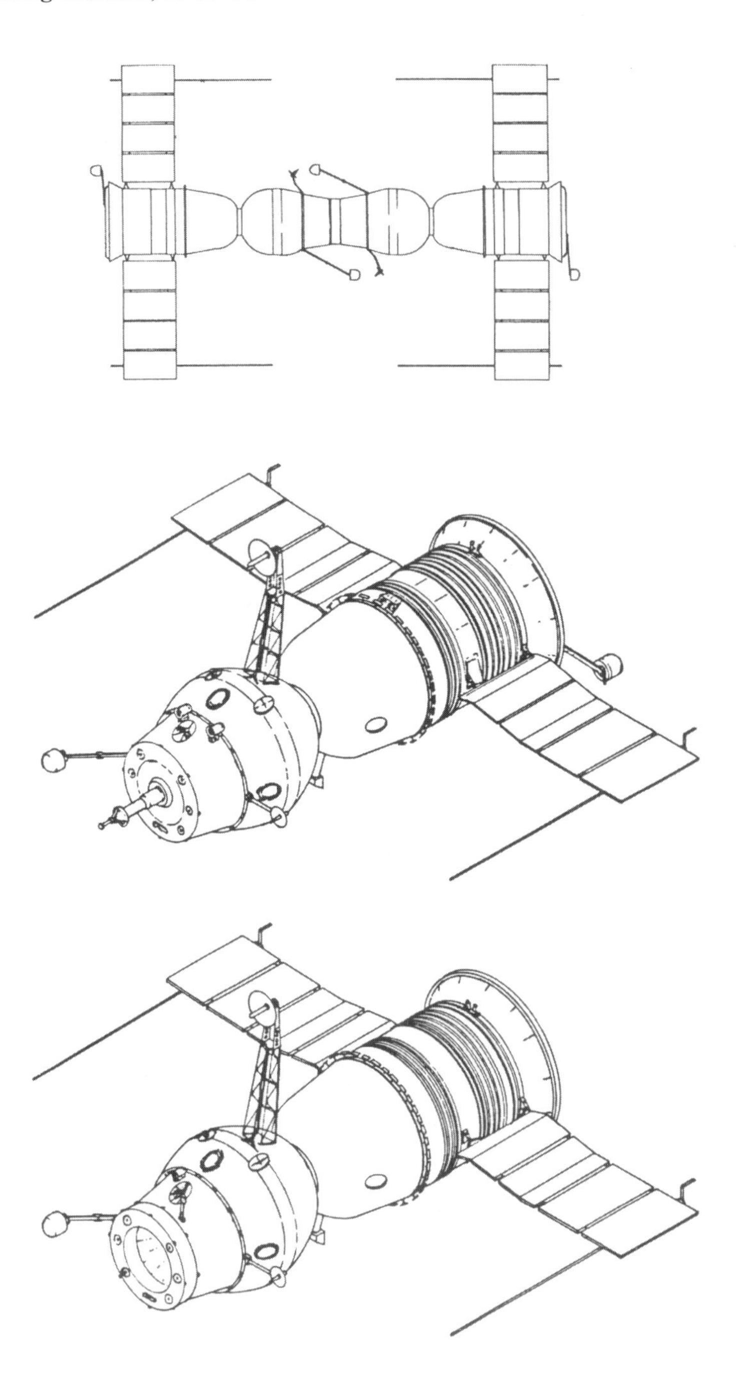

The Soyuz 7OK docking system, used from 1967 to 1969, but successful only on the Soyuz 4/5 mission in January 1969. (Courtesy Ralph Gibbons.)

the US was the series of Gemini flights, which provided flight experience in these techniques before the move to Apollo. By the time the Soyuz flights were finally launched, the hardware and objectives were being surpassed by both Apollo and Russia's own developing lunar programme. The Yastreb (Hawk) EVA suits used on Soyuz 5 were an improvement over the Berkut (Golden Eagle) suit used by Leonov on Voskhod 2,[8] but development of the Orlan EVA suit and Krechet-94 lunar EVA suits had surpassed the operational requirement for the Yastreb design. However, the EVA would still prove to be a valuable experience, as the Soviets had recorded only one EVA – by Leonov, logging just 24 minutes with 12 minutes actually spent outside, compared with 12.5 hours of experience by five Gemini astronauts during nine EVA periods.[9]

The suits – which were stored in a compartment in the OM – consisted of a multi-layered garment with the helmet and gloves attached by bayonet catches to the pressure garment. The helmet featured a separate Sun-visor for eye protection, the shoes had rigid soles, and the back of the suit had a reinforced area for the life support system back-pack with interior temperature regulation and a two-hour supply of oxygen. In practise, however, the LSS had to be attached to the legs instead of the back, when initial EVA simulations showed that the 66-cm diameter exit hatch could not accommodate a cosmonaut in a pressurise suit with the LSS on his back. A larger hatch of 70 cm was promised on later Soyuz OM, but was not ready by the time this mission flew. The cosmonauts also used safety line tethers.

The trio of cosmonauts found it difficult to don the suits in the cramped OM, but they eventually succeeded. Once they were suited, Volynov returned to the DM and closed the hatches, and depressurisation of the OM began. The EVA began 11 minutes later than planned, and some minor tasks had to be cancelled (the planned completion time of the EVA could not be altered, because the cosmonauts had to be inside the OM of Soyuz 4 before the spacecraft entered orbital darkness). Khrunov was the first out of the open hatch: 'The hatch opened, and a stream of sunlight burst in... I saw the Earth, the horizon and the black sky, and I had the same feeling I had experienced before my first parachute jump... I freely admit I felt all the anxiety of an athlete on the starting line. I emerged from the spacecraft without difficulty and looked around. I was amazed by the marvellous, magnificent spectacle of two spacecraft linked together high above the Earth. I could make out every tiny detail on their surfaces. They glittered brilliantly as they reflected the sunlight. Right in front of my eyes was Soyuz 4. When I had taken a good long look at this wonderful sight – the glittering spacecraft against the background of the Earth and the dark sky – I began to move.'

Khrunov's first task was to mount a film camera to record his own EVA. He then retrieved the TV camera (that had recorded the docking) from the outside of Soyuz 5, and took it over to install it on Soyuz 4 to record the transfer of Yeliseyev. Using the handrails, he gradually moved across the surface of the spacecraft hand over hand, and when he reached Soyuz 4, Shatalov (automatically) opened the hatch of the OM from the DM. As Yeliseyev prepared to conduct his transfer, he retrieved the film camera that had recorded Khrunov's EVA. He then tried to store it in a compartment in the Soyuz 5 OM, but could not close the compartment door correctly. It was then time for Yeliseyev to complete his transfer to Soyuz 4, and the TV pictures recorded the other camera floating out of the still open hatch on Soyuz 5, which Volynov was not meant to have closed until both men were safely inside Soyuz 4. Due to the loss of the camera and the images of Khrunov, the only pictures of the historic EVA were the ghostly images of Yeliseyev on the TV film.

With both men inside the OM, the hatch was closed, and repressurisation begun. Volynov finally closed the hatch, and then repressurised the OM on Soyuz 5. The EVA had lasted for an hour, with the two men being outside for a total of 37 minutes. Once pressure had been restored, the inner hatch of Soyuz 4 was opened and a happy reunion was held, with Shatalov and his two new crew-members toasting each other with blackcurrant juice (rather than the traditional vodka). The two cosmonauts had also brought Shatalov some mail from his wife and from Kamanin, as well as a newspaper featuring the launch of Soyuz 4 two days earlier. After completing their EVA and removing their suits, the three men prepared to undock from Soyuz 5.

The spacecraft had been docked for 4 hrs 33 min 49 sec when the connections between the two OM were severed and the two craft drifted apart. The next morning, Soyuz 4 safely re-entered (with Shatalov becoming the first cosmonaut to maintain a running commentary during descent, by using a VHF antenna embedded in the hatch of the DM). The uneventful landing took place 40 km north-east of Karaganda (and 40-km off target), in a clear sky and light winds – but in temperatures around –30° C.

A dangerous re-entry

Volynov continued in orbit for another 24 hours, before his attempted re-entry and landing. His recovery, however, was to prove far more arduous than Shatalov experienced. Temperatures at the landing zone were 5 degrees lower, but clear skies and light winds were still expected. On orbit, prior to landing, Volynov practised manual orientation of the spacecraft over a nine-minute period (the time between orbital sunrise and de-orbit burn on the re-entry orbit). He reported that he could not complete the task in such a short period of time, but despite this he was authorised to attempt manual orientation for re-entry on the next orbit. However, he was also given the commands for a second automatic attempt should he not complete the manual operation the first time.

Indeed, following the first attempt, Volynov reported that again he had not completed the manoeuvres successfully, and controllers expected to tell him to attempt automatic landing procedures on the next orbit. But when they heard from Volynov, it was clear that re-entry had already begun, and that the spacecraft was following a ballistic trajectory and violently tumbling. As he entered the period of black-out, worried controllers thought that another cosmonaut would be lost during recovery of a Soyuz.

Unbeknown to the ground, the Propulsion Module had failed to separate cleanly from the DM (unlike the OM), which caused the Soyuz to begin its re-entry 'back to front', with the inner hatch, instead of the heat shield, facing forward. Inside the DM, Volynov – who was not wearing a pressure suit – had heard the explosive charges fire to separate the modules, but out of the window he was alarmed to still see the antenna that was attached to the solar arrays of the PM. He reported the situation to the ground, but could do nothing except sit, and ride it out. He could clearly smell burning rubber due to the searing heat melting the gasket of the forward hatch, and expected it to fail at any moment. Deciding against saying goodbye to his family, he ripped the pages from his log-book (which detailed the rendezvous and docking events), rolled them up tightly, and stuffed them into his seat, hoping that they would survive. He then began recording his experiences on the onboard tape recorder, to assist in the identification of the failure should he not survive the landing. Despite this terrifying ordeal, he remained calm throughout. (Only a fraction of the surface density of the spacecraft is covered by re-entry insulation, as the majority is located at its base in the heat shield. At the hatch, there is only 1 inch of insulation, whereas the ablative area of the heat shield is 6 inches thick, with an expected burnaway of 3 inches.)

At one point, the Service Module propellant tanks exploded, and the resultant pressure wave forced the front hatch inwards and upwards as it strained against the pressure which seemed about to rip it off. During the ordeal, Volynov repeatedly uttered 'no panic' as, following a ballistic re-entry, he experienced 9 g. Normally during re-entry, the hydrogen peroxide thrusters would fire to produce lift to the capsule to reduce g forces, but on Soyuz 5 the instrument readings indicated that although the valves were open, nothing was happening. All propellant had been used at the beginning of retro-fire as the computer tried, without success, to correctly orientate the DM. The situation worsened, and Volynov thought that the end was near as he felt the increasing heat and was pulled against the seat harness as the ship tumbled towards destruction.

The connections were finally severed by atmospheric friction, allowing the PM to fall away and the DM to swing around to point its heat shield to the direction of flight and take the brunt of re-entry heating at 5,000° C. Volynov then realised that the PM had incinerated and separated from the DM, and that he had escaped death and would survive. But when the parachute system activated just above the ground, he became aware that the parachute had begun to tangle and that he might yet face Komarov's fate; but fortunately the tangle unfreed, and as he approached landing, the canopies deployed and the soft-landing rockets worked. However, the hard landing caused him to break free of his harness and fly across the cabin, and he tasted blood as he broke several front teeth in his upper jaw.

After landing, in the eerie silence Volynov could hear the hissing of the spacecraft as it cooled in the snow. He had landed 600 km from the intended landing site, and so rescue was some time away. The outside temperature was –36° C, and inside the spacecraft he soon began to feel bitterly cold. Radar had indicated that the capsule was down and far off course, and the controllers were still not sure if the cosmonaut was alive. Volynov knew that he had barely survived a second brush with death, but also that he would die if he stayed inside the capsule. Opening the forward hatch and looking around the landing site, 200 km from Kustanay, he spotted a vertical line of smoke in the distance; and after walking several kilometres he found a peasant hut, where the occupants welcomed him and kept him warm until the rescue team arrived.

On landing, only the specially formed couch and soft-landing shock-absorbers had saved Volynov from further injury, but he was so badly shaken that after the medal award ceremony he was admitted to hospital and was removed from flight status for two years. He later described how he had felt during the descent: 'There was no fear, but rather a deep and very clear desire to live on when there was no chance left.'

At the landing site, search teams located the DM and, to their surprise, found it empty; but they finally found Volynov by tracking the trail of blood in the snow where he had spat. (This incident remained secret until 1996.[10]) Volynov took some time to recover from the ordeal, but completed a second spaceflight in 1976.

Post-flight evaluation indicated that the separation devices between the DM and PM had failed to operate correctly, and did so only when the attitude control thrusters of the hydrogen peroxide tanks finally exploded due to the heat of re-entry. Volynov owed his life to the titanium structure that secured the forward hatch in place. Although the exact cause of the problem was never pinpointed, underpowered explosive changes seemed to have contributed to the problem – as experienced on Vostok 1, 2 and 5, and Voskhod 2.

Despite the close call during Volynov's re-entry, the Soyuz 4/5 mission was an outstanding success – especially when considering the events over the previous two-and-a-half years – although it had little operational value. The Soviets indicated that the mission clearly demonstrated space rescue capability, and opened the way for the creation of the first space stations,

THE TROIKA MISSION

With the success of the long-delayed Soyuz 4/5 docking and crew transfer mission, it was time to discuss the future direction of Soyuz. During late 1966, officials had planned to follow the Soyuz 1/2 manned docking flights with a repeat performance on Soyuz 3/4, and then a solo Soyuz 5 mission that would have had a programme of other research fields including space welding experiments. However, after Apollo 8 successfully sent American astronauts around the Moon during Christmas 1968, there was a lack of agreement over what should follow. In January 1969 it was suggested that a 30-day Soyuz mission could be undertaken – but this suggestion ignored. Then, in February 1969 the schedule was revised in the latest plans from TsKBEM: spacecraft no.14 (Soyuz 6) would fly first, with two cosmonauts, on a seven-day solo flight in April or May 1969; then, during August or September, spacecraft no.15 (Soyuz 7) and no.16 (Soyuz 8) would fly a seven-day docking mission, with five cosmonauts, and with the two spacecraft remaining docked for three days. These missions would then be followed by flights of spacecraft no.17–20, through to April 1970.

In reviewing these plans, Ustinov decided that the solo Soyuz mission was too conservative in its objectives, and he requested a 'more solid' mission plan. In his frustration, Kamanin wrote in his diary: 'Ustinov is very well aware that there is nothing 'more solid'. There are no more ships ready to fly, no clear flight plans, and not even a concrete programme for the forthcoming flight. We have reached complete absurdity and record short-sightedness... There is not a single person in this country who can tell when the next space mission will take place.'

Apparently, there was some move by Mishin (who had conceived the troika flight in late February) to incorporate Ustinov's plans, and at a meeting of the Soyuz State Commission on 25 April, the idea of flying three spacecraft at the same time was proposed. This was an adaptation of an old plan by Korolyov for flying three Vostoks after Titov's 24-hour space mission in 1961, which was abandoned in favour of a duel flight flown in August 1961. In the new plans, spacecraft no.14, 15 and 16 would fly on a three-vehicle mission in August. Spacecraft no.14 (Soyuz 6) would also be used to evaluate the Vulkan welding experiment and film the approach and docking of spacecraft no.15 (Soyuz 7) and no.16 (Soyuz 8), which would remain docked for three days. Each spacecraft would remain in orbit for 4–5 days. A tentative plan for follow-on missions also envisaged spacecraft no.17 (Soyuz 9) and no.18 (Soyuz 10) testing the new Kontact docking equipment and life support systems in November 1969, in support of the L3 lunar landing programme. They would also attempt to set a new Soviet endurance record of 15–16 days, to surpass the American 14-day record set by Gemini 7 in December 1964, and the Soviet record of five days set in June 1963. Spacecraft no.19 (Soyuz 11) and no.20 (Soyuz 12) would then fly a repeat mission in February or March 1970. These final four missions would also feature the much-delayed operations with the Kontakt docking hardware.

In a further meeting, on 7 June, Mishin and Ustinov discussed these plans, which gained a greater sense of urgency for the Soviets after the American success with Apollo 11 in July. However, it soon became apparent that there was a degree of concern and caution in the Soviet leadership about the possibility of a major accident during such a daring mission so soon afterwards.

Kontakt crewing and missions

During 1969–70, Soviet mission planners were already looking ahead to the era of the space station, and there was strong competition between design bureaux, especially concerning the rendezvous system to be used. Previous Soyuz missions had used the IGLA system, which had not always been successful, and many favoured a new system called Kontakt. It was decided to fly two Soyuz craft and to test the system with a docking mission, as with the Igla system. The crews chosen were:

Passive spacecraft prime crew	G.T. Dobrovolsky and V. Sevastyanov
Back-up crew	V.G. Lazarev and O.G. Makarov
Active spacecraft prime crew	A.V. Filipchenko and G.M. Grechko
Back-up crew	L.V. Vorobyov and V.I. Yazdovsky

These crews underwent considerable training, but due to delays in the delivery of the system the missions fell too close to the launch of Salyut, and it was eventually decided to use the Igla system. Kontakt, however, retained its supporters, and might well have been the better system; but the crews were stood down.

Preparing the spacecraft

On 8 August, the readiness of the three spacecraft was discussed at a meeting of the State Commission, and despite the fact that between forty and sixty faults had been discovered in all three spacecraft, the launch of Soyuz 6 was cleared for 4–6 October, with those of Soyuz 7 and 8 at 24 hour intervals.

On 24 September, preparations for the mission launches on 5, 6, and 7 October were put on hold when Ustinov summoned State Commission chairman Kerimov back to Moscow to inform him that no formal permission for the mission had been received, and this might have delayed the flights until November. Writing in his diary, Kamanin vented his dismay about a poorly-led space programme: '[Ustinov and Smirnov's] interference in the preparation of manned spaceflights shortly before the launches causes only harm. The indecisiveness of our highest leaders will delay the launch of the ships by at least a week.' He also expressed the hope that the bureaucracy of red tape would not create 'other consequences'.

Delay in the mission placed further pressure on the single Soyuz simulator, which was in urgent need of maintenance, and each of the vehicles was nearing the end of its recommended service life and needed to be launched soon to ensure optimum safety and reliability. Finally, on 30 September – after nearly a week of doubt about whether or not the missions would proceed – the Politburo approved the launch, and preparations continued. At a meeting on 1 October, the Soyuz State Commission scheduled the launches for 11, 12 and 13 October, and a week later these dates were confirmed. On 8 October, Soyuz 6 was rolled out to the pad at Area 31.

Launch after launch

On 11 October 1969, the launch of Soyuz 6 from Area 31 took place, after a smooth countdown, as strong whirlwinds whipped across the steppes. The vehicle followed a nominal ascent to orbit, and shortly after orbital insertion the crew configured their spacecraft for orbital flight. At the time, nothing was mentioned concerning the imminent launch of a further two spacecraft – only that the mission of Soyuz 6 was testing the onboard systems of 'an improved version of Soyuz' (in that it did not have docking equipment installed) and the spacecraft's manual guidance control, orientation and stabilisation capabilities, and that the crew would carry out scientific and welding experiments in space (see below). Twenty-four hours later, Soyuz 7 left Baikonur (this time from Site 1) and headed into orbit for a programme of 'joint manoeuvres' with Soyuz 6. Then, on 13 October, with light rain falling at the cosmodrome, Soyuz 8 was on the same pad as had been Soyuz 6, awaiting its turn to journey into space. Once Soyuz 8 reached orbit there was a record three manned spacecraft and seven cosmonauts in orbit at the same time.[11]

The plan was to have Soyuz 8 rendezvous and dock with Soyuz 7, to duplicate the experiment carried out on Soyuz 4/5; but this time the major difference was that Soyuz 7 was the passive vehicle and was placed in orbit prior to the launch of the active vehicle, Soyuz 8. Shatalov had planned to conduct a manual approach from the moment Soyuz 7 came within visual range of Soyuz 8, and Kamanin later wrote that this proposal would save both time and propellant, and would have applications in 'intercepting enemy satellites'. But prior to launch, Mishin rejected the idea.

Frustration with docking

On 14 October 1969, there was much anticipation as to whether Soyuz 8 would dock with Soyuz 7, and whether the cosmonauts onboard Soyuz 6 would film and photograph the historic event from a distance of 50 metres. But the first problem had already occurred three days earlier, soon after the launch of Soyuz 6. Shortly after orbital insertion, the failure of the automated pressurisation of the Approach and Orientation Engines' fuel tanks seriously threatened plans to manoeuvre Soyuz 6 to record the docking of the other two spacecraft. During the third orbit, however, Shonin was able to manually pressurise the tanks by using a switch in the DM, thus allowing the rendezvous to take place.

Soyuz 8's docking approach to Soyuz 7 began at 250 km, with a series of orbital manoeuvres that led to the Igla rendezvous system onboard both spacecraft acquiring the opposite spacecraft's signals. It was planned that, at 100 metres, Shatalov would take over manual control and bring Soyuz 8 to a docking with Soyuz 7, where they would have remained for the next three days. Unfortunately, at 1 km, the Igla system onboard Soyuz 8 failed to lock on to its Soyuz 7 target. The cosmonauts asked permission to take over manual control and attempt a docking from within 1,500 metres before the two spacecraft drifted apart; but by the time Mishin had decided to approve this action, the spacecraft were 3 km apart, and were too far away from each other for a second attempt that day.

On the following day, 15 October, mission control determined that the two spacecraft should have enough fuel on board to manoeuvre into a position to attempt a second docking approach. Over two orbits, small corrections finally brought the two spacecraft to the 1,700-metre point, which allowed them to close in for Shatalov take over with a manual approach. If the Igla system had been working, the main engines could have been used to close the distance; but without it, Shatalov could use only the smaller DPO approach and orientation engines. The cosmonauts therefore had to rely on visual cues, and there was little chance of success. In addition, Yeliseyev, onboard Soyuz 8, found it difficult to locate Soyuz 7 through the OM portholes against the Earth (the same problem as encountered by the Gemini 9 astronauts when approaching the Agena target in June 1966.[12] After four small burns of the DPO, Shatalov was still unaware of his range from Soyuz 7 or the orientation of his spacecraft, and was once again forced to abandon the attempt to dock; and consequently, Soyuz 8 and Soyuz 7 passed by each other. (TASS news releases afterwards indicated a closest approach of 500 metres, but did not mention a failed docking attempt.)

Later that day (according to Kamanin), Soyuz 6 manoeuvred to within 800 metres of Soyuz 7; but without an Igla system on board, a closer approach was not possible. Instead, Shonin and Kubasov used information up-linked from mission control, together with the onboard sextant, to conduct a dress rehearsal rendezvous operation with 'an unco-operative object' in orbit. Following his experience on Soyuz 4, Shatalov had suggested that a second control panel for the DPO thrusters be located in the OM, to allow the Commander to observe through the four portholes instead of through the periscope in the DM. Soyuz 6 carried this modification, although no docking equipment or Igla system was installed. It was merely to allow Shonin to perform a close approach to the other spacecraft. (An article in *Novosti Kosmonavtiki* claims that Soyuz 6 did not succeed in making any close approaches to the other two vehicles, although at one point the cosmonauts were able to see Soyuz 8.)

According to Kamanin's diary (translated by Soviet space specialist Bart Hendrickx), two other approaches were attempted by Soyuz 8 during the following day, but ballistics experts introduced errors into the orbital computations, and both attempts failed. On 16 October, TASS reported that Shatalov had completed two manual corrections of Soyuz 8's orbit during the forty-ninth and fifty-first orbit of his spacecraft. With a string of disappointments and failures highlighting the Soyuz programme's first three years of flight operations, it was Mishin's unenviable task to inform Ustinov and Brezhnev that once again the mission would fall short of their expectations.

Three landings on three days

On 16 October, Soyuz 6 returned to Earth – without (to everyone's relief) a repeat of the Soyuz 5 incident that almost cost Volynov's life. The post-flight investigation of that mission had highlighted several modifications that were required in the separation system of the DM and OM, and in August these had been successfully tested during the recovery of Zond 7 after its circumlunar mission. Soyuz 6 landed near Karaganda, Kazakhstan, and the cosmonauts emerged onto the bitterly cold and windy steppes.

Onboard Soyuz 7 on 17 October, the display for the automated landing system was accidentally activated (a step that should have occurred following separation of the DM from the OM), and this raised concern that perhaps the landing sequence might be activated earlier than planned. However, all was well, and the spacecraft landed without incident. Again the cold snow-covered steppes greeted the trio of cosmonauts. According to Soviet accounts, Volkov was the most enthusiastic at being welcomed by the local people and the recovery team, Fillipchenko carried on with his post-flight checks, and Gorbatko was embarrassed to be signing autographs on cigarette packets.

Now alone in orbit, the crew of Soyuz 8 conducted a successful communications link with the newly commissioned *Kosmonaut Vladimir Komarov* tracking and communications vessel – the latest addition to the space tracking fleet. The crew also used a communications link via a Molynia 1 communications satellite while out of direct communication link with ground stations in the Soviet Union, which was an important demonstration of improved orbital coverage. On 18 October, Soyuz 8 landed safely. Post-flight images of the landing showed the cosmonauts to be tired, weary and unshaven, but happy at completing their second spaceflight in ten months. In true Soviet reporting style, Shatalov was stated to be overjoyed at the success of the mission and how close each spacecraft had been in space and in their landing. 'They landed like airplanes,' the Soviet press reported him as saying, and he later added: 'We made a flight into the future – and the future belongs to orbital stations. We will have to make extensive investigations of orbital space... It is especially important to study the effects of prolonged weightlessness on human beings.'

Despite the snow-storms, all three spacecraft had landed safely and on target. Later, at the Vnukovo airport, Moscow, shortly after the landing of Soyuz 8, the traditional post-flight red carpet was again rolled out, but this time seven cosmonauts walked along the carpet to be greeted as heroes of what the Soviet press claimed was the latest Soviet space triumph. The operational phase of the group flight was over, but the investigation continued into why the Igla system (to be used in the space station programme) had failed.

Why Igla failed

Initial reports indicated that docking was never the purpose of Soyuz 7 and Soyuz 8, but Western observers never believed this. For many years the official line was that a significant amount of important work in space rendezvous had been achieved on the group flight, and emphasised that three separate manned spacecraft were operated in orbit. In 1970, Shatalov was interviewed by Dutch space journalist Peter Smolders. The Commander of the group flight followed the 'official' version that docking was not the plan, but rather overcoming the difficulty of controlling three ships at once (which some Americans had said would be a difficult operation). Smolders asked if Soyuz 7 and 8 had a docking mechanism, and Shatalov replied: 'No. Soyuz 7 and 8 were not equipped for docking. It was not necessary. We had already done this in January, and before that it had been done entirely by automatic control with Cosmos satellites.' Like many Western reporters, Smolders followed the official

On 18 October 1969, shortly after landing, Soyuz 8 cosmonauts Yeliseyev and Shatalov stand next to their spacecraft.

version that Soyuz 7 and 8 could not dock, that such an operation was never intended, and that Western rumours of the failed launch of a central docking unit were unfounded. However, it has subsequently been revealed that docking *was* the intention, but that the docking system had failed.

Writing in his diary of the events of October 1969, Kamanin was critical of the excessive use of automation on Soviet spacecraft: 'The unsuccessful attempts to perform a manual docking [without the Igla] have opened our eyes to the major shortcomings in the flight control systems used by Soyuz.' He recorded that successful docking relied on the flawless operation of automated systems, and that if these failed, the cosmonauts did not have a reliable means of control. He also emphasised that Shatalov and Yeliseyev consitituted by far the most experienced crew in Soyuz docking techniques, as they had trained together for some time and had served on Soyuz 4 and Soyuz 5. If they could not dock to Soyuz 7, this proved the inadequacies of the system at that time, and the apparent lack of minimal equipment available to the flight crew.

A commission, headed by Boris Chertok, was therefore set up to investigate why Igla had failed. This commission determined that on Soyuz 8, frequencies had 'shifted' because the piezoelectric quartz crystals used in the transmitter and receivers were affected by a thermostat failure, and they could not be maintained at the correct temperature for nominal operation.

In retrospect, it seems there was no major disquiet because the primary objective of the flight had not been achieved. Operationally, little was added to Soviet flight experience from the missions, apart from the control of three separate spacecraft in similar orbits. Ustinov clearly considered the troika flight to be a Soviet response to the Apollo 11 lunar landing the previous July, but the failure of the docking removed any chance of creating another Soviet 'space spectacular'. The day after the landing of Soyuz 8, the Soyuz State Commission glossed over the failure, and merely concluded: 'Much useful information [has been gleaned] about manual control and autonomous navigation, confirming the need for the cosmonauts to take a more active party in carrying out the flight programme in manned spaceships.'

SCIENCE ON SOYUZ

The main focus of the first eight Soyuz missions was to master the technique of rendezvous and docking, which the Americans had seen as a major objective of the Gemini programme,[13] with applications in both Earth-orbital and lunar operations. A second, longer-term objective was to gradually increase the duration of orbital spaceflight by using a series of space stations in Earth orbit for either military or scientific reasons. It was clearly seen that the Soyuz could be adapted to assist in this mode, as well as in lunar exploration, and so Soyuz could also be used as a test bed to provide a research capabilities in support of space stations, to evaluate equipment or procedures, or for an extended-duration mission as a precursor to the launch of a scientific research facility.

During 1967–69, several of the early Soyuz missions provided opportunities to test or evaluate equipment that might be used on extended flights in Earth orbit on space stations. Two of the missions flown at that time (Soyuz 6 and Soyuz 9) were solo Soyuz missions not related to rendezvous or docking tasks, but had potential application in the development of Soviet space stations in the 1970s.

Supplementary objectives

The first real opportunities to conduct scientific studies from a Soyuz spacecraft mostly arose after the rendezvous and docking operations had been conducted in 1968 and 1969. The first of these supplementary activities began on 28 October 1968. After attempting to dock with Soyuz 2, Georgi Beregovoi – onboard Soyuz 3 – carried out a modest programme of scientific research. He completed observations of planets and stars, detected typhoon and cyclone storm centres, reported on forest and jungle fires and the brightness of the Earth's surface, and photographed cloud and snow cover and the horizon of the Earth in both twilight and daylight,.

Following the undocking of Soyuz 4 and 5 on 16 January 1969, both crews conducted a number of scientific studies, including the evaluation of new equipment such as the RSS-1 spectrograph used for geophysical studies. On Soyuz 5, the instrument was used for a spectrometric experiment on the twilight aureole over the spectral range 400–650 nm at an altitude of 240 km, and medical and biological instruments were also tested. Khrunov also conducted experiments relating to the passage of radio waves though the ionosphere, and both spacecraft carried exterior mounted targets (each comprising fourteen plates constructed from one sheet of aluminium) to measure the presence of helium 3 and titanium. Astronomical investigations featured photography of the night sky directly opposite the Sun, and the study of developing comet tails, while Earth-related studies included the observation and photography of terrestrial cloud cover, the formation of storms, snow and ice cover, and the forms of glaciers.

Similar research programmes were conducted during the Soyuz 7 and 8 missions. Onboard Soyuz 7, the cosmonauts carried out photography of the Earth and of stellar objects in a variety of spectral bands, and completed some of the first detailed remote sensing exercises from a Soviet manned spacecraft. In particular, they focused on geological areas, with the intention of discovering reserves of raw materials for the national economy. Using the RSS-2 spectrometer, Soyuz 7 cosmonauts Gorbatko and Volkov undertook photography and spectrophotometry of the twilight aureole of Earth, cloud cover, and underlying terrain. This research included a session, on 13 October (while flying over the Arabian peninsular), in connection with ground-based observations and a pair of specially equipped Li-2 scientific aircraft flying at an altitude of 2.7 km. A further session was completed during the spacecraft's eighteenth-seventh orbit, over north-east Africa, with the spacecraft at an altitude of 218 km. The crews of Soyuz 7 and 8 also conducted several medical experiments, and photographed the Earth.

SCIENCE ON SOYUZ 6

When Soyuz 6 reached orbit, the initial announcements of the objectives for the mission were expounded much more than during previous flights. In addition to the normal tests and evolution of Soyuz control systems (at this stage, nothing was said about the flight of Soyuz 7 and 8), the crew was to conduct scientific observations and photography, over extensive areas of Earth, which could be of geographical or geological value, and also had to conduct atmospheric research, complete a programme of biomedical research, study the Earth for agricultural objectives, and conduct welding experiments in space. On the fourteenth orbit the crew evaluated a new sextant, and Kubasov took Earth terrain photographs while over the Caspian Sea and Volga river deltas, and photographs of the huge forests in central Russia. The crew also recorded cloud movements over Kazakhstan and a strong whirlwind in the same area – the data from which contributed in defining the landing areas for the spacecraft. A typhoon off the Mexican coast was tracked, as was a hurricane to the west of the country and tropical storms in the Atlantic and Indian Oceans; and medical tests – including several experiments and psychological tests – evaluated the working capacity of the cosmonauts in the microgravity environment.

Soyuz 6 had never been intended as a docking mission, and was not equipped with the Igla rendezvous systems or docking hardware, which was too heavy to be fitted. However, the presence of the Orbital Module allowed the inclusion of specialised scientific experiments and hardware to extend the flight objectives; and for Soyuz 6, the first welding experiments in space.

Vulkan welding experiments in space

The welding operation was completed on 16 October, a few hours before the re-entry. The Institute of Electrical Welding (IES), located in Kiev, had developed the 50-kg equipment, and the experiment had first been proposed to Korolyov by Boris Paton, the Director of the institute six years earlier in 1963. The plan was to fly one of the institute's engineers – Vladimir Fartushny – on a three-man solo Soyuz mission, but mass limits prevented the addition of a second Flight Engineer on Soyuz 6.

To complete the experiment, the OM had to be sealed and depressurised, with the operation being conducted automatically while the crew remained in the DM. The unit (Vulkan) was a cylinder divided into two sections – one containing the command and power systems, and the second holding the three welding devices to be evaluated. Three methods were used: low pressure compressed arc welding, electron beam welding, and arc welding using a consumable electrode. An electron gun was used to perform the welding, using samples of titanium, aluminium alloys and stainless steel.

On the seventy-seventh orbit, the operation began with the closing of the interconnecting hatch between the modules. Kubasov then flipped the switches to begin the welding operation, and then followed the fully automated operation on an indicator panel in the DM, while the data was directly transmitted to ground receiving stations. Academician Paton later claimed that a whole new era of metallurgy in space had begun with the experiment, and that the operation had revealed that 'a stable processing of welding metals by methods of melting is feasible [in space]. The miniaturised welding unit tested in the experiment proved highly reliable and efficient.' This was the official line for the next twenty-one years, until in 1990 it was revealed that the welding operation was a near catastrophe that could have claimed the lives of the crew, when a hole was almost burned through the hull of the OM.

It was thought that the low-pressure compressed arc had inadvertently aimed a beam at the inner compartment flooring and almost burned through. Unaware of this, the cosmonauts completed the experiment by turning off the unit and depressurising the module to retrieve the samples. Upon entering the OM they noted, to their astonishment, the damage to the OM structure, and fearing depressurisation they quickly returned to the DM and sealed the hatches. Kubasov, however, realised that the OM was structurally sound, and quickly re-entered the module to retrieve the samples before sealing off the OM to prepare for re-entry and landing later that day.

SOYUZ 9: A SPACE MARATHON

At a meeting of the Chief Designers on 23 October 1969, following the Soyuz troika flight, there were calls to expand the manned spaceflight programme to support the lunar effort, as well as further development of both EVA and rendezvous and docking techniques and attempts at extended-duration flights. For more than a year, Kamanin had been asking for the construction of an additional ten vehicles (up to flight unit no.20), and hoped to convince military officials that the additional vehicles might be used to bridge the gap until the Almaz and Soyuz VI were ready. Only four Soyuz 7K-OK vehicles (nos.17–20) were still available, and these were assigned to test the Kontakt rendezvous system designed to support L3 lunar operations during August–September 1970. However, in December 1969 these plans began to change when orders came to Kamanin to prepare for a 17–20-day solo flight – partly in celebration of the hundredth anniversary of the birth of Lenin. This followed shortly after the plans for an all-female Soyuz crew had been abandoned in early September 1969, and the female cosmonaut team (who had been selected in March 1962) were disbanded on 1 October 1969.

In January 1969, Afanasyev first proposed the idea of flying a 30-day Soyuz mission in an attempt to respond to the Apollo programme's achievements. Early planning of Kontakt missions had indicated durations of 15–16 days, and in one case a mission of up to 20 days was foreseen. On 6 December, a meeting of the TsKBEM discussed further mission planning and the possibility of flying only a solo long-duration mission that would require little time for preparation and upgrading of life support. There was also a desire to surpass the American Gemini 7's 14-day space record. Voskhod 3 should have achieved this in 1966, but the mission was cancelled in favour of the Soyuz manned programme. It seemed that the Kontakt missions would be delayed until 1971, and, fearful that the lunar programme would not produce any real results, there were moves to create a scientific space station programme from elements of the Chelomei Almaz station and the Soyuz spacecraft, which would be called Zarya. The long-duration mission – to be called Soyuz 9 upon reaching orbit – was regarded as a pathfinder for this space station programme, but due to delays it would not take place before June 1970.

Falcons in flight

The mission was planned to last for 18 days, but if the condition of the two cosmonauts deteriorated then it would be terminated. The docking equipment (but not the housing) of the vehicle to be used for the mission (spacecraft no.17) was removed, and a life support system – three or four times heavier than was normally carried on the early missions – was installed. Also installed inside the OM was a regeneration and thermal control system – a prototype for production systems being developed for the space station programme. For exercise and restraint, additional fixings were installed in the OM for use by the crew during exercise or other work. In the DM, the third couch was replaced by a storage rack, for extra equipment and scientific instrumentation and as a secure place for the return of exposed film and experiment results. Refinements in the seat shock absorber system were also included on this vehicle to support the cosmonauts after almost three weeks in space, which in 1970 was a major advance towards long-duration spaceflight. A consequence of these improvements was a build-up of carbon dioxide levels, which required the inclusion of additional carbon dioxide scrubbers inside the already cramped Soyuz.

The crew of the longest Soyuz mission in solo flight: the Soyuz 9 cosmonauts Sevastyanov and Nikolayev.

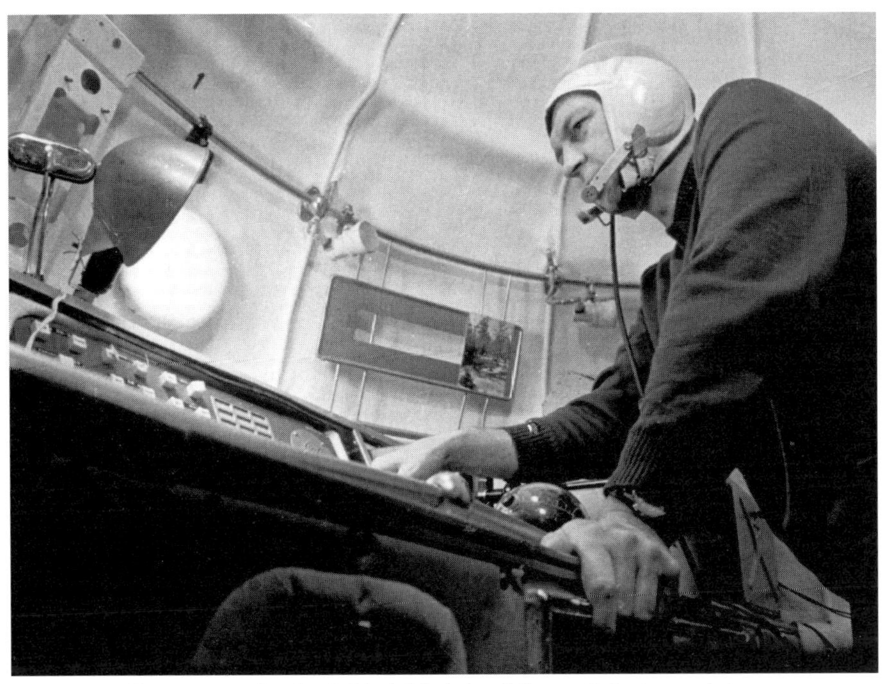

Sevastyanov in the Orbital Module of Soyuz 9 during training.

Soyuz 9 was launched on 1 June 1970, and returned to Earth eighteen days later on 19 June, to set a new world endurance record of 17 days 16 hrs 59 min, and 285 orbits of Earth.[14]

By far the largest programme of research conducted on Soyuz 9 was that related to the biomedical condition of the cosmonauts. This included their adaptation to, and working performance in, zero g, and their capability to recover after the flight. It was planned that the crew would spend two regular one-hour periods each day exercising and evaluating load-bearing suits, which were designed to provide baseline medical data and in-flight evaluation of procedures and equipment being developed for resident crews on space stations. Unfortunately, the crew found themselves so busy that they began evading the exercise programme and, despite a rebuke from mission control, they never really returned to the schedule. As a consequence, they suffered when they re-encountered Earth gravity. It was a valuable lesson for mission planners and crew trainers – but a painful lesson for the crew.

The research programme onboard Soyuz 9 also included a series of Earth resources experiments and observations, promarily linked to various fields of photography in conjunction with ground, aircraft and satellite programmes. The crew also evaluated manual methods of navigating in space by using known stars viewed through a sextant, and completed a range of astrophysical experiments. Another programme of research – listed under engineering – related to the evaluation of the Soyuz as a long-duration crew transport vessel to and from a space station, and its habitability over an extended period of time.

Orbital evaluation

The crew was to evaluate the habitability of the Soyuz, and establish a working day of 16 hours (inclusive of meals) and eight hours of sleep and rest. In the space station programme, the Soyuz would need to be placed in orbital storage for several weeks or months while the resident crew worked on the station, and would then be again powered up for the return to Earth. As well as continuing to contribute to an understanding of the rigours of spaceflight on the Soyuz design while supporting a human crew, Soyuz 9 also provided valuable information on the ability of Soyuz to endure a prolonged stay in orbit.

During the first few days of the mission there was some concern that the twin four-panel solar arrays were producing too much electricity, as the spacecraft was in almost constant sunlight. The device used to regulate the supply of electricity gathered by the solar arrays and fed to the onboard batteries was malfunctioning, and required the crew to manually cycle the system on or off as required. This could be carried out only fifty times, and there was concern that the mission would have to be terminated after only eight days. On 4 June, therefore, it was decided to spin the spacecraft around its longitudinal axis in order to maintain even temperatures. The Soyuz was spun at 2.5 turns per minute (15° per sec) to produce even distribution over its surfaces (similar to the American 'bar-b-que mode' that the Apollo spacecraft completed *en route* to the Moon). This minimised exposure to the Sun; but it drained the batteries, and the crew had to be awoken two hours before the end of their sleep periods to recharge the batteries by turning Soyuz to face the Sun. It was also found that the roll resulted in high temperatures on some of the spacecraft storage tanks, and it was therefore decided to suspend the roll for set periods. Fortunately, the problem of regulating solar array power remedied itself later in the mission.

The carbon dioxide scrubber was also not working as planned, and on 10 June the atmospheric pressure level was close to maximum at 900 mm. The crew was told to reduce the internal temperature from 21° C to 18° C, which reduced the pressure to 870 mm. On 17 June, after 15 days 5 hrs, the crew surpassed the Gemini 7 record by the required IAF percentage in order to qualify for a new official record. Later the same day, the Soyuz State Commission met to decide on the final duration of the flight. There had been some call for extending the flight by two days to 20 days, even though onboard supplies were sufficient for only 18 days; but the crew said that they had enough supplies for a two-day extension. However, the consensus was to return the crew after the 18-day target was achieved, with a landing on 19 June.

In all, the flight was a success and a confidence boost for the Soviets for planning longer missions on space stations. For the Soyuz designers, the flight had collected valuable data on the performance of the spacecraft over a long period of time – precisely as required for flights to space stations. In order to achieve this over a longer period of time, additional amendments to the basic Soyuz design would be required to allow protracted residency in orbit. In essence, Soyuz was about to become a crew ferry vehicle to support space station missions, at a time when the lunar programme was still under development. This was a prudent move that eventually proved decisive for the future of both Soyuz and Soviet manned spaceflight. However, there was still the flexibility to use specially-equipped Soyuz spacecraft on independent research missions, and several of these would fly during 1973–1976.

SUMMARY

After many years of development, and despite a troubled beginning and numerous setbacks, the Soyuz spacecraft finally flew in Earth orbit and proved Korolyov's concept of a three-module design. Between 1966 and 1970, seventeen spacecraft of the 7K-OK variant were flown on a variety of missions designed to qualify the spacecraft, to evaluate the first Soviet rendezvous and docking techniques and EVA transfer capability, and to support a programme of scientific investigations on missions lasting between 48 hours and 18 days. These spacecraft were the pioneers of the manned Soyuz programme and all that followed throughout the next four decades. Together, they accumulated more than 71 days in orbital flight time. At the end of the series the Soviet manned lunar programme was floundering, and the programme was being redirected towards the creation of a permanent manned space station supported by new variants of the Soyuz.

REFERENCES

1. Hall, Rex D. and Shayler, David J., *The Rocket Men*, Springer–Praxis (2001), pp.126–128.
2. *Ibid.*, pp.131–132; Shayler, David J., *Disasters and Accidents in Manned Spaceflight*, Springer–Praxis (2000), pp.76–77.
3. Shayler, David J., *Disasters and Accidents in Manned Spaceflight*, pp.369–387.
4. Rebrov, Mikhail F., *Kosmicheskie Katastrofy* [*Space Catastrophes*], Exprint, Moscow, 1994. This reports the restoration of communications after black-out, but is probably incorrect.
5. Shayler, David J., *Disasters and Accidents in Manned Spaceflight*, pp.85–88.
6. Hall, Rex D. and Shayler, David J., *The Rocket Men*, pp.190–191.
7. 'Potholes on the Starry Road: an inverview with cosmonaut Konstantin Feoktistov', *Trud*, April 2002.
8. Hall, Rex D. and Shayler, David J., *The Rocket Men*, pp.236–243.
9. Shayler, David J., *Gemini: Steps to the Moon*, Springer–Praxis (2001), pp.272–302.
10. For a Western account, see Oberg, James, 'Soyuz 5's Flaming Return', *Flight Journal*, June 2002.
11. For an example of Soviet 'real-time' press coverage of the mission, see 'Three Soyuz Spaceships in Group Flight', *Soviet News*, No. 5112 (14 October 1969), pp.13–15.
12. Shayler, David J., *Gemini: Steps to the Moon*, pp.256–257.
13. *Ibid.*, pp.225–267.
14. Shayler, David J., 'Flight of the Falcons: the 18-Day Space Marathon of Soyuz 9', *Journal of the British Interplanetary Society*, **54**, (2001), 27–46.

Soyuz variants: docking missions and hardware

Spacecraft (serial number) name	Design designation	Launch date	International designation	Docking date	Undocking date	Target spacecraft	Landing/ decay date (port)	Flight duration dd:hh:mm:ss
Key: (U), unmanned; (A), active; (P), passive								
11F615 series (Soyuz basic manned spacecraft for test flights)								
(1) ? (U)	7K-OK (P)	1966 Dec 14	*Launch pad explosion*				1966 Dec 14	
(2) Cosmos 133* (U)	7K-OK (A)	1966 Nov 28	1966-107A				1966 Nov 30	01:21:21:??
(3) Cosmos 140 (U)	7K-OK (P)	1967 Feb 7	1967-009A				1967 Feb 9	01:23:29:??
(4) Soyuz 1	7K-OK (A)	1967 Apr 23	1967-037A			Soyuz '2'	1967 Apr 24	01:02:47:52
(5) Soyuz '2'	7K-OK (P)	*1967 Apr 24 planned; cancelled*					Soyuz 1	
(5) Cosmos 188 (U)	7K-OK (P)	1967 Oct 30	1967-107A	1967 Oct 30	1967 Oct 30	Cosmos 186	1967 Nov 2	03:01:38:??
(6) Cosmos 186 (U)	7K-OK (A)	1967 Oct 27	1967-105A	1967 Oct 30	1967 Oct 30	Cosmos 188	1967 Oct 31	03:22:42:??
(7) Cosmos 213 (U)	7K-OK (P)	1968 Apr 15	1968-030A	1968 Apr 15	1968 Apr 15	Cosmos 212	1968 Apr 20	05:00:56:??
(8) Cosmos 212 (U)	7K-OK (A)	1968 Apr 14	1968-029A	1968 Apr 15	1968 Apr 15	Cosmos 213	1968 Apr 19	04:22:50:??
(9) Cosmos 238 (U)	7K-OK (P)	1968 Aug 28	1968-072A				1968 Sep 1	03:23:04:??
(10) Soyuz 3	7K-OK (A)	1968 Oct 26	1968-094A	*Failed to dock with target spacecraft*		Soyuz 2	1968 Oct 30	03:22:50:45
(11) Soyuz 2 (U)	7K-OK (P)	1968 Oct 25	1968-093A			Soyuz 3	1968 Oct 28	02:22:25:??
(12) Soyuz 4	7K-OK (A)	1969 Jan 14	1969-004A	1969 Jan 16	1969 Jan 16	Soyuz 5	1969 Jan 17	02:23:20:47
(13) Soyuz 5	7K-OK (P)	1969 Jan 15	1969-005A	1969 Jan 16	1969 Jan 16	Soyuz 4	1969 Jan 18	03:00:54:15
(14) Soyuz 6	7K-OK	1969 Oct 11	1969-085A				1969 Oct 16	04:22:42:47
(15) Soyuz 7	7K-OK (P)	1969 Oct 12	1969-086A			Soyuz 8	1969 Oct 17	04:22:40:23
(16) Soyuz 8	7K-OK (A)	1969 Oct 13	1969-087A	*Failed to dock with target spacecraft*		Soyuz 7	1969 Oct 18	04:22:50:49
(17) Soyuz 9	7K-OK	1970 Jun 1	1970-041A				1969 Jun 19	17:16:58:55
(18) (Kontakt)	7K-OK (A)							
(19) (Kontakt)	7K-OK (P)							
(20) Unused	7K-OK							
(21) Unused	7K-OK							

*The intended target vehicle was 7K-OK no. 1, but the launch of that spacecraft (what would probably have become Cosmos 134) was cancelled when Cosmos 133 developed problems in orbit. The launch of the passive vehicle had been planned for 29 November 1966.

The Soyuz ferry, 1971–81

One of the essentials in the operation of the space station programme (DOS) was the development of a crew transport system. Soyuz fitted this role perfectly, and in 1969 – while the 'original Soyuz 'spacecraft were flying, and plans to fly the Zond lunar variant continued – work began on the space station ferry. The first variant entered service in 1971 during the Salyut 1 programme and, with improvements, supported Soviet space station operations through to Salyut 6 until 1981, when a new variant, Soyuz T, emerged to take over the role of crew transport.

A FERRY FOR SALYUT

Work on the ferry began at TsKBEM late in 1969. The new spacecraft would be a redesigned 7K-OK, now designated 7K-T (11F615A8), and by early 1970 a new draft plan for the spacecraft had been produced by Department 231 at the design bureau. One of the major modifications was the inclusion of an internal docking and transfer system. The Soyuz would have an active probe assembly to mate with the passive drogue on the forward part of the Salyut, and the crew would transfer through a 0.8-metre hatch. The development of such an internal docking and transfer system is thought to have originated within Kozlov's Department 3 at OKB-1, in the framework of Soyuz R. The 7K-TK/11F72 variant of Soyuz for delivering crews to Soyuz R was also designed to have an internal transfer system. The life support system was simplified, as the spacecraft had to support only a three-man crew for the flight to the space station (up to two days) and the return to Earth (one day). When it was docked to the station (up to 60 days) it would not be required to support the crew and would be laid off, as all consumables and life support were onboard the space station. The Igla rendezvous system was relocated to the front of the spacecraft, inside the OM, and the removal of one of the radio command links enabled the toroidal compartment around the aft of the PM to be removed to save weight. At 6,700 kg the new ferry was about 50 kg heavier than the 7K-OK, but the vehicle could return only 20 kg of scientific specimens or equipment in the DM (with a full complement of three cosmonauts without pressure suits), suggesting that the design was at the upper limit of the capability of the launch vehicle.

One departure from earlier variants of this spacecraft was that there would be no unmanned test launches prior to man-rating the vehicle, reflecting the use of the proven equipment and sub-systems on the Soyuz 7K-OK; and already in development was a more advanced ferry craft, based on the 7K-S design, which would have greater capacity to support long-duration spaceflights and a larger payload return. This spacecraft was being developed in conjunction with a second-generation space station with two docking ports, but did not appear until 1979 (as Soyuz T). For the initial Salyut operations, the 7K-T would be the vehicle used to transport cosmonauts to and from the stations.

SALYUT FERRY OPERATIONS, 1971

On 19 April 1971, Salyut 1 – the world's first space station – was launched on a Proton rocket. It was designed to support at least two resident visits, and for this programme a cadre of cosmonauts had been formed to crew the space station and had begun training in the summer of 1970. The Salyut had been developed from the hull of a military Almaz station and elements of the Soyuz spacecraft in order to ensure that a Soviet space station was in orbit before the American Skylab. The Soyuz crew ferry featured the new internal docking and transfer hatch, but still relied on the solar arrays on the PM to supplement the power supply on the Salyut (which was also routed from adapted Soyuz solar arrays). More powerful arrays would be available on later stations, eliminating the requirement for the Soyuz to fly with solar arrays on such short flights.

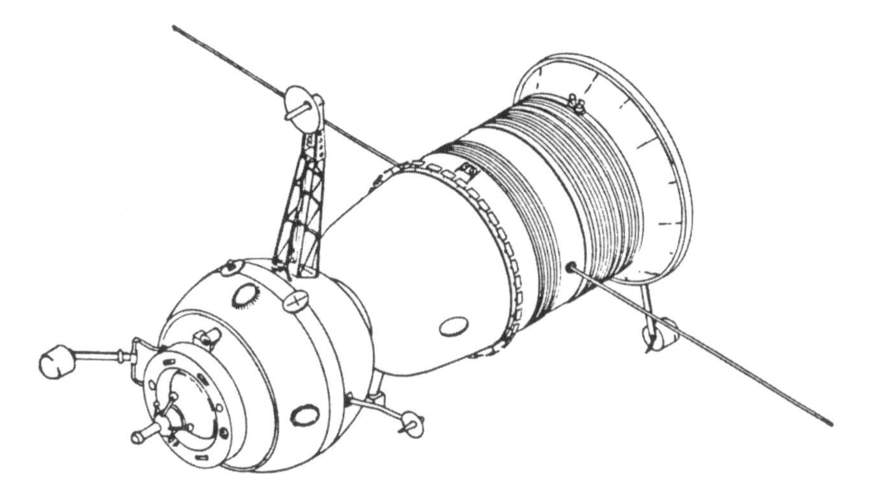

Soyuz space station ferry craft (1973–81). (Courtesy Ralph Gibbons.)

The crews of Soyuz 10 and Soyuz 11 prior to the launch of the first space station in 1971. (*Front row, left–right*) Leonov, Yeliseyev, Shatalov, Rukavishnikov and Kubasov; (*Back row, left–right*) Kolodin, Dobrovolsky, Volkov and Patsayev.

Salyut 1 crewing

The first crews were assigned to train for the first civilian space station in mid-1970. The mission planners identified four three-man crews, drawn from cosmonauts from the cancelled lunar programme and the recently flown Soyuz troika missions:

Crew 1	G.S. Shonin, A. Yeliseyev and N. Rukavishnikov
Crew 2	A.A. Leonov, V.N. Kubasov and P.I. Kolodin
Crew 3	V.A. Shatalov, V.N. Volkov and V.I. Patsayev
Crew 4	G.T. Dobrovolsky, V.I. Sevastyanov and A.P. Voronov

The first change came when Shonin was replaced because of a major drink problem. Shatalov was then moved from crew 3 to crew 1, and Dobrovolsky from crew 4 to crew 3, while Gubarev came into the support crew command position. The Shatalov crew flew Soyuz 10, and the Leonov crew was replaced as the prime crew on Soyuz 11 when Kubasov failed the pre-flight medical examination. The Dobrovolsky crew replaced them, but it was planned that if the Soyuz 11 mission was a success, then the Leonov crew, backed up by the Gubarev crew, would fly Soyuz 12 to Salyut. After the Soyuz 11 disaster, all crews were stood down.

Soyuz 10: the first ferry mission

The first manned mission to a space station, designated Soyuz 10 (call-sign, Granit – Granite), began with the launch on 23 April 1971, four days after the unmanned space station was placed in orbit. It was the second attempt to lift the crew off the ground. (During the first attempt on the previous day, the count-down had reached

T–1 minute when it was halted when one of the launch masts failed to retract.) All went well, and with its three-man crew, Soyuz 10 reached orbit to begin the chase to Salyut 1. Shortly after launch of the station, it was discovered that the cover of the OST-1 telescope had not jettisoned, and that two of the life support system ventilation units were malfunctioning, but this would not prevent manning of the station. The flight to the station would take a day.

During the approach, Soyuz 10 encountered difficulties in maintaining its orbit to intersect Salyut's orbit. On the fourth orbit there was a programming error in the command for a correction burn, and on the next orbit there was insufficient time to prepare for the burn, and it had to be abandoned. The optical surfaces of the ionic orientation system of the Soyuz had apparently become contaminated, and it had to be overridden. Shatalov then took manual control to position his spacecraft, and performed the manouevre flawlessly. However, six of Salyut's eight ventilation systems were not operating, and this raising concern over the quality of the atmosphere inside the station.

Early the following morning, Shatalov – the only cosmonaut in the cadre who had experience of docking in space – activated the Igla system to control the approach of Soyuz to 180–200 metres from Salyut's forward docking port. At this point he gently brought Soyuz 10 up to Salyut and, at 04:47 hours MT on 24 April, achieved a soft docking – the first link-up of a manned spacecraft with a space station. The next step was to initiate hard docking by withdrawing the probe on the Soyuz to draw the two spacecraft together, and to activate the docking latches in the ring of the docking adapter. However, some ten minutes after the initial docking, Shatalov reported that the hard docking indicator light on his control panel had not illuminated, indicating to the crew that they were not firmly attached to the station. A check of ground telemetry revealed that the two spacecraft were still soft docked, with a 9-cm gap between them. Shatalov fired the Soyuz manoeuvring engines to try to force the two spacecraft together, but without success.

On the fourth orbit after docking, Shatalov was ordered to undock Soyuz and again attempt hard docking. However, when he tried to do so the Soyuz refused to budge, indicating a serious problem. There were two options available. First, the crew could have entered the OM and dismantled the docking device from the Soyuz side. This was an emergency procedure in which explosive bolts would be used to separate the probe from the rest of the active docking mechanism, leaving it behind in the docking collar. It did not require any manual intervention by the crew, as they simply had to initiate the command to fire the bolts. Alternatively, they could jettison the OM and leave it attached to the Salyut. Either option would render the station inoperable, as there was only a single port where a ferry to dock. In addition, the crew had only about 40 hours of flight time left on the Soyuz, and they needed to save some of it for the return to Earth. During the following orbit, Shatalov tried a second time to undock his Soyuz – and this time the spacecraft gently slipped its berth and flew free of the station, prompting a round of applause in mission control.

While Shatalov maintained station-keeping near Salyut, ground control debated their options by reviewing the state of the spacecraft, the Salyut, the available air, the level of propellant, and the amount of power. It was decided that the only option was to abandon the docking attempt and to bring the crew home in an emergency recovery. With the news that their mission had been terminated, Shatalov and his crew flew around the station and photographed the docking port for post-flight analysis before initiating a separation burn to begin the recovery process.

During the early hours of 25 April, Soyuz 10 was on its way home, heading for Kazakhstan, and the first night landing in the Soviet programme. Inside the Soyuz, Yeliseyev and Rukavishnikov were looking out of the side portholes of the vehicle as it swung under the parachute, and noticed the large expanse of water towards which they seemed to be heading. Both questioned Shatalov about what to do; but Shatalov merely shrugged his shoulders, indicating there was nothing they *could* do except brace themselves and wait.[1] The spacecraft landed 120 km north-west of the town of Karaganda in Kazakhstan, missing a lake by 50 metres. Instead of an extended mission of up to 30 days, the crew had logged only 1 day 23 hrs 46 min 54 sec in flight.

Expecting a new space spectacular to compensate for losing the Moon race, the Soviets were once again organising an enquiry into what went wrong. The press statements, of course, praised the total success of the mission as a demonstration of the ability to dock to the station, and claimed that crew transfer was never the plan.

On 10 May the investigation team reported their findings concerning the docking failure. The crew had reported that following the soft docking, the Soyuz thrusters had continued to fire for 30 seconds, causing the Soyuz to swing violently after retraction of the probe, and placing a force equal to 160–200 kg on the shock absorbers in the docking mechanism. Part of the investigation included tests on a ground-based docking system, and it was found that the shock absorbers could accept no more than 130 kg (which itself was 60% more than the design limit). It was also determined that a similar situation of excess firing of the DPO thrusters had probably accounted for the failure of Cosmos 186/188 to achieve hard docking in October 1967 (see p.139). The recommendations of the investigation board included the strengthening of the shock absorbers to twice the upper limit (260 kg), and additional controls in the Soyuz to allow the Commander to manually control the retraction of the probe and the action of the thrusters.

Soyuz 11: triumph and tragedy

The disappointment caused by the Soyuz 10 failure did not dampen the planning for the future use of Salyut. Mishin apparently proposed two missions to complete the programme, although it is doubtful that this was a serious proposition. The first would be launched on 4 June, with two cosmonauts and EVA suits, allowing one of them to inspect the docking by EVA, and a second mission would be launched on 18 July. Kamanin rejected the idea, stating that none of the Salyut 1 training group cosmonauts had trained for EVA, and that the EVA suits under development would not be ready in time. It was therefore decided that any future mission would be aimed at reviving the station, and that its duration would depend on the state of the station and the level of consumables on board. Hopefully, two 30-day missions could still be mounted, but for planning purposes the first, Soyuz 11 (call-sign Yantar – Amber), was planned at 25 days, with a launch on 6 June 1971.

The original Soyuz 11 crew undergo instruction in Star City. (*Left–right*) Kubasov, trainer, Leonov and Kolodin.

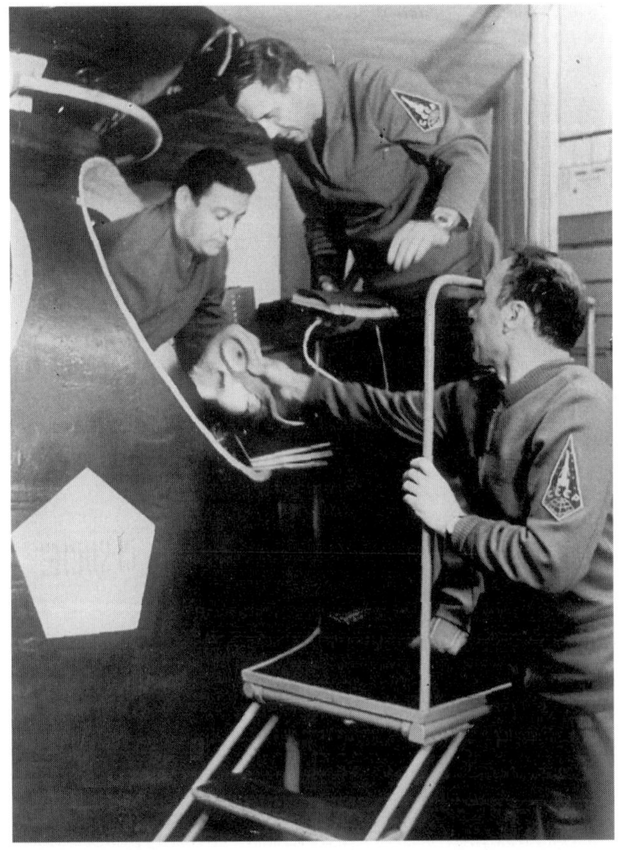

Volkov, Dobrovolsky and Patsayev complete a check-out of the Soyuz Descent Module compartment during the final stages of training for their mission.

Due to the illness of Kubasov, the back-up crew of Dobrovolsky, Patsayev and Volkov were launched to the station, as a complete crew, after only four months of training. However, despite their inexperience (only Volkov had flown – on Soyuz 7 in 1969) they achieved a successful docking and transfer into the Salyut, where they remained for three weeks. The trio of cosmonauts became the first to live and work on a space station and despite numerous in-flight difficulties they completed a successful residency. For those three weeks, the Soyuz (spacecraft no.32) remained docked to the front of the station in a dormant mode, ready for any emergency evaluation and for eventual return to Earth.

It had been decided not to mount a 30-day flight but to settle for a more conservative 23 days and, having broken the 18-day record of Soyuz 9, the mission was terminated on 30 June. The crew completed mothballing the Salyut for a possible second residency, packed the Soyuz with return material, transferred into the Soyuz, and closed the hatches. On the Soyuz display, the 'Hatch Open' indicator was on, referring to the inner hatch between the DM and OM, and not the hatch between the OM and Salyut. This caused some concern, as the inner hatch would be the only one separating the unsuited crew from the vacuum of space. Believing the sensor was in error, it was suggested that a piece of paper should be used to block off the sensor; but instead, Dobrovolsky placed a plaster over the sensor and tried the hatch closure sequence again. This time the light went off, and the pressure checks proceeded satisfactorily. With the hatches secure, the spacecraft undocked and was flown around the Salyut, with Patsayev taking photographs. That completed, Volkov reported that the 'Return' indicator light was on. At ground control, the CapCom replied that this was normal, and that 'Communications are ending... Good luck' – to which came the reply: 'Thank you... Be seeing you... I am starting orientation.' As Soyuz 11 headed to Earth, communications went out of range and contact was cut off... and no more was ever heard from the cosmonauts.

The crew had been told to provide, as they descended, a running commentary by using the VHF antennae located in the DM hatch and the parachute lines. In addition, due to their extended time in space and their expected difficulties in acclimatising to 1 g, they were ordered not to make any sudden movements after landing, and not to attempt to open the hatch themselves, but to wait for the recovery crews. When there was neither confirmation of de-orbit burn, nor any commentary from the crew prior to the plasma black-out caused by re-entry, the controllers tuned to the Soyuz frequencies usually used in orbit, in case the Soyuz had not made the burn. About 15 minutes after the start of the burn to take Soyuz 11 out of orbit, radar reports indicated that the DM had been picked up about 2,000 km from the planned landing site. The much-relieved controllers then expected to hear the crew over short-wave and VHF lines, at about 7 km altitude, as more reports were coming in that Soyuz 11 had been seen descending under its parachute. But surprisingly, all that was heard from the spacecraft were beacon transmissions, before Soyuz 11 apparently landed normally at 02.16.52 hours MT, approximately 200 km east of Dzhezkazgan. Almost immediately, four helicopters landed close by to assist the crew out of the capsule and to begin the celebrations of their record-breaking flight of 23 days 18 hrs 21 min 43 sec.

When the recovery crews opened the hatch, they found the three cosmonauts lifeless in their seats. They took them out of the capsule and attempted to revive them, but to no avail. Meanwhile, news of the tragedy was sent to the State Commission. From examination of the cosmonauts' injuries, it was deduced that they had died from decompression. On inspecting the DM, it was found that the radio transmitters had been turned off and that the shoulder straps of the seat harnesses were all loosened. Dobrovolsky had become entangled in the straps. All seemed normal inside the capsule, except that one of the two valves used for ventilation or pressure equalisation was open, which strongly indicated rapid decompression at a great height. Using the recovered onboard memory device (Mir), the sequence of events of the descent was determined, and with it the probable actions of the doomed crew. The DM and the OM had separated at a height of more than 150 km, and at that moment the pressure in the DM dropped to a near-vacuum in a period of 30–40 seconds, during which time the valve opened. The loss of cabin pressure was clearly due to the premature opening of the valve as the two modules separated; and as the crew were not wearing pressure suits, there was little time to do anything to same themselves.

The engineers' concern was that while it was almost certain what had happened, it was not so obvious *why* it had happened. Many tests were performed, and valve failures were recorded only with the introduction of situations outside normal operations, and in conjunction with other factors.

In his diary, Kamanin provided his own interpretation of the actions of the crew during the sequence of events that followed a punctual and successful re-entry burn. When the three modules separated, the twelve explosive bolts holding the DM to the PM generated too much power and inadvertently dislodged a ball joint from its position, after which the ball joint opened the ventilation valve too early, thus allowing the internal atmosphere to escape into the vacuum of space. According to Kamanin, the crew would have quickly become aware of the drop in pressure, and Dobrovolsky – assuming it to be due to the 'faulty' hatch – could have released his seat belts and checked the hatch, to find it sealed and secure. The crew would probably have heard the air whistling out of the spacecraft, but would have been unable to pinpoint it. To isolate the noise, Volkov and Patsayev apparently turned off all of the radio transmitters, and found the source of the noise under the Commander's central seat – the location of the ventilation valve. Dobrovolsky and Patsayev then appear to have tried to manually close the valve as the air escaped from their spacecraft; but with time running out, Dobrovolsky had managed to hurriedly refasten his harness before both fell back into the seats. The decompression was so rapid that the three men would have lost the ability to work within 10–15 seconds, and would have certainly been in agony for about 20–30 seconds. They died about 50 seconds after the failure of the valve, and the internal pressure in the DM dropped from 920 mm to zero in only 112 seconds.

It is reasonable to assume that the cosmonauts, as they attempted to find the leak, were aware that they were about to die. During an interview in 1990, Mishin indicated that they could have plugged the leak by simply placing a finger over the valve – if they had had enough time. Since the valve was not supposed to be activated outside the atmosphere, there was no reason to assume that the crew would have reacted any quicker to seal it. Their training was to open the valve much later. On 17 August the final version of the accident investigation report was signed off, and with it, several recommendations to ensure that such a tragedy could not happen again. The valve needed to be more stable in respect to shock loads; there needed to be manual quick-action chokes to close the valves within seconds; and space suits were to be used during all phases of flight on Soyuz during which depressurisation was a possibility.

Reoccupation of Salyut was impossible, as the implementation of the recommendations would take longer than the lifetime of the consumables and supplies on Salyut. Nine days after the loss of Soyuz 11 cosmonauts, all further crews were stood down. Despite efforts to keep Salyut in orbit until it could be visited by another crew, the station consumables were depleted by August 1971, and on 11 October the station was commanded to a destructive re-entry over the Pacific Ocean.

SOYUZ, SALYUTS AND SPACE SUITS, 1971–73

The accident investigation into the loss of the Soyuz 11 cosmonauts indicated that individual pressure garments for the crew should be introduced into future Soyuz flights, for launch and re-entry. Central Committee Secretary Dmitri Ustinov was determined that no cosmonaut would fly into space without a pressure suit, and plainly stated this to Mishin and his deputy Bushayev. The Zvezda Machinery Building Plant (under its chief designer, Gai Severin) was therefore tasked with accelerating its efforts to design the new pressure garment, designated Sokol K. Twenty years after the accident, Mishin wrote of his disagreement in developing this added safety element to Soyuz. He continued to argue that the best way to ensure safety for the crew of a multi-seat spacecraft was by ensuring the reliability of spacecraft systems. The addition of suits complicated the design and reduced space, and the additional weight necessitated the reduction of the crew capacity to two, and so research programmes and experiments also had to be reduced or removed. The workload of future two-person station crews was to be increased until the reintroduction of three-person resident crew.

Despite the loss of the Soyuz 11 crew, planning continued for future flights. During 1971, the Soviets and Americans discussed a possible joint mission involving a Soyuz and an Apollo (see p.205), and by August 1971 a provisional schedule had been drawn up for the next three DOS space stations. Each would have a six-month lifetime, with four months support for crews: DOS 2 would be launched during the first quarter of 1972, and three or four crews would be launched on Soyuz 7K-T spacecraft; DOS 3 would follow during the fourth quarter of 1972 – again, with three or four resident crews flying 7K-T spacecraft; and DOS 4 would be launched during the fourth quarter of 1973, and would be manned by four crews flying the new improved ferry (the 7K-S) and using a new rendezvous system designated Lira.

It is not clear why the next Salyut was slipped into the third quarter of 1972, but it was probably related to the re-qualification of the Soyuz and the development of the

Sokol K1 pressure garments. In order to test the changes made to the Soyuz and the new pressure suits, an automated Soyuz test flight was inserted into the flight programme. Launched on 26 June 1972, Cosmos 496 remained on orbit until 2 July, and was a successful flight. Little is known of the flight programme, apart from one orbital manoeuvre. The new equipment to support the suited crew weighed 100 kg, and was also tested on the flight. Cosmos 496 was the same vehicle that was supposed to have taken the Soyuz 12 crew to Salyut 1. It was originally vehicle no.33, but was renamed 33A (automatic) when it was decided to use it for an unmanned test flight. This may also be the reason why it still had solar panels, although it seems to have been decided to remove them some time during 1972.

The new Salyut training group

In July 1972, a new Salyut (DOS 2) was scheduled for launch, with the first crew planned to launch to the station during August and a second crew in October. Following the Soyuz 11 accident, the Soyuz required a numbers of changes – primarily the reduction of the crew to two because of the new requirement to wear space suits on launch and landing. The next Salyut was planned to fly in 1972, and so crews were formed in November:

Crew 1	A.A. Leonov and V.N. Kubasov
Crew 2	V.G. Lazarev and O.G. Makarov
Crew 3	A.A. Gubarev and G.M. Grechko
Crew 4	P.I. Klimuk and V.I. Sevastyanov

When the launch of the Salyut failed, these crews transferred to crew the next Salyut, due for launch in 1973; and when that also failed, the training group was broken up, and the crews were assigned to the ASTP, the interim flight of Soyuz 12, and the solo mission of Soyuz 13.

Lost Salyuts

The success of the Cosmos 496 mission restored some confidence in the Soviet programme – but the next Salyut was to present further disappointment. Launched on 29 July 1972, DOS 2 soared into the sky on top of a Proton launch vehicle, but after 162 seconds one of the four second-stage engines failed, causing the control systems of the second stage to fail, and preventing orbital insertion. It fell into the Pacific Ocean, and as it had not entered orbit it did not receive a Salyut designation or a Cosmos number. The two crews intended to man the station were stood down, and despite discussions about the launch of a solo Soyuz flight to test the new space suits and man-rate the Soyuz improvements later in the year, the mission was never launched.

This was a bad time for the Soviets. The Americans had succeeded again and again with Apollo, and their space station – Skylab – was on course for its planned launch in the spring of 1973. But while problems were dogging the Salyut programme, the military Almaz programme was slowly progressing towards launch in 1973. Taking advantage of the delay to manned flights, Mishin and his design bureau decided to prepare the next generation of DOS stations. Their requirements included improved solar arrays, improved systems to increase the lifetime from 90 days to 180 days, and the introduction of the Almaz military station from Chelomei's bureau so that they could complement each other and compete for headlines with the American Skylab during 1973. During 1972 and 1973, changes in the Soyuz 7K-T ferry were also incorporated to provide an upgraded ferry to both DOS and Almaz stations.

Crews for the Almaz (military) orbital station

In September 1966 the first cosmonauts were assigned to the Almaz programme, and more cosmonauts were assigned in January 1968. Throughout this period, mission plans were frequently reviewed and modified, and in 1970 – with the launch of the first military Salyut planned for 1972 – three crews were named:

Crew 1 A.P. Fyodorov, L.S. Demin and V.Y. Preobrazhansky
Crew 2 O.A. Yakovlev, V.M. Zholobov and E.N. Stepanov
Crew 3 V.D. Zudov, Y.N. Glazkov and M.I. Lisun

Following the Soyuz 11 disaster these crews were stood down, and the crewing for Almaz was reduced to two in line with the new requirement to wear space suits on launch and recovery. The first cosmonauts resumed training in late 1971, with the Salyut still on line for launch in 1973. In November 1971, the five crews named were: Popovich and Demin; Volynov and Khludeyev; Gorbatko and Zholobov; Fyodorov and Artyukhin; and Sarafanov and Stepanov.

Orbital operations, 1973

During the second half of 1971 it was decided to use the 7K-T for Almaz (previously, only the TKS was expected to fly to Almaz). The Almaz version of the 7K-T was assigned the index 11F615A9, and one of the chief differences, compared with the DOS version (11F615A8), was that shortly before docking, the Igla antenna was tilted slightly backwards to prevent it from colliding with the station's solar panels and antennae.

It was determined that the Soyuz need fly for no more than two days, and that for power it could rely solely on chemical batteries, which, after docking, could be recharged by the larger solar arrays on DOS or Almaz. The removal of the twin arrays on the Soyuz saved about 100 kg, reducing the mass of the ferry to 6,800 kg, although it was still flying with only two space-suited cosmonauts.

On 3 April a new space station was launched – the first Almaz, not a DOS station – and upon entering orbit it was designated Salyut 2 to disguise its true military nature. The first manned crew should have launched on 13 April, but were delayed to 8 May due to qualification problems with the Soyuz parachute system. Unfortunately, the crew were never launched to the Almaz as, although initially all seemed well, by the thirteenth day of flight, problems began to develop when the main engine of the station was fired, causing ruptures in the main hull and the loss of a solar array and radio transponders. On 28 May 1973, the orbit of the Almaz decayed into the atmosphere. With serious problems affecting the military station, all manned flights were cancelled, and those having trained specifically for a military-based mission would have to await the launch of the next Almaz station in 1974.

Meanwhile, on 11 May 1973 the 'civilian' DOS 3 was launched, with the first crew planned to launch to the new station on 14 May – the very day on which NASA would launch Skylab. Once again, the spectre of disappointment that had plagued the Soviet programme for so many years struck on the very first orbit, when problems with the attitude control system were recorded. It was later determined that this was due to a faulty ion sensor that caused irregular working of the thrusters. This rapidly depleted the reserves of attitude control gas – so quickly (1,500 times faster than normal) that the controllers initially refused to believe the data; but when they did, it was too late to save the station. Moreover, the loss of the station in orbit was so rapid that it never received an official Salyut designation, being identified as Cosmos 557. On 22 May the orbit decayed without a crew being launched.

Prior to the failure of Cosmos 557, the vehicles planned to fly to DOS 3 were nos.34, 35 (originally planned for DOS 2) and 36, while nos.37, 38 and 39 were planned for DOS 4. After the failure of Cosmos 557 it was decided to fly all three DOS 3 vehicles on unmanned test flights. Nos.34 and 35 were modified for their new roles, and were renamed 34A and 35A (automatic); in June 1973, no.36 flew, without modifications, as Cosmos 573; in September 1973, no.37 (the 'freshest' vehicle) flew as Soyuz 12; and in January 1974, no.34A flew, as Cosmos 613, on a 60-day endurance mission. Vehicle 35A was never launched, and was instead used to test the modified emergency escape system for the ASTP version of Soyuz.

A two-day test flight

After the launch of the first Salyut in April 1971, four space stations were launched in two years. Three of these stations were lost due to malfunctions during launch or in orbit, and the two manned missions were not successful because the first Salyut crew failed to transfer to the station, and the second crew died during recovery. With no prospect of a station in orbit before 1974, therefore, the Soviets took the opportunity to pursue other aspects of the programme. They were already in agreement with the Americans to fly a joint Soyuz mission with an Apollo in 1975, and crews from the cancelled civilian DOS stations had been reassigned to the Apollo–Soyuz Test Project in May 1973 (shortly after the loss of Cosmos 557/DOS 3), while all assigned Almaz crews were in training for the next military station to be launched in 1974. In the hiatus, Mishin decided to finally test fly various elements intended for the space station programme once the new stations were ready. The new Soyuz ferry would first be flown automatically, and then with a two-man crew, with pressure suits, for a short test flight (Soyuz 12); and he also added a third 'solo Soyuz mission' (Soyuz 13) to fly some of the experiments and instruments that had been abandoned due to the loss of the two civilian DOS stations. A fourth unmanned test flight was also included to test the capability of a Soyuz to remain in space for two months and simulate a docked mission with Salyut or Almaz.

The unmanned Soyuz – designated Cosmos 573 – was launched on 15 June, flew for 2 days 9 min, and completed one orbital manoeuvre. The new spacecraft was the first to fly without the twin solar arrays, and relied on chemical batteries for

Soyuz 12 crew Makarov and Lazarev discuss their flight plans with Georgi Beregovoi during a break in simulator training for the first Soyuz ferry mission in 1973.

electrical power; and as there was no docked spacecraft to recharge the batteries, the mission was limited to two days. This limited lifetime of the batteries would subsequently be an important factor in future operational flights of the vehicle. The OM of these 'ferry' vehicles was changed from a work and experimental area to a cargo area to ferry supplies and equipment to a station. If required, it could be used as an airlock for EVA, although this has never been put into operation since the Soyuz 5/4 transfer. The use of the internal docking probe and transfer hatch shortened the length of this Soyuz to 7.5 metres, and the OM was shortened to 2.65 metres with the removal of the large docking housing collar carried on the earlier 'basic Soyuz'. The success of this mission qualified the system for manned flight, without requiring any further unmanned tests.

The manned test of the ferry completed the pre-station testing. Launched on 27 September 1973, Soyuz 12 – the first manned ferry to fly without solar arrays, and crewed by Lazarev and Makarov – completed a flight of 1 day 23 hrs 15 min 32 sec. The duration of the flight was stated at launch, to ensure that there was no misunderstanding concerning its return after such a short flight. During the first few orbits, the crew followed the profile of both Cosmos 496 and 573, and probably duplicated the initial sequence of events for approaching a station, using the Igla system. Very few scientific experiments (undisclosed biological investigations) were carried on the flight, although Makarov used an LKSA multispectral camera to take

Earth resources photographs, while Lazarev used standard camera equipment to record the same targets for comparison. Apart from man-rating the new Soyuz, one of the major objectives of the mission was to test the Sokol K1 pressure suit.

The Sokol (Falcon) pressure suit

The Sokol K (Kosmichesky – Space Falcon) survival pressure suit was developed at Zvezda Plant 918, under the direction of Chief Designer Gai Severin, as a lightweight suit individually tailored for each cosmonaut, bearing their initials, and compatible with the Kazbek shock-absorbing seat liners. The designers reviewed their earlier work on the Vostok (SK-1 – Sokol) Intravehicular Activity (IVA) garment (1961–63) to develop the suit in an abbreviated design time.[2] The prototype was produced in July 1971, further modifications were introduced between August 1971 and March 1972, and ground tests were completed in 1972.

The suit was designed to protect the cosmonaut in the event of a sudden depressurisation of the DM, operating through the life support system of the spacecraft. First worn during the Soyuz 12 (K1) mission, it was upgraded to comply with the Soyuz T (KV-2), TM (KV-2) and TMA configurations, and has been used by every cosmonaut on a Soyuz spacecraft since 1973. It is used during launch, docking, undocking and recovery operations, operates at 41 kPa in the main pressure mode and 26.7 kPa in the back-up mode, and is designed for mobility in either mode. It weighs no more than 8–10 kg, can be donned in just a few minutes, and can reach operational condition (pressurised) in 30 seconds.

In the event of decompression on the Soyuz, the suit is automatically isolated from the cabin environment and supplied directly with either pure oxygen or an oxygen-rich mixture from a supply in the cabin or from self-contained systems. This life support system also produces the required pressure levels in the suit, and removes body heat, moisture, and expired carbon dioxide. The suit has a soft helmet that can be opened and pushed back over the head when not in use, a removable communications skull-cap, disposable pressure-sealed gloves, large over-boots to protect the soles of the feet during the walk to the spacecraft (these boots are removed before entry into the spacecraft), and strengthened soles of the pressure layer when inside the spacecraft. It is also provided with several pockets for gloves, pens and other personal items.

The suit is donned by placing in the legs, standing up, ducking through the neck area, putting in the arms, and then zipping up the chest. It has watertight seals and inflatable devices to keep the cosmonaut afloat in the event of water egress, and the structural systems include sizing and adjustment elements to match the individual wearer.

The Sokol can also be used in the emergency transfer of cosmonauts from one spacecraft to another, with the suit being fed by small hoses connected to the spacecraft's LSS or to a small portable back-pack. The transfer could, in theory, take place through docking units, through an airlock, or by EVA, although no such operation involving a Sokol suit has ever taken place. For a contingency EVA transfer, additional protective elements (layers) have to be added, and these are stored onboard the Soyuz. The suit can support a cosmonaut on the spacecraft's LSS until such time as a safe landing can be achieved.[3] During the Soyuz 12 mission, one of the cosmonauts' tasks was to partially decompress the spacecraft cabin while wearing the suits. During the second flight day, 'serious defects' in the LSS were reported, followed by a unidentified failure in the attitude control system of the Soyuz. The consequences of these 'failures' are unclear, but they apparently did not affect the completion of the two-day mission, and the spacecraft landed without further incident. Although perhaps not a total success, the flight boosted confidence in flight planning and the changes introduced after the Soyuz 11 mission. It was also the first Soviet flight, since Soyuz 9 in 1970, that had achieved all of its pre-flight objectives. It was now time to put the Soyuz ferry into operational use.

SALYUT 3 FERRY MISSIONS, 1974

On 25 June 1974, the second military Almaz station was finally launched as Salyut 3. The docking location differed from the civilian DOS stations, with the Soyuz docking at an aft port. The design of Almaz was conceived well before that of DOS, and included the location of the main propulsion system and docking port on the aft of the vehicle. There was also a requirement for a data return capsule to be located at the front of the station, and this was ejected to Earth recovery at the end of the Almaz mission. As Almaz and DOS engine compartments were developed at different design bureaux, a reason for using the Soyuz engine compartment on the original DOS design was perhaps to allow for a much earlier launch.

On 27 May 1974, an unmanned test of the Almaz ferry version was flown as Cosmos 656, on a mission which lasted 2 days 2 hrs 7 min. This was no.61 – the vehicle that should have flown the first manned mission to Almaz 1/Salyut 2. No.62 – also originally intended for Salyut 2 – was subsequently flown as Soyuz 14, and no.63 – Soyuz 15 – was specifically built for Salyut 3.

Salyut 3 crewing

In early 1973, the following crews were named for missions to Salyut 2; but when that mission failed they were transferred to Salyut 3, which flew in 1974. Experienced cosmonauts Shonin and Khrunov acted as group commander and training supervisor respectively:

Crew 1	P.R. Popovich and Y.P Artyukhin
Crew 2	B.V. Volynov and V.M. Zholobov
Crew 3	G.V. Sarafanov and L.S. Demin
Crew 4	V.D. Zudov and V.I. Rozhdestvensky

The first mission to Almaz/Salyut 3 was Soyuz 14 (call-sign, Berkut – Golden Eagle), launched on 3 July and returning on 19 July. At 1,000 metres from the station, the automated Igla rendezvous system was used to bring the Soyuz 14 spacecraft to within 100 metres, at which point Popovich took over control and completed a manual docking. Commander Popovich and Flight Engineer Artyukhin lived onboard the Almaz for 14 days, and completed a successful recovery after 16 days in space. It was the first wholly successful Soviet space station mission, three years after the loss of the Soyuz 11 cosmonauts. It was not a long mission, but it was a welcome success for the Soviets. The Americans had launched Skylab in May 1973, and despite early in-flight problems with the station, the trio of three-man crews established records of 28, 59 and 84 days, and collected volumes of scientific and research data and information on extended-duration spaceflights. It is ironic that, after establishing this lead in space station operations, and given the difficulties of the Soviet programme, the Americans would not again live on a space station until the Shuttle/Mir programme of 1995–97, which was a precursor to the International Space Station. Beginning with Soyuz 14, the Soviets took the lead in gaining long-duration space experience.

The Soyuz 15 docking failure

On 26 August 1974 a second crew, onboard Soyuz 15 (call-sign, Dunay – Danube), was launched towards the Salyut 3 station. The launch, ascent and first few orbits proceeded according to the flight plan, but it seems that the crew had difficulty on their first night in space, in either not completing their sleep period fully or in not having slept at all. Due to a shortage of sleep caused by the early start to their second day in space (the rendezvous day), the day was very long and arduous. Several firings of the Soyuz engine were required to bring them towards the Salyut on their sixteenth orbit, but as the engine fired, the crew noted that it was performing exactly opposite to what was intended: when Soyuz should have been accelerating, it slowed down, and *vice versa*. The consequence of this was that Soyuz 15 flew past Salyut 3 at a distance of only 7 metres; and during the second docking attempt, the spacecraft passed the station at 30–50 metres. After checking all the spacecraft's systems and instrument settings, including tests with directly input commands, the crew requested permission to attempt a manual docking approach – but ground control was reluctant to authorise such and attempt. As Soyuz 15 passed out of range, comments from Sarafanov indicated that he was determined to make the Soyuz dock with Salyut (although it is not clear whether he attempted manual docking). Then, as Soyuz 15 again returned to within communications range, they were ordered to stop any further attempts at docking and to prepare for a night landing – due again to the limitations of the chemical batteries.

As the spacecraft manoeuvred to the required attitude for the de-orbit burn, a spike in the electrical current was recorded, prompting the crew to calmly suggest that all they needed now was an onboard fire – and that maybe an ambulance should be sent to the landing site – just in case! However, they were successfully recovered without any further incidents, despite thunder and lightning over the landing site.

The Soviets were quick to mask the disappointment of the mission, and press releases indicated that Soyuz 15 was really a manned test of fully automated rendezvous and docking equipment planned for future robotic supply ships, and that manual docking was never included in the flight plan. Indeed, had they docked, then the crew would not have transferred to Salyut but simply undocked and returned to Earth. Other reports stated that one of the objectives of the flight was to test the psychological compatibility of space crew-members with vastly different ages

(Commander Sarafanov was 32, while Flight Engineer Dyomin was 48 – and both were on their first spaceflight). On 3 September an official government commission was established to determine the cause of the failure, and, despite excellent performances, the crew were officially reprimanded for 'cutting off the flight programme'. It also emerged that it was probably the Igla docking system that was at fault; but NPO Energiya engineers refused to believe that assessment, and so the cosmonauts took the blame. This, moreover, affected their future careers, as neither ever again flew in space.[4]

Twenty-five years were to pass before (in 1999) new information was revealed by the NPO Deputy Chief Designer, Boris Chertok, who was a member of the investigation board. Chertok noted that the Igla system had failed, and was also initiating false commands. Therefore, when Soyuz 15 was 350 metres from Salyut, Igla 'thought' it was 20 km away, and turned on the engines as it would on a long-range approach. Consequently Soyuz, when passing Salyut at a distance of 7 metres, was travelling at a relative velocity of 72 kph. Had the vehicle struck Salyut it would certainly have killed the crew; but this did not happen, because at 20 km the approach pattern had induced a small amount of lateral drift which misaligned the two spacecraft. After the two failed automated approaches, the crew were ordered to shut down Igla and to return to Earth, as there was insufficient fuel on board to allow Sarafanov to manually approach and dock with Salyut for a third attempt. According to official Energiya publications, the crew were held responsible for not shutting off Igla after it first malfunctioned, and for not assessing the clearly dangerous situation in which they had placed themselves by flying so close to Salyut. Despite evidence to suggest that Igla caused the situation, the crew-members' lack of spaceflight experience did not help their case, and they are still tarnished with the 'official reprimand'.

Salyut 3 could probably have hosted another crew, but there were simply no more Soyuz vehicles available to fly to the station, and on 23 September the 'original programme' of the station was completed with the return of the small recoverable capsule containing exposed film from the mission. On 24 January 1975 the unmanned Almaz was de-orbited on command over the Pacific Ocean and destroyed on re-entry. Shortly earlier, on 26 December 1974, Salyut 4 had been placed in orbit. This was the fourth 'civilian' space station DOS, and would be visited by crews during 1975.

SALYUT 4 FERRY MISSIONS, 1975

The new DOS space station, Salyut 4, was similar to Salyut 1, but with a number of improvements to allow it to support crews for at least six months in total. The station featured solar arrays (60 square metres) that could rotate to track the Sun and generate 4kW of power, but it still used a Soyuz-type propulsion module and a single forward docking port. The majority of the research programme and experiments were scientific rather than military in nature, and covered the fields of solar physics, astronomy, space sciences, Earth observations, technology development, life sciences, and medicine.

Salyut 4 crewing

The new training group for the Salyut 4 mission was formed in early 1974, with the following crews:

Crew 1	A.A. Gubarev and G.M. Grechko
Crew 2	V.G. Lazarev and O.G. Makarov
Crew 3	P.I. Klimuk and V.I. Sevastyanov
Crew 4	V.V. Kovolyonok and Y.A. Ponomaryov

Crew 1 flew Soyuz 17, and set a duration record of 28 days. They were backed up by crew 2, who lost their mission when their Soyuz suffered a mission abort. Crew 3 then flew a 64-day mission on Soyuz 18.

New steps toward permanent occupation

The first mission to Salyut 4 was Soyuz 17 (call-sign, Zenit – Zenith), launched on 11 January 1975. The launch was nominal, and on the second day of flight the Igla system was activated at a distance of 4 km from the Salyut. At 100 metres, Commander Alexei Gubarev – assisted by Flight Engineer Georgi Grechko – took over manual control, and achieved a perfect docking about 28 hours after launch. When the crew opened the transfer hatch into their new home, they found a note left by processing technicians at Baikonur: 'Wipe your feet'. Two days after docking, the crew installed a ventilation hose through the open hatches from the Salyut into the Soyuz, to use the station's environmental control system to ventilate their spacecraft while they remained docked to the station. For the next four weeks they worked onboard the station while Soyuz remained attached in a dormant mode, ready to act as a crew rescue vehicle should the need arise. The crew began packing the results of their expedition into Soyuz two days before undocking on 9 February. Soyuz 17 landed 110 km north-east of Tselinograd, with a cloud ceiling of 250 metres, visibility of 500 metres, wind gusts of 72 kmh, and in a snowstorm. The crew had surpassed Soyuz 11's record by flying a 30-day mission, and although there were two Skylab missions in front of them, in terms of the endurance, they had achieved a notable success for a Soviet space station – a landmark in a long series of missions that would gradually increase the duration of crew residencies from 30 days up to 14 months over the ensuing twenty years.

The 'April 5 anomaly'

The next crew was supposed to spend 60 days onboard the station, to continue the work begun by the Zenit crew. The pair of cosmonauts (Commander Vasili Lazarev, and Flight Engineer Oleg Makarov) was the same team that had flown Soyuz 12 to test the new ferry in September 1973. Their mission was planned to be Soyuz 18, and their call-sign was Ural (Urals). The mission was set for launch on 5 April and a return in early June, and on 15 July would be followed by the launch of the Soyuz 19 spacecraft, the Soviet part of the ASTP mission. The Soviets had introduced a new variant of the R-7 booster for ASTP, but for the Salyut mission, the older model of the R-7 would be used.

It was a bright and hot spring day at the cosmodrome, with a recorded temperature of 25° C and blinding sunlight. A little after 14:03 MT, the R-7 ignited and lifted of the pad to begin the planned 60-day mission to Salyut 4. As the rocket ascended, the crew was informed by mission control of the changing parameters of their trajectory, the pitch and yaw, and the pressure in the propellant tanks. At 120 seconds, the four strap-on Bloks of the R-7 were depleted and explosively separated from the 'stack', followed 30 seconds later by the jettisoning of the launch escape system and shroud. Suddenly, brilliant sunlight filled the DM, and at 180 seconds, ground control confirmed that everything was normal. With the escape tower gone, if an abort occurred the crew would have to make a ballistic re-entry to Earth. Onboard Soyuz, the crew relaxed as the first events of ascent proceeded like clockwork. After all, every cosmonaut had rode into space onboard a version of the R-7, which was a proven and reliable vehicle with many dozens of manned and unmanned launches to its credit. Both cosmonauts had previous launch experience, on Soyuz 12, and so this time there seemed no reason to suspect that anything would be different.

At T + 288 seconds, the crew expected the separation of the central core and the ignition of the upper stage to push them on and into orbit. Inside Soyuz, Lazarev felt the vehicle pitch and roll, and reported a heavier pitch than on Soyuz 12. At that moment the Sun suddenly disappeared from sight, and a loud siren sounded. On the instrument panel, the red 'Booster Failure' light flickered on. For a few moments the crew wondered what was happening as the sound of the booster engine stopped, and for second or two they became weightless as the forward velocity faltered.

What should have occurred was the firing of two sets of pyrotechnic charges to separate the spent central core. The two stages are connected by a lattice structure that is attached by six latches at the upper and lower points. This structure separates seconds after the central core, and splits into three elements that drop away from the ascending rocket, thus allowing the upper stage to fire. after which its exhaust gases vent through the lattice structure as the core stage drops way. On this mission, however, this did not happen. Shortly after this incident, the Soviets released details of the failure as part of the ASTP agreement with NASA. An excessive degree of vibration caused the relay in half of the upper sequencer to close down and to signal three of the six latches to fire prematurely, with the lower core and upper stages still firmly attached. This was activated only seconds prior to the planned separation, but with the latches armed. The connection that triggered was in the same location as the electrical link between the upper and lower segments of the structure. Therefore, when the electrical contacts were severed due to the premature explosions, so also were all links to the lower latches, causing an uneven linkage between the core and upper stage as the vehicle continued to climb.

When the upper stage of the booster ignited, the core stage was still attached, as the rest of the latches had failed to separate. The R-7 continued to climb, but it was dragging with it an empty and heavy central core. After just four seconds of the upper-stage burn, the dramatic increase in the atmospheric drag from an unstable booster caused a violation of the 10-degree safety limit in the flight path. The onboard gyros detected this, and automatically activated the abort programme.

Inside Soyuz, the cosmonauts realised that they were faced with a major failure.

Lazarev flipped off the siren, as it was distracting him, but everything was happing incredibly quickly as they scanned the instruments to determine the cause of the problem. Lazarev later wrote: 'The state of agitation indicates a lack of self confidence', I kept repeating, with this one thought dominating my mind; 'What is happening? What is coming next?' For a moment the question remained unanswered. There was the suppressive feeling of uncertainly or anxiety... or maybe fear.'[5]

Despite their years of training, hundreds of hours in the simulator, and a short spaceflight in their logbooks, the cosmonauts were still faced with the fear of the unknown. To save themselves they had to rely on automatic devices to work in timed sequence, and it was necessary that they avoid any mistakes which might worsen the situation. All those hours of training time and 'off-nominal' emergency simulations was beginning to pay off as they calmly went through their procedures. Then came another brisk jolt, as explosive charges separated the Soyuz from the booster and parted their Descent Module from the Orbital and Propulsion Modules. At an altitude of 90 miles, the abort signal was received by the spacecraft computer, and the separation from the upper stage was triggered by the firing of the propulsion system in the PM to take it to a safe distance before separating the three modules of the Soyuz.

Minutes after launch, two surprised cosmonauts began their descent to Earth, and Lazarev later recalled: 'We began to experience a creeping and unpleasant pull of gravity... It increased rapidly, and its rate was much greater than I had expected... Some invisible force pressed me into my seat and filled my eyelids with lead,,, Breathing was becoming increasingly more difficult. The loads pressing hard on us deprived us of the ability to speak, [and] 'ate up' all the sounds, leaving only guttural wheezing and puffing. We resisted the loads as best we could.'

The cosmonauts were held in their couches, which were designed to help absorb 3–4 g during a normal Soyuz entry. On this flight they had to endure 14–15 g, with a peak at 21.3 g. Lazarev later stated that when the second peak of 6 g was passed, the crew failed to notice it. The uncertainty of what had happened remained, but what was certain was that the cosmonauts were coming back down to Earth.

Lazarev radioed a request for information about where they were heading, but the airwaves remained silent. Makarov then checked his charts to determine where they might be heading. Normally, landing would take place on the steppes of Kazakhstan, but this time they were heading further east. They were particularly worried that they might head into mainland China, which at that time was not on good terms with the Soviet Union. The flight path crossed to the north-west of the town of Sinkiang, which had already been in the news, as two Soviet helicopters had landed there in error and their pilots had been captured by a Chinese patrol. The crew heard the firing of the parachute container hatch, and felt the jolts of the drogue and main chutes deploying. Lazarev was quietly expressing his confidence in the system and its operation, when Makarov apparently asked him what he was muttering about.

The cosmonauts landed in a rugged region of western Siberia, south-west of the town of Gorno-Altaisk, more than 1,600 miles from the cosmodrome and a safe 500 or so miles north of the Chinese border, well inside Soviet territory. The craft hit the side of a snow-covered mountain and began to roll down its side; and as it approached a sheer drop, the parachute lines snagged on scrub trees and stopped it. Lazarev later recalled the landing: 'Oleg calculated the landing place almost precisely. We landed a bit to the side of the place he had indicated. It was painfully disappointing and somewhat unpleasant. We had prepared ourselves for the mission for so long only to wind up like this... The very fact of failure was quite discouraging. [There was] a light jolt, and all of a sudden the spacecraft began to slowly turn around as though we had landed on water... The porthole, black with soot, suddenly became transparent [due to the friction of the snow as they slid], and I saw a trunk of a tree. Yes, this is the Earth.' Lazarev jettisoned the parachute lines, and the capsule came to rest. The two cosmonauts were bruised, shaken and dizzy, and instead of a flight of 60 days or 1,500 hours, they had landed – after travelling to an altitude of 192 km, and 1,574 km downrange – just 21 min 27 sec after launch.

In his account of the accident,[6] Makarov records that their first thought was that they could land in one of three places: Japan, China, or the mountains. However, they soon realised that it would not be Japan, because they were in daylight, and Japan was already in darkness at that time. Klimuk, who was CapCom, radioed the expected landing site to the crew 20–30 seconds after the failure, but for some reason the crew did not hear ground control after the failure (even though the ground could hear them). They were weightless for about 400 seconds after the emergency separation. Makarov further says that under the g-forces which they experienced they could easily have first lost vision and then consciousness; and although this did not happen, they experienced black-and-white vision and then tunnel vision. Post-flight medical tests showed they had not suffered any injuries as a result of the high g-forces. (It was not unknown for trainees to suffer haemorrhages in the back after being exposed to only 8 g during centrifuge tests.)

According to Makarov they had three options concerning the parachute immediately after landing: leave both parachute lines attached to the capsule (but then it is 'controlled' by the wind); jettison it completely (but then the capsule would roll down the slope); or leave one of the lines attached so that half the parachute would be released and could act as an anchor. They opted for the last mode – which, under the circumstances, happened to be the best choice.

The crew released then their harnesses and cracked the hatch, to be greeted by a cold breeze. The conditions at the landing site – an overcast sky with low cloud, a biting wind, and a great deal of snow – were very much different from those of less than 30 minutes earlier. Makarov states that the temperature was –7° C, compared with +25° C at the cosmodrome! The snow was about 1.5 metres deep and they could not walk in it, and they therefore had to crawl in it to enable them to make a fire; but, Makarov notes, this was nothing new to Lazarev, as he was Siberian. Indeed, when Lazarev emerged from the spacecraft he found himself chest deep in powdery snowdrifts. It was also becoming dark, and so they gathered their survival gear and awaited the rescue teams. Because of intense activity throughout the twenty minutes of flight, they were hot and perspiring – but they soon began to shiver. They therefore donned their warm survival gear, together with the Forel (Trout) emergency gear (used in water) for added warmth. As they were organising themselves outside the spacecraft, they saw that the hot capsule had begun to melt the snow and was slipping towards the 500–600-foot drop. Luckily, it held firm.

Fearing they might have landed in China, Lazarev also decided to burn some documents related to a military experiment he was supposed to perform in orbit – an experiment so secret that even *he* did not know its purpose. (It apparently involved testing the usefulness of the naked eye in reconnaissance from orbit.) Soon they heard the voices of the rescue teams on the radio, and after asking for their position they were relieved to learn that they were on Soviet territory.

The capsule landed not far from the small town of Aleisk – at an altitude of 1,200 metres, on the slopes of a mountain called Teremok-3 – and the rescue teams spotted them about half an hour later. The first plan was to drop parachutists (including doctors) from one or more of the aircraft circling overhead; but Lazarev, an experienced parachutist, radioed that this would be too dangerous, and insisted that they would remain on the mountain for the night. Overnight, plans were devised to lower a ladder from a helicopter, but when this was attempted the next morning it proved very difficult to keep the helicopter stable at this altitude, and the plan was therefore retained only as a back-up option. The next plan was to land a helicopter, with a group of rescuers, on the frozen river Uba near the foot of the mountain, and to then climb to the stranded Soyuz. Unfortunately, the team – which did not include any experienced mountain climbers – soon found itself buried under a self-induced avalanche, and had to be rescued by another group of rescuers. Meanwhile, an MI-4 helicopter, with geologists (from an unknown location) on board, was hovering over the capsule, but with the Air Force not permitting the rescue of the cosmonauts with a ladder, a forest guide had to be lowered to join the two cosmonauts. A little later, another helicopter, with a tail number 74, arrived – seemingly from nowhere. It had apparently been despatched from a Siberian military district (although it is not clear who ordered it to join the rescue), and was piloted by a young man named Sultan-Galiyev, who, after the Afghan war, was to be honoured as a Hero of the Soviet Union. It was this helicopter that eventually managed to winch the cosmonauts (and the forest guide) on board, and the DM was recovered some time later.

Both men were reported to be in good health after their ordeal. Russian officials refrained from breaking the news of the abort until 7 April, after the men had been rescued, and then informed the Americans of the incident. Makarov has said that after the flight, some Americans believed that they had died in the accident – which is why they were once ordered to play football with some Americans to convince them that they were still alive. The accident necessitated amendments to prevent a repeat, and so all subsequent Soyuz launchers were fitted with a series of changes to the circuitry of stage separation. If one relay were to fire prematurely, then the rest would be automatically triggered to ensure clean separation of the stages.

A replacement mission

The aborted mission was not assigned a flight number, and the Soviets simply designated it 'the April 5 anomaly'. It was decided that the back-up crew for the aborted mission should be assigned to fly the replacement mission, which was designated Soyuz 18 (call-sign, Kavkas – Caucasus). In the West, this became known as Soyuz 18-B, while the aborted flight was labelled 18-A. Launched on 24 May, the two-man crew rode a smooth R-7 flight into orbit and began the approach to Salyut over the next two days. As the Soyuz approached Salyut 4 automatically, the spacecraft entered Earth's shadow some 800 metres from the station. Commander Klimuk had to turn on the spotlight on the Soyuz, and after a few minutes spent in trying to find Salyut, they finally docked manually from 100 metres to become the first crew to reoccupy a space station – a major milestone in extending the operational lifetime of the station. The Kavkaz crew continued the scientific research programme begun by the Zenits, and, on 26 July, finally completed their mission after 63 days in space. They also tested a new orientation system that would enable subsequent Soyuz craft to perform precise night landings.

When Klimuk and Sevastyanov were launched they were expected to stay in orbit for 28 days, and it was only during their flight that it was decided to extend their mission by 35 days. This was primarily because Salyut 4 had not yet outlived its usefulness, but also because no further Soyuz vehicles were available to carry an additional crew to it, as all the vehicles intended for it – nos.38, 39 and 40 – had been used.

In America, the timing of the flight raised some questions concerning the Soviets' ability to simultaneously conduct two separate manned missions. However, the Salyut crew remained in orbit throughout the duration of the joint ASTP mission, and this success demonstrated a growing maturity in Soviet space operations after several years of frustration, disappointment, delays and tragedy. The American contribution to the ASTP was the last time for six years that astronauts would fly in space. During that time the Soviets would increase their station operations, beginning with the second Almaz station (Salyut 5) launched in 1976. Salyut 4 would receive one more spacecraft: the unmanned Soyuz 20, launched on 17 November and recovered on 16 February 1976. Soyuz 20 was primarily intended to evaluate whether the 7K-T could remain in a docked state for 90 days. The 7K-T had been modified to extend its docked lifetime from 60 to 90 days, and Soyuz 20 tested this. The flight was probably also applicable to Progress (including the slower 48-hour approach). Soyuz 20 (vehicle no.64) was originally intended to carry the first crew to Salyut 5 (it had a serial number in the '60 series' – as did all Almaz/Soyuz ferries – and the index 11F615A9), but when it was decided to 'borrow' it for the endurance test with Salyut 4, NPO Energiya compensated by assigning vehicle no.41 – a DOS ferry version intended as the first ferry to Salyut 6 – as the first spacecraft to fly to Salyut 5 (what became Soyuz 21). Salyut 4 continued to orbit until it was commanded to a destructive re-entry on 3 February 1977, two years and a month after its launch.

SALYUT 5 FERRY MISSIONS, 1976–77

The third Almaz vehicle was launched on 22 June 1976, and received the designation Salyut 5 once it was safely in orbit. There had been rumours in the West that Salyut 5 would be equipped with dual docking ports, but with a single port in the rear and a return capsule at the front, it resembled Salyut 3. It was to be the final military-orientated station manned by cosmonauts (although this was not known at that time).

Salyut 5 crewing

After the missions to Salyut 3, the Soviets announced four more crews to train for Salyut 5, due in 1976:

Crew 1	B.V. Volynov and V.M. Zholobov
Crew 2	V.D. Zudov and V.I. Rozhdestvensky
Crew 3	V.V. Gorbatko and Y.N. Glazkov
Crew 4	A.N. Berezovoi and M.I. Lisun

After the first three missions it was intended to fly a fourth mission – Soyuz 25 – with the Berezovoi crew, but no Soyuz was available before the station decayed, and so the mission did not take place. No further Soyuz missions were planned to fly to a military Salyut. Almaz training continued, but those cosmonauts were to fly on the TKS craft (see p.240).

An acrid odour

Two weeks after the launch of the station, the first crew was launched onboard Soyuz 21 on 6 July. Ascent was nominal, but upon reaching orbit one of the Igla antennae was not deployed, although it operated as planned for the first few orbits. Despite modifications to the system since Soyuz 15, there were clearly still a few problems. Visual contact was achieved at 350 metres, but as Soyuz approached to 270 metres, the velocity between the spacecraft and the station suddenly increased to more than 2 metres per second. This time the crew immediately asked to switch to manual control, but ground controllers told them to wait, as the approach was proceeding normally. As soon as the braking indicator light went out, the crew reported that there was no longer an increase in velocity, but that they were rapidly losing the line of sight to the station. At 70 metres, Volynov assumed manual control before they passed by the station in a repeat of Soyuz 15's approach. His skills were demonstrated in a successful docking 25 hours after leaving Earth. Five hours later they were inside the station, to begin what was expected to be a 54–66-day residency.

The crew conducted a range of military-orientated experiments and observations, as well as several scientific and biological studies. However, on 17 August, as they were working, the station alarm suddenly sounded, all the lights went out, and several systems failed. Passing over the night side of the Earth, they shut off the alarm to find a station void of sound. It then passed out of communication range, leaving the crew alone to discover what had happened. There was no indication of a loss of pressure (as during the Soyuz 11 accident), but they found that the life support system had stopped working and that attitude control had been lost, leaving the station drifting in orbit. It took a difficult two hours to restart the systems, but there were still problems. After this emergency, the health of Flight Engineer Vitaly Zholobov – who was experiencing his first flight into space – began to deteriorate, and became worse as each day passed. The medication, it seemed, was of no use. Press reports suggested that he was suffering from sensory deprivation due to prolonged isolation. On 23 August – with no indication of improvement in Zholobov's health – the decision was made to return the crew at the earliest opportunity; that is, during the following 24 hours. Volynov carried Zholobov into the capsule and strapped him in.

The undocking from Almaz/Salyut 5, on 24 August, almost proved to be another setback in the Soviet programme. As Volynov tried to undock his spacecraft, the latch failed to release properly, and as he fired the jets to move away, the docking gear jammed, resulting in the spacecraft being undocked but still linked to the station. As the spacecraft moved out of range of communications, only the first set of emergency procedures was received onboard the Soyuz. Volynov tried to disconnect a second time, but only managed to loosen the latches slightly. He looked across to his ill colleague, who seemed horrified at the second unsuccessful attempt to detach from the station.[7] For a complete orbit (90 minutes) the two spacecraft could not undock, but the latches finally let go of Soyuz when the second set of commands was received.

Because Soyuz 21 returned early it was outside the normal recovery window, and as it descended it encountered strong gusty winds, which caused an uneven firing of the soft landing rockets. Around midnight local time, it landed hard (and bounced several times) in the field of a Karl Marx collective farm, 200 km south-west of Kokchetav, in Kazakhstan. Inside the capsule, both cosmonauts were hanging by their restraint harnesses; and as the early return had interrupted their exercise routine, they were both quite weak. Having survived a difficult landing on his first mission (Soyuz 5), Volynov again found himself in an awkward situation. With some effort he opened the hatch and crawled out, but was unable to stand without help, while Zholobov remained in the capsule, unable to follow his commander as his Sokol helmet had caught on an obstruction in the capsule. Volynov then crawled to his Flight Engineer and helped him out. Both men were too weak to move far from the capsule and expected to spend the night there; but just forty minutes later, the recovery team arrived.

Zholobov's illness – and, to a lesser extent, Volynov's undermined condition – was apparently caused by nitric acid fumes which had probably leaked from the station's propellant tanks. Other reports have indicated that the crew broke their physical training exercise programme, suffered from a lack of sleep, and were not sufficiently supported from the ground. Whatever the reasons, and despite returning to fitness, neither of the cosmonauts flew in space again.

Fluctuations in the flight plan

The next crew destined for Salyut 5 consisted of rookie cosmonauts Commander Vyacheslav Zudov and Flight Engineer Valeri Rozhdestvensky (call-sign, Rodon) on Soyuz 23 (Soyuz 22 was a solo Soyuz mission; see p.217). Alhough the official Soyuz 21 post-flight investigation had determined that there was still air in the Salyut, there was concern about the state of the internal atmosphere. The crew of Soyuz 23 were given gas masks for initial entry into the station, and specialised equipment to take samples and analyse 'toxic' atmosphere to determine whether it was safe for further occupation.

On 14 October 1976, the mission began. First, the transfer bus broke down on the way to the launch pad. Then, during the ascent, the R-7 deviated from its planned flight profile to such an extent that it verged on signalling a launch abort, which resulted in a orbital insertion lower than planned. This was due to heavy winds at the launch site, which caused the rocket to slightly veer off course during lift-off. By the sixteenth orbit, the Soyuz had begun its final approach to Salyut 5. At 7 km the crew placed Soyuz in its automatic rendezvous mode, but at 4.5 km from the station they reported 'strong lateral fluctuations' in the Soyuz. At 1,600 metres the oscillations had increased to the point where Soyuz began turning away from the Salyut, although onboard instrumentation did not indicate that this was happening. At 500 metres the spacecraft was still turning away from the Salyut and was travelling too fast for a safe docking, and so the crew turned off the approach programme.

The crew were not happy with the situation, as they could clearly see Salyut close by. While trying to determine the cause of the failure, they requested a second attempt at docking, as the Soyuz had now ceased its rotational motion. Ground control, however, was aware of the low levels of attitude control propellant onboard as a result of the reduced orbital insertion, and knew that there was only enough fuel for one docking and one landing attempt. Mission rules dictate that sufficient propellant should be held in reserve for two de-orbit burns, and although the cosmonauts were confident of a manual docking, they were told that a second approach would not be possible. They were also told to remove their Sokol suits, and to rest. With the Soyuz having passed the landing opportunity for that day, the crew was told to shut down all non-essential equipment, including the radio, in order to conserve power. As their Soyuz did not carry solar arrays, the available power reserves were severely limited, and they were further disadvantaged by having to wait for the next day's landing window. For the first time on a Soviet manned mission, the news announcements revealed that a docking had been cancelled due to malfunctioning equipment.

A Soyuz splash-down

At Zudov's home, his family gathered with colleagues to await the news of the safe return of the crew. The night landing was targeted for the normal Soyuz landing area, but Shatalov told the crew to remain in their seats after landing, as the conditions in the recovery area – high winds and blizzards – were far from ideal. Due to the DM's limited battery supply, there was very little that could be done to change the landing site; but Soyuz could land in a variety of terrains and had already demonstrated its versatility on several missions – most recently the Soyuz 18-A launch abort in 1975 – and so there were high hopes of a successful landing.

More than twenty-five years have passed since the dramatic flight of Soyuz 23, and accounts of the recovery events still conflict. It was not until 1984 that a journalistic description of the splash-down was published, in *Literaturnaya Gazeta*;[8] but a more detailed description of the rescue effort has recently been published in a book by Iosif Davydov,[9] who for a long time was in charge of survival training and who actually took part in the rescue. This account is probably more reliable.

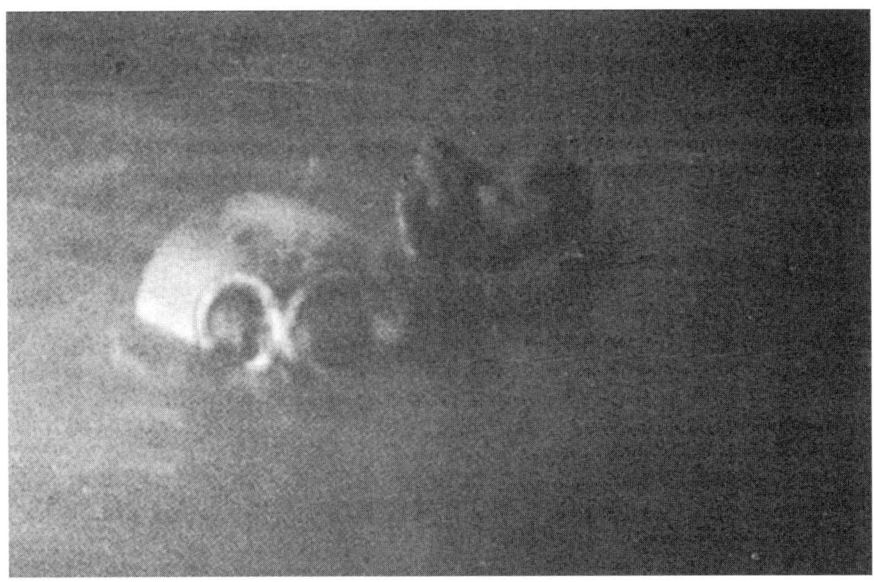

During the recovery of Soyuz 23 on 17 October 1976, the Descent Module floats in the icy waters of Lake Tengiz. (Courtesy British Interplanetary Society and D Haeseler.)

The de-orbit burn, capsule separation and re-entry took place normally, but the high winds contributed to a 121-km overshoot from the planned landing near Arkalyk, and a descent in fog at a temperature of –22° C. Inside the capsule, the crew braced themselves for the landing as they swung beneath the parachute. At 164 metres, Zudov told Rozhdestvensky to brace himself for the retro-rocket firing and the hard thud as they hit the ground; but to their surprise, the capsule splashed down in the sludgy and salty ice of the 32-km wide Lake Tengiz, about 140 km west of Arkalyk, in Kazakhstan. The splash-down took place about 8 km from the northern shore, after which the main parachute dragged the capsule through the water for some distance. Lying on its side, the capsule's forward exit hatch could not be opened without allowing the freezing water to flood in, and so fresh air was not available. The crew had only two hours of oxygen remaining, and had to rely on a pressure equalisation vent that was fortunately still above the water line, but which could only support them for another five hours or so.

Despite being hot from entry, the capsule rapidly cooled in the freezing waters of the lake. The two cosmonauts quickly removed their pressure garments and put on their flight suits. Exhausted from their efforts in the small capsule, they then decided to eat some of their limited rations while awaiting what they expected would be a prompt rescue. Unfortunately, the fog obscured the descent capsule's light beacon from the rescue helicopters. After fifteen minutes, salt corrosion activated the pyrotechnic charges of the reserve parachute container. The parachute then deployed and filled with icy water, which dragged it to the bottom of the shallow lake. Fortunately, the lake was not very deep, as the crew would have drowned if the capsule had slipped below the water. Zudov was still strapped to his couch, suspended above Rozhdestvensky as the capsule turned 'upside down' due to the submerged parachute acting as a sea anchor.

In order to conserve oxygen, the cosmonauts stopped talking and moving, but this only hindered attempts to find the capsule, as the rescue teams had difficulty in establishing communications from the crew and had no search signal to focus upon. The capsule was found by chance by one helicopter, piloted by Lieut-Colonel Bogatyrev, and carrying the leader of the rescue operation (Silayev), a KGB representative (Chekin), and Davydov. Flying in winds of 20 metres per second, Bogatyrev skilfully descended to 4–5 metres above the lake and used his powerful searchlight to pinpoint the capsule. The article published in 1984 states there were no signs of life from the crew but Davydov writes that his helicopter was in contact with the cosmonauts, and ordered them to don their water survival suits. Inside Soyuz 23, ice had begun to form on the inner walls, and the internal temperature had begun to drop rapidly. Neither were exterior conditions improving, with snow falling heavily and the fog thickening. Unfortunately, the rescue team had not brought rubber boats. Davydov was ready to be winched down, but because of the snow-storm the pilot could not stabilise the helicopter. A further attempt to hover above the capsule also failed, and they were forced to fly back to shore because they were low on fuel.

Other helicopters encountered the same difficulties, but others managed to deliver amphibious vehicles to the shore, although they were unable to negotiate the numerous bogs and marshes surrounding the lake; and even the deployment of life rafts was hampered by numerous blocks of ice and sludge on the surface. With all available land, water and air routes blocked by the weather and conditions, it was decided to wait for the first light of dawn before attempting to rescue the crew. Although Soyuz was 'sea-worthy', there was about a 40-minute limit to the onboard power supplies, and so the cosmonauts shut down all internal power except for a small light. Onboard rations were sufficient, although cold. When the reserve parachute was accidentally activated, it acted as an anchor and caused the capsule to lean over on its side, as a result of which the pressure equalisation valves entered the water. Two hours later, the cosmonauts began to reveal signs of oxygen starvation. Over the radio they could be heard breathing heavily, after which their voices became hoarse and were hardly recognisable. Zudov reported they had donned their water survival suits and were ready to leave the capsule; but everyone outside knew that if the hatch were to be opened, water would enter the capsule and the cosmonauts would drown. As morning approached, Rozhdestvenskiy reported that Zudov had lost consciousness due to a lack of oxygen. Somewhat later, contact with the crew was lost as the frozen antenna also entered the water. One of the helicopters was, however, carrying rubber boats and water survival gear, and overnight some of the rescuers tried to reach the capsule in boats from the shore. Most of them became stuck, but one of them – Mi-6 helicopter pilot Nikolai Chernavskiy – reached the capsule, although there was little he could do to help the cosmonauts.

In Yevpatoria, the controllers and officials at flight control had received three communications reporting the progress (or rather, lack of progress) at the rescue site. The initial report indicated that the spacecraft had indeed landed in the lake, and that all-terrain vehicles had been dispatched to affect a rescue within the hour. Then a second report recorded adverse weather conditions and the failure of the helicopters to lift the capsule from the water. During the early hours of 17 October, a third report stated that the all-terrain vehicles had also been unable to negotiate the path to the capsule, and that all attempts had been abandoned until dawn. In the control centre, there was considerable concern that the crew could freeze to death before anyone could reach them.

The recovery of a Soyuz refrigerator

The next day, plans for the recovery included preparing a heavy lift helicopter (Mi-8) capable of lifting 20 tonnes, and a lighter Mi-6 helicopter to carry rescue frogmen. Inside the capsule, the cosmonauts were still wearing their emergency water survival suits in order to keep warm. They also turned on the exterior lights to assist the helicopters in locating the capsule in the improved but still severe conditions at dawn.

As morning broke, the temperature was –22° C, and the weather had cleared. However, due to a lack of fuel, all the helicopters of the Arkalyk rescue force were stranded near the shore of the lake. It was at this point that Nikolai Kondratyev (the commander of the Karaganda rescue force) arrived with his helicopter. He then picked up Davydov, a co-pilot (Nefedov), a diver and a doctor, and flew to the capsule. The diver was winched down with a rubber boat, but reported that it was impossible to open the hatch, although he was able to communicate with Zudov and Rozhdestvenskiy by tapping on the wall of the capsule. It seemed that the only way to rescue the cosmonauts was to tow the capsule to the shore. However, this was forbidden, as at that time the technique had not been perfected, although it had been successfully carried out with a crew onboard a mock-up Soyuz near Feodosiya. But as this was the only way of rescuing the crew, it was decided not to waste time by

The Soyuz 23 Descent Module, with two cold cosmonauts on board, is dragged towards the shore of the partly frozen Lake Tengiz. The nearly-collapsed parachute is still attached by lines. (Courtesy British Interplanetary Society and D. Haeseler.)

asking permission over the radio, and the attempt proceeded. It was a difficult operation, as the helicopter was nearly dragged down by the reserve parachute when it emerged from the water. The towing caused severe buffeting during a demanding 45-minute trip for both the helicopter crew and for the cosmonauts in the capsule. The cosmonauts, moreover, almost suffocated in the attempt to rescue them, as 5 km from the shore the capsule nearly sank. Eventually, however, almost eleven hours after the landing the capsule was finally hauled onto dry land. The rescuers had all but given up hope that the crew were alive, and were somewhat surprised when Rozhdestvensky opened the hatch and revealed that both of them had survived their ordeal.

News of the safe recovery was issued at 07:00 hours MT on 17 October. The crew underwent a series of medical checks and took a few days rest before flying home, and when they returned to TsPK on 26 October they were received by leading officials. Shatalov praised the courage of the crew and that of helicopter pilot Kondratyev, who was awarded the Order of the Red Star for his piloting skills.

An investigation into the docking failure focused on the large oscillations in the signals of the Igla rendezvous system. A reconstruction of events indicated that the crew may have acted reasonably, but that their actions could have violated earlier instructions. Former cosmonaut Alexei Yeliseyev believed that the crew were to blame for not attempting a second docking approach during the thirty-third orbit, for which they had trained. It was determined that when the spacecraft had acquired the Salyut, the lateral movement light in the cabin came on, indicating that all such motion had ceased. However, as the engines used to stop lateral motion were not turned on, the spacecraft continued to swing around. The crew reported feeling the motion, but their instruments actually told them they were stationary. The cosmonauts pondered on the situation for another two minutes, and realised that if they abandoned the approach, then the mission would have to be aborted due to low propellant levels as a result of excessive use in the early phase of the approach. The crew waited until the last moment to abandon the docking, ignoring the four points during the approach when they could have terminated the operation. Fortunately for the crew, the indicator lights on the control panel had recorded a normal approach. Clearly, it was again the Igla system that was at fault.

During a subsequent meeting with Afanasyev, Armen S. Mnatsakanyan (the designer of the Igla system) could not explain why ground controllers had not detected the difficulty, and stated that instead of trying to use Igla with its inherent problems, work should progress towards the introduction of the Kurs docking system already under development. It was concluded that the problem on Soyuz 23 was caused by excessive vibration of the boom on which the gyro-stabilisation antennae were located. The official report, dated 2 December 1976, severely reprimanded Mnatsakanyan, and stated that if sufficient measures were not implemented to prevent such an occurrence on future missions, he would be removed from his post. On 10 December, Mnatsakanyan refused the opportunity to resign, and on 6 January he was fired from his position as Director and Chief Designer of NIITP.

The crew was praised for acting 'in a confident way, efficiently discharging their

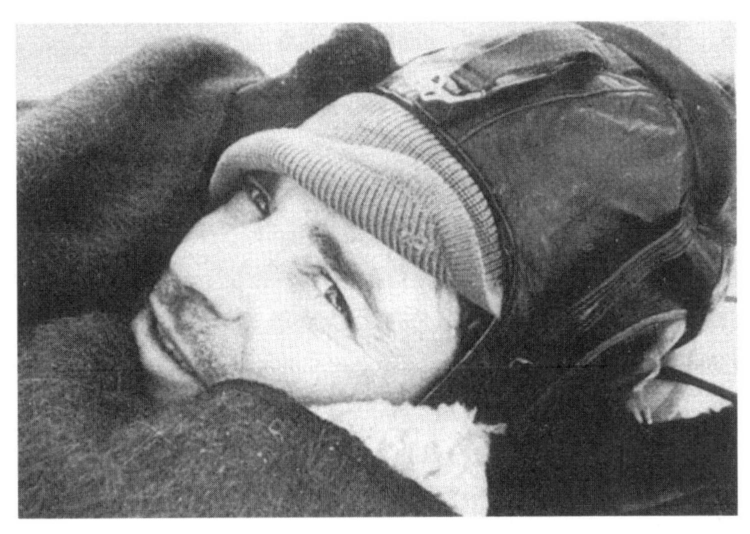

Cosmonaut Zudov – cold, tired, and very relieved after recovery from the capsule of Soyuz 23. (Courtesy British Interplanetary Society and D. Haeseler.)

duties.' Neither cosmonaut appeared to suffer any long-term effects from their ordeal, nor did they ever fly a second mission.

The last military Soyuz crew

Meanwhile, Salyut 5 carried on orbiting the Earth and continued an automatic programme of scientific studies while another crew was prepared to visit the station. The crew of Viktor Gorbatko and Yuri Glazkov were to spend 18 days on the station, to solve the problem unanswered by the failure of Soyuz 23 to dock with the station – whether the atmosphere had indeed been poisoned towards the end of the Soyuz 21 mission.

Launch took place on 7 February, and everything went according to the flight plan until Soyuz 24 was just 80 metres from the docking port of Salyut 5 on 8 February. At this point, Igla failed yet again, prompting Gorbatko to take over manual control and effect a successful docking. However, the crew remained inside their Soyuz for a six-hour sleep period, and it was not until 9 February that they floated into the station. They wore breathing apparatus and carefully took samples of the air as they went, and eventually declared the atmosphere to be clear, and without odours. After a final check of the data, they were told to remove their facemasks and work as normal for the duration of the mission.

The mission concluded on 25 February, with a landing in strong winds, snow, and temperatures of –17° C. Soyuz 24 did not land in the planned recovery zone, and it took some time for the rescue crews to reach the capsule. On landing, the capsule rolled a few times and ended up on its side, inflicting minor injuries on both cosmonauts. After waiting in an uncomfortable position for the rescue crews, Gorbatko and Glazkov decided to unstrap themselves and leave the capsule to await rescue; but once outside they found it bitterly cold, and opted to return inside. An hour later, the recovery forces arrived to attend to the crew and recover the spacecraft. It was determined that, as a result of the experiences on several missions (notably Soyuz 18-1, Soyuz 23 and this mission), improvements in survival equipment and protective clothing would be introduced on future missions.

This mission was the final Soyuz manned mission to the Almaz-type space station. For a short while after the return of the Soyuz 24 crew, it was planned to launch cosmonauts Anatoly Berezovoi and Mikhail Lisun to Salyut 5 for a 15-day stay onboard Soyuz 25. However, at a meeting of the State Commission in March 1977, Glushko reported that it would take two months to construct and test an extra 7K-T spacecraft, and then another two months at the cosmodrome for it to be man-rated. (Construction of this vehicle, no.67, began – but it eventually flew to Salyut 6 as Soyuz 30.) This necessitated keeping the Salyut in orbit for four months, which would consume about 250 kg of propellant to sustain its orbital attitude and altitude. By the time the Soyuz 25 crew arrived at Salyut 5, there would have been a 70-kg shortage of propellant for their mission. Although the Progress tanker was in development for Salyut 6, there was no capability on Salyut 5 to resupply the fuel tanks, and so the State Commission cancelled the plans for Soyuz 25/Salyut 5, and left the station in an automated mode for the remainder of its orbital lifetime. Salyut 5 successfully de-orbited over the Pacific Ocen on 8 August 1977, clearing the way for the next station – the first of a new generation.

THE SOLO SOYUZ MISSIONS, 1973–76

During the period 1973–76, there were six flights of Soyuz spacecraft that were not intended to dock with a Salyut or Almaz space station. Soyuz 13 was a 7K-T (no.33) variant that in December 1973 flew a scientific research mission to test some of the experiments and procedures intended for the cancelled Salyut missions of 1972 and 1973 and for forthcoming station operations. During 1974–75, four flights (two unmanned and two manned) were included as part of the first international manned spaceflight programme: the Apollo–Soyuz Test Project. These spacecraft were designated 7K-TM. Another index for 7K-TM was 11F615A12, and an alternative name for the vehicle was Soyuz M.

The final flight using back-up ASTP hardware was Soyuz 22 (7K-TM no.74) which in September 1976 flew an Earth resources mission as a precursor to a series of nine Interkosmos flights to Salyut 6 between 1978 and 1981. The six flights between 1973 and 1976 could be termed 'Solo Soyuz missions', as they were not flown in conjunction with any other Soviet manned spacecraft or space station.

The early scientific missions

The earliest example of a scientific Soyuz mission was Soyuz 6, which flew in October 1969 and did not carry any docking equipment. Flown in conjunction with Soyuz 7 and Soyuz 8 – which were intended to dock – the Soyuz 6 crew performed the first welding experiments in space in a flight delayed from earlier that year. In June 1970 there was a second scientific Soyuz flight when Soyuz 9 flew for a record

18 days and established a baseline of medical data for use on later long-duration Salyut missions (see p.165). The scientific solo Soyuz missions of 1969–76 were as follows:

Soyuz	*Type*	*Launch*	*Landing*	*Crew*	*Duration*	*Purpose*
6	7K-OK	1969 Oct 11	1969 Oct 16	2	04d 22h	Space welding equipment tests
9	7K-OK	1970 Jun 1	1970 Jun 19	2	17d 16h	Biomedical research
13	7K-T/AF	1973 Dec 18	1973 Dec 26	2	07d 20h	Astrophysical and biological research
22	7K-TM/F	1976 Sep 15	1976 Sep 23	2	07d 21h	Earth resources observations

Soyuz 13 crewing

In May 1973 the Soviet programme encountered a major setback when a civilian Salyut-type (DOS-type) spacecraft designated Cosmos 557 failed soon after entering orbit. Soviet mission planners therefore decided to recover the situation by flying two solo Soyuz missions. The first of these was a two-day ferry Soyuz test (Soyuz 12; see p.182), and the second was the first flight of the Soyuz craft capable of independent flight. The Soyuz orbital module was adapted to fly scientific instruments on a week-long flight, and two Salyut-trained crews were nominated.

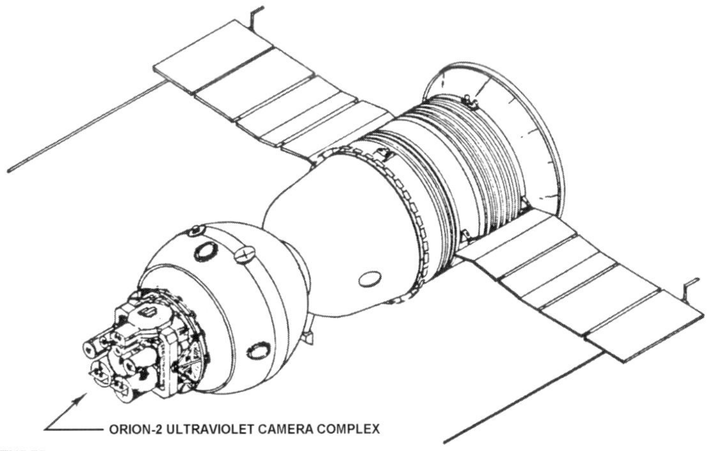

Soyuz 13 with the Orion telescope system installed in the nose of the spacecraft. (Courtesy Ralph Gibbons.)

The prime crew was Colonel Lev. V. Vorobyov, with Flight Engineer Valeri I. Yazdovsky. Their back-ups were initially Major P.I. Klimuk and Flight Engineer V.I. Sevastyanov; but due to medical problems, Sevastyanov was quickly replaced by V.V. Lebedev. The team therefore consisted of rookies.

Only three or four days before the mission, however, Vorobyov and Yazdovsky were replaced because they were apparently incompatible for effective working together in space. They were replaced by their back-ups, and no other cosmonauts were assigned to the mission, and so the replacement Soyuz 13 crew completed their training and flew the mission without a back-up crew. However, the real reason (which has been revealed only recently) was that the original Soyuz 13 crew-members were both very principled cosmonauts who did not suffer fools gladly, and spoke their mind. They consequently made enemies – who decided that they should not fly under any circumstances.

The Soyuz 13 mission

After the failure of the DOS 2 launch in July 1972, the original plan was to fly a manned solo test flight of the modified 7K-T vehicle using spacecraft no.34 (originally intended to fly to DOS 2). The prime crew consisted of Leonov and Kubasov, with Lazarev and Makarov as their back-ups. However, as these crews were undergoing training for the mission in August–September 1972, it was decided to cancel the mission and to instead fly a dedicated Soyuz solo science mission to re-fly the Orion-2 telescope that had been lost on DOS 2. (DOS 3 was a modified station which would carry different scientific instruments, and so the only way to re-fly Orion-2 was on a solo Soyuz.)

Vehicle no.34 was consequently reserved for the first flight to DOS 3, and a new 7K-T was specifically built for the Orion-2 mission. The dedicated solo vehicle for

The original solo Soyuz 13 prime crew: Vorobyov and Yazdovsky.

Orion-2 was assigned as no.33 – a number which had already been assigned to Cosmos 496 (the 7K-T test flight of June–July 1972). There were therefore two vehicles no.33: Cosmos 496 (33A) and Soyuz 13 (33). After the failure of DOS 3 (Cosmos 557), it was decided to fly a manned solo test flight of 7K-T. This would use vehicle no.37, which was originally built to fly to DOS 4. This would now fly before the Orion-2 mission. The Soyuz 12 mission – as it was flown by Lazarev and Makarov in September 1973 – was actually conceived well after the Soyuz 13/Orion-2 mission of Klimuk and Lebedev. The timing of the Soyuz 13 mission in December 1973 was also apparently connected with the appearance of comet Kohoutek.

Soyuz 13 was launched on 18 December 1973, and upon entering orbit it became the first Soviet manned spacecraft to be in orbit at the same time as an American mission. The three Skylab 4 astronauts had been launched on 16 November, and had just completed the first month of their three-month record-breaking mission. The two Soyuz 13 cosmonauts would be spending only a week in orbit, and neither crew reported sighting the other spacecraft, despite being in similar orbits. This Soyuz also marked the first use of a new mission control centre in Kaliningrad.

The Soyuz used on this flight was fitted with solar arrays for power, but the docking equipment had been removed and replaced by the Orion-2 UV telescope (a follow-on to the unit flown on Salyut 1, and identical to that lost in the DOS 2 launch failure) and an X-ray camera which seems to have been part of the Orion-2 complex. Lebedev operated the instruments from the OM, uncovering the apertures for 1–20 minutes to obtain the observations, while Klimuk controlled the spacecraft's orientation from the DM. The RSS-2 spectrograph used NASA-supplied film for a range of stellar studies, and during the course of their mission the crew obtained 10,000 spectrograms of more than 3,000 stars. They also performed medical and biological experiments, including the Levkay-3 experiment, which took measurements of blood distribution to the brain in weightlessness to determine its relationship to space sickness. In the Oazis-2 biological experiment, chlorella and duckweed were used to demonstrate a closed-cycle life support system, in which the waste products of one type of bacteria consitituted the initial material used by a different type of bacteria to accumulate protein mass up to thirty-five times. The crew also conducted navigational exercises and observations of the Earth and its atmosphere.

The mission was completed with a successful landing – despite concerns over a heavy snowstorm at the landing area – after 7 days 20 hrs 55 min. The short but successful mission of Soyuz 13, and that of Soyuz 12 three months before, provided a much-needed boost of confidence for the Soviets, and as the joint American/Soviet docking mission was being developed, it demonstrated their capability in recovering from the Soyuz 11 tragedy.

THE SOYUZ–APOLLO EXPERIMENTAL FLIGHT (EPAS), 1969–75

The idea of a joint mission with the Americans evolved from a letter and then discussions between Anatoliy Blagonravov, Chairman of the Soviet Academy of

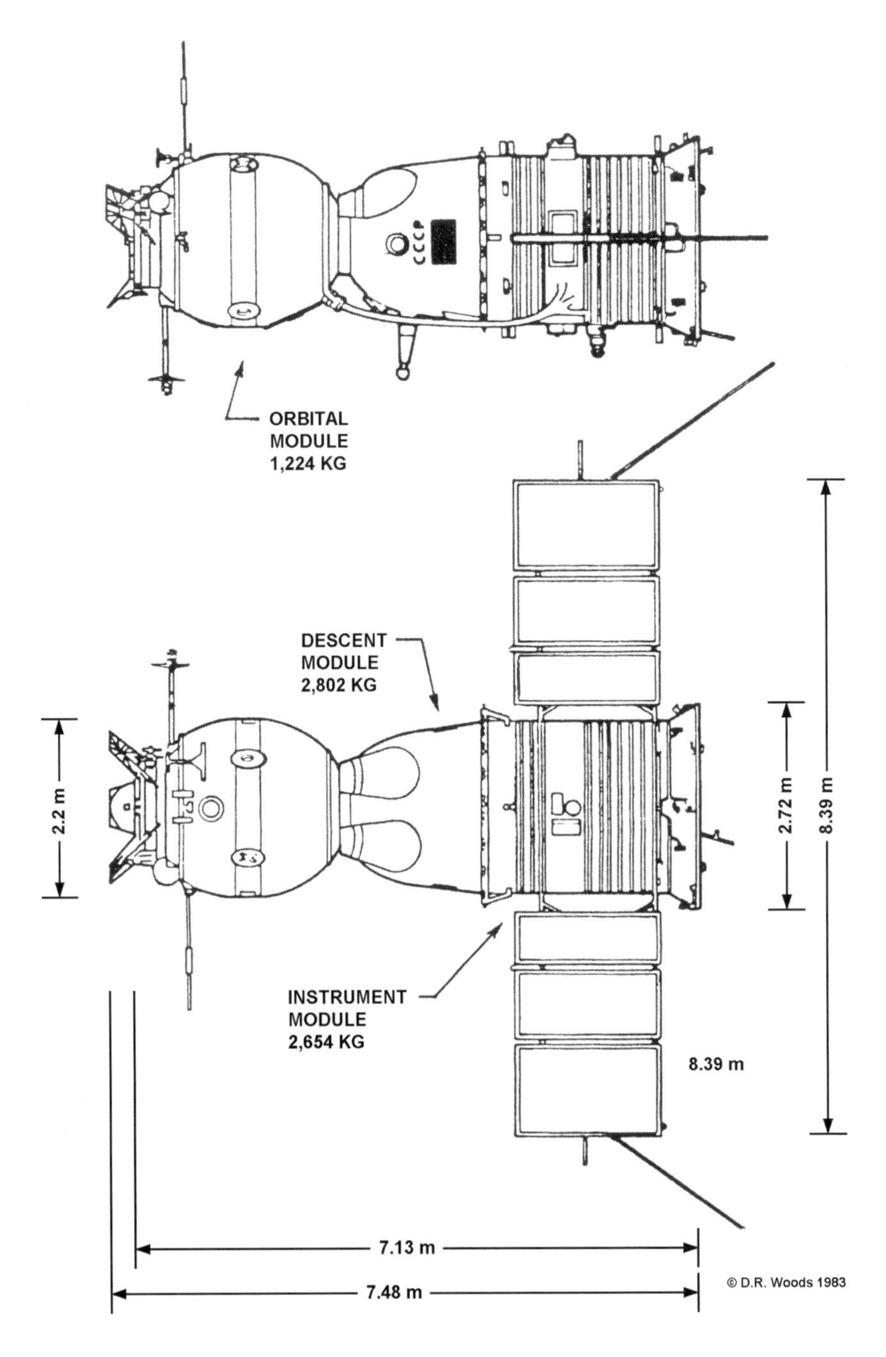

Soyuz as used during EPAS (ASTP) in 1974–75. (Courtesy D.R. Woods.)

Sciences Commission on Exploration and Use of Space, and the third NASA Administrator, Thomas Paine, during 1969. Paine believed that international cooperation in the peaceful exploration of space was the way forward after Apollo had achieved its manned lunar landing goal, and he attempted to renew interest within NASA for a joint docking mission with the Soviets.[10] After three years of further official discussions, proposals, meetings and agreements, there emerged, on 24 May 1972, a five-year agreement between the Soviet Union and the United States to cooperate in the fields of science and technology for the peaceful exploration of space. Of the wide-ranging proposals, by far the most dramatic was the rendezvous and docking of a Soviet Soyuz with an American Apollo CSM in 1975. In America this became known as the Apollo–Soyuz Test Project (ASTP), and in Russia it was called the Soyuz–Apollo Experimental Flight (EPAS).

Early proposals

Suggestions for a joint Soviet/American spaceflight had been reported throughout the 1960s, and in the early years this focused on a joint mission to the Moon.[11] By 1969 it was clear that such a cooperative venture would only be performed in Earth orbit. During 1970, one of the earliest ideas for a joint flight – once the compatibility of docking systems and spacecraft atmospheres had been resolved – was the sending of a Soyuz to an American Skylab space station for a joint research mission.[12]

In trying to achieve this, several hurdles would need to be overcome; most notably, how the docking of the Apollo and Soyuz to the Skylab Orbital Workshop would be accomplished. One suggestion was to dock a Soyuz to the already manned Skylab, which required the adaption of the Apollo docking probe to fit to a Soyuz, and either moving the Apollo to the radial port, or for the crew to separate and fly Apollo in a station-keeping mode to record the docking of Soyuz with Skylab before

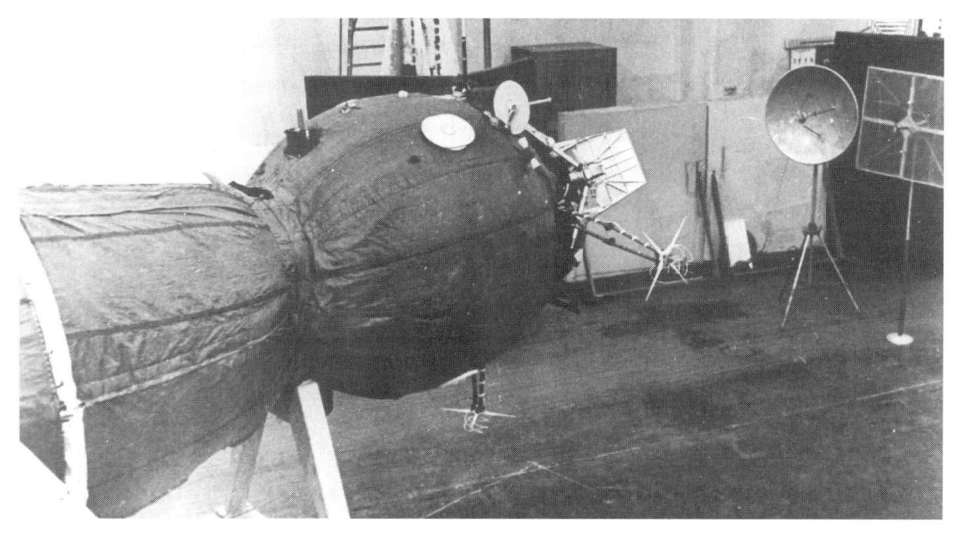

Soyuz undergoes rendezvous and radar qualification tests.

The EPAS (or ASTP) Soyuz 19 spacecraft during ground launch processing at Baikonur.

themselves re-docking. This was a complicated sequence of events to add to the problem of atmosphere compatibility. Another suggestion was to allow the Soyuz crew to occupy Skylab *after* the American crew had departed (a scenario that was feasible, but politically undesirable for the Americans). A simpler plan proposed the direct docking of Apollo with a Soyuz, without the use of Skylab; but the Soviets considered that this would be regarded as nothing more than another space stunt. This indicated their desire to fly to Skylab, but they did not mention that their own space station was much closer to launch.

In April 1971, talks were progressing towards an agreement, but not a defined mission, when the Soviets launched their first Salyut space station. Suddenly, America was faced with a Soviet space station capability two years before it planned to launch Skylab, and they therefore suggested that since a Salyut was in orbit, the proposed mission should focus on an Apollo/Salyut/Soyuz docking mission in order to demonstrate a truly international space station capability. As a follow-on mission, a Soyuz/Skylab/Apollo mission might be considered, although with the funding for follow-on Skylab under doubt, such a mission was never really a possibility.

During the summer of 1971, discussions continued about an Apollo/Salyut mission, together with plans for a Docking Module to facilitate the changes in atmosphere and docking systems. But when the Soyuz 11 crew were lost in a landing accident, the impetus for such a mission – together with that for a Soyuz/Skylab mission – slowly declined. In April 1972, a month before the agreements would be signed, the Soviets told the Americans that a Salyut docking by an Apollo and a Soyuz, simultaneously, would be impossible. They stated that at that time it was not technically possible to incorporate a second docking port, and would also be economically prohibitive on the current Salyut design. There were indications that the Soviets did not believe that the Americans were abandoning Skylab and were in fact preparing to fly a second (military) Skylab,[13] although it was soon clear that this was not the case, and that Skylab would be a single space station with three manned missions. The Soviets did not reveal that they were working on a dual-port space station – probably because it would not be ready to fly for some years.

According to the NPO Energiya history, it the first consideration was to use the improved 7K-S (the later Soyuz T) for the ASTP (from about mid-1972 until the autumn of 1972), but this was not discussed with the Americans, and the idea was eventually discarded. Instead, the 'solo' 7K-TM version of the 7K-OK – re-introducing the solar panels – would be used, with a modified emergency escape system and upgraded version of the R-7 (see p.89). Also at this time (1972), the individually tailored Sokol suits, initially for use by two-man 7K-OK crews, were also being tested prior to their introduction on all future Soyuz flights. All of this hardware would be ready in time for the planned 1975 mission, whereas the Soyuz T would not not be finished before the deadlines.

The flight

The proposed EPAS mission would begin with the launch of a two-man Soyuz from the Baikonur cosmodrome into Earth orbit. Seven and a half hours later, the three-man Apollo would be launched using a Saturn 1B vehicle. Once in orbit, the Apollo crew would extract the Docking Module from the top of the S-IVB stage, in a move similar to extracting the LM on Apollo lunar missions. Over the next two days, Apollo would begin the orbital chase to the Soyuz, achieving a docking on the third flight day. The two spacecraft would remain docked for two days, during which (using the docking module to equalise the pressure to enable entry to each other's vehicle) the crews would perform a sequence of exchanges so that each crew-member would visit the other spacecraft at least once. At undocking, on flight day 5, the Apollo would perform manoeuvres around the Soyuz, and then re-dock with the Soviet craft docking system in the active role. After only a couple of hours the two craft would undock for the final time, and then drift apart. The Apollo would remain in orbit for a further five days, to conduct a programme of experiments. The Soyuz, however, would return on flight day 7.

In preparation for the EPAS, the Soviets agreed to fly an independent manned dress rehearsal mission using EPAS hardware. In total, six sets of flight hardware were prepared for the EPAS programme, including a back-up spacecraft and launcher already on the second Soyuz pad in case the primary mission failed to dock with Apollo, or the launch of Apollo had to be significantly delayed *after* the primary Soyuz vehicle had already been placed on orbit. The original plan was as follows:

Vehicle no.71	Unmanned test flight
Vehicle no.72	Manned test flight
Vehicle no.73	Manned test flight
Vehicle no.75	Vehicle to be used for the actual EPAS mission
Vehicle no.76	Back-up vehicle (on stand-by on the pad)
Vehicle no.74	Second back-up vehicle

At some point, this plan was changed in order to fly two unmanned flights and only one manned test flight.

EPAS crewing

In early June 1973 the Soviets assigned four crews (eight cosmonauts, including four rookies) to the EPAS (ASTP) missions; and, for the first time, identified the cosmonauts prior to flying. The crews were:

Crew 1	A.A. Leonov and V.N. Kubasov
Crew 2	A.V. Filipchenko and N.N. Rukavishnikov
Crew 3	V.A. Dzhanibekov and B.D. Andreyev
Crew 4	Y.I. Romanenko and A.S. Ivanchenkov

The Leonov/Kubasov crew was to fly to the 1973 failed Salyut mission as its initial crew, but were reassigned due to the importance of this project. In December 1974, the no.2 crew – Filipchenko and Rukavishnikov – flew the Soyuz 16 mission, with the no.4 crew – Romanenko and Ivanchenkov – as back-up; and in July 1975, the no.1 crew – Leonov and Kubasov – flew the Soyuz 19 mission, with the no.2 crew – Filipchenko and Rukavishnikov – as back-up. Another rookie cosmonaut – Valeri V. Illarionov – was identified when he served as a mission CapCom at Mission Control, Houston.

Cosmos 638 and Cosmos 672

On 3 April 1974, the first mission associated with EPAS – Cosmos 638 – was launched by an improved version of the R-7 booster designated 11A511U (also known as Soyuz U). The spacecraft flew an extended test flight of 9 days 21 hrs, and landed on 13 April 1974. During the mission, the new life support system, designed to support up to four crew-members in the Soyuz (for the series of crew transfers with the American astronauts), was evaluated, and extended solar arrays were employed to support the additional power requirements. The mission followed the orbital parameters being developed for the actual EPAS flight plan, and the OM was fitted with the universal docking system that would be used to dock with the

American Docking Module. During the re-entry of Cosmos 638 (7K-TM no.71) – which was a ballistic rather than a controlled re-entry – problems arose because a valve used to release air from the orbital module prior to the separation of the compartments had caused unexpected motions. It was therefore decided to fly the next 7K-TM vehicle (no.72) unmanned rather than with a crew, and so there would be only one manned dress-rehearsal flight rather than two.

The second unmanned test flight – of Cosmos 672 – was launched on 12 August 1974, and was a similar mission to Cosmos 638. The second vehicle also carried the EPAS docking system (which was jettisoned before the retro-fire burn), a modified life support system, and extended solar arrays; but this time it completed a duplicate EPAS mission duration of 5 days 22 hrs. The spacecraft was successfully recovered on 18 August.

Soyuz 16: dress rehearsal for the EPAS

Between 2 and 8 December 1974, the final precursor mission (5 days 22 hrs) for the EPAS was flown as Soyuz 16. It was crewed by Filipchenko and Rukavishnikov (call-sign, Buran – Snowstorm); although during a recent interview, Filipchenko said that he and Rukavishnikov had wanted their younger back-ups to fly Soyuz 16, but they were overruled. During the mission, the crew followed the orbital manoeuvres that the EPAS Soyuz would follow on the real mission, and to demonstrate the reduction of the transfer time between the spacecraft from two hours to one hour they adjusted the air from 760 mm pressure and 20% oxygen to 540 mm pressure

Soyuz 16 commander Anatoli Filipchenko inside the Orbital Module of Soyuz 16 during the mission of December 1974.

Flight Engineer Nikolai Rukavishnikov works at the Orbital Module control panel of Soyuz 16 during the mission of December 1974.

and 40% oxygen. The OM had an EPAS docking ring and a simulated US Docking Module ring attached to it to practise capture, latching and undocking operations in the vacuum of space. The crew completed several docking procedures during orbits 32–36 and again on orbit forty-eight, before jettisoning the device with explosive bolts, and demonstrating a fail-safe method should the Soyuz be unable to undock from the American Docking Module. Apart from their EPAS obligations, including simulated manoeuvres for the solar eclipse experiments with Apollo, the cosmonauts also performed numerous medical, biological and Earth observation experiments during their six days in orbit. They also determined that the Soyuz (like all spacecraft) expelled gases and a cloud of debris, such as ice particles and paint chips, followed the spacecraft in a trail that could affect observations and measurements – information valuable for future scientific missions. The spacecraft landed on frozen ground, and the mission was deemed a complete success, setting the stage for the actual EPAS mission seven months later.

Soyuz 19, and a handshake in space

When Soyuz 19 (call-sign, Soyuz – Union) was launched to begin the EPAS mission on 15 July 1975, it was the first Soviet manned launch to be announced in advance and the first to be televised live. Upon entering orbit, the mission also marked the first time that two crews from the same country were conducting separate missions. The Salyut 4 cosmonauts – Klimuk and Sevastyanov – were nearing the end of their two-month residency onboard the station. The insertion into orbit was, according to the Soviets, to within 2–3 km of the desired parameters, and attention then turned to the launch of the Apollo from Florida while the Soyuz cosmonauts Leonov and Kubasov settled down to checking and configuring their vehicle for orbital flight.

On the adjacent pad 31, vehicle 7K-TM (no.76) was ready in case of problems with the primary vehicle (no.75). It would have been launched if either vehicle no.75 had failed to dock with Apollo, or if the launch of Apollo had to be significantly delayed after vehicle no.75 had already been placed on orbit. In the event, the 7K-TM (no.76) was not required for the joint mission, as all primary hardware performed as planned.

The Apollo was successfully launched on time, and ten minutes later the three American astronauts – Commander Tom Stafford, CM Pilot Vance Brand, and DM Pilot Deke Slayton – were in orbit. At this point, the seven men – three on Apollo, two on Soyuz 19, and two on Salyut – set a new record for the number of humans in space at one time.

The Apollo then began a series of manoeuvres to bring the two spacecraft together for a docking on the thirty-sixth orbit of the Soyuz on 17 July. The following day, the Soyuz crew unsuccessfuly attempted to repair a faulty black-and-white TV camera. This camera did not work at all during the flight, and plans for transmitting TV pictures of the Apollo CSM during the flight had to be cancelled.

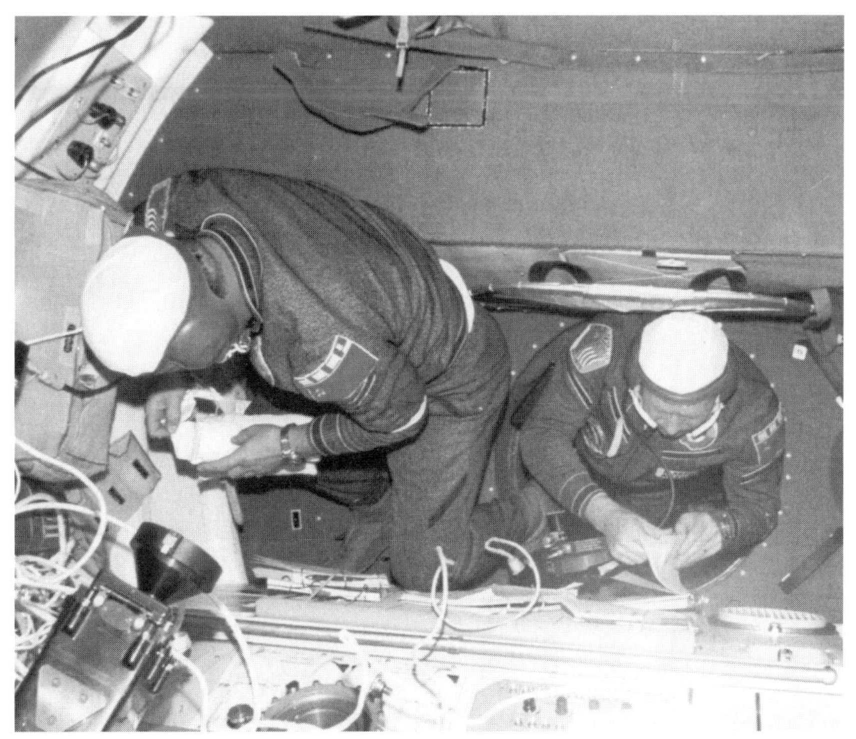

EPAS cosmonauts Leonov (*left*) and Kubasov (*right*) in the Soyuz 19 Orbital Module during the historic first international manned space mission with the American Apollo.

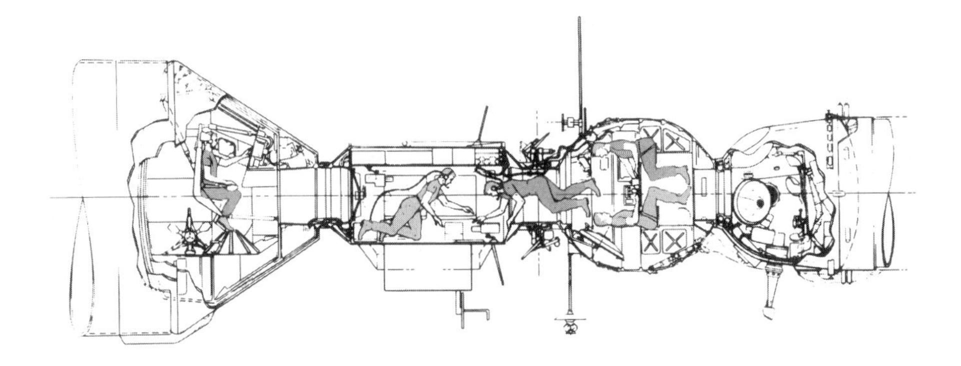

The crew transfer system used during the EPAS (ASTP) mission with the Americans in July 1975.

The day also saw a circulation burn, and a communication link with colleagues onboard Salyut 4.

On 17 July the Apollo crew closed in on the Soyuz to a point where, at 222 km, Kubasov turned on the range tone assembly to determine the exact range between the two spacecraft. After a small burn, Apollo closed in on Soyuz as the two crews spoke to each other – the Americans in Russian, and the Russians in English. After the Soyuz was re-orientated by 60° to match the approaching Apollo with the DM on its nose, Tom Stafford eased the two spacecraft together until the docking latches fired, sealing the two spacecraft together at 19.10 Moscow Time on 17 July 1975. Leonov reported: 'We have capture. OK, Soyuz and Apollo are shaking hands now.'

With the docking completed, the two crews began to prepare their respective spacecraft for the series of transfers. The hatches between the CSM and the DM were opened, but a brief alarm sounded because of an odour inside the module, which, however, soon dispersed. There then followed the first meeting of two crews in space. With Brand still in the CM, Stafford and Slayton entered the DM and closed the hatch to the CM behind them. They began to raise the pressure from 255 mm to 490 mm by introducing nitrogen to the CM's previous 78% oxygen atmosphere. On the other side of the hatches in the Soyuz, the pressure had been reduced to 500 mm prior to docking.

At 22:17:26 MT on 17 July, Stafford opened the final hatch to greet the two cosmonauts in the OM, calling his two colleagues over to shake hands as the joint spacecraft flew over the French city of Metz. The meeting was both personal and technical, and included mostly commemorative events, marking their unique mission and achievement. After 3.5 hours in the Soyuz, the two Americans returned to the Apollo with a reversal of the procedure they had to complete to enter Soyuz.

The following day, 18 July, began with a TV tour of Soyuz by Kubasov and one of Apollo by Brand. Despite the apparent ease of how everything appeared, it had taken considerable periods of negotiation and planning to arrive at this point. The Soviets were visibly overjoyed by the success of the flight, but many Americans seemed to regard the TV show as commonplace, and far from the days when millions were glued to their TV sets to watch men walk on the Moon. The day also saw the more complicated transfer of Brand to Soyuz, for lunch, and Leonov into Apollo. Then, as Brand returned to Apollo, Stafford and Leonov moved across to Soyuz, and then Kubasov transferred to Apollo. Each transfer included the checking of the atmospheric composition in each vehicle, and the opening and closing of three hatches (one at each end of the DM, and the forward Soyuz OM hatch; the CM forward hatch and the inner Soyuz hatch between the OM and DM remained open at all times during the docked period). Following the completion of more commemorative events and a number of small science demonstrations for the TV cameras, there began the final transfer to return the crews to their own spacecraft for the undocking.

During the transfers, Leonov spent 5 hrs 43 min in the DM and CM, while Kubasov logged 4 hrs 47 min in the American modules. In Soyuz, Stafford completed 7 hrs 10 min, Brand 6 hrs 30 min, and Slayton 1 hr 35 min. The next day, 19 July, after almost forty-four hours docked together, the two spacecraft separated. The Apollo crew then pulled away and placed their spacecraft between the Soyuz and the solar disc, creating an artificial solar eclipse (experiment MA-148) and allowing the cosmonauts to photograph the solar corona. Then, in a second major experiment (MA-509), the Apollo flew out of plane with Soyuz at 150 metres, 500 metres, and finally 1,000 metres. The cosmonauts projected monochromatic laser beams of light at reflectors mounted on Apollo, which were bounced back to a spectrometer on Soyuz to record the wavelengths of the returned light. Later analysis would reveal information on the quantities of atomic oxygen and atomic nitrogen at the altitudes at which the spacecraft were orbiting (approximately 210 km), in studies of UV absorption.

Following this experiment, Apollo returned to Soyuz, which acted as the passive partner to achieve a second docking, with Slayton at the Apollo controls. At first, Soyuz was visible against the blackness of space, but as they approached it moved against the back-drop of Earth, and Slayton lost sight of his target from 100 metres. With too much light flowing into the optical instrument, he was worried that he might hit Soyuz, but continued with the docking by instruments; or, as Slayton later commented, 'flying by the seats of the pants. I guess I got a little closer than they [the cosmonauts on Soyuz] or the ground anticipated.' After three hours, and with no transfer this time, the two spacecraft separated for the final time. The operation was activated by Leonov in command of the active spacecraft. A programme of fly-around manoeuvres of Apollo by Soyuz resulted in photodocumentation of the two spacecraft before Apollo performed a separation burn to separate the two spacecraft into a 217 × 219-km orbit. Apollo would remain in orbit until 24 July, to conduct what would be the final programme of experiments and observations onboard an American spacecraft before the launch of the Space Shuttle six years later.

On 20 July, the Soyuz crew spent the day conducting independent activities, including Earth and solar photography and a few smaller onboard biological experiments. The crew completed simulations of de-orbit procedures, together with a final live TV transmission to Russian viewers. The next day, 21 July, Soyuz 19 returned to Earth, and the landing watched live on Soviet TV for the first time.

A tradition continued by every returning Soyuz crew – signing the Descent Module. Here, EPAS cosmonauts Leonov and Kubasov add their signatures to the Soyuz 19 Descent Module at the end of the first international space mission.

Soyuz landed on its side, and a few minutes afterwards Kubasov and Leonov emerged, slightly shaky but otherwise in good health. They received the best wishes of the Apollo crew still flying high above them. The first joint manned spaceflight was over, and after a flight of 5 days 22 hrs, Soyuz 19 was back on Earth. Five days later it was followed by the pair of Salyut 4 cosmonauts after 63 days in space (see p.192).

Future cooperation

The initial agreement with the Russians, signed in 1972, lasted until 1977, and following the EPAS mission there were discussions about flying an American Shuttle to a Russian Salyut station in 1981.[14] During the 1970s this idea was quietly dropped, as relations between the Soviet Union and America became more strained in light of other international developments. It would be two decades before American astronauts and Soviet cosmonauts once again worked together in orbit. By the 1990s, the Russians were partners in the International Space Station, and in 1995, memories of the EPAS returned when the Shuttle *Atlantis* docked with Mir.

The Soviets probably gained more from the EPAS than did the Americans. In flying Soyuz 19, the Soviets demonstrated their ability to launch on time and complete a difficult mission with an international partner, fulfilling their earlier agreements (which the current Russian programme has difficulty in maintaining). This was a clear demonstration of the intent to open up the Soviet programme (albeit briefly and to a limited extent) to the Western world, to show the apparent might and scope of the Soviet programme at a time when then Americans were winding down the Apollo programme in favour of the Shuttle programme. For the Americans, this was a brief insight into the workings of the Soviet manned spaceflight programme, and they obtained key information about the Soyuz programme, the elements of its hardware, and – in the case of Soyuz 11 and 15, and the 5 April 1975 launch abort – its failures.

For the Soviets, the programme offered the chance to undertake a programme of international cooperative experiments, which prompted Premier Leonid Brezhnev to announce, prior to the first handshake in space on 17 July, that 'Soyuz/Apollo is a forerunner of future international orbital stations.' From the possibilities of cooperation with other countries (and with plans for the Shuttle to fly European science specialists as passengers on space laboratory missions) emerged the idea of the Interkosmos programme to fly guest cosmonauts from Eastern Bloc countries before the Americans flew Europeans on the Shuttle. The EPAS Soyuz flights of 1974–75 were as follows:

Soyuz	*Type*	*Launch*	*Landing*	*Crew*	*Duration*	*Purpose*
K 638	7K-TM	1974 Apr 3	1974 Apr 13	–	09d 21h	Unmanned test flight
K 672	7K-TM	1974 Aug 12	1974 Aug 18	–	05d 22h	Unmanned test flight
16	7K-TM	1974 Dec 2	1974 Dec 8	2	05d 22h	Manned Soyuz dress-rehearsal
19	7K-TM	1975 Jul 15	1975 Jul 21	2	05d 22h	Soviet EPAS mission
BUp 2	7K-TM	Converted to fly Soyuz 22 Earth resource mission in September 1976				

The first EPAS back-up vehicle (no.76), which had been on stand-by on pad 31, was not actually used for another mission. Since it had been fuelled for a possible mission (it should have been launched within about 75 days of being fuelled), it could not be kept in storage for much longer. After the successful launch of Soyuz 19 (no.75), vehicle no.76 was returned to the assembly building, after which its fuel tanks were drained and it was sent back to NPO Energiya. The DM was later used for vehicle 7K-T no.47 (Soyuz 31), but the OM and SM were used for a Soyuz that is now on display (with a mock-up DM) in the Energiya museum.

SOYUZ 22: THE LAST SOLO SOYUZ

The success of Salyuts 3, 4 and 5 indicated that after EPAS there would be no further need to fly solo Soyuz missions that were not part of a space station programme. In June 1976, cosmonaut Klimuk, Commander of the solo Soyuz 13 mission and second expedition to Salyut 4, indicated as much. However, there was one more flight in the planning that used the back-up unflown EPAS spacecraft and its launcher, and allowed the in-flight testing and evaluation of scientific equipment destined for the next civilian space station (Salyut 6). The equipment chosen was a 204-kg East German multispectral camera designated MKF-6, and its programme of observations from Soyuz 22 was called Raduga-1 (Rainbow-1). This unit replaced the EPAS docking system in the front hatch area of the OM, and was protected by an outside cover when not in use. A special 428-mm-diameter porthole was installed on Soyuz 22 to facilitate MKF-6 observations. Data could be recorded in four visible wavelengths and two infrared wavelengths, to a resolution of 20 metres, and the cosmonauts could photograph 800,000 square km in only ten minutes. For military observations, this resolution was not as good as from the Almaz stations, and relied on the cosmonauts' own visual observations to supplement the gathered data. However, for Earth resources photographs, the camera was very useful. It was not until 1974 that the basic technical requirements for the MKF-6 were decided upon. The actual camera was not ready for tests until April 1976, and a development model was therefore assigned to fly on Soyuz 22 prior to an improved version on Salyut 6.

Soyuz 22 crewing

In January 1976, the Soviets assigned six cosmonauts to train in three crews for a solo Soyuz mission, using the back-up ASTP craft fitted with an East German camera which was also planned to be used on the forthcoming Salyut 6 station. The crews were:

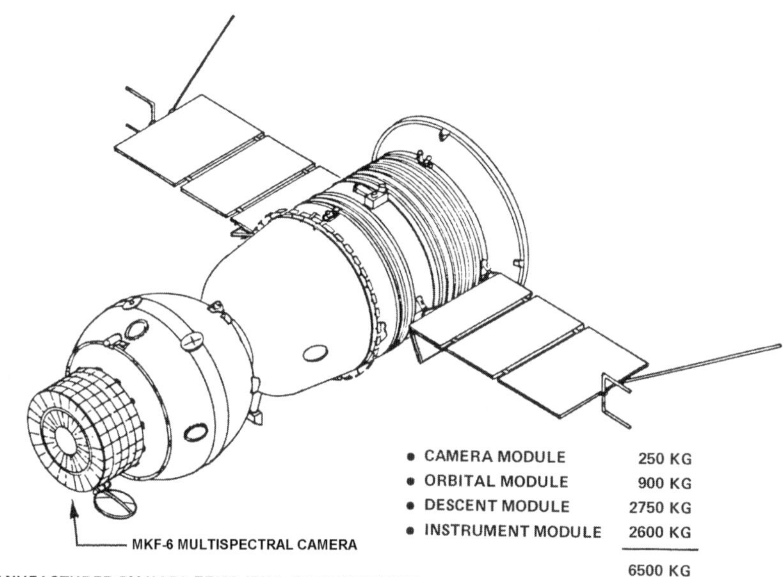

Soyuz 22 with the MKF-6 Earth Resources Camera installed in the nose of the Orbital Module. (Courtesy Ralph Gibbons.)

Prime crew	V.F. Bykovsky and V.V. Akyonov
Back-up crew	Y.V. Malyshev and G.M. Strekalov
Support crew	L.I. Popov and B.D. Andreyev

Before this assignment, some of these crews were working in the cosmonaut training group for the next Salyut mission; and Bykovsky – who two years later commanded the East German Interkosmos mission – had developed contacts with the camera specialists in Germany. When the mission flew in September the group was disbanded.

The Soyuz 22 mission

The final Soyuz to fly independent of a space station – Soyuz 22, crewed by Valeri Bykovsky and Vladimir Akyonov (call-sign, Yestreb) – was launched on 15 September 1976. The mission flew in an orbit of 185 × 296 km inclined at 64°.75 (much higher than the normal Salyut missions), and presented the opportunity to use the camera at higher altitudes than normally available from Soviet spacecraft. Although civilian in nature, here was a military application to the mission, which was flying at the time of a NATO exercise called Teamwork, being conducted in Norway. A delegation from East Germany witnessed the launch, and TASS news reports at the time indicated that cosmonauts from eastern bloc countries would soon join the Soviet cosmonauts on joint missions. From the Soyuz, the cosmonauts took photographs of the Soviet Union and East Germany, with Akyonov operating the camera from within the OM while Bykovsky orientated the Soyuz from the DM. During the fifteenth orbit, tests with the camera system were completed, including photographs of the Earth's horizon, the Moon, and the Baikal–Amur railway, as well as locations in Asia. Part of the Earth resources experiments was carried out in conjunction with photography of the same area from a high altitude AN-30 aircraft also equipped with MKF-6 cameras. After the flight, the photographs from space were compared with the higher-resolution photographs taken from the aircraft. By the end of the mission, the crew had taken 2,400 pictures, which were stored in the DM for return to Earth. Also onboard the Soyuz was the Biogravistat plant growth experiment, which used a centrifuge as an artificial aid to plant growth.

In a recent interview, Akyonov said that the crew photographed 20 million square km – half of which was Soviet territory – and that 95% of the images were of excellent quality. He also said that there were problems with changing out the film cassettes, and that their recommendations were taken into account in building the modified MKF-6M. The specialists who had built the camera had also privately asked them, before the flight, to recover the camera's colour filters (the camera itself would burn up on re-entry). In order to do that, the cosmonauts had to disassemble most of the camera (without mission control being aware) during one of their final nights in orbit. Soyuz 22 landed on 23 September 1976, after a flight of 7 days 21 hrs, and ended the solo Soyuz series.

A solo Soyuz series?

There were probably never any plans for a dedicated long-range solo Soyuz

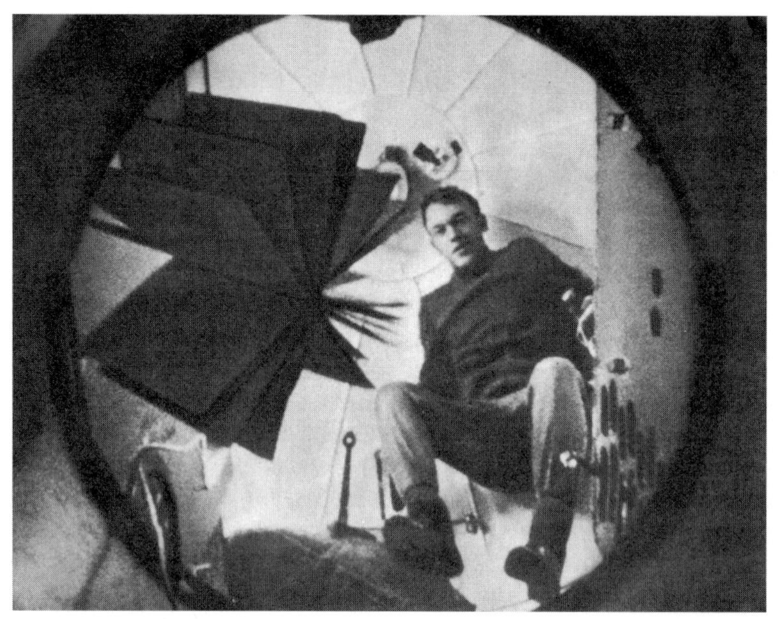

The additional volume in the OM on later Soyuz missions is evident in this photograph taken during the Soyuz 4/5 mission.

programme. The mission of this type that was planned well in advance was Soyuz 6. Based on the schedules in the Kamanin diaries, this was planned for even before Komarov's flight. To increase the propaganda effect, the decision to combine it with the Soyuz 7/8 mission was not made until a few months before the flight. Soyuz 9 was a scientific flight, but it used a vehicle originally assigned to the Kontakt programme. Soyuz 13 was also a dedicated scientific flight (even using a Soyuz vehicle specially built for this purpose), but it was a stop-gap mission to compensate for the failed DOS 2 (and primarily the loss of the Orion-2 telescope). All the others used ASTP vehicles. Only six vehicles were manufactured for ASTP, and there was therefore no prospect of a later continuation of this series.

SOYUZ FERRY MISSIONS TO SALYUT 6

The launch of the next space station, designated DOS 5 (Salyut 6), took place on 29 September 1977. DOS 5 featured two docking ports, an EVA hatch, and the capability for resupply by unmanned (Progress) ferry spacecraft, although none of this was identified at launch. Unlike the military-orientated Salyut 3 and 5, the programme was scientific in nature, and continued the work begun on Salyut 1 and 4.

Salyut 6 crewing

The new Salyut was to be a second-generation version with two docking ports, and because of the opportunity to exchange Soyuz craft, the intention was to set a number of long-duration records. Fourteen cosmonauts were trained, to form seven crews during 1975 for long-duration and visiting missions, with a first mission due in 1977. The crews for the first three long-duration missions were:

Crew 1 V.V. Kovolyonok and V.V. Ryumin
Crew 2 Y.I. Romanenko and A.S. Ivanchenkov
Crew 3 V.A. Lyakhov and G.M. Grechko

Crew 1 flew on Soyuz 25, but were unable to dock with the station. This was a major failure, as the mission had been planned to coincide with a major Communist Party conference, and the new second-generation Salyut 6 was waiting to be activated. The review resulted in a major change to the crewing of future missions. All subsequent had to have an experienced cosmonaut on board, and so all the existing training crews had to be changed:

Crew 1 V.V. Kovolyonok and G.M. Grechko
Crew 2 V.V. Kovolyonok and A.S. Ivanchenkov
Crew 3 V.A. Lyakhov and V.V. Ryumin
Crew 4 L.I. Popov and V.V. Lebedev

Crew 1 flew Soyuz 26, setting a record of 96 days in orbit; while crew 2, on Soyuz 29, exceeded this with a record of 140 days. This in turn was beaten by crew 3, who flew Soyuz 32 to Salyut on a record 175-day mission. Immediately prior to the launch of crew 4 on Soyuz 35, Lebedev injured his knee and was replaced by Ryumin.

Crew 4 L.I. Popov and V.V. Ryumin
Crew 5 V.D. Zudov and B.D. Andreyev

Again the record fell, as crew 4 set a duration of 185 days in orbit. Their back-up crew, Zudov and Andreyev, transferred to the Soyuz T programme and were planned to fly Soyuz T-4, but failed a medical examination prior to the mission.

Visiting missions to Salyut 6

The first planned visiting mission was to fly a Soyuz to the newly launched Salyut and dock at the second port. This would test the port and its rendezvous system. The first crews for this were drawn from the main Salyut 6 training group, with Dzhanibekov and Kolodin as the prime crew, and Lazarev and Makarov as back-up. This changed following the docking failure of Soyuz 25 and the State Commission's decision to include an experienced cosmonaut on every mission. As both members of the prime crew were rookies, they were split up – Dzhanibekov being paired with Makarov, and now being backed up by the next long-duration crew consisting of Kovolyonok and Ivanchenkov. Because of this change it was the last flight opportunity for Kolodin, who retired soon afterwards.

All future visiting missions to Salyut would also have to have an experienced Soviet cosmonaut as Commander, with an Interkosmos cosmonaut in the second seat. The Interkosmos programme was devised in early 1976, and certainly had a major propaganda element in flying citizens from other socialist countries. Other reasons were less obvious, but when pieces of hardware (as well as experiment packages, some of which stayed onboard the station for the long-duration teams to use) came from these countries, it was clear that the Soviets were saving substantial sums of money in key areas of science, and were using the skills of foreign scientists and research institutes to enhance the work on the station. Another reason for flying these missions was to exchange the Soyuz craft to enable the long-duration crews to remain in orbit for these record-breaking mission.

The first mission would fly with a Czechoslovakian cosmonaut on board. The Soviets assigned Gubarev as the prime crew's Commander, with Isaulov as his back-up – but again this had to change as a result of the failure of Soyuz 25. Gubarev was retained on the prime crew, but Isaulov was replaced by Rukavishnikov. It has never been made clear why he was selected, but it could have been because of the introduction of the new rules, as there were too few experienced Air Force Commanders available. Rukavishnikov was the first Soviet civilian to be named to command a manned craft. It also seemed that the back-up Commander was now required to miss two missions and command the third. One problem caused by these changes was that it denied seats to members of the cosmonaut team, which certainly caused much anger. This, together with the reduction to two crewmen, had a profound effect on crewing opportunities for Soviet cosmonauts, and several of them were denied a flight after many years of training.

Mission	*Prime crew*	*Back-up crew*
Soyuz 28 (original) Czech	Gubarev and Remek	Isaulov and Pelcak
Soyuz 28 (revised) Czech	Gubarev and Remek	Rukavishnikov and Pelcak
Soyuz 30 Poland	Klimuk and Hermeszewski	Kubasov and Jankowski
Soyuz 31 DDR	Bykovsky and Jahn	Gorbatko and Kollner
Soyuz 33 Bulgaria	Rukavishnikov and Ivanov	Romanenko and Aleksandrov
Soyuz 34 Hungary	(Cancelled due to the failure of the Soyuz 33 mission)	
Soyuz 36 Hungary	Kubasov and Farkas	Dzhanibekov and Magyari
Soyuz 37 Vietnam	Gorbatko and Pham Tuan	Bykovsky and Bui Thanh Liem
Soyuz 38 Cuba	Romanenko and Mendez	Khrunov and Lopez Falcon
Soyuz 39 Mongolia	Dzhanibekov and Gurragcha	Lyakhov and Ganzorig
Soyuz 40 Romania	Popov and Prunariu	Romanenko and Dediu

Soyuz 40 would be the last Soyuz ferry mission in this configuration. Command was offered to Khrunov, but after training with Prunariu for some time he was removed due to marital problems, and was replaced by Popov.

The Soyuz 25 docking failure

The launch of Soyuz 25 was timed to mark the twentieth anniversary of the launch of Sputnik, the world's first artificial satellite, and was launched from the same pad on 9 October 1977. Onboard Soyuz 25 were rookie cosmonauts Vladimir Kovalyonok and Valeri Ryumin (call-sign, Photon), who had trained to complete a three-month visit to the station. By the seventeenth orbit, the vehicle was aligned to dock with the forward port of the station, but despite several attempts to engage the docking latches, the contact light on the display panel in the Soyuz failed to come on, and the latches refused to engage in a hard docking. The crew reported that the approach force did not seem hard enough. They thought that they had achieved a soft docking, with the probe engaged in the drogue of the Salyut port, but the system would not retract the probe to bring the two vehicles together and activate the latches around the circumference of the docking ring. During the twentieth orbit, Kovalyonok was told fly in formation with the Salyut and await instructions while administrators and ground controllers discussed the options.

Soyuz 25 was apparently the first flight for some time during which manual docking once again became the primary mode (in the wake of the Igla failures on Soyuz 15 and Soyuz 23). Igla was activated at 25 km, but Kovalyonok assumed manual control at a range of 100 metres. There is, however, some conflict between the various reports concerning subsequent events. According to Boris Chertok's memoirs, at a range of 1 metre the cosmonauts noticed that the longitudinal axis of the Soyuz had deviated by 2°. The maximum allowable deviation was 4°, and they could therefore have docked; but they considered that the situation was not nominal, and decided to pull back the Soyuz. Chertok states that they misintepreted the situation because of inadequate simulations on the ground. Reportedly, the simulator did not provide an accurate image of the station when the deviation was more than 1°, and therefore during the actual docking attempt the cosmonauts saw, through their periscope, an image of the station different from the image they expected to see. The RKK Energiya history says that the cosmonauts thought the orientation of the *station* was not correct, and that they pulled the Soyuz back to a distance of 25 metres. Based on these two sources, it would appear that on this first attempt there was no actual contact between Soyuz 25 and Salyut 6. However, Yeliseyev (the flight director) writes in his memoirs that on the first attempt, the probe of Soyuz 25 failed to enter the drogue mechanism of Salyut's docking port. It does not appear, however, that the situation was in any way similar to that which occurred on the Soyuz 10 mission (see p.173). Either they pulled the Soyuz back *before* contact, or there *was* contact but without even a soft dock.

On the twenty-third orbit there was another attempt at hard docking, but again it was not achieved. With onboard supplies of power and fuel running low, there were no reserves left to try to dock with the aft port; and, moreover, it had been determined that the fault lay in the Soyuz docking equipment and not the Salyut port. Not wishing to inflict damage on the second port and risk putting the station out of commission before it was manned, it was decided to bring the crew home.

It is not really clear how many docking attempts Kovalyonok made after the initial attempt. Yeliseyev says that there was one more attempt; Chertok says there were two; and according to the RKK Energiya history, there were three – and no reasons for the failures are given. The RKK Energiya history relates that on the subsequent three attempts, 'the television system did not work properly'. It is possible that there was no actual contact between the two docking mechanisms until one of these later attempts.

In his memoirs, Yeliseyev reveals that a very dangerous situation arose *after* the failed docking attempts. For some reason the Soyuz – already low on fuel after the repeated docking failures – continued to hover dangerously close to the station, and any manoeuvre to bring it to a safe distance would have required precious feel needed for re-entry. Apparently, ground controllers were also not exactly sure of the mutual position of the two spacecraft, and were therefore wary of performing a manoeuvre which might have caused the Soyuz to collide with the station. In the event, Yeliseyev decided not to perform any manoeuvres, and instead hoped that natural forces would cause the two vehicles to drift apart. This eventually happened, but not until about two or three orbits later.

Soyuz 25 landed successfully on 11 October; but unfortunately, during part of the normal descent sequence, the docking equipment on the front of the OM was lost in the separation and destruction of the module, and so post-flight analysis of why the latches did not engage was not possible. In the post-flight investigation of the failure, it was decided that Soyuz 26 would be launched in December to dock with the aft port, and that the crew would perform an inspection EVA of the forward hatch area to assess any damage caused by the failed docking. If no serious damage was discovered, then Soyuz 27 would be launched to the forward port and the latter crew would return in Soyuz 26, freeing the rear port for the first unmanned resupply craft.

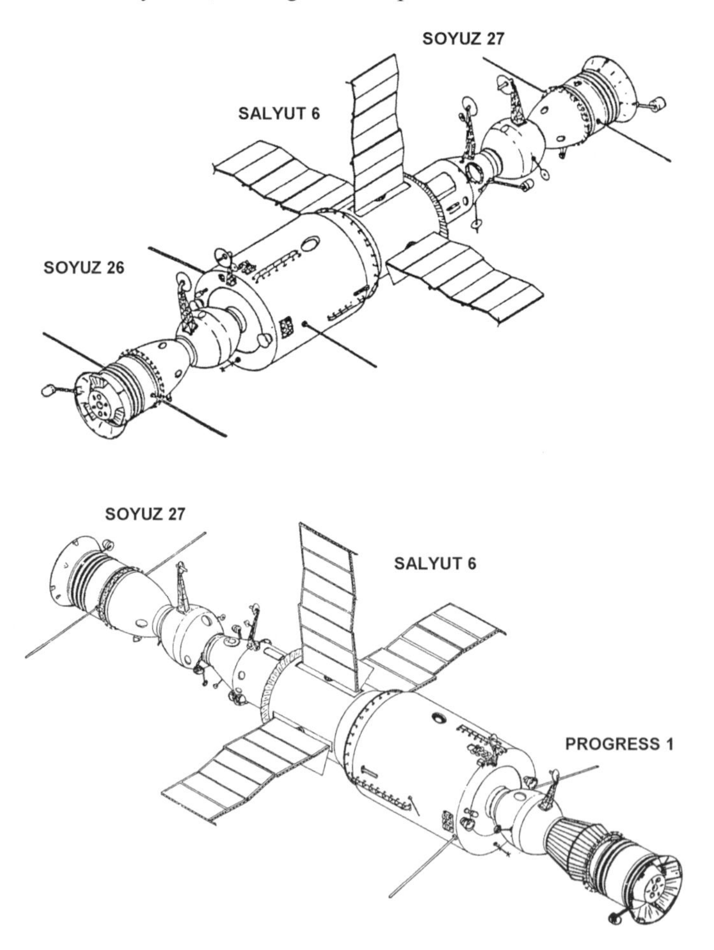

Soyuz and Progress docking configurations.

Success upon success

Soyuz 26, with cosmonauts Yuri Romanenko and Gregori Grechko (call-sign, Tamyr) on board, was launched on 10 December 1977, and docked to the rear port of Salyut 6, without incident, on 11 December. While reporting the successful manual docking, the Soviets for the first time revealed the presence of a second docking port at the rear of the Salyut. After three hours spent verifying the security of the docking, the crew entered the Salyut for what was to be a record-breaking 96-day mission. At mission control, former cosmonaut and now Salyut 6 flight director Alexei Yeliseyev reported that the crew had carried out the docking with exceptional accuracy, and both calmly and neatly; while Shatalov reported that the docking was 'very smooth, almost exemplary.' Leading spacecraft designer and former cosmonaut Konstantin Feoktistov explained that in the past, if a fault had developed in the Soyuz ferry while docked to the station, it would have necessitated the jettisoning of the spacecraft and the sending of a replacement. But now, with two docking ports, the station could simultaneously accommodate two spacecraft and expand the experimental programme with two crews working together. If the Soyuz was unable to undock, the second port offered greater flexibility for station operations. On 20 December, Grechko (supported by Romanenko from inside the unpressurised transfer compartment) performed the EVA to visually inspect the forward docking port, and during his 20-minute excursion, TV pictures were transmitted to mission control to reveal the state of the docking cone. It was reported that there were scratches on the docking cone where Soyuz 25 had attempted to dock, but that there was no serious or permanent damage.

As a result of the inspection EVA, plans were finalised for Soyuz 27 to attempt docking at the forward port. On 10 January 1978, Soyuz 27 was launched towards the Salyut station; and by the following day, Commander Vladimir Dzhanibekov and Flight Engineer Oleg Makarov (call-sign, Pamir) reported seeing the Salyut at 1.5 km. At 300 metres the lights on the Salyut were clearly seen, and at 10 metres the docking mechanism clearly revealed itself to the crew. Docking took place without incident – and for the first time, three spacecraft were linked together in space. For safety, Romanenko and Grechko had retreated to the safety of the Soyuz 26 DM in case of a collision or rapid depressurisation of the main Salyut compartments. All went smoothly, and three hours later the crews were inside the Salyut celebrating their success with a 'toast' of cold cherry juice squeezed from tubes.

Setting the standards

According to Feoktistov, the success in docking two ferry craft at the same space station opened up the possibilities of creating sophisticated engineering space complexes in Earth orbit. He also revealed the concern of some engineers about the possibility of creating a 'switch' effect that would cause a break between the spacecraft, but further calculations revealed that this was extremely unlikely. With the addition of a second docking port, the dynamic characteristics of the Salyut were seriously altered, and the mass, inertia and centre of gravity had to be recalculated. The increased length of the complex added to its mass, and it was therefore decided to test the dynamic loads on the whole complex in a programme of experiments called Rezonans (Resonance).

This research included mission control sending a metronome signal to Salyut as the cosmonauts ran on the treadmill or 'jumped' using bungee devices. Scientists at mission control looked for any vibrations that could lead to fatigue between the station and the attached ferry craft. In the worst-case scenario, appendages such as solar arrays and antennae could break away from the spacecraft. In order to monitor potential fatigue during future expansion and docking of spacecraft, the resonant frequencies would be amended. The results from the experiments carried out on Soyuz 26/Salyut 6/Soyuz 27 were encouraging, and led to confidence in the future operations of the Salyut station and the use of the second port for manned missions as well as unmanned resupply vehicles.

On 13 January the crews carried out what was to become a standard procedure on space station missions, and has continued with Salyut 7, Mir, and the ISS. They removed their individually moulded seat liners, and took them across to the other Soyuz, so that at the end of the mission, the crew of Soyuz 27 would fly home in Soyuz 26, and the crew of Soyuz 26 would fly home in Soyuz 27. Therefore, as the orbital lifetime of the Soyuz neared completion, a new crew could be launched to replace the resident crew already in orbit, or complete a short visiting mission to change out the older craft with a fresher vehicle to extend the duration of the mission, or eventually enable continuous permanent manning of the space station.

The docking of Soyuz 27 with the front port greatly relieved the mission planners on the ground, as it indicated that it was indeed the equipment on Soyuz 25, and not the equipment on Salyut, that had prevented that spacecraft from docking during previous October. On 16 January, the departure of the Soyuz 27 crew in Soyuz 26 freed the rear docking port for the first designated Progress resupply craft. The Progress was launched on 20 January, and docked to the rear port of Salyut 6 on 22 January (see p.253).

The Progress departed Salyut on 6 February, and again freed the rear port for another manned Soyuz – this time carrying the first Interkosmos cosmonaut from Czechoslovakia. Launched on 2 March, Soyuz 28 carried Soviet commander Alexei Gubarev and Czech cosmonaut Vladimir Remek (call-sign, Zenit), who successfully docked with Salyut's rear port, without incident, on 3 March. The first international crew undocked Soyuz 28 from the station a week later, on 10 March, and landed safely in a snow-covered field after a flight of 7 days 20 hrs. Six days later, the main Salyut crew vacated the station onboard Soyuz 27, and landed (again in a snow-covered field) 265 km west of Tselinograd. They had set a new space endurance record of 96 days 10 hs, surpassing the American Skylab 4 record of 84 days 1 hr 16 min, set in February 1974.

New beginnings

The success of the first expedition crew to Salyut 6 was followed three months later, on 15 June 1978, with the launch of Soyuz 29. Using the Igla rendezvous system, the crew – Commander Vladimir Kovalyonok and Flight Engineer Alexandr Ivanchenkov (call-sign, Photon) – docked with the forward port of Salyut, without

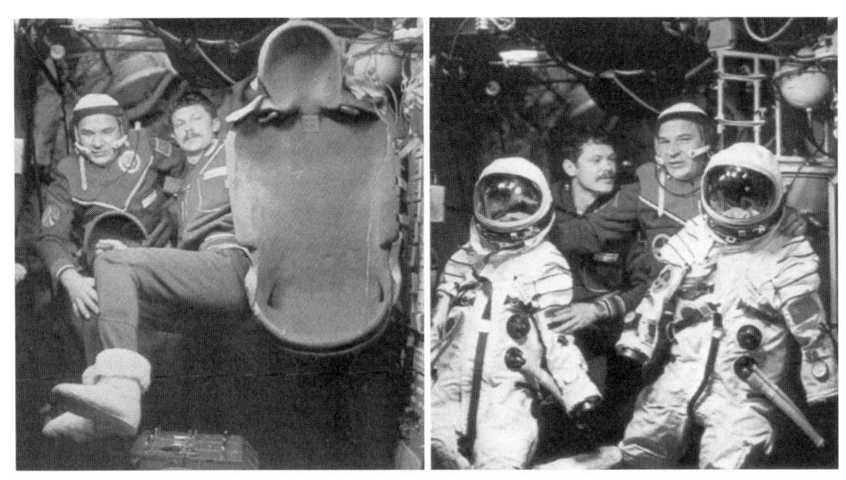

Frequent operations during exchanges of crews and spacecraft during Soyuz operations include the removal and exchange of seat liners and Sokol suits from one Soyuz to another. This was first completed during the Soyuz 27/26 mission in early 1978, and here the operation is shown during the Hungarian mission to Salyut (Soyuz 36) in 1980.

incident, on 16 June. Again, they broke the duration record with a mission of 139 day 14 hrs 48 min, and finally landing on 2 November 1978. During this residency, the crew received three unmanned Progress vehicles and two international crews. Soyuz 30, carrying Pyotr Klimuk and Polish cosmonaut Miroslaw Hermaszewski (call-sign Kavkas), was launched on 27 June 1978, and two days later docked at the aft port of the Salyut. On 5 July, this crew undocked their spacecraft and landed safely after a week of joint experiments. On 26 August – following the flights of Progress 2 and Progress 3 – the two-man Soyuz 31 mission was launched, carrying veteran cosmonaut Valeri Bykovsky and East German cosmonaut Sigmund Jahn (call-sign, Yastreb). They docked at the aft port on 27 August. The Soviet/GDR crew returned in the older Soyuz 29, leaving the fresher Soyuz 31 for the resident crew, and so during their week onboard the station they had exchanged the seat liners ('lodgements') between the two spacecraft. On 2 September – after Soyuz 29 had been in orbit for seventy days, and as the time for undocking approached – the cosmonauts tested the onboard systems and the main engines.

In addition to the seat liners, the centring weights also had to be transferred. Located beneath the flight seats, these are specially calibrated to take into account the weight of the cosmonauts and the equipment to be returned, in order to accurately determine the centre of gravity used in the ballistic determination of the re-entry and landing. The slightest miscalculation would cause an overshoot or undershoot of considerable distance, and might lead to serious injury to the crew. On 3 September, Soyuz 29 left the mooring port at the front of Salyut 6 and headed back to Earth. On the Salyut, the rear docking port needed to be freed to accept a new Progress resupply craft, and on 7 September the resident crew entered Soyuz 31 to carry out an operation which was later to become routine on Salyut, Mir and the ISS: the undocking and re-docking of a Soyuz at different locations.

Placing Salyut in a gravity-gradient mode, the crew entered Soyuz 31 at the aft of the station and sealed the hatches, after which they undocked and backed away 100–200 metres to maintain a station-keeping position. Mission control was then supposed to command a firing of Salyut thrusters to turn the station 180° to face its forward port towards the waiting Soyuz. However, as Soyuz 31 undocked, orbital mechanics connected to the gravity-gradient mode imparted a 90-degree pitch manoeuvre on the station, and so the thrusters needed only to turn Salyut the remaining 90°. The crew then moved their Soyuz in for a re-docking with the front port of the station, freeing the rear port for the fourth Progress vehicle to dock near the end of their mission.

Because of the success of Soyuz 26–31 and the Progress vehicles, in the space of twelve months, six manned Soyuz spacecraft and four resupply craft (ten vehicles) had successfully launched, docked with the same Salyut, and completed their planned mission objectives without serious incident. In addition, all the crews had been recovered normally at the end of their planned missions, thus doubling the success rate of the Salyut programme operations of 1971–77. Prospects were good for 1979, as a new resident crew prepared for launch and visits by other international crews were planned.

Six months in orbit

On 25 February 1979, the next resident crew for Salyut 6 – Commander Vladimir Lyakhov and Flight Engineer Valeri Ryumin (call-sign, Proton) – were launched onboard Soyuz 32 for a planned 6.5-month occupancy and further visits by Interkosmos cosmonauts. The docking with the forward port took place, without incident, on 26 February, and the crew soon began reactivating the station by placing an air hose between the station and the Soyuz to allow a flow of air into the mothballed ferry craft. One of the objectives of the crew was to evaluate the condition of the station in order to support both their own long mission and further crewing into 1980.

By April the cosmonauts had settled into their routine onboard the station, and were looking forward to the arrival of their first visitors onboard Soyuz 33. These would be the fourth Interkosmos crew of Soviet Commander Nikolai Rukavishnikov and Bulgarian Cosmonaut Researcher Georgi Ivanov, who were to spend a week onboard the station. The launch occurred in one of the highest winds recorded for a Soviet launch – 40 kmh; but this did not hinder the ascent, and within ten minutes Soyuz 33 was safely placed in orbit. The crew then spent their first three orbits configuring the Soyuz for orbital flight by deploying appendages, establishing communications with ground- and ocean-based communication stations, and checking the Igla approach and rendezvous system. During the fourth orbit they took off their Sokol suits, opened the connecting hatch to the OM, and prepared a meal, before firing the orientation engines on the fourth and fifth orbits to begin the long flight towards the Salyut.

By the next day, during the seventeenth orbit, five orbital corrections had placed Soyuz 33 on the correct approach path to intersect the orbit of Salyut. On the large display screen at the flight control centre, a pale blue light represented Salyut 6, and

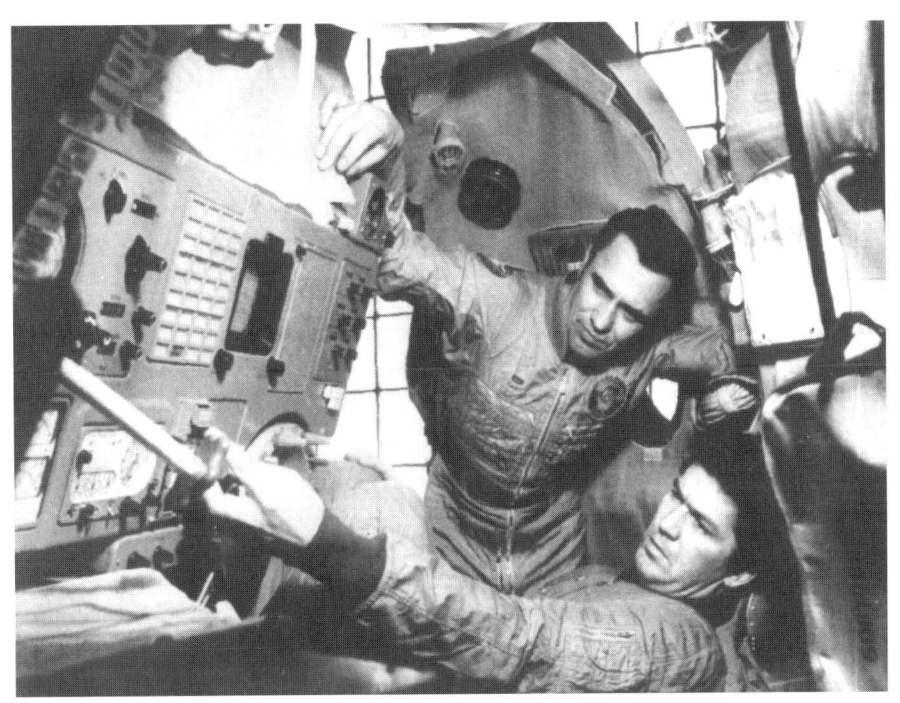

The Soyuz 35 crew inside the original Soyuz ferry simulator. Whilst strapped in the restraint couch, cosmonauts use a pointer stick to reach the control panel.

a red dot revealed Soyuz 33 closing in on the station. The Soyuz crew reported that they were well, and that all of their spacecraft's systems were functioning normally. They were ready for the final docking approach as the two spacecraft entered the same communications zone, allowing the flight comtrol centre to talk to both crews simultaneously.

Instructions were sent to the Soyuz crew to complete the final approach for a docking on the eighteenth orbit. As the two spacecraft approached to within 3 km, Rukavishnikov reported that all systems were functioning normally. It was then that both ground control and the crew noted 'deviations in the regular operating mode of the approach-correcting propulsion unit of the Soyuz 33 engine'. Rukavishnikov later commented that onboard the Soyuz, the crew noted something wrong with the functioning of the engine. It was supposed to fire for six seconds but shut down after only three seconds of erratic firing. Abnormal vibration – felt by both men strapped inside the DM – then shook the Soyuz, and the docking was aborted.[15] The problem was subsequently traced to the gas generator feeding the main engine's turbo-pump.

A failure in the engine

The crew had only read-outs of the functioning of the engine, and no data to diagnose the fault. Engine firing and shut-down was controlled from the flight control centre, with the cosmonauts using a stop-watch and reporting the operation to the ground. This engine would also be used for the de-orbit burn. Despite pleas from the crew to attempt a second approach, flight control wanted to ensure the safety of both the Soyuz 33 crew and the Salyut crew, and decided to follow the mission rules to bring the crew home as soon as possible by using the back-up engine. The back-up engine could have been used to attempt a docking with the Salyut; but this was too risky, because there would then be no back-up for de-orbit if the docking attempt failed.

Although obviously bitterly disappointed at not being able to complete their mission, the crew adhered to their training and prepared for an early descent. As the landing had been initiated due to an emergency, it would take place in the western area of the landing zone at the first opportunity, although this would not be until the following day. Despite their predicament, the Soyuz crew were still able to conduct some Earth observations before eating and settling down for a second night in space.

The following day – 12 April – the crew donned their Sokol suits, sealed the hatch between the DM and OM, and strapped themselves into their couches ready for landing. The engine was fired for 213 seconds to initiate the descent trajectory, and this was followed by the separation of the OM and PM. The burn lasted 25 seconds longer than normal, and Rukavishnikov had to manually shut it down; and because the back-up engine had been used, they were heading for a ballistic entry. During the 530 seconds of descent, the loads experienced by the crew reached 8–10 g instead of the more normal 3–4 g. Both cosmonauts reported that breathing was difficult for several minutes as the loads pressed against their bodies, but despite the difficulty they were able to talk to each other. Rukavishnikov later commented that as the vehicle descended he felt as if he were inside the flame of a blowtorch, with external temperatures of 3,000° C and a great deal of vibration and noise.[16]

After a period of tension in the flight control centre as they waited to re-establish communications with the crew after re-entry, the red-hot capsule was spotted by one of the pilots of the approaching recovery aircraft, and shortly afterwards communication was established. The spacecraft landed safely, but as normal it rolled on its side. By the time the recovery crews arrived, the cosmonauts had exited the spacecraft and were standing close by. With the crew safely home, attention then turned towards determining the cause of the engine failure. Unfortunately, the PM had been destroyed as planned during re-entry, and so the engineers had to analyse the telemetry data from the flight, which indicated that the primary engine combustion chamber pressure was lower than normal during the final approach burn.

Both the RKK Energiya history and Yeliseyev's memoirs state that the engine did not provide enough thrust and that the deceleration impulse was less than planned. Bart Hendrickx suggests that perhaps the automatic systems tried to compensate for the inadequate thrust by firing the engine longer than planned, but the crew of course would not be aware of this, and shut it down. Yeliseyev also states that Soyuz 33 came down close to the planned landing point, but that this was sheer coincidence. Because of the insufficient deceleration, it took longer for the Soyuz to enter the atmosphere, but this was compensated by the steeper re-entry angle. Rukavishnikov's account of the extended engine burn was also reported by Mikhail

Rebrov in 1996.[17] (The accurate reporting of Russian accidents and incidents is difficult, because there is so much conflicting information. What *would* be of use are the post-flight reports of the investigation boards rather than personal reminiscences; but it is doubtful when, if at all, these will be made available.)

Back on Salyut, the crew resigned themselves to a longer wait for visitors. Ryumin later wrote in his diary that both he and Lyakhov were watching the approach of the Soyuz with its blunt end forward, waiting for the braking burn that never came. The engine ignited and flickered, and they saw the change of engine exhaust colour and an abrupt shut-off. As Soyuz 33 sped past Salyut instead of slowing down, they realised that a serious problem had occurred on their colleagues' craft, and they therefore discussed how they might be able to use their own Soyuz 32 to assist them should they become stranded in orbit. This also led to concerns about their own engine onboard Soyuz 32, and the fact that the next mission, Soyuz 34, would be used to replace Soyuz 32 for the resident crew. Due to doubts about the Soyuz engine, the launch of Soyuz 34 was postponed pending the inquiry into the Soyuz 33 incident.

Soyuz 33 Commander Rukavishnikov later said that although the flight had lasted only two days, it had seemed like a month. It was particularly frustrating for Rukavishnikov, who was a member of the Soyuz 10 crew that had failed to enter Salyut 1. With the loss of Salyuts before crews could reach them in 1972 and 1973, he had been reassigned from the Salyut training group to the ASTP training group before returning to Salyut training, only to be disappointed in once again failing to reach a space station.

The Soyuz 33 crew also faced the possibility of being marooned in orbit with little hope of rescue. Calculations taken on the day of the incident revealed that natural decay would result in the re-entry of Soyuz 33 ten days later, but there were only five days of breathable oxygen onboard the spacecraft, and just one day of electrical power. Rukavishnikov realised that if the main engine was damaged, then perhaps the back-up engine might also be affected. The priority was to return the crew to Earth, as time and consumables were low. According to the crew, the night was long, and neither of them had much sleep. Rukavishnikov felt the burden of command, in the responsibility for his mission, his spacecraft and his crew. In order to take their minds off the situation, they opened some of the gifts that they had intended to give to the main crew. The contents were not revealed, but they reportedly 'fortified themselves'.

In determining the cause of the engine failure, the Salyut crew's comments on the performance helped pinpoint the cause of the problem. The engine failed after only three seconds of its seventh burn of the mission. It appeared that a pressure sensor in the rocket's combustion chamber had terminated the burn when it detected off-normal performance, and this in turn prevented fuel from being pumped into an engine that was not performing correctly. The suspect part that failed had been tested 8,000 times without failure and the Soyuz engine had performed 2,000 firings in space during flights begun in 1967. Amendments made to the engine would be incorporated in future flight versions.

These improvements would be tested on the unmanned Soyuz 34, which was launched on 6 June 1979. The spacecraft and its systems performed flawlessly, and docked automatically with the aft port of Salyut on 9 June, delivering further biological samples for the experiment programme. The older Soyuz 32 was filled with experiment samples and data, and undocked from Salyut, without a crew, on 13 June. The Soyuz carried 50 kg of experiment results, and 130 kg of replaced instruments from the station, to be inspected back on Earth. There were also samples from the Kristall furnaces, film rolls from cameras, biological specimens, and other used equipment. After three independent orbits, Soyuz 32 re-entered, and landed safely. The following day, the crew moved Soyuz 34 from the aft port to the forward port as Salyut turned 180°, clearing the rear port for further Progress supply craft. Two months later, Soyuz 34 became the vehicle that ferried the Proton crew back to Earth, landing without incident on 19 August 1979 after a 175-day flight. This crew also became the first to use the ground recovery system – a framework of platforms and a slide – to exit the upright spacecraft.

The end of an era

The next resident crew – Leonid Popov and Valeri Ryumin (call-sign, Dneiper) – was launched to Salyut 6 on 9 April 1980, and docked the next day. Soyuz 36 was then launched on 26 May, carrying the fifth Interkosmos crew of Valeri Kubasov and Hungarian cosmonaut Bertalan Farkas (call-sign, Orion). This was the mission delayed from July 1979, but on 27 May 1980 they finally arrived on the Salyut for a week of joint experiments. The international crew filled the OM of their Soyuz 35 return spacecraft with unwanted rubbish, to be incinerated when the OM was de-orbited. On 3 June they undocked from Salyut, and completed a successful landing

A rare view (from Apollo during the ASTP in July 1975) of the rear of a Soyuz in orbital flight, showing the KDTU engine location in the aft of the Propulsion Module.

later that day, although the soft-landing engines failed because the altimeter failed to send a signal to the engines. Neither cosmonaut was injured by the landing. The next day, the resident crew moved the new Soyuz 36 from the aft port to the forward port of Salyut.

The next mission was the first manned flight of the Soyuz T spacecraft (see p.284), followed by two further Interkosmos missions flying the older Soyuz spacecraft. Soyuz 37 carried the sixth Interkosmos crew of Viktor Gorbatko and Vietnamese cosmonaut Pham Tuan, who were launched on 23 July and docked with the aft port of Salyut a day later. On 28 July the international crew moved their personal belongings and couch liners across to Soyuz 36 for the return to Earth, which was accomplished without incident on 31 July. The following day – 1 August – the resident crew once again moved a newer vehicle from the aft to the forward port of the Salyut. The final mission to this resident crew was launched onboard Soyuz 38 on 18 September. The crew – Commander Yuri Romanenko and Cuban cosmonaut Arnaldo Tamayo-Mendez (call-sign, Tamyr) – docked to the aft port on 20 September, and spent a week onboard the station. On this occasion a spacecraft exchange was not planned, and so they undocked and landed in Soyuz 38 on 26 September. Despite having to land in the dark to allow another Progress to be launched in the same launch window, this was one of the most accurate landings in the Soviet programme – only 3 km from the target point. On 11 October the resident crew landed in Soyuz 37, after a record flight of 185 days.

The next two manned launches were Soyuz T mission. Soyuz T-3 (see p.294) was a maintenance mission designed to evaluate the station for a further residency flight in 1981, and Soyuz T-4 (see p.294) subsequently flew that residency mission for 74 days – the final long-duration mission to the record-breaking Salyut 6 station. During the T-4 mission, the final two original Soyuz ferry craft flew to the station.

On 22 March 1981, Soyuz 39 was launched, carrying the eighth international crew of Commander Vladimir Dzhanibekov and Mongolian cosmonaut Judgeremidiyin Gurragcha. They docked with the aft port of Salyut on 23 March. As the main crew had arrived onboard the modernised Soyuz T, there was need to change out the new spacecraft with an older model, and so the Soyuz 38 visitors' landing, on 30 March, took place in fog and rain, although the crew was quickly located and recovered without incident. The ninth international crew – Leonid Popov and Romanian cosmonaut Dumitriu Prunariu – were the last crew to fly a Soyuz of the ferry type and original designation. Soyuz 40 was launched on 14 May 1981, and docked with the aft port of Salyut 6 on 15 May. The final original Soyuz began its return to Earth on 22 May, and after landing, Popov praised the work of the Soyuz designers in the reliability of the original series of ferry craft, as the Soyuz T spacecraft took over the role of station ferry. The work of these designers in overcoming the problems of the early 7K-T craft was clearly demonstrated with the successes at the Salyut 6 station, with only two manned 7K-T missions (Soyuz 25 and Soyuz 33) failing to dock out of a total of sixteen attempts.

SUMMARY

In its role as a space station ferry between 1971 and 1981, thirty-three Soyuz 7K-T spacecraft logged more than 1,176 days in space in both its unmanned and manned roles in support of Salyut stations and solo flight operations. Despite early problems with the Igla docking system, the 7K-T vehicle proved its role as a space station ferry craft and its versatility in supporting other objectives while the improved Soyuz T and TM spacecraft were under development.

REFERENCES

1 Private conversation between Vladimir Shatalov and D.J. Shayler, Northampton, England, 19 May 2002.
2 Hall, Rex D. and Shayler, David J., *The Rocket Men*, Springer–Praxis (2001), pp.91–94.
3 Severin, G.I., Abrahamov, I.P. and V.I. Svertshek, 'Crewman Rescue Equipment in Manned Space Missions: Aspects of Application', presented at the 38th Congress of the IAF, 10-17 October 1987, Brighton, UK; *Aviation Week and Space Technology*, 26 October 1987, p.75; Soyuz KB2 Ultra light Pressure suit for Soyuz TM, Aviaexport USSR press release (undated).
4 Siddiqi, Asif, 'The Almaz Space Station Complex: a History, 1964-1992. Part 1, 1964–1976', *Journal of the British Interplanetary Society*, **54**, (2001), 389–416.
5 Lazarev, V., *Orbits of Peace and Progress, 1984, A Contingency Situation*, pp.194–200.
6 *Novosti Kosmonavtika*, No.6, 2000.
7 Siddiqi, Asif, 'The Almaz Space Station Complex: a History, 1964-1992. Part 2, 1976–1992', *Journal of the British Interplanetary Society*, **54**, (2002), 35–67.
8 *Ibid.*, pp.39–43; Haeseler, D., 'The Soviets only water recovery', *Spaceflight*, **37**, (August 1995), 283; Shayler, David J., *Accidents and Disasters in Manned Spaceflight*, Springer–Praxis, 2000, pp.364–367.
9 Davydov, Iodif, *Triumph and Tragedy of Soviet Cosmonautics, through the Eyes of a Tester,* Globus, 2000.
10 Ezell, Edward and Ezell, Linda, *The Partnership: a History of the Apollo Soyuz Test Project*, NASA SP-4209, 1978, pp.1–14.
11 Hall, Rex D. and Shayler, David J., *The Rocket Men*, Springer–Praxis (2001).
12 Shayler, David J., *Skylab: America's Space Station*, Springer–Praxis (2001).
13 Oberg, James, *Red Star in Orbit*, p.143.
14 Shayler, David J., 'The Proposed USSR Salyut and US Shuttle Docking Mission, circa 1981', *Journal of the British Interplanetary Society*, **44**, (1991), 553–562.
15 Shayler, David J., *Accidents and Disasters in Manned Spaceflight.*
16 Blagov, V., 'A Day after the Blast Off', in *Orbits of Peace and Progress*, P. Popovich (*ed.*), Mir Publishers, 1998.
17 Rebrov, Mikhail, *Space Catastrophes*, Eksprint, Moscow, 1994.

Soyuz variants: ferry missions and hardware

Spacecraft (serial number) name	Design designation	Launch date	International designation	Docking date	Undocking date	Target spacecraft (port)	Landing/ decay date	Flight duration dd:hh:mm:ss
Key: (U), unmanned								
11F615 A8 series (Soyuz transport manned spacecraft for DOS (Salyut) space stations)								
(31) Soyuz 10	7K-T	1971 Apr 22	1971-034A	1971 Apr 24	*Unable to hard dock*	Salyut 1 (Fwd)	1971 Apr 24	01:23:45:54
(32) Soyuz 11	7K-T	1971 Jun 6	1971-053A	1971 Jun 7	1971 Jun 29	Salyut 1 (Fwd)	1971 Jun 30	23:18:21:43
(33) Soyuz '12A'	7K-T	*1971 planned; cancelled*				Salyut 1 (Fwd)		
(33A) Cosmos 496 (U)	7K-T	1972 Jun 26	1972-045A				1972 Jul 2	05:23:02:00
(34) Soyuz '12B'	7K-T	*1972 Jul planned; cancelled*				Salyut *launch failure*		
(34) Soyuz '12D'	7K-T	*1973 May planned; cancelled*				Cosmos 557		
(33) Soyuz 13	7K-T	1973 Dec 18	1973-103A				1973 Dec 26	07:20:55:35
(34A) Cosmos 613 (U)	7K-T	1973 Nov 30	1973-096A				1974 Jan 29	60:00:09:??
(35) (Unused)	7K-T	*Intended for DOS 2 and DOS 3; adapted for unmanned test flight, but never flew*						
(36) Cosmos 573	7K-T	1973 Jun 15	1973-041A				1973 Jun 17	02:00:04:00
(37) Soyuz 12	7K-T	1973 Sep 27	1973-067A				1973 Sep 29	01:23:15:32
(38) Soyuz 17	7K-T	1975 Jan 10	1975-001A	1975 Jan 12	1975 Feb 9	Salyut 4 (Fwd)	1975 Feb 9	29:13:19:45
(39) Soyuz 18-1	7K-T	1975 Apr 5	*Ballistic trajectory (180 km) qualified as a spaceflight*			Salyut 4 (Fwd)	1975 Apr 5	00:00:21:27
(40) Soyuz 18	7K-T	1975 May 24	1975-044A	1975 May 25	1975 Jul 26	Salyut 4 (Fwd)	1975 Jul 26	62:23:20:08
(41)	7K-T	*Flew as Soyuz 21*						
(42) Soyuz 25	7K-T	1977 Oct 9	1977-099A	1977 Oct 10	Unable to hard dock	Salyut 6 (Fwd)	1977 Oct 11	02:00:44:45
(43) Soyuz 26	7K-T	1977 Dec 10	1977-113A	1977 Dec 11	1978 Jan 16	Salyut 6 (Aft)	1978 Jan 16	37:10:06:19
(44) Soyuz 27	7K-T	1978 Jan 10	1978-003A	1978 Jan 11	1978 Mar 16	Salyut 6 (Fwd)	1978 Mar 16	64:22:52:47
(45) Soyuz 28	7K-T	1978 Mar 2	1978-023A	1978 Mar 3	1978 Mar 10	Salyut 6 (Aft)	1978 Mar 10	07:22:16:30
(46) Soyuz 29	7K-T	1978 Jun 15	1978-061A	1978 Jun 16	1978 Sep 3	Salyut 6 (Fwd)	1978 Sep 3	79:15:23:49
(47) Soyuz 31	7K-T	1978 Aug 26	1978-081A	1978 Aug 27	1978 Sep 7	Salyut 6 (Aft)		

					1978 Sep 7	1978 Nov 2	Salyut 6 (Fwd)	1978 Nov 2	67:20:12:47
(48)	Soyuz 32	7K-T	1979 Feb 25	1979-018A	1979 Feb 26	1979 Jun 13	Salyut 6 (Fwd)	1978 Jun 13	108:04:24:37
(49)	Soyuz 33	7K-T	1979 Apr 10	1979-029A	*Soyuz engine failure; docking aborted*		Salyut 6 (Aft)	1979 Apr 12	01:23:01:06
(50)	Soyuz '34A'	7K-T	*1979 Jun planned; cancelled*				Salyut 6		
(50)	Soyuz 34 (U)	7K-T	1979 Jun 6	1979-049A	1979 Jun 8	1979 Jun 14	Salyut 6 (Aft)		
					1979 Jun 14	1979 Aug 19	Salyut 6 (Fwd)	1979 Aug 19	73:18:16:45
(51)	Soyuz 35	7K-T	1980 Apr 9	1980-027A	1980 Apr 10	1979 Jun 3	Salyut 6 (Fwd)	1980 Jun 3	55:01:28:01
(52)	Soyuz 36	7K-T	1980 May 26	1980-041A	1980 May 27	1980 Jun 4	Salyut 6 (Aft)		
					1980 Jun 4	1980 Jul 31	Salyut 6 (Fwd)	1980 Jul 31	65:20:54:23
(53)	Soyuz 37	7K-T	1980 Jul 23	1980-064A	1980 Jul 24	1980 Aug 1	Salyut 6 (Aft)		
					1980 Aug 1	1980 Oct 11	Salyut 6 (Fwd)	1980 Oct 11	79:15:16:54
(54)	Soyuz 38	7K-T	1980 Sep 18	1980-075A	1980 Sep 19	1980 Sep 26	Salyut 6 (Aft)	1980 Sep 26	07:20:43:24
(55)	Soyuz 39	7K-T	1981 Mar 22	1981-029A	1981 Mar 23	1981 Mar 30	Salyut 6 (Fwd)	1981 Mar 30	07:20:42:03
(56)	Soyuz 40	7K-T	1981 May 14	1981-042A	1980 May 15	1981 May 22	Salyut 6 (Aft)	1981 May 22	07:20:41:52
(61)	Soyuz 20 (U)	7K-T	1975 Nov 17	1975-106A	1975 Nov 19	1976 Feb 15	Salyut 4 (Aft)	1976 Feb 16	90:11:52:00
(67)	Soyuz 30	7K-T	1978 Jun 27	1978-065A	1978 Jun 28	1978 Jul 5	Salyut 6 (Aft)	1978 Jul 5	07:22:02:59

11F615 A9 (Soyuz transport manned spacecraft for OPS Almaz)

(41)	Soyuz 21	7K-T	1976 Jul 6	1976-064A	1976 Jul 7	1976 Aug 24	Salyut 5 (Aft)	1976 Aug 24	49:06:23:32
(61)	Soyuz '12C'	7K-T	*1973 Apr planned/cancelled*				Salyut 2 (Aft)		
(61)	Cosmos 656	7K-T (U)	1974 May 27	1974-036A				1974 May 29	02:02:07:05
(62)	Soyuz 14	7K-T	1974 Jul 3	1974-051A	1974 Jul 4	1974 Jul 19	Salyut 3 (Aft)	1974 Jul 19	15:17:30:28
(63)	Soyuz 15	7K-T	1974 Aug 26	1974-067A	*Unable to dock with target spacecraft*		Salyut 3 (Aft)	1974 Aug 28	02:00:12:11
(65)	Soyuz 23	7K-T	1976 Oct 14	1976-100A	*Unable to dock with target spacecraft*		Salyut 5 (Aft)	1976 Oct 16	02:00:06:35
(66)	Soyuz 24	7K-T	1977 Feb 7	1977-008A	1977 Feb 8	1977 Feb 25	Salyut 5 (Aft)	1977 Feb 25	17:17:25:58

11F615 A12 (Soyuz manned spacecraft for (ASTP) ApolloSoyuz Test Project)

(71)	Cosmos 638	7K-TM (U)	1974 Apr 3	1974-018A				1974 Apr 13	09:21:35:30
(72)	Cosmos 672	7K-TM (U)	1974 Aug 12	1974-064A				1974 Aug 18	05:22:40:00

Spacecraft (serial number) name		Design designation	Launch date	International designation	Docking date	Undocking date	Target spacecraft (port)	Landing/ decay date	Flight duration dd:hh:mm:ss
(73)	Soyuz 16	7K-TM	1974 Dec 2	1974-096A	*Tested ASTP docking system on nose of OM*			1974 Dec 8	05:22:23:35
(74)	Soyuz 22	7K-TM	1976 Sep 15	1976-093A				1976 Sep 23	07:21:52:17
(75)	Soyuz 19	7K-TM	1975 Jul 15	1975-065A	1975 Jul 17	1975 Jul 19	US Apollo (18)		
					1975 Jul 19	1975 Jul 19	US Apollo (18)	1975 Jul 21	05:22:30:51
(76)	Soyuz '19A'	*This was the unused ASTP standby vehicle on the pad. The Descent Module was later used for vehicle no. 47, and the Orbital Module and Propulsion Module were subsequently used for a Soyuz mock-up at the RKK Energiya museum.*							

Progress, 1978–

In January 1978 a new robotic version of Soyuz was introduced as part of the manned space station programme. Called Progress, it outwardly resembled the familiar Soyuz design, but featured completely new systems and components designed to support logistical unmanned resupply of station consumables, crew provisions and equipment. This was a significant development in the Soviet space programme, and a key element in prolonging the operational use of Salyut beyond the six-month useful lifetime of the early stations. It would also prove to be one of the most fundamental elements in allowing the Soviets to achieve permanent occupation of a space station over several years. As with Soyuz, Progress has been upgraded over the years, and twenty-five years after its maiden flight the latest version continues to play a crucial role in ISS operations.

THE DEVELOPMENT OF PROGRESS

To obtain maximum benefits from a space station it should remain in orbit for as long as possible. Even at the dawn of the space age there had been suggestions that space platforms could be resupplied by smaller spacecraft on a regular 'shuttle' service to and from Earth. During the 1960s, in design studies for early American space stations NASA evaluated the use of the Gemini or Apollo CSM for such a role, before deciding upon a new manned spacecraft – the Space Shuttle – to ferry supplies and crews to and from orbital space bases that would also be serviced by inter-orbital space tugs. Unfortunately, the funding for such an infrastructure was not forthcoming, and sole reliance on the manned shuttle as a cargo logisitics vehicle has proven to be both costly and complex.

In the Soviet Union during the late 1960s, the Soyuz was being flight-proven in a series of manned and unmanned test flights to support both lunar and space station operations. It was soon realised that a resupply vehicle based on the flight-proven Soyuz could adequately support long-term space station operations without the need for a costly new vehicle. It would also be much simpler to develop as an unmanned craft, alleviating the complexities and launch weight penalties of life support, crew rescue and recovery sub-systems. By removing the launch escape system, crew provisions, and a re-entry heat shield, there would be a considerable weight saving that would provide greater capacity for cargo deliveries to the station. In addition, any rubbish accumulated in the station could easily be transferred to the empty cargo hold. The craft could then be commanded to destructive re-entry in the atmosphere, alleviating the problem of what to do with all the rubbish accumulated on orbit over a long period of time.

During the early 1970s, therefore, the Soviets decided to develop a space station cargo ship by taking the basic Soyuz spacecraft, launch shroud and R-7, and modifying it for a new unmanned support role. Work on the vehicle began in mid-1973, and by February 1974 a draft plan was issued for what would later be identified as Progress. The reason for calling it Progress is unclear. The Russians recycle older names for different projects, but in this case it might imply significant progress in extending space station operations almost indefinitely. The prime manufacturer, TsKBEM (formerly OKB-1), gave the new design the designation 7K-TG (transportnyy gruzovoi – transport cargo) and the initial design was assigned article number 11F615A15.

Military cargo ships

A cargo ship was also considered for Mishin's Soyuz VI military space station. It was to be called 7K-SG (or 11F735), and would have been based on 7K-S (the later Soyuz T). Some of the ideas for 7K-SG may have been incorporated into the first-generation Progress, but it was actually built on the basis of 7K-T (the 'Soyuz 12' type). Progress never inherited any systems from Soyuz T, and the first-generation Progress was based on the 'Soyuz 12'-type spacecraft, Progress M was based on Soyuz TM, and it therefore appears that the Soyuz T phase in Progress development was skipped (not taking 7K-SG into account).

The TKS vehicles for Chelomei's Almaz stations were also, in essence, cargo ships (aside from delivering crews). Work on these began in the late 1960s – long before Progress. The TKS (Transport Logistics Spacecraft) vehicles consisted of two spacecraft joined together, and were to be launched by three-stage Proton vehicles. Merkur was a conical re-entry capsule for up to three cosmonauts. Joined to it until re-entry was a slim forward section containing the propulsion system. The whole capsule was reusable, and the aft end was covered by a re-entry heat shield. Attached to the aft of Merkur was the Functional Cargo Block (FGB), which housed a pressurised compartment and, in its aft end, a probe unit for docking to Almaz space stations. Crew access was via a hatch in Merkur's heat shield, similar to the American MOL/Gemini design. The FGB was filled with cargo and supplies for the station, and the Merkur capsule was used by the three-person crew for access, to and from orbit. The launch weight was about 19,000 kg, and it had two 400-kg main thrust engines, an overall length of 17.51 metres, a diameter of 4.15 metres, and a span of 60 square metres across the solar array wing span. The Merkur capsule was 3 metres in diameter and 2 metres in height. The crew launched and landed in Merkur, but for docking they operated the spacecraft from an aft control post in the FGB.[1] Tests of the TKS took place on Cosmos 881/882 in 1976, Cosmos 929 in 1977–78,

Cosmos 997/998 in 1978, Cosmos 1100/1101 in 1979, Cosmos 1267 in 1982, and Cosmos 1443 in 1983. Modified TKS/Merkur capsules flew as Cosmos 1686, and remained docked to Salyut 7 from 1985 to 1991.

Automated docking tests

Flying unmanned versions of manned craft was nothing new to the Soviets, who had converted Vostok spacecraft for an unmanned military reconnaissance role from 1962.[2] Both the Americans and the Soviets had also flight-tested manned spacecraft in unmanned roles, but the cargo craft was something radically different.

The vehicle would be prepared in a similar way to the manned Soyuz, using modified Soyuz/R-7 facilities at Tyuratam Cosmodrome. It would be flown in an automated approach and docking with the station, initially controlled from ground control (although on later versions of Progress, the cosmonauts could control the docking themselves). The vehicle would have to have the capacity to automatically transfer fresh supplies of propellant and air into the station storage tanks, as well as a compartment for dry cargo that the station crew would unload over several days. Once the freighter was emptied, the crew could then fill it with rubbish. After the sealing of the hatches, the craft would be automatically undocked from the station, and later commanded to a predetermined destructive re-entry over the Pacific Ocean, to incinerate itself and its cargo in the atmosphere.

This automated docking capability had been demonstrated some years earlier. In 1967, Soyuz flight-testing hardware was under development for the manned lunar programme, part of which involved perfecting the techniques of rendezvous and docking of both manned and unmanned vehicles. This technique (described on p.139) was first used for the docking of Cosmos 186 and Cosmos 188 in October 1967, and the exercise was repeated with Cosmos 212 and Cosmos 213 in April 1968 (see p.142). Although related to the manned programme, it provided the Soviets with experience in automated docking in orbit, although it was not repeated until Soyuz 20 docked with Salyut 4 in November 1975. The next automated docking took place during the mission of the first Progress, which docked with Salyut 6 in January 1978.

Progress precursor test flights

Although Soyuz 20 (see p.193) was a test of the in-orbit storage capability of the Soyuz ferry, senior spacecraft designer and former cosmonaut Konstantin Feoktistov has subsequently stated that it was also a test of the automated transport systems intended for the Progress series. In addition, the docking of the manned Soyuz 27 to the Soyuz 26/Salyut 6 combination demonstrated the ability to safely attach a second craft to an occupied station, although the resident Salyut crew retired to the Soyuz DM in case anything went wrong with the docking approach of Soyuz 27. This helped qualify the launch of Progress 1 by providing data on integrity and structural strength when joining more than two vehicles together in orbit – an important piece of information, as Progress was planned to transfer propellant across to the station while the crew was on board.

The role of Progress

After the first Progress launch, Feoktistov explained the reasoning behind the introduction of the resupply craft, the development of which coincided with the development of the first Salyut stations during the early 1970s. 'Even at that time we were aware that in order to ensure the functioning of space stations over prolonged periods of time, both manned launches and cargo flights would be necessary.' Feoktistov also pointed out the additional advantages of such a cargo craft in its ability to carry sensitive biological payloads that have to be used for only a short time after they arrive at the station, or replacement photographic film to prevent the deterioration caused by long-term storage. The Americans experienced the restrictions of not having a regular resupply capability during the Skylab programme in 1973–74. Each of the three Command Modules had to be filled with additional supplies to replace or supplement the cargo and consumables launched with the unmanned station.[3]

As a space station mission continues, a new discovery or procedure might be revealed that requires a completely new set of hardware to be delivered to the station. Alternatively, old or faulty hardware might need to be exchanged for spare or upgraded parts. Feoktistov also emphasised the ability to restock onboard manoeuvring fuel supplies, air for the artificial atmosphere (replacing the quantities lost by repeated hatch openings and EVA airlock operations), or the restocking of carbon dioxide absorbers to clean the air. He quoted a daily use of expendable materials such as clothing, linen, food, medicines and water that equated to about 30 kg per person per day. Without Progress this would require several tonnes of cargo to be launched with the station in order to fully support the planned mission; and as the number of cosmonauts in orbit increases, so does the amount of logisitics required. Senior Salyut 6 flight controller and former cosmonaut Alexei Yeliseyev added that sending up cargo by unmanned automated spacecraft was by far the best way to prolong the active functioning of an orbital station, and that although there remained many hurdles to overcome before Konstantin Tsiolkovsky and Sergei Korolyov's forecast of a permanent human presence in space was finally realised, the flight of the first Progress represented a major step towards that goal.

Progress variants

Before describing the flight operations of Progress, it is worth explaining the variants used over a period of twenty-five years from 1978. Use of the Progress required the target station to incorporate a docking port featuring connections and communication links to transfer fuel and air from the freighter's tanks to those on board the station. The early Salyut and Almaz stations were not equipped for this, but by 1973 designs had been begun on the development of a station with more than one docking facility to allow for supply transfer. The development of Progress was therefore conducted in parallel to the 'second-generation' space stations.

On 20 January 1978, the first Progress-designated vehicle was launched to Salyut 6 with great success, demonstrating its capability to resupply the orbital station. Between 1978 and the end of the Salyut 6 programme in 1981, twelve Progress vehicles were used to support station operations. In April 1982, Salyut 7 replaced the

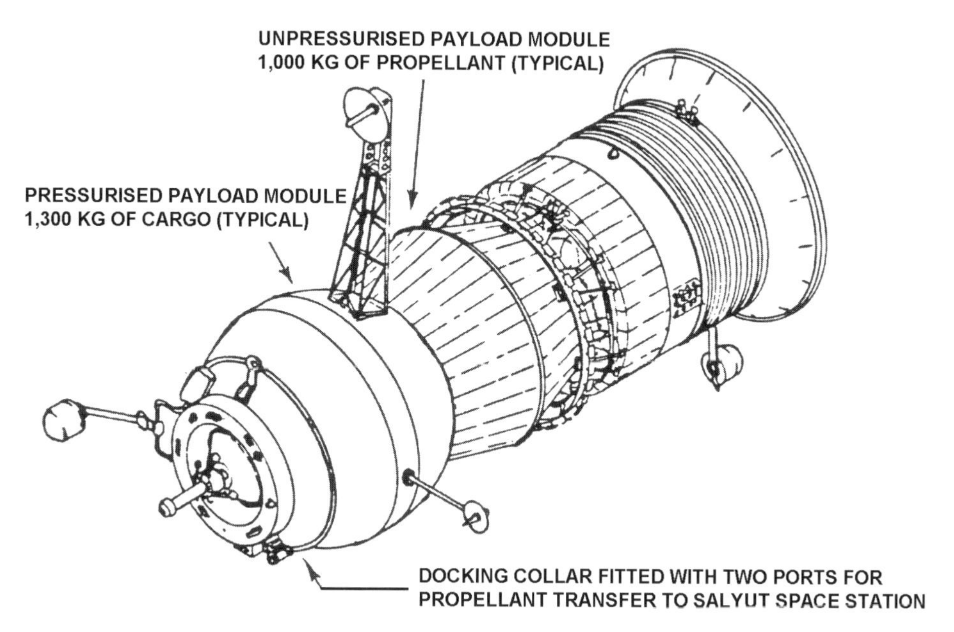

Progress (1–40) resupply freighter. The later Progress M and M1 had an additional pair of wing-like solar arrays. (Courtesy Ralph Gibbons.)

ageing Salyut 6, and over the next three years another thirteen Progress vehicles (including Cosmos 1669) were used to support the new station. The longevity of the Progress design was clearly demonstrated during the fifteen-year career of the space complex Mir where, between March 1986 and May 1990, another eighteen 'first-generation' Progress freighters delivered supplies to the station (see the Table on p.272).

During August 1989, an upgraded version of the Progress (Progress M – Modification), capable of delivering a larger payload, was introduced, and a new article number 11F615A55 and designation of 7K-TGM (again, the additional M refers to the modification of the original design) was assigned to it (see the Table on p.272). A total of forty-three Progress Ms were launched to the Mir station, and a further four were assigned to early resupply missions to the ISS. A second upgrade to Progress (for greater propellant delivery) began operations in 2000 when the first of the Progress M1 vehicles was launched to both Mir and the ISS. It is not clear why no new designator was issued to this variant; however, it has been observed that on occasions, different designators have been assigned to similar Soviet/Russian vehicles with almost no obvious physical changes. Equally, some spacecraft that are evidently different have retained the same designators!

Other, unflown variants (see p.407) in the series include Progress MT (11F615A75) and M2 (11F615A77), which were originally Zenit-launched vehicles developed to support the abandoned Mir 2 space station, with a mission potential far greater than just a tanker/freighter. When funding restrictions forced the cancellation of the Progress MT and M2 to Mir 2, the M2 design was reassigned to the ISS before it too was cancelled. Then, an alternative Progress design (MM) was proposed, but this also had to be scaled down to the more modest Progress MS, which is intended to support future ISS operations. Progress M3 – essentially an upgraded Soyuz with an NK-33 engine in the core stage – was a short-lived plan for the Yamal heavy-lift launch vehicle.

By January 2003, almost a hundred Progress, Progress M and Progress M1 vehicles had been sent to four different space stations (Salyut 6, Salyut 7, Mir, and the ISS) over a 25-year period. This remarkable record will undoubtedly continue in support of ISS operations in the foreseeable future.

PROGRESS HARDWARE

The design of the Progress vehicle relied to a great extent on existing Soyuz hardware in order to use the manned vehicle processing infrastructure, the R-7 launch vehicle, the Soyuz launch shroud, and the launch facilities at Baikonur. By using compatible Soyuz rendezvous and docking systems, as well as other off-the-shelf hardware for the modified instrument compartment, both time and substantial funds were saved by not having to develop new hardware and systems.

The overall length of the Progress has been quoted at 7.9 metres, with a maximum diameter of 2.7 metres; and the total launch weight has been quoted as 7,020–7,240 kg, with a cargo capacity of 2,300–2,500 kg. The independent design life was approximately three days, with a 30-day orbital storage capability while docked to the Salyut station.

Almost every source cites different masses and dimensions. Glushko's encyclopaedia,[4] however, is one of the most authoritative, and lists the following overall dimensions and masses: mass of fully loaded vehicle, about 7 tonnes; cargo mass, 2.3 tonnes (1.3 tonnes in the Cargo Module, and 1 tonne in the propellant compartment); length, 7 metres; maximum diameter, 2.72 metres; autonomous flight capability, four days, as part of station up to two months. The RKK Energiya history cites an overall mass of 7,020 kg (although this is slightly different for each vehicle).

Progress 1–19 were all launched by means of a Soyuz 11A511U R-7 variant; Progress 20 used the Soyuz U-2 (11A511U2) variant for the first time, for the greater payload lift capability intended to support Mir operations; and Progress 25–42 – the remaining first-generation vehicles – were all launched on the U2 variant on resupply flights to Mir. The early Progress launches also utilised a Soyuz-type shroud, with the launch escape system included to maintain the flight-proven aerodynamics during ascent through the lower atmosphere. The escape motors and deployable stabilisers were, however, not required, and were removed, although the shroud separation motors were retained as on the Soyuz configuration.

The Cargo Module (CM)

The Orbital Module was modified to carry dry cargo to the station and to provide a location for the rubbish bags being dumped. The *Kosmonavtika Entsiklopedia* cites the following dimensions for the Cargo Module: maximum diameter, 2.2 metres; length (including docking unit), 3.15 metres; internal volume, 6.6 cubic metres (although this includes volume that could not actually be used by the cosmonauts). The same book quotes 6.6 cubic metres as the internal volume of the Soyuz Orbital Module, of which only 4 cubic metres (about the same as on Progress) could be used by the cosmonauts. Unlike on Soyuz, there was no rear hatch in the CM. The module's structural mass was 2,520 kg, and it could deliver a payload mass of approximately 2,500 kg. The Cargo Module structure was formed from two hemispheres welded to a short intermediate section, in much the same way as the Soyuz OM. The docking equipment and forward entry hatch occupied the front of the module (see below), and inside was a latticework of racks for various dry cargos such as foodstuffs, clothing, replacement equipment and new experiments for transfer into the station. Large items were connected to the supporting framework, while small items were located in small bins. For easy removal, the fastenings were simple half-turn bolts.

Unloading took several days, and once completed, the cosmonauts would begin filling the compartment with redundant equipment, rubbish bags and waste packaging. In some instances (depending on fuel reserves), the main engines of the Progress were used to boost the station to a higher orbit to reduce atmospheric drag. After undocking, the departing Progress was allowed to drift away from the vicinity of the station, before firing its engines to lower its orbit to around 200 km. Entry usually took place a few hours after undocking, and normally as the spacecraft passed over the Soviet Union. The command for retro-fire would be issued, and the Progress would enter the atmosphere to destruction over the Pacific Ocean.

Loading and unloading Progress

Crew training for loading and unloading the Progress follows a defined format, and observes certain key procedures and safety requirements. Prior to the transfer of cargo, the crew prepare cotton gloves, goggles, respirators and waste bags, and don a special tool-belt containing unlocking tools to remove the cargo. Each transfer is logged in a transfer operations document, and the instructions are provided on decals attached to containers and equipment. Normally, one crew-member works inside the Progress, and a second remains in the station side of the access tunnel. To find the items in sequence, a locator diagram is provided in each Progress.

The crew wear protective gear (gloves, goggles and other items) when required, and are advised to avoid unrestrained drifting of logisitics when passing them from one set of hands to another. This is not always adhered to, as it is quicker to transfer lighter items by sailing them through the connecting hatches into the main station area. Crew-members are also told to prevent items from hitting the structure (especially the control panels) of the Progress or station, and are quick to ensure that their hands, arms and legs are kept out of the way as larger items pass through the small hatches.

The unbolting of units begins with the loosening all custom bolts and then the sequential unbolting of them to prevent the payload skewing. The crews are also advised to make use of all available restraints, and to tether all tools. As the crew

Yuri Romanenko unloads the first Progress freighter during his 96-day mission to Salyut 6 in 1978. The tube supplies air from the Salyut to the small spacecraft, allowing the cosmonauts to work comfortably inside the confines of the freighter. Photographs of Progress unloading and loading operations are rare, due to the small working volume and the time it takes to complete the operation.

release the larger items, the bolts are then returned to the vacant mounting holes, or in some cases to the collapsed support frames. Fasteners and disposable material used in securing the cargo are stored in the waste bags to prevent additional clutter in the small confines of the freighter.

The crew are trained to at all times avoid blocking the depress valve located in the Cargo Module when loading the Progress with used equipment after the unloading operation has been completed. Equipment destined for plane IV must be the last to be stowed on the vehicle. All items relocated into the Progress cargo compartment must observe the mass-centring characteristics of the vehicle, and loading is conducted under instruction from mission control. (On the ISS, to ensure all items are positioned correctly, videotaping provides visual confirmation that all items for disposal are correctly restrained.) This is conducted throughout loading, and is supplemented by crew commentary and then down-linked to mission control. All tools are then re-stowed, and a checklist is completed to ensure that nothing is missing from the equipment list. Food left-overs are stored in one container for food rations, and are placed in a designated area. The crew follows an up-linked message from mission control on what should be stowed in the removable and non-removable containers in the Progress, including used equipment and solid waste products.

In 1996, American astronaut John Blaha wrote of the excitement of early morning entry into the Progress for the first time: 'We stayed up a few extra minutes as we searched for our crew packages. Once we found our packages, it was like Christmas and your birthday all rolled together when you are five years old. We really had a lot of fun reading mail, laughing, opening presents, eating fresh tomatoes, cheese, and so on. It was an experience I will always remember.'

American astronaut Jerry Linenger was onboard Mir for only one Progress resupply mission in 1997, and later wrote of the pleasure of receiving a package of 'family goodies' among the more mundane cargo: 'When the Progress arrived, we quickly unlocked equipment and repair parts in search of the packages. Once found, and munching on fresh apples that had also arrived on the Progress, we individually retreated from our work and sneaked off to private sections of the space station, eager to peruse the box contents.' These 'goodies' can include photographs of the family, birthday or celebration cards, taped or written massages, small souvenirs, long letters, newspaper clippings, and treats such as sweets and snacks outside the normal food inventory.

The constant unbolting and bolting of items is an arduous task. Once the new items are in the main station, the crew then has to relocate them wherever there is room. As the station ages (as did Mir), it becomes a challenge to find room for the new logisitics, and a struggle to move the waste material past the new supplies and into Progress. The biggest difficulty arises when reloading the Progress with unwanted equipment. Each Progress is fitted with a latticework designed to restrain the equipment being delivered to the station, and sometimes this does not fit the items for disposal. With everything floating, it is almost impossible to pack the Progress as well as it was packed on the ground. On Mir, the ground controllers constantly changed their minds about what should or should not be removed. After being told that an item was to be removed, the crews often found themselves retrieving the worn-out item a few days later and cannibalising it for spares. Naturally, that item was invariably behind half a dozen bolted-down items of hardware which they had since put in front of it. When working in Progress it can be cold and damp, and the excitement of receiving fresh food and packages from home is soon forgotten during the chore of removing the new stores, stowing them, and repackaging all the rubbish.[5]

The docking system

The docking equipment located on the forward hemisphere of the Progress CM resembled the lightweight probe and drogue unit featured on the Soyuz ferry spacecraft. The docking collar was modified to include two propellant ducting mating connectors to match those on the opposite collar on the Salyut, for transfer of the UDMH and nitrogen tetroxide to the station's storage tanks. The Igla rendezvous system (see p.139) was used on the early Progress vehicles, and incorporated deployable antennae and a tower-mounted dish, two infrared local vertical (horizon) sensors, and two ion guidance sensors (as opposed to single units on the Soyuz). The installation of dual sensors on Progress provided redundancy for the fully automated spacecraft.

Flight time from orbital insertion to docking with the station was about two days, allowing for additional analysis of orbital parameters following each orbital manoeuvre by the Progress during the approach. Two TV cameras located near the docking unit broadcast stereo views, and superimposed closing rates of the docking approach for controllers at a control room separate from that handling Salyut operations. Alternative views from the station docking port were also available. The Progress carried spotlights to illuminate the docking target if docking took place during the night-side pass. It was initially planned that the cosmonauts retire to the Soyuz DM during the docking, in case of additional impact damage; but the success of docking Soyuz 27 to Salyut 6 a few days prior to the first Progress allowed the cosmonauts to remain in the station and man the controls, in case they needed to move the station to compensate for any deviations in the docking approach. A feature of the docking system on Progress was the ability to remove the docking hatch and probe system to ease movement of logisitics to and from the cargo compartment.

The Refuelling Module (RM)

On Progress, the Refuelling Module was located where the Descent Module would be on Soyuz. The compartment was 2.1 metres long, and had a diameter of 1.7 metres and an overall mass of 1,846 kg. Inside an unpressurised conical housing were two tanks containing the UDMH and nitrogen tetroxide. In order to avoid accidental spillage of the highly corrosive and poisonous fuels into the habitable compartments of the cargo ship, all station propellant feed lines were fitted along the exterior of the CM to the docking collar, and thus into the feed system leading to the storage tanks located in the aft compartment of the station.

The fuel transfer process began either under the control of the onboard crew, or automatically from mission control. The latter allowed refuelling of the station without interrupting the work schedule of the crew, or during periods when Salyut was left unmanned between long-term missions.

After verification of the integrity of each transfer line, the Salyut fuel tanks were closed, and the operational pressure of 220 atm reduced by introducing nitrogen from the Salyut pressure feeding system back into the nitrogen storage tanks. Each storage tank was spherical, and was divided into two parts by an internal membrane – one part holding the propellant, and the other containing the pressurising gas introduced into the tank as the propellant is transferred. This part of the operation was conducted over several days in order to reduce the drain on the Salyut electrical system caused by the operation of the nitrogen compressor. Once the internal pressure was down to 3 atm, UDMH was pressure-fed from the Progress tanks to the Salyut tanks by means of the cargo vessel's own onboard nitrogen supplies located in the propellant module. With the freighter's tanks pressurised to 8 atm, fuel was forced from the tanks through the connecting plumbing and into the Salyut tanks, thus forcing the nitrogen back into the storage tanks as the Salyut tanks were refilled. Once the UDMH had been transferred, the operation was repeated with the nitrogen tetroxide, using the separate set of pipes and connections. The total propellant transfer operation also took several hours, because the 1 kW of power

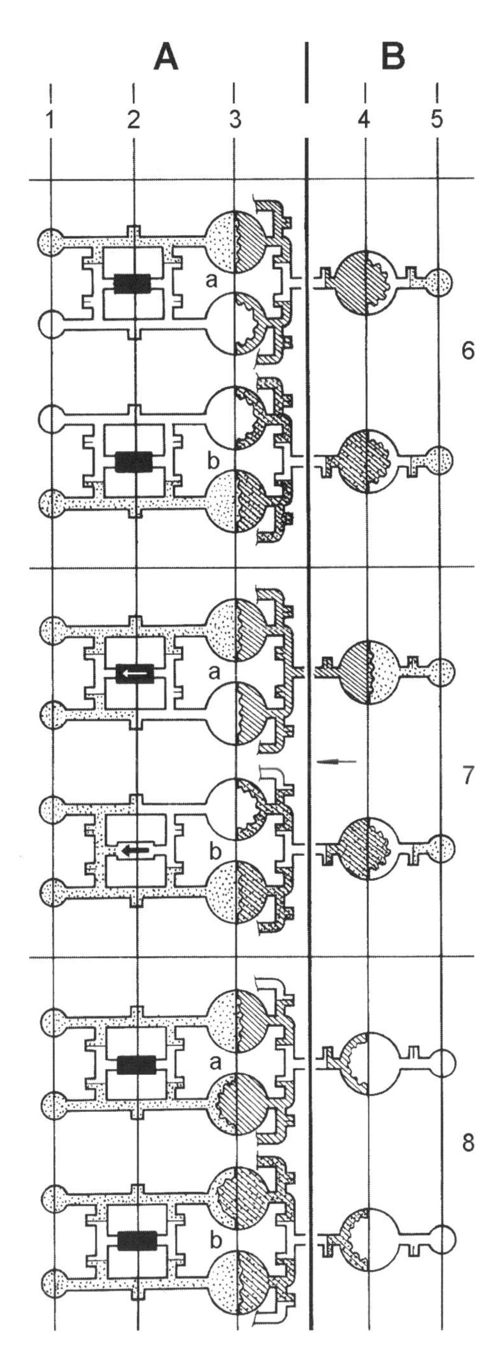

The Progress refuelling system. A, Salyut 6; B, Progress; a, fuel tanks; b, oxidiser tanks; 1, Salyut high-pressure nitrogen supply; 2, propellant tank nitrogen pumps; 3, Salyut tanks with internal partitions; 5, Progress high-pressure nitrogen supply; 6, system prior to refuelling; 7, system refuelling from Progress to Salyut tanks; 8, post-refuelling status.

consumed by the pumping action drained the Salyut power reserves and necessitated periodic recharging of the station's batteries. When the transfer operations were completed, both sets of lines were purged with nitrogen to prevent residual propellant spillage when the Progress undocked at the end of its mission. The propellant compartment also housed the Rodnik tanks, which contained water to be pumped over to the station.

The Propulsion Module (Service Module)

The third element of all Progress vehicles was the aft Propulsion Module. This resembled the Soyuz PM, but featured several modifications and improvements over the early Soyuz ferry version, and served as a test bed for the PM design that would be used on the manned Soyuz T vehicle. With a length of 3.1 metres it was longer than the normal PM used on Soyuz, because flight control equipment that was normally installed in the Soyuz Orbital and Descent Modules had to be relocated in the Progress design. The maximum diameter was 2.72 metres, and the overall mass was approximately 2,654 kg.

The internal arrangements of the PM included fourteen 10-kg and eight 1-kg hydrogen peroxide thrusters for attitude control during independent flight. The fuel storage tanks for these were installed inside the PM. The main engine on the vehicle was the KTDU-35 system – a 417-kgf thrust engine fuelled by nitric acid/hydrazine propellant for a specific impulse of 282 sec, at 200 metres per sec ΔV.

All first-generation Progress ferries relied on chemical batteries for independent power supply, and did not feature the large twin solar array panels carried on early and upgraded models of the Soyuz. This limited the independent operational life of the early Progress to only a few days. Fortunately, each vehicle was able to achieve its docking mission within that time, before battery power was exhausted.

Progress M upgrades

The first generation Progress spacecraft (nos.1–42) all flew to the second-generation space stations Salyut 6 and 7. The next step in Soviet space station development was the much larger Mir complex, which would require a greater resupply capacity than was offered by the early Progress. Work on modernising the vehicle began in 1986, with the draft plan completed in May 1986. From August 1989, the Progress M (Modernised) vehicle replaced the earlier design.

Overall, the three-module design was retained, but the Propulsion Module was adopted from the Soyuz TM manned craft, as was the improved Kurs rendezvous and docking system that replaced the docking tower on the earlier vehicle. The engine installation was identical to that of Soyuz-TM, with a KTDU-80 (also called S5.80) main engine burning UDMH and nitrogen tetroxide.

The increased cargo capability was not that considerable. The first volume of the RKK Energiya history quotes averages of 2,300 kg for the first-generation Progress and 2,400 kg for Progress M, while the second volume quotes 2,480 kg for Progress M – about 100–200 kg more than the earlier vehicle. Independent flight time was extended by installing two solar arrays with a span of 10.6 metres and an area of 10.0 square metres. This produced a 10% increase to 30 days of independent flight time and orbital storage of up to six months, and offered greater flexibility in orbital operations. However Progress M-17 stayed in orbit for almost *one year*, although this was not originally planned (see p.260).

With Mir, there was the added flexibility to dock Progress with either the forward or aft (Kvant) docking ports, both of which had the capability to transfer fuel to the main tanks. This allowed the Soyuz to remain where it was rather than be moved to accommodate the incoming Progress. Another innovation was the ability to move excess propellant (up to 200 kg) from the Progress M SM tanks to the Mir system, or to transfer propellant from Mir back to the SM tanks to raise the station's orbit using the Progress SM engines. The installation of the solar arrays produced 0.6 kW of electrical power. This was useful during periods of free flight, but was also incorporated into Mir's main electrical network to supplement the electricity generated by the station. There was also considerable weight-saving, including the use of a new launch shroud that did not require a dummy launch escape system. Progress M could therefore deliver an increased payload mass to Mir.

Raduga return capsules

On 9 Progress M flights, a new ballistic return capsule was introduced to offer an intermediate sample return capacity. This capsule – designated Raduga (Rainbow) – was approved in July 1988 and introduced in September 1990. It allowed the return of small but valuable payloads from Mir, indicating future commercial possibilities. Carried in the dry cargo compartment, it could be filled with 150 kg of payload (exposed film, magnetic data tapes, samples from processing furnaces, and other items), and was planned to fly two or three times a year. Official figures from RKK Energiya quote the dimensions of the capsule: 1.4 metres long, 0.78 metres in diameter, and a total mass of 350 kg, including the return payload of 150 kg.

The launch penalty for using this unit was 100 kg of overall launch cargo capacity, and this could have been one reason for the termination of the series, in that the loss of cargo delivery capability could not be justified by the volume which the empty return capsule occupied – especially given the relatively small amount that was returned. The maximum return capacity of Soyuz TM was 50 kg when a three-man crew was flown, and 120 kg when a two-man crew was flown. The same RKK Energiya document also mentions a dedicated cargo return vehicle (without crew) that could return 1,000–1,500 kg. (The second volume of the Energiya history includes a drawing of it.) It was essentially an unmanned Soyuz using both Soyuz TM and Progress M hardware, and would carry up supplies in its Orbital Module and return cargo in the Descent Module. It appears that this was studied as a possible alternative to Raduga during the early 1990s.[6] Following the de-orbit burn, the capsule (having no propulsion system of its own) was separated at a height of 120 km, and while the Progress was destroyed, the Raduga made a ballistic re-entry and parachuted to recovery in Kazakhstan.

NPO Energiya also proposed various versions of the Raduga capsule as autonomous spacecraft attached to a propulsion module. One of these featured in a proposal for the recoverable Japanese Express spacecraft. However, the Japanese selected not the NPO Energiya but the Khrunichev proposal.

The Raduga return capsule and parachute container used on nine Progress mission to return samples from Mir. (Courtesy Andy Salmon.)

By 1992, and with the emergence of the Shuttle/Mir programme, the availability of greater cargo return capability from the American Shuttle, and the fact that one capsule was lost during entry probably, contributed to the termination of the Raduga series. With the ISS in orbit, and the availability of the two-way Multi-Purpose Logistics Module onboard the Shuttle, the small Raduga system was no longer required.

Progress M1 upgrades

The inclusion of Russian station hardware in the ISS programme offered greater utilisation of Progress vehicles in resupplying the station. The main reason for the development of Progress M1 was to carry more propellant supplies for Mir 2 orbital reboost, and the same role could be applied to the ISS. The Progress M1 series was the first of several designs to support the (later cancelled) Mir 2 operations. With the delayed demise of Mir, there was also a need to support the overlap with the creation of the ISS. The remaining Progress M and initial M1 vehicles were therefore assigned to both stations for a short time. The Progress M1 payload capacity is still within that of the standard Soyuz launch vehicle, as it has eight rather than four propellant tanks in its refuelling compartment. However, because of the the four additional tanks, there is no space for the Rodnik water tanks, and on Progress M1 the water is carried in containers in the cargo compartment. When compared with Progress M it has *less* overall cargo capacity, but it carries more propellant. The second volume of the Energiya history quotes the following data:

	Progress M	*Progress M1*
Overall cargo mass (kg)	2,480	2,280
Dry cargo (kg)	1,800	1,800
Propellant (kg)		
(in the refuelling compartment)	870	1,950
Gas (kg)	50	40
Water (kg)	420	220

The disposal of Progress

Normally, a few hours after undocking from the station, the Progress, full of rubbish and unwanted hardware, is commanded by ground control to fire its braking engine to begin a de-orbit manoeuvre. Unlike Soyuz, none of the three spacecraft components are separated, and all are incinerated during the fiery re-entry, usually 'over the Pacific Ocean'. During the re-entry of Progress M27, the ITAR–TASS news agency reported that the permanent destruction site for disposing of non-recoverable Russian spacecraft lay at coordinates 3,600 km south-east of Wellington, New Zealand.

PROGRESS FLIGHT OPERATIONS

The first Progress was launched to the Salyut 6 space station on 20 January 1978. Over the next twenty-five years, a total of one hundred Progress-type vehicles – forty-three Progress (including Cosmos 1669), forty-seven Progress M, and ten Progress M1 – were dispatched to Salyut 6, Salyut 7, Mir, and the ISS. Each of these craft carried different cargos, although most followed the same basic flight profile.

Pirs (Pier) – the docking and airlock facility on the ISS – is not specifically a Progress M1 type. The only feature which it has in common with Progress is the propulsion compartment, and this is identical on Progress M and Progress M1. (Pirs is described below.)

The exact payload weights (see the Table on p.272) reveal similarities in overall capacity, but these can be further divided into the amount of fuel, consumables (air and water), and dry goods, which depend on the individual mission requirements. The stated payload of Progress 1 (2,300 kg) has been broken down to 1,000 kg of propellant and oxygen and 1,300 kg of general cargo and experiments, and it can therefore be calculated how frequently a Progress needed to call at the station. In general, it has been determined that each cosmonaut consumes about 15–30 kg of material each day. A two-man residency crew would therefore require supplies of 30–60 kg per mission day, and a three man crew would require 45–90 kg per mission day, without contingencies and additional hardware and experiments. A two-man crew (the average Salyut residency number) would require a Progress to dock with the station every three to six weeks throughout the mission, which would place a further load on spacecraft construction, ground preparation and launch infrastructure to meet that demand.

However, over the years it has become clear that payload figures and flight frequency are very flexible, and can vary considerably, depending on the requirements at the time. On several occasions, the amount of fuel transferred has been minimal, while the dry cargo has significantly increased. During other periods, resupply between Progress craft has been less frequent in order to use up supplies onboard the station.

PROGRESS MISSIONS TO SALYUT 6

The first of twelve Progress flights to Salyut 6 took about two days to reach the station and dock with it. Soon after docking, the cosmonauts opened the hatches and began to unload the dry cargo. Meanwhile, preparations continued for the transfer of fuel across to the station, with several days of checking onboard systems and verifying the connections and processes to prepare the storage tanks for operation. The transfer of fuel was a complete success, and was followed a few days later by the transfer of oxygen into the station's storage tanks. And because of this success, Salyut's own engines could be used to adjust its orbital parameters as required, without the requirement of a large vehicle to be docked to the station at all times to provide manoeuvring thrust. Following the fuel transfer, the crew filled the Cargo Module with rubbish and unwanted material. With the hatches resealed, the Progress was undocked – but its mission was not over. During its first orbit after undocking it was allowed to drift about 13 km from the station to test the back-up automated search and approach system, qualifying it for operational use on later missions. No re-docking was planned on this occasion, and so the test approach was interrupted, allowing the craft to move away from the Salyut to a predetermined point where the engines could be ignited to begin the de-orbit profile. This profile would become the standard mission profile for Progress: delivery of the primary cargo and disposal of waste, plus specific tests and investigations as required. (Details of the Progress missions are listed in the Table on p.275)

Varied cargoes

The next two Progress flights to the station provided an early demonstration of the variety of payloads which the craft could deliver under a similar flight profile. Progress 2 supplied only about 600 kg of propellants, while Progress 3 appears to have transferred no propellant at all. Progress 3 followed on closely from the undocking and de-orbiting of Progress 2, and was intended to quickly provide additional scientific equipment and supplies to the station's reserves as the second resident crew carried out their programme.

The series of Interkosmos flights – of representatives of the Soviet bloc east European countries spending a week onboard the station – began on Salyut 6, and over the years this expanded to include several international cooperative missions. These have been supported by Progress spacecraft supplying additional instruments, hardware and experiments for a forthcoming Interkosmos or international visiting mission. Ivanchenkov's wife sent a box of chocolates, via Progress 4, to the crew, who found that they were filled with brandy. She had substituted the original plain chocolates for liquor-filled chocolates in a normal chocolate box, so that they would not be noticed during loading. However, despite the luxury, the crew stated that they preferred the fresh food items much more.[7]

On 16 March 1979, a report on the third resident crew revealed a problem with the Salyut propulsion systems which was noticed at the end of the second residency in late 1978. Tests performed during the unmanned autonomous operation between the residencies revealed a faulty membrane in the propellant storage tanks, which needed to be isolated from the rest of the fuelling system if the stations operational status were to be maintained. With Progress 5 docked, the combination was slowly spun on its transverse axis to separate the fuel and nitrogen by centrifugal force, allowing most of the fuel to be transferred to another tank on the station and the residual fuel and nitrogen to be transferred back into a spare tank on Progress 5. The freighter was used to stop the rotation, and by 30 March the transfer of new fuel to the remaining Salyut tanks was complete. On the same day, the engines of the cargo ship were used to raise the orbit of the complex – an operation that was becoming a standard feature of many Progress missions, and a much-needed bonus, considering the problems with Salyut's primary propulsion system.

The KRT-10 antenna experiment

On Progress 7, an additional experiment was performed during undocking on 18 July 1979. This time the outer hatch of the Salyut was removed, while the inner hatch was sealed and attached to the rear of the docking tunnel. Near the closed hatch was a folded radio telescope – designated KRT-10 – with a mass of 200 kg (including support structures). The folded instrument – which the cosmonauts had earlier prepared – lay through the tunnel into the Cargo Module of the Progress. As the freighter pulled away, so the antenna unfolded and deployed to a full 120-metre diameter, its operation relayed by TV from onboard the Progress. Its job done, Progress was programmed for de-orbit the day after undocking. The KRT-10 was thought by Western observers to signal the end of the Salyut 6 programme, as it occupied the rear port used for resupplying the station; but on 9 August 1979 it was commanded to separate from the station, its research programme completed. However, separation was not totally successful, and on 15 August it required an unscheduled EVA to free the port for further docking operations.

Progress 9 completed a new operation when it used the Rodnik system to transfer water by directly pumping 180 kg to the station's reserve tanks. During earlier flights, water supplies had to be unloaded by hand, in 5-kg water containers, which took time, and occupied cargo space in the Progress and storage space in the Salyut.

PROGRESS MISSIONS TO SALYUT 7

Salyut springs a leak

On 9 September 1983, during the transfer of propellant from Progress 17 to Salyut 7, an event occurred that would have a major impact on future Salyut 7 operations, but which was not immediately reported by the Soviets. One of the Salyut propellant lines sprung a leak, and plans were formulated to conduct a series of EVAs to try to locate, isolate and repair the faulty ducting, allowing the Salyut to continue its planned programme.

In 1984, Progress 20 delivered spare parts and tools for the repair of the station's propulsion system, including some parts that were located on the outside of the freighter's hull and required retrieval during the series of EVAs. The Cargo Module of this Progress had also been modified to include foot restraints, on an extension that was unfolded and locked in place by ground command shortly after docking with the rear port of Salyut. The two EVA cosmonauts were able to attach themselves here, as it provided firm support while they worked.

Progress 21 and Progress 24 delivered the second and third sets of additional panels to be attached to the solar array structures (the first set had been delivered by Cosmos 1443). Progress 24 also supplied replacement parts for the stricken Salyut during repair operations in 1985. The undocking of Progress 22 on 15 July 1984 consisted of a simple push by the docking system springs rather than with a short burst of the engines, as it was thought that the exhaust plumes might be damaging the solar arrays.

Although twelve designated Progress missions were used in support of Salyut 7 operations, a thirteenth craft (Cosmos 1669) was in reality a Progress vehicle. Initially, Western analysts suggested that this was the first flight of a free-flying science platform based on the Progress design, to be use in conjunction with the expected next station (Salyut 8). Initial telemetry indicated that the Igla rendezvous antenna had not deployed, which would have made docking impossible (there was a similar problem on Soyuz T-8, and on that occasion not even the crew could dock the Soyuz), and it was therefore decided to assign it a Cosmos cover name rather than name it Progress 25. Later analysis revealed that the antenna *had* deployed as planned, and that the telemetry readings had resulted from a faulty sensor. By that time, however, TASS had already announced the mission as Cosmos 1669.

PROGRESS MISSIONS TO MIR

With twenty-five Progress spacecraft supporting two Salyut stations over seven years, the Soviets had evolved a reliable system for sustaining a station in orbit for successive crews. The true value of the Progress freighter series came with the Mir complex, when no fewer than sixty-four freighters were launched to the station between 1986 and 2001.

First-generation Progress

For the first four years of Mir operations, the original Progress vehicles (Progress 25–42) were flown to Mir. The first of these assigned to Mir was launched soon after the base module to supplement the basic twenty days of rations carried on the core station. Progress 28 featured the deployment, after undocking, of a 60-metre antenna for geophysical experiments. (According to some reports, it was a prototype design for future larger space structures.) The Progress 28 experiment was part of the Model programme (also flown on Progress 11, 14 and 18). A second, similar experiment was also conducted on Progress 40 as part of the unrelated Krab programme (see p.261).

In 1987, the Kvant astrophysical module was docked to the rear port of Mir. However, Kvant could still accept Progress freighters as well as manned Soyuz craft, by docking with the module's aft port. Duplicated fixtures and fittings for docking and consumables transfer were incorporated into the design. These extended through the module into the core module by way of the rear docking connections used before Kvant was permanently attached to Mir.

After undocking on 10 November 1987, Progress 32 demonstrated a new ability by performing independent manoeuvring tests controlled from the ground, and re-docking with Mir 1.5 hours (1 orbit) later. This was to evaluate the reduction in the use of onboard supplies during manoeuvring and docking, and the ability to station-keep to free up the port and then re-dock at a later stage.

Twenty flights later, in May 1990, the final mission of the original version, Progress 42, flew to Mir. By 1990 the new improved version, Progress M, was in operational use, but this final flight of the original type of freighter (which usually employed the Igla rendezvous) delayed both use of an upgraded Kurs rendezvous system and integration of the more powerful Salyut 5B orientation control computer that was delivered with the second module, Kvant 2. The Salyut 5B computer was required because the complexity of the Mir station had outgrown the capacity of the Argon 16B (original) central computer on Mir. Since the older Progress model could only interface with the Argon 16B, the complete phase-in of the Salyut 5B was postponed (and together with other problems with the integration of the new computer, this delayed the launch of Kristall).

Buran ejection seat tests

The Buran ejector seat (K-36M, 11F35) was installed in an experimental droppable compartment placed instead of the LES engine on top of the payload shroud. (On Progress launches, a standard Soyuz shroud with an LES engine was used, but the engine lacked the powder charge). During the launcher's initial ascent phase, the tower was fired and the ejector seat was released at a speed reaching Mach 4.1 (according to Zvezda engineers who designed the seat). The experiment was flown on five Progress craft nos.38, 39, 40, 41 and 42 from September 1988 to May 1990.

These experiments were unique, as no others have been flown using this method. The Soviets were testing the seats which would be used on manned flights of the Buran (Snow-storm) shuttle, which was fitted with them to enable the crew to blast clear if there was a problem during the ascent and descent phase of the mission.[8]

Progress M takes over

The first Progress M, launched on 23 August 1989, was notable not only for its new payload capacity, but also because it docked with the forward port of Mir instead of the rear port, which revealed that the new space station had duplicate sets of propulsion transfer systems. This produced far greater redundancy than on either Salyut 6 or Salyut 7, and was a feature that was to become very useful throughout the long life of the station.

Schedules and setbacks

Throughout the twenty-five-year operational history of Progress resupply missions, the majority of the hundred or more vehicles launched and performed flawlessly. However, there have inevitably been occurrences that have threatened the smooth delivery of supplies to operational stations, and some of these from the Mir era are recalled here.

The launch of Progress M-22 on 16 March 1994 had to be rescheduled for 19 March. However, heavy snowfalls and a fire at the MIK assembly building forced a further delay to 22 March. The fire was in an annex to one of the main buildings, and although five rooms were gutted, there was no long-term damage. It was apparently made worse by an electrical shortage at the cosmodrome, and by snow blocking road and rail links to the site. Press reports indicated that such events were becoming more frequent in the cash-starved Russian programme. Eighteen months later, on 27

Progress M approaches for docking with the space station, to deliver a cargo of propellant and other supplies.

September 1995, it was reported that a flat-back rail freight car, used to transport the Soyuz launch vehicles to the pad, had been derailed. Initial reports stated that this would seriously affect the launch of the next Progress M, but this was soon denied. The M-29 vehicle was indeed rolled out and launched as planned.

The following year, two failures of the Soyuz U launcher in May and June delayed the launch of Progress M-32 by several weeks 'for additional checks'. The failures arose from weakening of the same glue that was used on earlier versions to bond the launch shrouds, resulting in failure of those shrouds under aerodynamic forces. On 25 July, the count-down had reached T–45 sec when automated checks revealed a drain valve malfunction. At the time, a NASA report pointed to a pressure sensor failure on the first stage, and the Russians indicated the pre-launch malfunction of a drainage safety valve. Whatever the problem, the situation became worse when other problems were discovered with the booster on the pad. As a result, it was rolled back to the MIK assembly building, and the Progress was swapped out to a second launcher undergoing final processing at the complex. After a hurried turnaround, Progress M-32 was launched on 31 July, with an abbreviated stay to allow docking of the EO-22 crew on 19 August. Following the departure of the EO-23 crew on 2 September, Progress M-32 re-docked with Mir – the first time that a Progress had been stored in an orbit away from a station for more than two weeks while main crews were exchanged on the station.

On 19 November 1996, Progress M-33 was launched on a Soyuz U launcher, reportedly with several substandard components in some of the rocket assemblies. Nevertheless, the spacecraft docked with Mir on 20 November. On 6 October 1998, news that Progress M-40 would be delayed until 25 October was publicly released. The new hold was due to the lack of funding to purchase the Progress carrier rocket (Soyuz U) from the Samara construction facility. The Progress was scheduled to deliver time-critical experiments and supplies, and so its launch was critical. It was, however, launched on time, and two days later it docked with Mir.

Additional hardware, experiments and research objectives

The improvements incorporated in the design of Progress provided greater payload mass, extended duration docked to a station, and capacity for independent flight. This allowed the Soviets to expand the flight envelope of the freighter to include a number of payloads intended to demonstrate greater scientific opportunity (for potential commercial marketing of the Progress vehicle), and as a test bed for development of future space technologies and procedures.

In December 1989, Progress M-2 to Mir delivered a protein crystal growth experiment provided by Payload Systems – a private US company co-founded by Spacelab 1 Payload Specialist Byron Lichtenberg. In September 1990, when Progress M-4 undocked from Mir, the resident cosmonauts installed a plasma production system (on the freighter's docking unit) from which plasma was released over a three-day period. This was observed and recorded by the Mir crew.

An inflatable balloon was deployed from Progress M-8, but unfortunately the integrity of the balloon failed, and it did not fully inflate. Two weeks later, the limp satellite decayed.

On 4 February 1993, after 97 days attached to Mir, the M-15 freighter undocked from the station. Twelve minutes later the Progress was still only 160 metres from Mir when a prototype solar reflector was deployed. The Znamya 2 (Banner) Kevlar sheet reflector took three minutes to unfurl out of a special container attached to the front of the Progress. The structure was manufactured in eight triangular sections, and was automatically spun – initially to 95 rpm, and later to only 14 rpm. This forced the sections to remain relatively flat to create a near-uniform disc 20 metres in diameter. Progress M-15 was then moved 12.1 km from the station, 4 hrs 38 min after undocking, to begin an experiment called New Light. For six minutes, Znamya reflected a spot of sunlight up to 30 km in diameter on the surface of the Earth, until the spacecraft crossed the Earth's terminator, blocking out direct sunlight. It was almost immediately ejected, allowing the Progress to be flown in a series of manoeuvres controlled by the cosmonauts from Mir two days later to test the TORU remote control docking system onboard Mir. On Progress M-40 in February 1999, a test of Znamya 2.5 also failed. The Znamya experiment has potential for illuminating regions of the Earth in prolonged winter darkness, and for future solar sail propulsion techniques; but it is dependent upon successful flight testing and adequate funding – which does not appear forthcoming.

The Progress M-14 spacecraft was specially modified to carry the framework of the 700-kg VDU (Vynosnaya Dvigatel'naya Ustanovka – External Engine Unit) Mir propulsion block used for attitude control of the complex. The VDU was externally mounted on a special framework between the Cargo Module and the Propulsion Module. On 3 September 1992, the crew removed the VDU by EVA, and then installed it on the end of the Sofora beam that extended from the Kvant module. In March 1998, Progress M-38 delivered a second VDU.

Progress M-17 was used in a unplanned longevity test of onboard systems, both in docked and storage orbit flight. It remained docked to Mir for 132 days, and was kept in free flight for 204 days during a mission lasting 336 days. For some unkown reason, Progress M-17 did not have enough propellant left for a nominal de-orbit burn after undocking from Mir. One source sstates that this was because it had required more fuel to reach Mir (after an inaccurate orbit insertion), and another source suggests that it was because the vehicle performed many manoeuvres while docked to Mir (it stayed docked unusually long, because Soyuz TM-16 was parked at the Kristall module). Only 80 kg of propellant was left in the Propulsion Module, whereas 200 kg was required for a nominal de-orbit burn. This is why it was decided to leave Progress M-17 in orbit and let natural orbit decay gradually bring it closer to Earth, until only a gentle push from the engines would be sufficient to de-orbit it.

Several other experiments were flown on Progress spacecraft after undocking. At the time, some of these were not announced by the Russians because they had military applications or were not entirely successful. However, the information is now available,[9] and a brief account follows.

Kant–Sirius A radar system, with a surface of about 8 square metres, for detecting ships and submarines. Flown on Progress 17 and Progress 22.

Model A series of 20-metre-diameter circular-shaped antennae (loop aerials) for testing a new space communication system using ultra-slow-frequency radio waves. The purpose of the experiments was two-fold: first to test the deployment of large structures in space; and secondly to test the propagation of ultra-low-frequency radio waves, both between two objects in space (Progress and the space station) and between Progress and receiving stations on the ground. Each Progress carried two such antennae (as represented in a picture in the Energiya history). Flown on four missions, the first three tested only space-to-ground propagation, while the fourth tested both space-to-ground and space-to-space propagation.

Progress 11 Only one of the two antennae unfurled.

Progress 14 Both antennae unfurled, but did not assume a circular shape.

Progress 18 This had improved antennae (Model 2), but again did not assume a circular shape.

Progress 28 Also a Model 2 type, but this time successful. This also tested the propagation of ultra-low-frequency radio waves between Progress and Mir. For this purpose a special receiver was extended from the base block's airlock (usually used to dispose of rubbish) at the end of a 10-metre long boom. During two days of autonomous flight, the Progress performed twenty communications sessions using the system both with Mir and with ground stations. There were also plans to use the receiver on Mir to pick up ultra-low-frequency radio waves beamed up from the ground (in an experiment called Sekventa), but the results are not available. The hope was that these experiments would pave the way for similar 100-metre-diameter antennae. (The applications or benefits of this communication system are not clear. There is no indication of Ministry of Defence involvement, but there were possibly military applications.)

Svet (Light) A 600-kg instrument package, flown on Progress 30, for testing a communications system in the optical waveband. It did not required crew participation. More than thirty communications sessions were performed between Progress and two ships in the Pacific Ocean and Atlantic Ocean. The signal was first detected by a device located 50 metres under water.

Krab (Crab) Two 20-metre-diameter circular structures, made of a form-remembering alloy, deployed from the propellant compartment of Progress 40. The experiment – which began after Progress 40 had moved some 70–80 metres from Mir – was flown for two days, and was filmed with a camera onboard Progress. (Further details are supplied in David Harland's book about Mir.[10])

Small satellites

Over the years, the Progress freighters have provided the opportunity to deliver small satellites subsequently deployed by the crew (Iskra-HAM radio satellites) via the station air-lock into short-term orbits. Other commemorative small 'satellites' delivered by Progress, but deployed during EVA, included the MAK-1 and MAK-2 upper atmosphere research satellites deployed from Mir's airlock on 17 June 1991 and 20 November 1992, and a small replica Sputnik deployed in late 1997 to celebrate the fortieth anniversary of the launch of the world's first artificial satellite. In addition to these station-deployed satellites, several were deployed from the Progress vehicles.

Progress M-27 delivered a German 215-mm, 20-kg laser geodetic satellite covered with sixty reflectors, to be targeted by laser stations located in Germany, Austria and the UK. Designated GFZ-1, it was spring-ejected out of Mir's base block waste airlock on 19 April 1995.

In December 1997 a German satellite was prepared for deployment from Progress M-36 after it undocked from Mir. The 72-kg, 62 cm × 56-cm hexagonal prism contained manoeuvring jets and a camera for several manoeuvres around the Progress and Mir. The Inspector satellite was installed in place of the docking unit on the Progress by the EO-24 crew, to be used in twenty-nine hours of tests as a precursor to a similar satellite to be flown for use on the ISS. The crew used a lap-top computer to control the satellite, which provided visual inspection video across the surface of the spacecraft in order to reduce future time spent on EVA inspections and operations.

The Progress undocked from Mir on 17 December, and 95 minutes later, Inspector was jettisoned from the front of Progress. Unfortunately, not all of the small satellite objectives were completed. The failure of a star-tracker guidance system and onboard software prevented it from flying elliptical orbits around Mir. Several TV images of the station were received, but on 18 December a separation manoeuvre by the crew ended the experiment. In order to avoid a collision, Inspector was moved from behind the station to the front, to precede it in orbit.

These small satellite ejections continued into ISS operations. On 19 March 2002, Progress M1-7 undocked from the ISS, and just over four hours later, a 20.5-kg Kolibri-2000 (Humming-bird) satellite was ejected from the front of the Progress. The plan was for students in Russia and Australia to use the satellite for particle field research in Earth orbit for four months.

The Raduga missions

Progress M-5 was launched on 27 September 1990, and docked with Mir on 29 September. It was the first freighter to be equipped with the Raduga payload return system, and returned 150 kg of samples to Earth. Raduga 9 landed in the Orenburg region, and was the final mission to recover a capsule from Progress.

Raduga	*Launch*	*Landing*	*Duration (d:h:m)*	
1 (M-5)	1990 Sep 27	1990 Nov 28	62:00:26	
2 (M-7)	1991 Mar 19	1991 May 7	49:03:34?	Capsule lost during re-entry
3 (M-9)	1991 Aug 20	1991 Sep 30	40:09:24	
4 (M-10)	1991 Oct 17	1992 Jan 20	55:12:01	
5 (M-14)	1992 Aug 15	1992 Oct 22	66:23:54	
6 (M-18)	1993 May 25	1993 Jul 4	43:10:23	
7 (M-19)	1993 Aug 10	1993 Oct 13	63:01:59	
8 (M-20)	1993 Oct 11	1993 Nov 21	40:11:33	
9 (M-23)	1994 May 22	1994 Jul 2	41:10:38	Last Raduga mission

The data for landing times are contradictory, and the exact mission durations are not certain. The official RKK Energiya data on the mass of returned payload for the first six Raduga capsules is as follows:

Raduga 1	113.1 kg
Raduga 2	94.1 kg
Raduga 3	147.0 kg
Raduga 4	121.8 kg
Raduga 5	106.1 kg
Raduga 6	95.3 kg

Rendezvous and docking operations and incidents

Over the years, a number of difficulties have arisen in docking Progress vehicles to Mir. Progress 26 required three days to reach the station, although the reason for the delay is not clear. On 21 March 1998, Progress M-7 was attempting to dock with the aft port when the craft suddenly broke off its approach just 500 metres from the station. Two days later there was a second attempt, but when it was only 20 metres from the rear of the station, controllers at TsUP noted a 'catastrophic error' in the approach, and once more terminated the docking. The freighter was only 5–7 metres from the station when it flew by, narrowly avoiding crashing into antennae and solar arrays. To test the docking approach again, the cosmonauts boarded their Soyuz TM-11 craft, undocked from the forward port, and manoeuvred to dock with the rear port. They used the Kurs automatic approach system rather than the manual approach, and the flight deviated away from the docking port, indicating a problem with the Mir Kurs hardware and not the Progress. The cosmonauts achieved a manual docking, while the Progress was taken out of its station-keeping orbit and commanded to dock successfully with the front port. The decision to include fuel transfer and docking capabilities at two ports in the design of Mir was vindicated during this mission.

Progress M-10 also experienced docking difficulty in a delayed docking of two days. The aborted docking attempt on, 19 October 1991, was reportedly due to a ground control failure. An initial attempt on 21 October also failed, but the attempt on the next orbit was successful. Progress M-13 also required two docking attempts, as an initial attempt, on 2 July 1992, was aborted at a range of 172 metres.

Funding restrictions during 1994 delayed the launch of Progress M-24 – the first of two flights planned to resupply EO-16. The second was cancelled, and its was cargo combined with Progress M-24 and Soyuz TM-19. When at last the Progress was launched on 25 August, it failed to dock automatically on the front port on 27 August, and drifted 150 km in front of the station while new rendezvous software was loaded into its computers. Three days later, a second automated approach also failed, and the freighter twice struck the forward docking port before floating away. Evaluation on the ground determined that there was onboard propellant sufficient for two more attempts; but this time the approach would be handled by the resident cosmonauts, and not the automated systems. On 2 September, Malenchenko took control of the Progress from the TORU control panel located inside Mir, and succeeded in docking. A week later, Malenchenko and Musabayev performed an EVA to inspect the damage which the collision had caused the Kvant. Several statements have indicated the probable cause of this problem, ranging from software or Kurs electronics failures on the Progress, to a failure of equipment in TsUP on the ground. On 5 October 1994, the spacecraft performed a programmed re-entry, and so the hardware evidence was lost.

On 4 March 1997, Vasily Tsibliyev attempted to re-dock Progress M-33 with Mir's aft port, but the cosmonaut lost his TV picture and reported that the cargo ship failed to respond correctly to the TORU system inside Mir, with the freighter flying past about 225 metres away. Over the next few orbits, the distance between the freighter and the station increased, preventing any further re-docking attempts. Therefore, on 11 March the Progress was commanded to re-enter and destruct. With the next Progress vehicle, Tsibliyev would not be so lucky.[11]

The Progress M-34 collision

On 8 April 1997, Progress M-34 docked with the aft Kvant port, as programmed. On 24 June the vehicle was undocked in order to perform a re-docking test controlled from inside the station by one of the cosmonauts using a newly developed remote-control docking approach. At the time, Russia was trying to source alternative sub-systems for its space hardware from inside the Russian federation, instead of purchasing expensive components from the former USSR states – in this case, the Ukraine – which it could no longer afford. The cosmonaut called upon to control this approach was Commander Vasily Tsibliyev, who had received only minimal training on this technique. On 25 June it was clear to Lazutkin and astronaut Michael Foale that their Commander was not comfortable with this attempted docking, as he had not trained for the procedure for some months. Tsibilyev was also concerned that his only view of the approach trajectory was from the TV camera onboard the Progress, displayed on the TV screen in front of him, and that he would need visual cues from Lazutkin and Foale, looking out of the Mir portholes. Unfortunately, the views from the portholes were quite restricted, and when Progress approached, the cosmonauts realised that they could not see the spacecraft as clearly as expected.

Tsibliyev had not been provided with information on when to begin the approach or in which position the spacecraft should be in relation to the Earth, and Progress was first spotted against the back-drop of a cloud-covered Earth, and was difficult to distinguish. Lazutkin suddenly shouted that the freighter was approaching much faster than supposed, and was much closer to the station than he assumed. Tsibliyev immediately initiated the braking manoeuvre; but it was too late and Progress missed the docking port and passed Mir by inches. Normally, docking speed was approximately 1 metre per sec, but the Progress was travelling ten times that speed as it struck the solar array on the Spektr module. It punctured a hole in the pressure shell and then bounced into a solar array, which it also damaged. With alarms indicating a serious pressure leak, the crew raced to seal off the Spektr module from the rest of the station, and at the same time prepared Soyuz for possible evacuation (see p.357). On a scale of 1 to 7 – in which 7 represented an event requiring immediate evacuation – this collision rated as 5. Soyuz TM-25 was the only element of Mir not directly affected by the collision, and as it had its own independent power supplies it was powered up as soon as possible in case it was required. The struggle to save Mir and to restore the station to full operational status would take several weeks; but Spektr would be sealed off from the rest of the station, and would never again be used.

On 26 June, Michael Foale (whose sleeping berth had been in Spektr) mentioned in a radio report that he had lost most of his personal equipment in the sealed-off Spektr, and requested replacements to be sent up on the next Progress, which had been delayed while plans were formed to restore electrical power to Mir. At the same time, plans were devised to attach a permanent hatch to seal off the Spektr module. Once tested, the necessary equipment was placed in the Progress M-35 freighter and launched to the station on 5 July, to dock successfully two days later. Its cameras also recorded the damage to the Spektr arrays. After the collision, Progress M-34 had continued past the station. Over the next few days, until 2 July, it was subjected to a programme of automated tests by mission control, after which ground commands were sent to the freighter, commanding it to destructive re-entry. A few days later, on 8 July, Foale (like Blaha and Linenger before him) commented on the pleasure of unloading his packages from Progress: 'We are now unpacking Progress the smell was heavenly, of fresh apples. Have opened two work packages for me and found my hygiene stuff... and chocolate.'

The problems of docking Progress to Mir did not end with M-34. The initial re-docking of Progress M-35 on 17 August 1997 was cancelled; and there was also trouble during the second attempt on 18 August, forcing Solovyov to use TORU to bring in the vehicle.

The de-orbiting of Mir

By 1999 it was clear that the useful life of Mir was coming to the end, and with the first elements of ISS in orbit, the need to divert Russian funds, hardware and resources into the ISS was stronger than Russia's desire to make Mir a commercially attractive proposal. After ten years of continual operation, the twenty-seventh resident crew departed the station in August 1999. Despite a short, privately funded mission to the station in 2000, manned operations came to an end. The next task was to control the descent of the large station into the atmosphere to destroy it over the Pacific Ocean rather than over a populated area.

Despite attempts to continue Mir operations with private funding and plans for 'tourist' deals, the Council of Chief Designers decided, on 3 October 2000, that Mir operations should be ended and that the station be brought out of orbit. It required two Progress craft to de-orbit the station, and Progress M-43 and M1-5 were assigned to that task. M-43 was used to raise the orbit of Mir to ensure that it remained in orbit until at least February 2001. On 24 January 2001, Progress M1-5 became the final spacecraft to dock to the station, although it was six days late, due to the loss of control of the station. Just over a month after the fifteenth anniversary of the launch of the core module of Mir (20 February), during the early hours of 23 March, Progress M1-5's attitude control engines were ignited for a 1,293.8-second to begin the first orbit-lowering sequence of burns to bring Mir down. Ninety minutes later, the attitude control engines were reignited for 1,444.5 seconds. About five hours after the beginning of the sequence, this was followed by combination burns of the attitude control (1,166.5 seconds) and main propulsion systems (700 seconds) until the fuel on the Progress was depleted. The performance of the Progress main engine appeared to be slightly higher than predicted, leading to a minor shortfall (but still within the target area) of the planned debris impact footprint.

PROGRESS M1 OPERATIONS

Progress M1 craft were now being built and financed to fly to the newly-planned ISS. Construction plans continued, but hardware was not launched on schedule, and due to ground storage problems, three Progress M1s – M1-1, M1-2 and M1-5 – were dispatched to Mir. This caused some difficulties with ISS partners, who questioned the Russians' ability to simultaneously supply two operational stations and yet still meet their ISS obligations.

Progress M1 at the International Space Station

On 8 August 2000, Progress flights to the ISS began with the docking of Progress M1-3 to the ISS Zvezda module, prior to the arrival of the first resident crew in October. The STS-106 mission docked to the ISS in September, and one of Yuri Malenchenko's roles as a Mission Specialist on that flight was to supervise the unloading of the docked Progress and the stowage of its cargo pending the arrival of the ISS-1 crew. NASA continues to confuse the identification of ISS Progress missions by labelling them Progress 1, Progress 2, Progress 3... and so on.

On 18 November 2000, Yuri Glazkov docked the second ISS Progress (M1-4) at

A Progress freighter approaches the space station docking port.

the nadir port of the Zarya module, using the TORU system from within the ISS. It took two attempts to dock the freighter, because the Kurs automated system failed on the first attempt. The freighter moved so erratically 100 metres from the station that Gidzenko took over manual control to dampen out the motions and dock manually. Unfortunately, 5 metres from the docking port, lighting angles and suspected icing on the periscope lens of the Progress TV camera forced him to abort the approach. He backed the Progress out to 35 metres and, after waiting for the lens to clear, reapproached and docked successfully with ISS. On 1 December, Progress M1-4 was temporarily undocked and stored in orbit some 250 km away, while STS-97 docked at the station. After STS-97 departed, Gidzenko re-docked Progress M1-4, using TORU, on 26 December.

On 23 May, Progress M1-6 became the first cargo vessel to dock with the ISS by employing the Microgravity Acceleration Measurement System (MAMS) to measure vibration disturbance onboard the station during docking. Then, on 28 November, despite an apparently routine approach, there appeared to be problems with the docking of Progress M1-7, in that it was unable to achieve a hard docking even though it secured a soft docking. Rewriting the daily activity plans, the crew's schedule was changed from unloading the Progress to other duties, including preparations for an EVA to investigate a U-shaped object seen in the replayed video of the docking approach of the Progress. It seemed to be preventing hard docking, and was possibly an o-ring or cable left on the docking cone during loading of Progress M-45 in November. That EVA, on 3 December, was conducted by Vladimir Dezhurov and Mikhail Tyurin, who reported that a twisted sealing ring from the active docking unit of Progress M-45 was caught in the port, preventing the docking of Progress M1-7. Controllers at TsUP sent a command for the Progress to extend its docking probe, providing 395 mm of clearance for the cosmonauts to use a cutting tool (fabricated from spares inside the station) to snip the blockage in two places and remove the debris. With the cosmonauts clear, Progress was commanded to retract the probe – and this time the docking hook engaged for a successful hard docking.

After Progress M-46 undocked from the ISS on 24 September 2002, it did not immediately manoeuvre for re-entry, and instead made an engine burn to place it into a 401 × 360-km orbit, to operate in free-flyer mode. The Russians reported that it was using its onboard camera to monitor natural disasters, and also the man-made fire which had been started shortly earlier and had covered parts of Moscow in smoke for a number of days. It re-entered the atmosphere on the 14 October 2002, during orbit 1,718. Progress M1-9 was launched on 25 September, but instead of docking with the ISS two days later (the normal procedure), it kept station with it so that tests could be carried out (in automatic mode) on the Kurs rendezvous system on Zvezda and the Progress. These tests confirmed the reliability of the system, and Progress docked on 29 September. Its 854 kg of fuel consituted only about half a load, which probably reflected requirement rather than capacity.

(On 2 February 2003, the launch of Progress M-47 to the ISS – the day after the loss of STS-107 *Columbia* – demonstrated the commitment to ISS operations. It carried 2.5 tonnes of foodstuff (enough for three months), water, fuel, and parcels of gifts for the crew. It docked on 4 February, the first of three planned Progress dockings manifested for 2003.)

APPLICATION OF PROGRESS HARDWARE

A number of projects were evaluated, drawing upon flight-proven Progress hardware to support a range of objectives, many of which failed to leave the drawing board.

Gamma

On 11 July 1990, an astrophysical research mission began with the launch of a Progress Propulsion Module supporting a large gamma-ray experiment package: Gamma. The vehicle was an evolutionary design of space station modules based on the Soyuz spacecraft (see p.19) that evolved during the 1960s and 1970s. However, Gamma was not intended to dock with a space station, and carried a telescope housing in the Cargo Module area where the docking system would have been.

The history of this satellite began during the mid-1960s, with studies in using Soyuz as the basis of add-on space station modules. Work on the instrumentation began in 1972, with the French joining the science programme in 1974. On 17 February 1976, Gamma (19KA30) was included in a government resolution that sanctioned the DOS 7 and DOS 8 (Mir) space stations. It was a man-tended free-flyer, to be serviced by dedicated solo Soyuz missions that would rendezvous and dock every six months, to remove film cassettes to and repair or upgrade instruments. By 1982 the weight of the satellite had increased, and most Soyuz production was required for Mir, and so all man-tended instrumentation was removed, to be replaced with automated systems. It was set for launch in 1984, but a series of technical problemes delayed the launch until 1990 – around twenty-five years after the idea had first been conceived.

The dimensions of Gamma – 7.7 metres in length, and 2.7 metres in diameter – were similar to those of Progress, allowing it to fit in the launch shroud. Its mass was 7,320 kg, and it carried 1,700 kg of scientific equipment, of which 1,500 kg constituted the telescopes. The Gamma research programme – a cooperative venture between the USSR, France and Poland – included the Gamma-1 telescope (50 MeV – 6 GeV range), to which was linked the Telezvezda star-tracker (with a 6-degree field of view, and a sensitivity to fifth magnitude); the Disc M telescope (20 keV – 5 MeV); and the Pulsar X-2 telescope (2–25 keV), mounted coaxially on the vehicle for simultaneous observations of any given region of the celestial field. The solar arrays were driven by electric motors (unlike the Progress), with a total area of 36.5 square metres and a maximum power level of 3.5 kW. The spacecraft was placed into a 382×387-km orbit inclined at $51^\circ.6$. At the end of its mission on 28 February 1992, Gamma was de-orbited, and was destroyed on re-entry after 1,388 days in orbit.

The scientific programme was hindered by a power failure in the Gamma-1 telescope, which restricted angular resolution to only 10° throughout the entire mission. Research continued, however, with studies of the Vega pulsar, the central region of the Galaxy, the Cygnus binaries, and the Heming gamma-ray sources in

Taurus and Her X-1. It was also hoped that Gamma would locate COS-B sources and gather information on high-energy emissions during a period of maximum solar activity. Unfortunately, during its two-year mission Gamma 1 failed to produce any significant scientific results.

Aelita

There were plans at NPO/Energiya for a series of Progress-based scientific satellites, but none seem to have advanced beyond paper studies. One of them – Aelita – was intended to launch a cryogenically-cooled infrared telescope.

An Earth observation satellite

The basic design of these vehicles was also to be reconfigured to support remote sensing packages, but none developed past the design study stage. One of them – developed by RKK Energiya during the early 1990s – featured a 10-tonne low Earth orbit (400–800 km) platform, Sun synchronous, with a package totalling 1.4 tonnes with 2kW available power. The package of instruments included a side-looking radar, a TV camera, a scanning radiometer, and a video spectrometer. The plan was to operate this platform for 3–5 years, and to return data directly to ground stations or via relay satellites. The project, which received no funding, was to be launched by Zenit booster no earlier than 1996.

Plans for Soviet Star Wars

Also planned was a series of five Progress-derived spacecraft that were supposed to be flown as part of the Soviet Union's equivalent of the Star Wars programme (which was approved in 1976). As part of this programme, Energiya worked on two types of DOS-based 'battle stations', which would be only periodically visited by crews for periods of up to seven days. These stations would have had much more propellant than an ordinary DOS for manoeuvring. One type would be equipped with laser weapons to destroy targets in low Earth orbit, and another – Kaskad (17F111) –would carry missiles to destroy targets in medium and geostationary orbits.[12] In 1998, an article published in *Novosti Kosmonavtiki* revealed that Energiya had plans to use Progress-type vehicles to test the anti-satellite missiles that were to be deployed on Kaskad. The construction of five such vehicles, nos.129–133, began at the Energiya factory, but when the plan was cancelled they were rebuilt as ordinary Progress resupply ships (with different serial numbers).

THE ISS DOCKING COMPARTMENT

In design studies for potential missions of the Buran space shuttle, there was provision for a docking adapter to allow the shuttle to link up with other manned vehicles in orbit (in much the same way that the NASA Shuttle docked to Mir during 1995–97). A flight-qualified version was constructed for a planned docking between Buran 2 and a Soyuz TM, but it was never flown. The Buran docking adapter consisted of two elements. The first was a spherical section (2.55 metres in diameter) bolted to the keel of the orbiter, with two hatches for access to the Buran crew compartment and to a Spacelab-type research module, or to provide an EVA exit and entry into the payload bay. On top of this was the cylindrical tunnel (2.2 metres in diameter), with, on top of that, an APAS docking system for mating with a second vehicle.

For Mir 2, a modified version of the docking adapter would have flown on a Progress M tug which would separate from the adapter after delivery to the station. Here, the central spherical section of the Buran docking module was retained, with the forward part of an Orbital/Cargo Module added to the uppermost part of the structure, including an access hatch into the compartment. Part of the Buran cylindrical tunnel at the aft end allowed EVA, and docking by other spacecraft. This version was reassigned from Mir 2 to the ISS as the Docking Compartment, launched in 2001. On Mir 2 the function of the Docking Module would be to act as an airlock and (initially at least) to provide an APAS docking port for Buran.[13]

Docking Compartment 1

Docking Compartment 1 (DC1), also called Flight ISS 4R (fourth Russian), and known as Pirs (Pier), was launched on 14 September 2001. It weighed 3,600 kg, and was 4.85 metres long and 2.6 metres in diameter. It was docked to the nadir (Earth-facing) port on Zvezda on 17 September. The Service Module section of this vehicle (M-SO1 – Stykovochnyy Otsek), connected by a short intermediate section to the DC, was pyrotechnically severed from Pirs on 26 September, and was de-orbited the same day. DC1 also carried 800 kg of supplies that were later transferred into

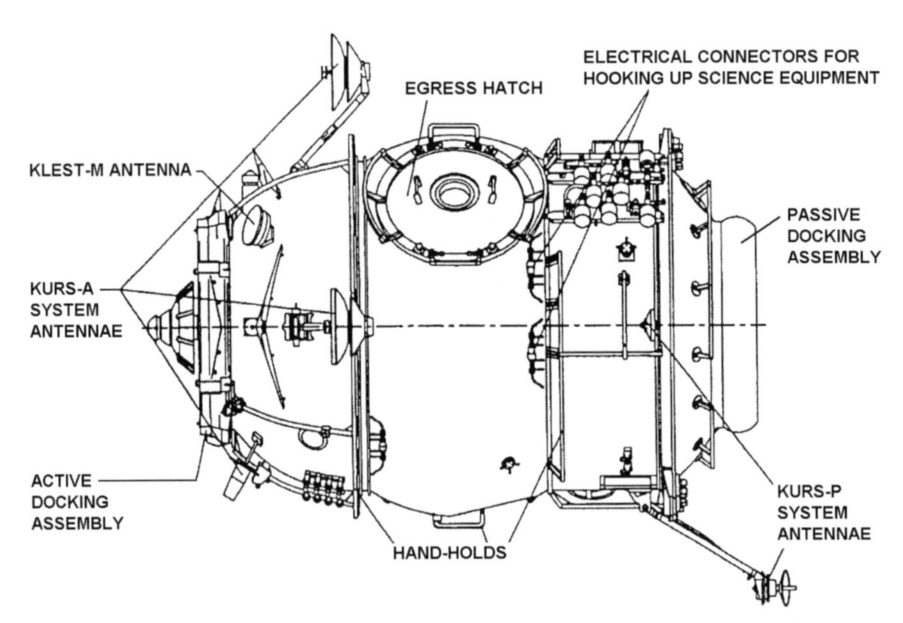

Pirs (Pier) docking and airlock module installed on the ISS during September 2001. (Courtesy NASA).

the station. This included a second Strela cargo boom (GStM-2) and an extra Orlan M pressure suit. *Novosti Kosmonavtiki* states that the mass of Pirs (with cargo) is 3,676 kg, and that it has a length of 4.907 metres and a maximum diameter of 2.55 metres.

The frontal docking port includes an active SSVP-M system (a hybrid of probe/drogue and androgynous systems), and the rear port features a passive SSVP (standard probe/drogue) system capable of receiving either Soyuz or Progress M1 vehicles, amd with the capacity to transfer propellants, water and air to Zvezda's storage tanks. This reduces the need to reposition the Soyuz to free docking ports for new or replacement vehicles.

Construction of the hardware began at RKK Energiya during 1988, and was completed late in 2000. After mating with the Propulsion Module, a series of electrical interface tests was completed during the first six months of 2001. The hardware completed its pre-shipment test on 11 July, and on 16 July it arrived at Baikonur for launch processing and mating to its Soyuz U launcher vehicle. In all, three Pirs articles were built: the flight article (DC1), a mock-up to be used in the hydro lab for EVA simulations, and a 1-g trainer for crew familiarisation. Pirs was located at the docking port (Zvezda nadir) that was intended for the much larger Universal Docking Module, which in the original planning would see Pirs discarded to allow docking of the UDM. A new Docking Compartment would then have been launched, to be docked to one of the UDM's radial ports. Current (2002) planning will see Pirs transferred from the Zvezda nadir to the Zvezda zenith docking port, and the Science Power Platform will later be mounted on top of Pirs. It will therefore no longer be necessary to build a replacement Pirs Docking Compartment. On 22 June 2001, Yuri Grigoryev, Deputy General Designer at Energiya, stated that Pirs was utilised over the larger, Proton-launched UDM due to lower costs.[14] The Centre for Analysis of Strategies and Technologies (CAST) in Moscow stated that the launch of a Proton cost $70–75 million per mission, compared with $55 million to launch a Soyuz U vehicle.

Pirs-based EVAs from the ISS, 2001–02

The location of Pirs also provides additional docking clearance from the main body of the Zvezda and additional elements that will be located on the hull of the station over the years. It also serves an EVA airlock for Russian-based EVAs (using only Orlan suits), with a 1-metre EVA hatch either side of the spherical section. The choice of hatch depends on which way the EVA crew is heading across the station. Each hatch opens inwards, which – although reducing the strain on the hinges when opened (a problem which damaged the Kvant 2 hatch on Mir in July 1990) – restricts the internal volume (13 cubic metres) of the airlock.

Pirs EVA	*EVA date*	*Duration (h:m)*	*ISS EVA*	*EVA crew*	*Expedition/ EVA*
1	2001 Oct 8	04:58	27th	Dezhurov/Tyurin	EO-3/1
	Erected antennae and docking target on Pirs module; first EVA from Pirs airlock; first EVA from the ISS without the Shuttle docked; hundredth Soviet/Russian EVA since Voskhod 2 in March 1965				

2 | 2001 Oct 14 05:52 | 28th | Dezhurov/Tyurin | EO-3/2
Placed Kromka sample detector on Zvezda, and MPAC and SEED experiments; recovered Russian flag and erected commercial logos.

3 | 2001 Nov 12 05:04 | 29th | Dezhurov/Tyurin | EO-3/3
Routed cables from Pirs to ISS interior; checked solar arrays.

4 | 2001 Dec 3 02:46 | 30th | Dezhurov/Tyurin | EO-3/4
Unplanned EVA to remove rubber o-ring obstruction from Pirs docking module to allow docking of Progress M1-7.

5 | 2002 Jan 14 06:03 | 32nd | Onufrienko/Walz | EO-4/1
Completed assembly of Strela unit; installed ham radio antenna.

6 | 2002 Jan 25 05:59 | 33rd | Onufrienko/Bursch | EO-4/2
Installed six thruster-plume deflectors; mounted four experiment packages and retrieved one; attached tether guides and ham radio antenna.

7 | 2002 Aug 16 04:25 | 42nd | Korzun/Whitson | EO-5/1
After a 1 hr 43 min delay due to misconfigured oxygen regulation valve in their Orlan suits, the pair moved the first six (or a planned twenty-three) micrometeoroid debris shields from their temporary stowage location on the ISS to their permanent location around Zvezda; due to delayed start of EVA, the planned refurbishment of Kromka experiment was postponed to a later EVA.

8 | 2002 Aug 26 06:00 | 43rd | Korzun/Treschev | EO-5/2
Installed EVA work-frame on Zarya; installed Japanese material samples, tether aids for future EVAs and two ham radio antennae.

PROGRESS CARGO MASS

The following shows the total cargo mass flown by Progress vehicles to Salyut 6, Salyut 7, Mir and the ISS.

Progress, 1978–90						*Progress M, 1989–2003*							
Salyut 6		Salyut 7		Mir		Mir		Mir		Mir		ISS	
1	2,300	13	2,116	25	2,482	M-1	2,682	M-16	2,598	M-31	2,410	M-44	2,542
2		14	1,981	26	2,405	M-2	2,726	M-17	2,604	M-32	2,402	M-45	>2,530
3		15	1,969	27	2,406	M-3	2,643	M-18	2,192	M-33	2,200	M-46	2,580
4	1,902	16	2,136	28	2,084	M-4	2,689	M-19	2,249	M-34	2,430	M-47	2,568
5	2,300	17	1,968	29	2,227	M-5	2,594	M-20	2,210	M-35	2,425		
6		18	1,879	30	1,856	M-6	2,546	M-21	2,385	M-36	2,501		
7	2,130	19	2,094	31	2,441	M-7	2,542	M-22	2,363	M-37	2,492		
8		20	2,376	32	2,341	M-8	2,693	M-23	2,207	M-38	2,377		
9		21	2,080	33	2,082	M-9	2,730	M-24	2,355	M-39	2,437		
10		22	2,126	34	2,324	M-10	2,624	M-25	2,380	M-40	2,552		
11		23	2,267	35	2,283	M-11	2,576	M-26	2,388	M-41	2,438		
12	>1,600	24	1,977	36	2,237	M-12	2,748	M-27	2,390	M-42	2,419		
		K1669	2,254	37	2,305	M-13	2,752	M-28	2,379	M-43	2,174		
				38	2,282	M-14	2,532	M-29	2,382				
				39	2,242	M-15	2,558	M-30	2,325				
				40	1,993								
				41	2,238								
				42	2,409								

Progress M1, 2000–							*Total*	*Average*
Mir		ISS		ISS				
						Progress to Salyut 6	10,232	2,046
						Progress to Salyut 7	27,223	2,094
M1-1	2,576	M1-3	2,434	M1-7	2,468	Progress to Mir	40,637	2,258
M1-2	2,271	M1-4	2,420	M1-8	2,407	Progress M to Mir	106,299	2,472
M1-5	?2,500	M1-6	2,478	M1-9	?	Progress M1 to Mir	7,347	2,449

SUMMARY

For more than twenty-five years the Progress variant of Soyuz has sustained operations on four space stations by delivering dry cargo, gas and liquids, and by facilitating the disposal of unwanted equipmemt and rubbish. It has also supported EVA operations and individual experiments, and continues to play a vital role on the ISS in maintaining resupply and reboost in conjunction with the American Shuttle system. Like Soyuz, it will remain an integral part of manned spaceflight operations for at least the next decade, and probably for the duration of the ISS.

The current cost of a Progress is $23 million – $6 million for the spacecraft, and $17 million for the launch vehicle. Each Progress takes about eighteen months to construct, but the process can be accelerated if necessary. Following the *Columbia* disaster, the rate of Progress launches was reviewed to ensure continued resupply of the ISS.

REFERENCES

1 David S.F. Portree, *Mir Hardware Heritage*, NASA RP-1357, March 1995: Part 3, 'Space Station Modules', pp.155–161.
2 Hall, Rex D. and Shayler, David J., *The Rocket Men*, Springer–Praxis (2001), pp.265–275.
3 Shayler, David J., *Skylab: America's Space Station*, Springer–Praxis (2001).
4 Glushko, Valentin P., *Kosmonavtika Entsiklopedia*, Soviet Encyclopaedia, 1985.
5 Linenger, Jerry M. *Off the Planet*, McGraw Hill (2000), pp.154–162.
6 See *Spaceflight*, August 1990, p.255.
7 Oberg, James, *Pioneering Space*, Newkirk, p.360.
8 *Aviation Week and Space Technology*, 10 June 1991, p.44.
9 RKK Energiya history, **1**, p.345–349.
10 Harland, David, *The Mir Space Station*, Wiley–Praxis, {year?}, p.168.
11 Burrough, Bryan, *Dragonfly*, Harper Collins, New York, 1998, pp.156–165.
12 RKK Energiya history, **1**, p.419.
13 Hendrickx, Bart, 'From Mir 2 to the ISS Russian Segment', *The International Space Station: from Imagination to Reality*, British Interplanetary Society, 2002, pp.30–34.
14 *Space News*, 2 July 2001.

Progress missions and hardware

Spacecraft (serial number) name	Design designation	Launch date	International designation	Docking date	Undocking date	Target spacecraft/port	Landing/ decay date	Flight duration dd:hh:mm:ss
Unmanned freighter for space station								
11F615 A15 (Progress automated transport cargo spacecraft for DOS								
(101) Progress 2	7K-TG	1978 Jul 7	1978-070A	1978 Jul 9	1978 Aug 2	Salyut 6 (Aft)	1978 Aug 4	27:14:04:51
(102) Progress 1	7K-TG	1978 Jan 20	1978-008A	1978 Jan 22	1978 Feb 6	Salyut 6 (Aft)	1978 Feb 8	18:17:35:20
(103) Progress 3	7K-TG	1978 Aug 8	1978-077A	1978 Aug 10	1978 Aug 21	Salyut 6 (Aft)	1978 Aug 23	15:18:58:38
(104) Progress 5	7K-TG	1979 Mar 12	1979-022A	1979 Mar 14	1978 Apr 3	Salyut 6 (Aft)	1979 Apr 5	23:19:16:32
(105) Progress 4	7K-TG	1978 Oct 3	1978-090A	1978 Oct 6	1978 Oct 24	Salyut 6 (Aft)	1978 Oct 26	22:17:18:43
(106) Progress 6	7K-TG	1979 May 13	1979-039A	1979 May 15	1979 Jun 8	Salyut 6 (Aft)	1979 Jun 9	27:17:33:51
(107) Progress 7	7K-TG	1979 Jun 28	1979-059A	1979 Jun 30	1979 Jul 18	Salyut 6 (Aft)	1979 Jul 20	21:16:32:49
(108) Progress 8	7K-TG	1980 Mar 27	1980-024A	1980 Mar 29	1980 Apr 25	Salyut 6 (Aft)	1980 Apr 28	31:12:01:??
(109) Progress 9	7K-TG	1980 Apr 27	1980-033A	1980 Apr 29	1980 May 20	Salyut 6 (Aft)	1980 May 22	24:18:20:00
(110) Progress 10	7K-TG	1980 Jun 29	1980-055A	1980 Jul 1	1980 Jul 17	Salyut 6 (Aft)	1980 Jul 19	19:21:06:18
(111) Progress 11	7K-TG	1980 Sep 28	1980-079A	1980 Sep 30	1980 Dec 9	Salyut 6 (Aft)	1980 Dec 11	73:22:50:05
(112) Progress 15	7K-TG	1982 Sep 18	1982-094A	1982 Sep 20	1982 Oct 14	Salyut 7 (Aft)	1982 Oct 16	28:12:09:00
(113) Progress 12	7K-TG	1981 Jan 24	1981-007A	1981 Jan 26	1981 Mar 19	Salyut 6 (Aft)	1981 Mar 20	56:02:40:58
(114) Progress 13	7K-TG	1982 May 23	1982-047A	1982 May 25	1982 Jun 4	Salyut 7 (Aft)	1982 Jun 6	13:18:08:00
(115) Progress 16	7K-TG	1982 Oct 31	1982-107A	1982 Nov 2	1982 Dec 13	Salyut 7 (Aft)	1982 Dec 14	44:05:57:00
(116) Progress 21	7K-TG	1984-May 7	1984-042A	1984 May 10	1984 May 26	Salyut 7 (Aft)	1984 May 26	18:15:29:45
(117) Progress 14	7K-TG	1982 Jul 10	1982-070A	1982 Jul 12	1982 Aug 10	Salyut 7 (Aft)	1982 Aug 13	33:15:30:59
(118) Progress 18	7K-TG	1983 Oct 20	1983-106A	1983 Oct 22	1983 Nov 13	Salyut 7 (Aft)	1983 Nov 16	26:18:18:55
(119) Progress 17	7K-TG	1983 Aug 17	1983-085A	1983 Aug 19	1983 Sep 17	Salyut 7 (Aft)	1983 Sep 17	31:11:34:37

(120)	Progress 19	7K-TG	1984 Feb 21	1984-018A	1984 Feb 23	1984 Mar 31	Salyut 7 (Aft)	1984 Apr 1	40:11:31:55
(121)	Progress 20	7K-TG	1984 Apr 15	1984-038A	1984 Apr 17	1984 May 6	Salyut 7 (Aft)	1984 May 7	22:01:43:07
(122)	Progress 22	7K-TG	1984 May 28	1984-051A	1984 May 30	1984 Jul 15	Salyut 7 (Aft)	1984 Jul 15	48:03:54:08
(123)	Not flown	7K-TG	*Not used (?) – test vehicle, a non-flightworthy vehicle, or not built*						
(124)	Progress 23	7K-TG	1984 Aug 14	1984-086A	1984 Aug 16	1984 Aug 26	Salyut 7 (Aft)	1984 Aug 28	13:18:59:10
(125)	Progress 24	7K-TG	1985 Jun 21	1985-051A	1985 Jun 23	1985 Jul 15	Salyut 7 (Aft)	1985 Jul 15	24:23:20:00
(126)	Cosmos 1669	7K-TG	1985 Jul 19	1985-062A	1985 Jul 21	1985 Aug 28	Salyut 7 (Aft)	1985 Aug 30	
(127)	Progress 29	7K-TG	1987 Apr 21	1987-034A	1987 Apr 23	1987 May 11	Mir (Aft-K)	1987 May 11	19:16:36:59
(128)	Progress 30	7K-TG	1987 May 19	1987-044A	1987 May 21	1987 Jul 19	Mir (Aft-K)	1987 Jul 19	61:00:57:50
(129)		7K-TG	*Not used*	*129–133 were originally built for flight tests as part of the subsequently cancelled Soviet Star Wars programme. Elements of the vehicles were incorporated in later Progress craft.*					
(130)		7K-TG	*Not used*						
(131)		7K-TG	*Not used*						
(132)		7K-TG	*Not used*						
(133)		7K-TG	*Not used*						
(134)	Progress 25	7K-TG	1986 Mar 19	1986-023A	1986 Mar 21	1986 Apr 20	Mir (Aft)	1986 Apr 21	32:03:51:35
(135)	Progress 27	7K-TG	1987 Jan 16	1987-005A	1987 Jan 18	1987 Feb 23	Mir (Aft)	1987 Feb 25	40:09:10:22
(136)	Progress 26	7K-TG	1986 Apr 23	1986-032A	1986 Apr 26	1986 Jun 22	Mir (Aft)	1986 Jun 23	60:23:00:56
(137)	Progress 28	7K-TG	1987 Mar 3	1987-023A	1987 Mar 5	1987 Mar 26	Mir (Aft)	1987 Mar 28	24:15:46:56
(138)	Progress 31	7K-TG	1987 Aug 3	1987-066A	1987 Aug 5	1987 Sep 21	Mir (Aft-K)	1987 Sept 12	50:03:37:49
(139)	Progress 32	7K-TG	1987 Sep 23	1987-082A	1987 Sep 26	1987 Nov 10	Mir (Aft-K)		
					1987 Nov 10	1987 Nov 17	Mir (Aft-K)	1987 Nov 18	55:00:26:06
(140)	Progress 33	7K-TG	1987 Nov 20	1987-094A	1987 Nov 23	1987 Dec 19	Mir (Aft-K)	1987 Dec 19	28:03:08:48
(141)		7K-TG	*Not used (see no. 123)*						
(142)	Progress 34	7K-TG	1988 Jan 20	1988-003A	1988 Jan 23	1988 Mar 4	Mir (Aft-K)	1988 Mar 4	43:07:53:06
(143)	Progress 35	7K-TG	1988 Mar 23	1988-024A	1988 Mar 25	1988 May 5	Mir (Aft-K)	1988 May 5	42:09:56:30
(144)	Progress 36	7K-TG	1988 May 13	1988-038A	1988 May 15	1988 Jun 5	Mir (Aft-K)	1988 Jun 5	23:19:57:35

Spacecraft (serial number) name	Design designation	Launch date	International designation	Docking date	Undocking date	Target spacecraft/port	Landing/ decay date	Flight duration dd:hh:mm:ss
(145) Progress 37	7K-TG	1988 Jul 18	1988-061A	1988 Jul 20	1988 Aug 12	Mir (Aft-K)	1988 Aug 12	24:15:38:21
(146) Progress 38	7K-TG	1988 Sep 9	1988-083A	1988 Sep 12	1988 Nov 23	Mir (Aft-K)	1988 Nov 23	74:18:52:20
(147) Progress 39	7K-TG	1988 Dec 25	1988-114A	1988 Dec 27	1989 Feb 7	Mir (Aft-K)	1989 Feb 7	44:08:37:23
(148) Progress 40	7K-TG	1989 Feb 10	1989-008A	1989 Feb 12	1989 Mar 3	Mir (Aft-K)	1989 Mar 5	22:16:14:08
(149) Progress 41	7K-TG	1989 Mar 16	1989-023A	1989 Mar 18	1989 Apr 21	Mir (Aft-K)	1989 Apr 25	39:17:07:46
(150) Progress 42	7K-TG	1990 May 5	1990-041A	1990 May 7	1990 May 27	Mir (Aft-K)	1990 May 27	21:14:55:59
11F615 A55 (Progress M modified automated transport cargo spacecraft for DOS								
(201) Progress M	7K-TGM	1989 Aug 23	1989-066A	1989 Aug 25	1989 Dec 1	Mir (Fwd)	1989 Dec 1	100:07:02:28
(202) Progress M-2	7K-TGM	1989 Dec 20	1989-099A	1989 Dec 22	1990 Feb 9	Mir (Aft-K)	1990 Feb 9	51:03:36:10
(203) Progress M-3	7K-TGM	1990 Feb 28	1990-020A	1990 Mar 3	1990 Apr 27	Mir (Aft-K)	1990 Apr 29	58:00:39:03
(204) Progress M-4	7K-TGM	1990 Aug 15	1990-072A	1990 Aug 17	1990 Sep 17	Mir (Fwd)	1990 Sep 20	36:07:03:46
(205) Progress M-6	7K-TGM	1991 Jan 14	1991-022A	1991 Jan 16	1991 Mar 15	Mir (Aft-K)	1991 Mar 15	60:02:23:33
(206) Progress M-5	7K-TGM	1990 Sep 27	1990-085A	1990 Sep 29	1990 Nov 28	Mir (Fwd)	1990 Nov 28	61:23:46:46
(207) Progress M-8	7K-TGM	1991 May 30	1991-038A	1991 Jun 1	1991 Aug 15	Mir (Fwd)	1991 Aug 16	77:22:53:??
(208) Progress M-7	7K-TGM	1991 Mar 19	1991-020A	1991 Mar 28?	1991 May 6	Mir (Fwd)	1991 May 7	49:03:18:45
(209) Progress M-14	7K-TGM	1992 Aug 15	1992-055A	1992 Aug 18	1992 Oct 21	Mir (Aft-K)	1992 Oct 21	66:23:11:29
(210) Progress M-9	7K-TGM	1991 Aug 20	1991-057A	1991 Aug 23	1991 Sep 30	Mir (Fwd)	1991 Sep 30	40:21:22:??
(211) Progress M-10	7K-TGM	1991 Oct 17	1991-073A	1991 Oct 21	1992 Jan 20	Mir (Fwd)	1992 Jan 20	66:23:11:29
(212) Progress M-11	7K-TGM	1992 Jan 25	1992-004A	1992 Jan 27	1992 Mar 13	Mir (Fwd)	1992 Mar 13	48:07:57:??
(213) Progress M-12	7K-TGM	1992 Apr 19	1992-022A	1992 Apr 22	1992 Jun 27	Mir (Fwd)	1992 Jun 28	69:02:31:??
(214) Progress M-13	7K-TGM	1992 Jun 30	1992-035A	1992 Jul 4	1992 Jul 24	Mir (Fwd)	1992 Jul 24	23:15:21:??
(215) Progress M-15	7K-TGM	1992 Oct 27	1992-071A	1992 Oct 29	1993 Feb 4	Mir (Aft-K)	1993 Feb 7	102:13:23:??

(216)	Progress M16	7K-TGM	1993 Feb 21	1993-012A	1993 Feb 23	1993 Mar 26	Mir (Aft-K)		
					1993 Mar 26	1993 Mar 27	Mir (Aft-K)	1993 Mar 27	37:15:53:??
(217)	Progress M-17	7K-TGM	1993 Mar 31	1993-019A	1993 Apr 2	1993 Aug 11	Mir (Aft-K)	1994 Mar 3	336:23:54:??
(218)	Progress M-18	7K-TGM	1993 May 22	1993-034A	1993 May 24	1993 Jul 3	Mir (Fwd)	1993 Jul 4	43:10:41:??
(219)	Progress M-19	7K-TGM	1993 Aug 10	1993-052A	1993 Aug 13	1993 Oct 12	Mir (Aft-K)	1993 Oct 13	63:02:00:??
(220)	Progress M-20	7K-TGM	1993 Oct 11	1993-064A	1993 Oct 13	1993 Nov 21	Mir (Aft-K)	1993 Nov 21	40:12:03:??
(221)	Progress M-21	7K-TGM	1994 Jan 28	1994-005A	1994 Jan 30	1994 Mar 23	Mir (Aft-K)	1994 Mar 23	54:03:01:??
(222)	Progress M-22	7K-TGM	1994 Mar 22	1994-019A	1994 Mar 24	1994 May 23	Mir (Aft-K)	1994 May 23	61:23:58:??
(223)	Progress M-23	7K-TGM	1994 May 22	1994-031A	1994 May 24	1994 Jul 2	Mir (Aft-K)	1994 Jul 2	41:10:14:??
(224)	Progress M-24	7K-TGM	1994 Aug 25	1994-052A	1994 Sep 2	1994 Oct 4	Mir (Fwd)	1994 Oct 4	40:07:18:48
(225)	Progress M-25	7K-TGM	1994 Nov 11	1994-075A	1994 Nov 13	1995 Feb 16	Mir (Aft-K)	1995 Feb 16	97:08:44:02
(226)	Progress M-26	7K-TGM	1995 Feb 15	1995-075A	1995 Feb 17	1995 Mar 15	Mir (Aft-K)	1995 Mar 15	27:12:49:32
(227)	Progress M-27	7K-TGM	1995 Apr 9	1995-020A	1995 Apr 11	1995 May 22	Mir (Fwd)	1995 May 23	43:07:06:03
(228)	Progress M-28	7K-TGM	1995 Jul 20	1995-036A	1995 Jul 22	1995 Sep 4	Mir (Fwd)	1995 Sep 4	46:05:54:14
(229)	Progress M-29	7K-TGM	1995 Oct 8	1995-053A	1995 Oct 10	1995 Dec 19	Mir (Aft-K)	1995 Dec 19	71:20:35:20
(230)	Progress M-30	7K-TGM	1995 Dec 18	1995-070A	1995 Dec 20	1996 Feb 22	Mir (Aft-K)	1996 Feb 22	65:23:31:01
(231)	Progress M-31	7K-TGM	1996 May 5	1996-028A	1996 May 7	1996 Aug 1	Mir (Fwd)	1996 Aug 1	88:12:40:12
(232)	Progress M-32	7K-TGM	1996 Jul 31	1996-043A	1996 Aug 2	1996 Aug 18	Mir (Fwd)		
					1996 Sep 3	1996 Nov 20	Mir (Aft-K)	1996 Nov 21	112:02:42:19
(233)	Progress M-33	7K-TGM	1996 Nov 19	1996-066A	1996 Nov 22	1996 Feb 6	Mir (Aft-K)		
					1996 Mar 4	*Failure*	Mir (Aft-K)	1997 Mar 12	112:04:03:??
(234)	Progress M-34	7K-TGM	1997 Apr 6	1997-014A	1997 Apr 8	1997 Jun 24	Mir (Aft-K)		
					1997 Jun 25	*Failure; collision*	Mir (Aft-K)	1997 Jul 2	86:14:28:??
(235)	Progress M-35	7K-TGM	1997 Jul 5	1997-033A	1997 Jul 7	1997 Aug 6	Mir (Aft-K)		
					1997 Aug 18	1997 Oct 7	Mir (Aft-K)	1997 Oct 7	94:14:12:??
(236)	Progress M-37	7K-TGM	1997 Dec 20	1997-081A	1997 Dec 22	1998 Jan 30	Mir (Aft-K)		
					1998 Feb 23	1998 Mar 15	Mir (Aft-K)	1998 Mar 15	75:14:19:??

Spacecraft (serial number) name	Design designation	Launch date	International designation	Docking date	Undocking date	Target spacecraft/port	Landing/ decay date	Flight duration dd:hh:mm:ss
(237) Progress M-36	7K-TGM	1997 Oct 5	1997-058A	1997 Oct 8	1997 Dec 17	Mir (Aft-K)	1997 Dec 19	73:22:12:??
(238) Progress M-39	7K-TGM	1998 May 14	1998-031A	1998 May 16	1998 Aug 12	Mir (Aft-K)		
				1998 Sep 1	1998 Oct 25	Mir (Aft-K)	1998 Oct 29	167:06:02:??
(239) Progress M-40	7K-TGM	1998 Oct 25	1998-062A	1998 Oct 27	1999 Feb 4	Mir (Aft-K)	1999 Feb 5	101:04:33:??
(240) Progress M-38	7K-TGM	1998 Mar 14	1998-015A	1998 Mar 17	1998 May 15	Mir (Aft-K)	1998 May 15	61:23:41:??
(241) Progress M-41	7K-TGM	1999 Apr 2	1999-015A	1999 Apr 4	1999 Jul 17	Mir (Aft-K)	1999 Jul 17	105:23:51:??
(242) Progress M-42	7K-TGM	1999 Jul 16	1999-038A	1999 Jul 18	2000 Feb 2	Mir (Aft-K)	2000 Feb 2	201:13:32:??
(243) Progress M-43	7K-TGM	2000 Oct 16	2000-064A	2000 Oct 20	2001 Jan 25	Mir (Aft-K)	2001 Jan 29	103:05:31:??
(244) Progress M-44 (3P)	7K-TGM	2001 Feb 26	2000-008A	2001 Feb 28	2001 Apr 16	ISS Aft-Zvezda	2001 Apr 16	47:05:14:??
(245) Progress M-45 (5P)	7K-TGM	2001 Aug 21	2001-006A	2001 Aug 23	2001 Nov 22	ISS Aft-Zvezda	2001 Nov 22	93:??:??:??
(246) Progress M-46 (5P)	7K-TGM	2002 Jun 25	2002-033A	2002 Jun 29	2002 Sep 24	ISS Aft-Zvezda	2002 Oct 14	108:??:??:??
(247) Progress M-47 (10P)	7K-TGM	2003 Feb 2	2003-006A	2003 Feb 4		ISS Aft-Zvezda?		
(248) Progress M-48 (12P)	7K-TGM	2003				ISS		
11F615 A55 (Progress M1 (Modified) automated transport cargo spacecraft for Mir; ISS)								
(250) Progress M1-1		2000 Feb 1	2000-005A	2000 Feb 3	2000 Apr 26	Mir (Aft-K)	2000 Apr 26	84:12:40:??
(251) Progress M1-2		2000 Apr 25	2000-021A	2000 Apr 27	2000 Oct 15	Mir (Aft-K)	2000 Oct 16	174:??:??:??
(252) Progress M1-3 (1P)		2000 Aug 6	2000-044A	2000 Aug 8	2000 Nov 1	ISS Aft-Zvezda	2000 Nov 1	86:12:38:??

(253)	Progress M1-4 (2P)	2000 Nov 16	2000-073A	2000 Nov 18	2000 Dec 1	ISS Nadir-Zarya	–	
				2000 Dec 26	2001 Feb 8	ISS Nadir-Zarya	2001 Feb 8	84:12:07:??
(254)	Progress M1-5 7K-?	2001 Jan 24	2001-003A	2001 Jan 27		Mir (Aft-K)	2001 Mar 23	58:01:16:??
(255)	Progress M1-6 (4P)	2001 May 20	2001-012A	2001 May 23	2001 Aug 21	ISS Aft-Zvezda	2001 Aug 21	92:??:??:??
(256)	Progress M1-7 (6P)	2001 Nov 26	2001-051A	2001 Nov 28 *Soft*		ISS Aft-Zvezda		
				2001 Dec 3 *Hard*	2002 Mar 19	ISS Aft-Zvezda	2002 Mar 20	114:??:??:??
(257)	Progress M1-8 (7P)	2002 Mar 21	2002-013A	2002 Mar 24	2002 Jun 25	ISS Aft-Zvezda	2002 Jun 25	96:??:??:??
(258)	Progress M1-9 (9P)	2002 Sep 25	2002-045A		2002 Sep 29	2003 Feb 1	ISS Aft-Zvezda	2003 Feb 1
(259)	Progress M1-10 (11P)	2003 May	*Planned*				ISS	
(260)	Progress M1-11	2003 Nov				ISS		
(301)	Progress M-SO1 (4R) 7K-?	2001 Sep 14	2001-041A	2001 Sep 17	*Permanently docked*	ISS		

The Instrument Unit of Progress (301) re-entered, after twelve days, on 2001 Sept 26

Soyuz T, 1979–86

In December 1979 a new variant of the Soyuz emerged as the next generation of manned ferry craft to second-generation Salyut space stations. Called Soyuz T (Transport), it featured several upgrades to the previous ferry vehicle, and offered greater flexibly and reliability in support of the long-duration missions that were being planned for the space station programme

THE ROLE OF SOYUZ T

This variant of Soyuz was designed to operate as a ferry to orbital space stations. The addition of solar panels allowed for longer independent flight, so that if the initial docking should fail, it could be re-attempted. It also allowed the return to the three-person crews that would be essential for efficient operations on larger stations, and the crews would continue to wear Sokol space suits on the ascent and descent stages of each mission.

THE ORIGINS OF SOYUZ T

Soyuz T can be traced back to the 7K-S design authorised in 1968 as a manned ferry to the Soyuz VI military space station. The ferry survived the cancellation of that programme in February 1970, and in two versions: a two-man space station ferry craft, and a solo research vehicle. The S denoted 'special', which, in the terminology of those days, indicated that it was military. 7K-S was notable as the first Soyuz for which a probe/drogue docking mechanism with an internal transfer tunnel was developed, and 7K-T (the 'Soyuz 10 type'), which was developed later, inherited this design. It seems that after the cancellation of Soyuz VI, the focus shifted to solo missions in the interests of the Ministry of Defence. Plans for a transport version moved to the background. Even while Soyuz VI was still alive, there were plans for 'solo' versions of 7K-S; namely, 7K-S-I (11F733) for short-duration flights, and 7K-S-II (11F734) for long-duration flights, although priority was given to the transport version.

On 11 August 1972, a supplement to the original draft for the 7K-S ferry craft was issued for the construction of a series of flight and test models to run up to four unmanned test flights, followed by two manned test flights and two 'operational' test flights. One of the changes in the August 1972 plan was to reduce the crew from three to two. This was due to the Soyuz 11 disaster, which had a direct effect on the number of cosmonauts flown on both 7K-T and 7K-S. By May 1974, three test models of the new spacecraft had been fabricated and were in the final stage of production.

The summer months of 1974 were challenging for the Soviet space programme as it faced major reorganisation. On 19 May, all further launches of the N1 lunar vehicle were suspended, followed by suspension of work on the N1–L3 on 25 June, effectively ending the disappointing Soviet manned lunar programme. In addition, on 22 May 1974, TsKBEM (formerly OKB-1) merged with KB EnergoMash and formed a new organisation to be called NPO Energiya, with Valentin Glushko succeeding Vasily Mishin as Chief Designer.

On 21 June 1974 a decree was issued by the Military Industrial Commission (VPK) – a government body that oversaw the entire defence industry and implemented space policy. It called for the establishment of a State Commission (Chairman, A.G. Karas; Deputy Chairman, Gherman Titov) to oversee test flights of the vehicle, and also to speed up the development of a transport version. This was an important turning point in the programme. It seems that in 1974 the military lost interest in solo research flights of Soyuz T, and that the focus once again shifted to the use of Soyuz T as a ferry vehicle. Some of the experiments planned for the solo flights were transferred to the space station programme, and the article number 11F732, used from the beginning of the 7K-S programme, was not introduced at this point. Major overhauls of internal systems in the proposed new spacecraft, which

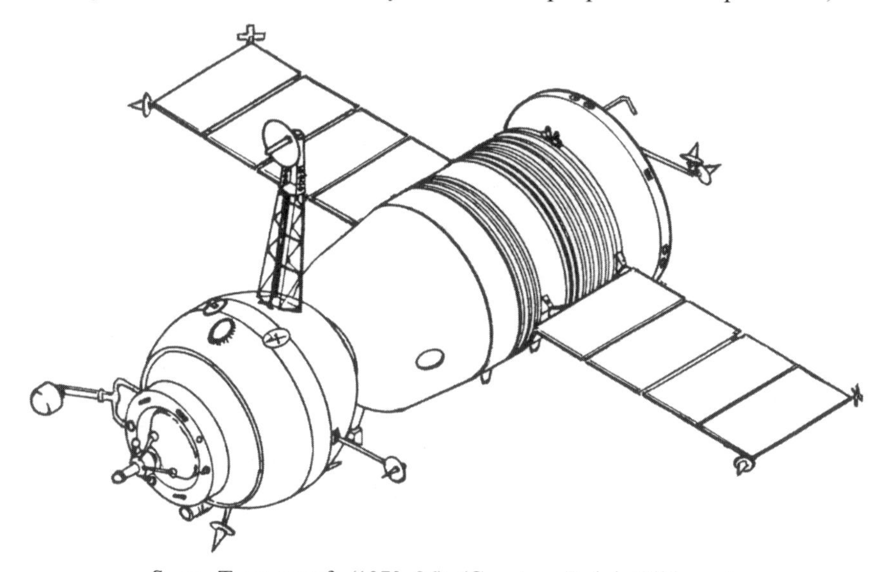

Soyuz T spacecraft (1979–86). (Courtesy Ralph Gibbons.)

were radically different from the original 7K-S, resulted in a design that was essentially a new spacecraft in an old shell. In 1975, a further revision in the design enabled a crew of up to three to be flown wearing Sokol KV-2 pressure garments – an improvement over those first introduced on Soyuz 12 in September 1973.

TESTING THE TECHNOLOGY, 1974–80

During the 1970s, Western analysts identified several Cosmos flights as possibly being related to the manned programme; but it was not until the first Soyuz T was identified in 1979 that some of these mysterious Cosmos missions could finally be assigned as development flights of this new variant of Soyuz. As the 7K-S test vehicles were already constructed and in storage at Baikonur, it was decided to fly three spacecraft under the Cosmos label to test some of the systems and technology that would flown operationally on Soyuz T. It was decided to fly the three 7K-S test vehicles already available – despite the decision having already been taken in June 1974 to reorient efforts to the ferry version. The three 7K-S vehicles that had already been built by that time did not yet incorporate the changes required for the transport version, although it was decided to launch them to test some of the systems.

The first of these was Cosmos 670 (7K-S no.1) launched on 6 August 1974 into a new inclination of 50.6° for Soviet manned spacecraft and after an unplanned ballistic re-entry was recovered after a three day flight on 8 August without having performed any manoeuvres. This flight occurred during a busy period at Baikonur, with final preparations for the launch of a second Soyuz (Cosmos 672) six days later on a test of ASTP systems, and a third (Soyuz 15) launched on a docking mission to the military Salyut (Almaz) 3 station on 26 August. Flying the test vehicle at 50.6° was new for manned spacecraft and could have been because the Soyuz launch vehicle had a slightly greater lifting capability.

The second test vehicle – Cosmos 772 (7K-S no.2) – was launched on 29 September 1975, and was also recovered after three days, although this time the vehicle performed a series of orbital manoeuvres. Finally, Cosmos 869 (7K-S no.3) was launched on 26 November 1976. The mission was planned to last for eight days, but due to problems this was extended to eighteen days (thus matching the duration record of Soyuz 9). Cosmos 869 completed a programme of extensive manoeuvres before it was recovered on 17 December.

There were no communications with the vehicle for about two days after orbit insertion. Analysis showed that a built-in safety system, developed specifically for the military solo version, had shut down all communications with the ground because of a certain sequence of commands sent up to the spacecraft early in the mission. As expected, the vehicle 'called back' to the ground after two days. However, there was also a problem with the infrared vertical sensor, and it was finally decided to extend the mission and to complete the flight programme as originally planned. The first of the Soyuz 7K-ST spacecraft, launched on 4 April 1978, was flown as Cosmos 1001 (7K-ST no.4); but the mission was plagued by troubles, and the vehicle landed eleven days later, on 15 April.

Originally there were to be three solo test flights of the 7K-ST transport version: two unmanned (4L and 5L – the 'L' denoting Lyotnaya (Flightworthy), as opposed to a boilerplate mock-up or a test model), and one manned (6L). Three crews began training for the solo test mission in 1978: Malyshev and Akyonov, Kizim and Makarov, and Lazarev and Strekalov). However, in late 1978 (several months after the Cosmos 1001 (4L) mission) several changes were made to the schedule. This coincided with Yuri Semyonov taking over from Konstantin Bushayev as the Chief Designer of Soyuz T (Bushayev had died in October 1978). Although 7K-ST was intended as a crew ferry, the emphasis in the test flights remained on its capability to perform military solo missions, and Semyonov wanted to eliminate this contradiction. He therefore changed the test flight programme, as follows: an unmanned solo endurance mission simulating a long-docked state (5L); an unmanned docking flight to Salyut 6 (6L); and two manned docking flights to Salyut 6 (7L and 8L). It is interesting that the military remained in charge of the entire unmanned test programme; despite the emphasis having shifted to the transport function in 1974. Control of the Soyuz T State Commission was turned over (by a decision of the Military Industrial Commission) from the Ministry of Defence to the Ministry of General Machine Building (Chairman, Kerim Kerimov) on 10 May 1979.

Cosmos 1074 (7K-ST no.5) was launched on 31 January 1979, and was the second and final flight of a Soyuz T craft under the Cosmos designation. The spacecraft extended the orbital storage test, and remained on orbit for 60 days. The mission was originally planned as a ninety-day longevity test, but due to technical problems it was curtailed on the sixty-first day.

Following these five test flights, the next spacecraft (7K-ST no.6) was assigned to complete a full systems and procedures test in conjunction with the Salyut 6 space station (then unmanned) late in 1979. It was assigned the mission identification Soyuz T.

Soyuz T: the inaugural mission

On 19 August 1979, the third resident crew of Salyut 6 (Lyakhov and Ryumin) vacated the space station at the end of a record-breaking 175 days, and landed in Soyuz 34. In September, Konstantin Feoktistov reported that an evaluation on the status of Salyut 6 was underway to determine if it could support further manned spaceflight. After two years in orbit, it had already supported expeditions of 96, 140 and 175 days, as well as four short visiting missions, seven Progress freighter missions and three EVAs, becoming by far the most successful Salyut station to date. New crews were in training to occupy Salyut in the spring, but first a new version of the Soyuz was about to be revealed to the world.

On 16 December 1979, 73 minutes after the unmanned Salyut flew over the cosmodrome, Soyuz T left the pad and headed for orbit on the final unmanned test flight. Two days later the spacecraft was commanded to approach the Salyut; but it overshot, and moved in from above and ahead of the Salyut instead of the planned 'behind and below' approach. It would be 24 hours before control manoeuvres placed Soyuz T back in the correct position to try again. The attempt on 19 December was flawless, and Soyuz T nudged into the forward port of Salyut 6. Six days later, on Christmas Day, the engines of the Soyuz were ignited to raise the station's orbit from 342 × 362 km to 370 × 382 km. For 95 days, Soyuz T remained docked to the Salyut while tests of its systems and storage capability were conducted, alongside tests of the Salyut for receiving a new crew. Undocking took place on 24 March, and this was followed by further independent tests over a 48-hour period, ending with recovery of the DM in Kazakhstan on 26 March, and qualifying the spacecraft for manned flight.

SOYUZ T UPGRADES

Outwardly, Soyuz T resembled its predecessor, with an OM containing the docking and internal transfer equipment, the DM for a crew of up to three wearing pressure garments, the PM for orbital manoeuvring, and a pair of solar arrays for electrical power in flight. The Soyuz T was designed for an operational life of up to 14 days without a full power-down if the docking failed in the first two days, and an orbital storage duration of 180 days before it required exchange with a fresh craft. The launch vehicle for these spacecraft was the Soyuz 11A511U, with the 11A511U2 being used from T-12 onwards (see p.89).

The Orbital Module

The OM changed very little from earlier versions, although it reportedly could be left docked to the Salyut station to use its internal compartment for extra storage space (5 cubic metres). This was apparently demonstrated during at least two missions (T-3 and T-4) to Salyut 6, where the OM was left attached to the Salyut docking port for a few hours, and then cast off, to be identified by US tracking networks. Alternatively, it could be jettisoned prior to the orbital descent burn, thus allowing a 10% reduction in propellant to about 250 kg at the end of the mission. This in turn allowed for a third crew-man for the first time since 1971 (with life support system and space suit), or two cosmonauts and 100 kg of extra cargo. The docking equipment was still a lightweight probe assembly, using the Igla rendezvous system.

Reports that the orbital modules of Soyuz T-3 and T-4 remained temporarily attached to Salyut 6 after undocking of the DM/PM have *never* been confirmed by the Russians, and some of the cosmonauts involved in those missions have been asked, on several occasions, if this was really true; but they denied it A possible explanation for these reports is that NORAD tracked the OM after the return of the DM, and thought that they had been separated from Salyut 6; whereas they had actually been separated from the DM/PM shortly before retro-fire. This raises the question of when this technique was introduced, as several Western accounts have indicated that several of the unmanned Cosmos test flights of Soyuz T had tested this apparent capability. If such a technique was introduced on T-3, then this misunderstanding would be entirely plausible, because before that NORAD had never seen an Orbital Module remain in orbit after retro-fire.

Further during the 1981 T-4 mission, Cosmos 1267 was launched on 25 April 1981, well before the return of T-4. But would the Russians have taken the risk of leaving the OM docked to Salyut 6 while Cosmos 1267 was already on its way to link up with the same docking port? Neither is it clear whether the OM separation before retro-fire was the reason (or certainly not the only reason) why a third crew-member could be included. It seems that this was mainly achieved by using lighter onboard systems, and other factors such as the use of the Soyuz U launch vehicle and reconfiguration of the Descent Module. On the Soyuz 12-type vehicles, the third seat was replaced by an emergency oxygen system in case of depressurisation, and on Soyuz T this was somehow reconfigured to allow the third seat to be reinstalled. The RKK Energiya history includes a detailed technical description of Soyuz T, but the OM separation before retro-fire is not mentioned. Finally, after the Soyuz TM-5 mishap in 1988, the old technique of detaching the OM *after* retro-fire was re-introduced, but this did not prevent three crew-members being flown on future TM missions.

The Descent Module

A notable feature incorporated into the Descent Module of Soyuz T was the new window covers, which could be jettisoned after re-entry to offer a clearer view during parachute descent and after landing. Crews had previously reported that the DM windows had been blackened from the fiery re-entry. The improved Sokol pressure suit weighed only 8 kg, incorporating more plastic for lightness and flexibility and a new helmet with increased visibility; and although not specifically designed for EVA operations, it could still protect a cosmonaut during an emergency EVA transfer from the Soyuz OM to a second spacecraft (although this never occurred on Soyuz T).[1]

Other provisions for the crew included an improved escape tower for additional safety margins and to cope with the larger cargo capacity. This escape tower was heavier than the earlier version, but it was intended to be jettisoned earlier (at T + 123 sec, rather than T + 160 sec). It used improved solid-fuel engines that were able to pull the DM/OM to a higher altitude in case of an on-the-pad abort. This in turn made it possible to use the main parachute rather than the (less reliable) back-up parachute during the descent of the DM. The escape system automatics even took into account the prevailing winds to ensure that the DM did not land back near the pad. In addition, a set of six soft-landing engines were incorporated over the previous four engines on the earlier Soyuz vehicle.

The Chayka (Seagull) flight control system of controls and displays featured integrated circuitry chips – a saving in both volume and weight. The system also included a BTSVK digital computer that was significantly more automated than earlier Soyuz systems, although in practise, on several occasions the cosmonauts had to override the automated systems and revert to manual operations. This computer – Argon 16 (it had 16 kbytes of RAM) – had been used on Salyut 4, and, under normal flight conditions, would reduce reliance on the ground computers and tracking stations. (The crew, however, retained the flight data books as back-up.) The Argon 16 displayed data simultaneously in the DM and at TsUP, and the reliability, capacity and speed of telemetry between Soyuz T and TsUP was also improved.

Another new addition was the crew compartment control panel, which included a cathode ray tube display. The world drive scope – on which the spacecraft's relative position over Earth was plotted – was relocated from the left-hand side to the right-hand side of the panel. The analogue sequencers stored in boxes on the left of the control panel were removed altogether, along with numerous steam-gauge-type dials. They were replaced by rows of push buttons or annunciation display panels. The cosmonauts still used the 'swizzle stick' to reach these controls when in their pressure suits and strapped in the couches. When Soyuz T appeared, the Western press compared the improvements and true 'piloting' capability with the more sophisticated Gemini and Apollo spacecraft. For attitude determination, the Soyuz still relied on the Vzor periscope system, and there was no '8-ball' horizon indicator as installed on the American craft.

The Propulsion Module

A significant change was incorporated into the propulsion sub-system of Soyuz T, so that it became similar to that flown on the Progress freighters. The smaller attitude control thrusters were fully integrated with the main propulsion system, and both systems drew from the same propellant supply (N204/UDMH), as some of the earlier Soyuz–Salyut docking failures were caused because propellant could not be transferred from the attitude control system to supplement the main propulsion system. Main engine thrust had been reduced to 315 kg, producing a vacuum thrust of 305 seconds, and there was a set of fourteen 14-kg attitude control thrusters and a second set of twelve 2.5-kg thrusters available for pitch, yaw and roll control. The decision to unify the RCS and primary engines eliminated the need for a separate back-up engine for descent orbit, as the RCS network could now perform this function.

On the earlier Soyuz, the attitude control thrusters and main propulsion system not only used different propellant supplies, but also different *types* of propellant. The attitude control thrusters used hydrogen peroxide, and the main propulsion system nitric acid/UDMH. On Soyuz T the main propulsion system was fed by pressure rather than turbo-pumps as in the previous Soyuz design. The only designator identified for the Soyuz T main engine is 11D426, and no KTDU designator has been identified.

The Soyuz T PM also included two twin solar arrays, similar to those flown during the ASTP. With a span of 10.6 metres, these were slightly smaller than those flown on Soyuz 1–11. Their 10-square-metre surface area generated 0.6 kW.

THE SOYUZ T TRAINING GROUP, 1973–81

By the summer of 1980 – after six unmanned test and evaluation flights and one brief manned test flight – Soyuz T-2 (7K-ST 7L), the next spacecraft on the manifest (7K-ST 8L), would inaugurate the operational flights with a short manned mission to Salyut 6, with the following Soyuz T 7K-ST 10L or Soyuz T-4 considered as the first operational flight.

Salyut 6 operations

In January 1974, four crews were formed to train for Soyuz 7K-S operations, each crew consisting of a pilot and an engineer: Kizim and Akyonov, Lyakhov and Voronov, Malyshev and Strekalov, and Popov and Burdayev. These crews continued to train until January 1976, when four members of this group were assigned to work on the solo scientific Soyuz 22 mission that had been planned at short notice. The remaining four continued to train as the Soyuz 7K-ST group. Pressure was put on the Air Force, as all four members were Air Force cosmonauts.

The four cosmonauts who were transferred to Soyuz 22 in January 1976 were Akyonov, Malyshev, Strekalov and Popov, leaving Kizim, Lyakhov, Voronov and Burdayev (all Air Force) in the Soyuz T group. In the spring of 1976, Voronov and Burdayev were bumped from the training group after failing their Soyuz T examinations at NPO Energiya. (They afterwards claimed that they had been intentionally failed because they were Air Force cosmonauts, and because NPO Energiya wanted its own engineers on Soyuz T). Also in 1976, Lyakhov was transferred to Salyut 6 resident crew training. In September 1976, three cosmonauts of the Soyuz 22 team (Akyonov, Malyshev and Strekalov) returned to the Soyuz T group, but Popov was transferred to Salyut 6 resident crew training (*not* to Interkosmos programme). He was placed on a support crew for Soyuz 25, together with Andreyev. So, in late 1976 the Soyuz T group consisted of Malyshev, Akyonov, Kizim, Makarov, Strekalov. Lazarev joined in 1977.

In 1978, three two-man crews were identified for training for the Soyuz T manned test programme: Malyshev and Akyonov, Kizim and Makarov, and Lazarev and Strekalov. These crews were selected for a solo manned mission using spacecraft 7L, but in late 1978 this was cancelled in favour of two manned flights to Salyut 6. The Malyshev/Akyonov and Kizim/Makarov crews remained assigned to the first manned flight, while the Lazarev/Strekalov crew (plus Polyakov) was assigned to the second flight.

These assignments reflected the policy (established in 1977, following the Soyuz 25 docking failure) for at least one crewman to have had flight experience. In the same year, two cosmonaut doctors, Polyakov and Potapov (sometimes called the ‘Pol Pot’ team), began training for cosmonaut research seats on the Soyuz T craft. This was the first evidence of the possibilities for using the third seat.

Then, in 1979, the crews for the first two tests mission were announced. For Soyuz T-2, the prime crew was Malyshev and Akyonov, with Kizim and Makarov as back-up. The Soyuz T-3 prime crew was Lazarev, Strekalov and Polyakov, with Isaulov, Rukavishnikov and Potapov as back-up.

Both missions were to be visiting missions to Salyut 6, and the Soyuz T-3 mission (with two doctors on the crew) would include a major medical research programme. However, in May–June 1980 the Soyuz T-3 crew was changed when the flight became a repair mission due to a problem with the thermal control system on Salyut 6. The new prime crew consisted of Kizim, Makarov and Feoktistov, with Lazarev, Savinykh and Polyakov as back-up. A third support crew was also announced: Isaulov, Rukavishnikov and Potapov, who, however, were disbanded in September 1980, two months *before* the mission flew. Feoktistov was brought into the crew to lead the repair part of the mission, but on 9 November 1980 he failed a medical examination and was replaced by Strekalov, who was an original Soyuz T training group member.

After the successful repair mission, it was decided to fly a last mission to Salyut 6. This provided an opportunity for another long-term test of the Soyuz T craft and a further occupation of the space station, and the last two Interkosmos visiting missions using the old Soyuz could also be flown to the old station. The cosmonauts assigned were Zudov and Andreyev as the prime crew, with Kovalyonok and Savinykh as back-up. The third crew was Isaulov and Lebedev, who were already assigned as crew no.1 on the new Salyut station. The prime crew was replaced before the flight due to medical issues, and the flight of Soyuz T-4 marked the beginning of operational use of the Soyuz T.

According to the *Novosti Kosmonavtika* cosmonaut book,[2] the crew was changed because Salyut 6 was nearing the end of its lifetime, and that it was decided to have a crew-member with Salyut 6 experience. In the biography of Kovalyonok it is also claimed that the decision was made 'as a result of the exams', so it appears that the Zudov/Andreyev crew failed them or performed relatively poorly, although there is no mention of medical problems. The crew exchange was decided by the State Commission in February 1981, which was also when Isaulov was dropped from the third crew, to be replaced by Anatoly Berezovoi.

THE SOYUZ T TRAINING GROUP, 1981–86

Salyut 7 operations

In 1981, the Soviets began forming the crews to fly to Salyut 7 in the new Soyuz T craft. It would be reasonable to assume that although the crews trained on Soyuz T, their main focus was the operation of the new space station.

One change in the way that crews were selected and trained was that they were named in sequence, several months apart. Soviet mission planners had discovered that crews peaked too early if the old pattern was used, and so the training group was formed over several months. In September 1981, Isaulov failed a medical and was replaced by Berezovoi, because it was not wished to break up existing crews who were already in advanced training.

According to the biography of Isaulov in the *Novosti Kosmonavtika* cosmonaut book,[2] he trained as a crew- member for the third crew of Soyuz T-4 until February 1981, and was then replaced by Berezovoi because of medical problems. In the biography of Berezovoi it is said that Berezovoi trained 'in a group for Salyut 7' from February to June 1981. In June he was assigned to the first Salyut 7 mission with Lebedev, and they began joint training in September 1981. In the biography of Lebedev, the latter is the earliest date mentioned in connection with Salyut 7. The initial three crews were:

Mission 1 A.N. Berezovoi and V.V. Lebedev
Mission 2 V. Titov and G. Strekalov
Mission 3 V. Dzhanibekov and A. Aleksandrov

As the programme called for very long-duration missions to the new station, visiting crews would be required to bring up the replacement Soyuz craft. Mission 1 would receive two crews. The visiting no.1 prime crew was Malyshev and Ivanchenkov, backed up by Kizim and V. Solovyov; while the visiting no.2 prime crew, Popov and Serebrov, was backed up by Vasyutin and Savinykh. The planners took advantage of the third seat by assigning two French cosmonauts – Chrétien and Baudry – to the first visiting crew (Soyuz T-6). For the next visiting crew (Soyuz T-7), mission planners chose three women – Savitskaya, Pronina and Kuleshova – to train for the mission. Before these missions flew, Malyshev was replaced by Dzhanibekov after failing a medical examination, in December 1981 Kuleshova – the original back-up on Soyuz T-7 – was replaced by Pronina, and in January 1982 Dzhanibekov, on mission 3, was replaced by Lyakhov.

The next expedition crew, Titov and Strekalov – who also had a heavy EVA schedule – were joined by a cosmonaut–researcher when it was also decided to fly a woman on a long-duration flight. Pronina was assigned to the crew, with Serebrov as her back-up, but due to the internal politics of the Soviet programme there was heavy pressure not to fly her, and in March 1983 – only a month before launch – she was replaced with Serebrov, with Savinykh his as back-up.

Several Western accounts suggest that Pronina was supposed to participate in the EVAs – but the EVAs were assigned to Titov and Strekalov (which is why they were recycled for Soyuz T-10 following the Soyuz T-8 docking failure). Strekalov said in an interview that the reason why Pronina was 'bumped' was because 'someone in the Central Committee' was afraid that if something untoward happened during EVA, Pronina would be left alone onboard Salyut 7, and would not know how to return – indicating a limit to the amount of Soyuz spacecraft training which Pronina had received. The *Novosti Kosmonavtika* cosmonaut book states that members of the Military Industrial Commission were opposed to women flying in space. The female cosmonaut programme of the early 1980s was really Glushko's idea, but he was not wholly supported.

In September 1982, Malyshev and Manarov were selected as one of the back-up crews for Soyuz T-8 (together with Titov and Strekalov, and Lyakhov and Aleksandrov, which would eventually have placed them in line for a prime crew assignment to mission 4).

One visiting mission for the third mission crew was also planned, and three crews were named to begin training: Kizim, V. Solovyov and Volk as the prime crew; Vasyutin, Savinykh and Levchenko as the back-up crew; and Viktorenko, Sevastyanov and Stankyavichus as the support crew. The third-seat occupants were all civilian test pilots on the Buran shuttle programme. Their mission was to gain flight experience and to test their reactions to space. Following landing, Volk would fly a Tu-154, equipped with Buran's flight control systems, from Baikonur to Moscow, and then a MiG-25 from Moscow back to Baikonur. However, this mission was cancelled, and the crews were disbanded following flight problems with Soyuz T-8 (see p.300).

After the failure of the second mission crew to occupy the space station, the crews were reorganised. Lyakhov and Aleksandrov crew were assigned to mission 2/1

(Soyuz T-9, EO-2/1), and V. Titov and Strekalov were recycled to fly mission 2/2 (Soyuz T-10 EO-2/2). This was because the original mission objectives for EO-2 (Soyuz T-8) were spread over two flights The other core crew in training was the existing pairing of Malyshev and Manarov, and they were joined by Viktorenko and Sevastyanov, with Kizim and Vasyutin nominated as potential Commanders. No visiting crews were planned to go to the Soyuz T-9 crew, and so the mission 2/2 crew would replace both them and the Soyuz T. This did not, however, proceed according to plan, as the Soyuz launch vehicle exploded on launch, and the crew was blasted away from the burning vehicle by the escape tower.

In 1983, veteran cosmonaut Feoktistov was asked to train for a long-duration mission, due to fly in 1984. The purpose was for him to participate in an experiment to determine how an elderly person would react to a long term in space. He would have been joined on the flight by a doctor, and for this purpose, Atkov and Polyakov began training in June 1983. However, Feoktistov developed a 'chronic illness, which became acute' and was stood down from training. A 'propaganda aspect' of Feoktistov's mission would have been to break the record for the oldest man to fly in space (which at that time was held by Deke Slayton). Glushko had produced the 'old-man-in-space' idea in 1979.

In September 1983, the crews for the third mission were named: Kizim, V. Solovyov and Atkov as the prime crew; Vasyutin, Savinykh and Polyakov as back-up; and Viktorenko and Sevastyanov as support crew. This mission was due to fly in 1984, and would be visited by two crews. The first of these was named at the same time as those for mission 3, and was the crew for the International Soviet–Indian mission. The prime crew was Malyshev, Rukavishnikov and Sharma, with Berezovoi, Grechko and Malhotra as back-up. Malyshev was replaced on the second visiting crew by Dzhanibekov, and on the Indian mission Rukavishnikov was replaced by Strekalov, for medical reasons. Grechko remained in the back-up position.

In December, the prime crew for the second visit was named as Dzhanibekov, Savitskaya and Volk. The mission would include the first space walk by a woman and the long-delayed flight by a Buran pilot. In February 1984 they were joined by their back-up crew – Vasyutin, Savinykh and Ivanova. Manarov, having lost Dzhanibekov as his Commander, was moved on to the Mir programme, and officially, Savitskaya's back-up (as Flight Engineer) was Savinykh, with Ivanova as Volk's back-up (as cosmonaut–researcher). Ivanova was not trained for EVA, and according to the *Novosti Kosmonavtika* cosmonaut book, Savitskaya was chosen for the EVA not only because of her spaceflight experience, but also because of her physical strength. Savinykh has also confirmed, in his space diary (privately published in Moscow in 1999), that Glushko consciously ordered the female EVA in order to upstage Kathy Sullivan's EVA on STS-41G, to be flown in October 1984.

The crews for mission 4 were named in September 1984: the prime crew of Vasyutin, Savinykh and A. Volkov was backed up by Viktorenko, Aleksandrov and Salei, with A. Solovyov, Serebrov and Moskalenko as the support crew. The third seat was occupied by a military pilot, because a military experiment was due to be launched to the station in a heavy module, and a number of military engineers had also been considered as potential crew-members. Only one visiting mission was planned – the all-female crew of Savitskaya, Ivanova and Dobrokvashina (who was a doctor) – and they began work in December 1984.

However, in February 1985 a problem developed onboard Salyut 7, and ground control lost contact with the station. This was a very serious problem, and so a group of four pilots – Dzhanibekov, Berezovoi, Popov and Lyakhov – were selected to train for a repair mission. The crew for the mission was named in March, with a crew of two – Dzhanibekov and Savinykh – flying the mission as Soyuz T-13. Popov and Aleksandrov were their back-up, and the mission was designated Ex 4/1.

In anticipation of the success of the first mission, the crews for mission 4/2 were also named: Vasyutin, Grechko and A. Volkov formed the prime crew; Viktorenko, Strekalov and Salei were the back-up; and A. Solovyov, Serebrov and Moskalenko were the support crew. After the swift return of mission 4/2 there was discussion about a further mission using the Soyuz T-14 back-up crews, but it was finally decided to fly a two-man mission, with the third seat being used to transfer cargo from the old station to the new Mir. This crew – which flew on the last Soyuz T (no.15) – consisted of Kizim and V. Solovyov, with Viktorenko and Aleksandrov as back-up. Finally, before Vasyutin's illness, the plan was for the all-female crew to visit the Vasyutin crew in March 1986 on board Soyuz T-15, with no more visits to Salyut 7 planned – because there were no more Soyuz T vehicles.

SOYUZ T FLIGHT OPERATIONS, 1980–81

Between June 1980 and July 1986, fourteen manned Soyuz T spacecraft reached Earth orbit. The first two (one unmanned; the other manned) met the qualification programme. The third opened the operational programme, and all were flown to the Salyut 6 station, with the remainder to Salyut 7, and with the last mission also completing the first manned visit to Mir. A further Soyuz T launch was aborted seconds before lift-off, in the first known launch-pad abort and activation of the launch escape system.

Soyuz T-2: the first manned mission

On 5 June 1980, a Soyuz T left the launch pad at Baikonur with cosmonauts on board for the first time. Commander Yuri Malyshev and Flight Engineer Vladimir Akyonov were heading for the Salyut 6 space station already occupied by the fourth resident crew (Leonid Popov and Valeri Ryumin). (When the first drawings of Soyuz T were released they only revealed two crew positions, and there was no mention, at that time, of a third seat.) The resident crew had been on board since April (and had already hosted a visit by the Hungarian Interkosmos crew), and repositioned Soyuz 36 from the aft to the forward docking port, to clear the way for the Soyuz T-2 to link up at the aft port.

Once Soyuz T-2 had separated from the carrier rocket and been placed in orbit, the solar arrays were deployed, and there then began the sequence of events to reach Salyut. After the integrity of the DM and OM had been confirmed, the cosmonauts released themselves from their seats, removed their suits, and opened the inner transfer hatch for more room inside the Soyuz. The docking probe was extended from its stowed launch position ready for capture, while the Argon computer accessed the amount of thrust required to place the spacecraft in the correct orbit. This information was displayed to the cosmonauts, who then hit the proceed switch. The computer then commanded the main ODU engine to fire to correct the orbital parameters in two stages. The aim of the flight was stated as continuing testing of the new spacecraft and its systems in a tentative short flight. One of these new systems allowed the crew to maintain the internal temperature of their Soyuz at a comfortable 20° C, and, for the first time on a Soviet craft, they could also monitor the internal pressure (787 mm) of the spacecraft.

Docking with the Salyut was also part of the flight plan, and unlike the previous unmanned Soyuz T flight, this time a standard 24-hour approach was followed. As the spacecraft approached, the crew tested the new systems of the Soyuz T, including the performance of the solar arrays. After receiving permission to proceed, Malyshev programmed the Argon computer to select the best approach from a selection of options based on real-time data, and to control the manoeuvres required to achieve that flight path. However, only 180 metres from the Salyut, first-time space-flyer Malyshev was unsure of the chosen approach (his heart rate at this point was stated as being 130 beats per minute, while that of veteran Aksyonov peaked at 97 beats per minute) – an approach for which neither he nor the ground controllers had trained. As this was the first docking of the craft with a crew on board, the Commander decided to override the automated approach and instead perform a manual docking at the aft port of Salyut, revealing the greater flexibility offered by the Soyuz T design. (Later analysis indicated that the computer programme had selected a flight path that would have achieved a successful docking, and that Malyshev was perhaps a little eager or over-cautious.)

A few hours later, the two cosmonauts joined their colleagues onboard the Salyut for three days of joint operations, including the unloading of supplies from the Soyuz T and the performance of biomedical tests and other experiments. These supplies were probably located in the third crew position, which may have been why only a two-person crew was flown, although it also offered the chance to evaluate the probable mass of a third crew-member.

On 9 June, Soyuz T-2 undocked from the Salyut, and a photographic survey of the station was conducted using the unified propulsion system fed to the RCS system. During previous operations the station had to be rotated by the onboard crew. After separating from the station, the Argon computer was again programmed to select the best of several options for descent. This was also the first time that two Soyuz crews had returned home from a Salyut in the same landing window – the Hungarian crew having landed in Soyuz 35 only six days earlier. The crew jettisoned the OM prior to de-orbit burn, and the descent trajectory selected by the Argon computer would depend upon real-time flight conditions, with Malyshev standing by to assume control if necessary. This time all went well, and about three hours after leaving the vicinity of the Salyut, Soyuz T-2 was on the way to a successful landing. After emerging from black-out, the crew also jettisoned the outer panes of the DM windows to obtain a clearer view of the landing area. Soyuz T-2 was judged a successful inaugural mission, with a flight of 3 days 22 hrs.

Soyuz T-3: the three-man maintenance mission

During the latter stages of the fourth residency onboard Salyut 6, the cosmonauts had to evaluate the state of the station and its capacity to support a further long visit. Several systems were identified as requiring attention, repair or replacement, and the seriousness of the problems with the thermal control system resulted in a mission change. As a result, the Soyuz T-3 crew was assigned a short 13-day maintenance and refurbishment mission, which was also an opportunity to test the new spacecraft systems over a longer period.

When Soyuz T-3 was launched on 27 November 1980, the announcement was not unexpected in the West; but what *was* a surprise was that for the first time since Soyuz 11 in 1971, a Soyuz carried three cosmonauts to orbit. It demonstrated the greater flexibly in the design of the new Soyuz T, and it offered the chance to crew Salyut with three cosmonauts, for greater productivity. The Soviets also revealed that crew composition on future flights would vary, and would not always include a third person (Cosmonaut-Researcher) who had no piloting functions. Sometimes, their place would be taken by extra supplies.

On Soyuz T-3, the Argon computer performed flawlessly (mission control referred to the system as the 'fourth crew-member'), without any of the problems experienced on Soyuz T-2, and from a distance of 5 km it achieved a fully automated docking. The three cosmonauts completed their assigned mission in preparing Salyut for one last extended-duration residency. The Soyuz undocked from Salyut on 10 December, and a few hours after the T-3 crew had departed to head for the landing, the OM was ejected, to re-enter the atmosphere to destruction on 26 January 1980. (Western reports of the OM being left behind, docked to Salyut, are unconfirmed by Russian sources.)

Soyuz T-4: the long-duration test

With Salyut 6 cleared for a further residency visit, the opportunity arose to test Soyuz T on a relatively long mission supporting a crew. In addition, due to delays in the Interkosmos programme as a result of the failed Soyuz 33 docking attempt, two further missions remained to be flown with the older Soyuz ferry craft. These were also assigned to visit Salyut 6 in 1981.

Soyuz T-4 launched on 12 March 1981, and the following day Commander Kovalyonok sighted Salyut, with his periscope, at 5 km distance. With the Argon computer guiding the spacecraft, the approach was halted 2,300 metres from Salyut, as the two craft passed out of communications range of mission control. The docking took place during the dark-side pass of the orbit, with the docking area illuminated by spotlights on the Salyut. During their stay, the two cosmonauts hosted visits by the final Soyuz ferry craft, Soyuz 39 and Soyuz 40. No vehicle exchange was possible, as neither of the two visiting crews had trained on Soyuz T, and they were not qualified to return one to Earth. Onboard the Salyut, a routine daily pattern was followed for the programme of experiments and activities.

One report stated that on 12 May (the sixty-first day in flight), the crew had dismantled the active docking unit on Soyuz T-4 and replaced it with a passive docking unit, allowing other craft to dock with it. In Western accounts this was originally interpreted as an experiment for future application in rescuing cosmonauts in orbit, possibly promoted by the near loss of the Soyuz 33 mission in 1979, although to date no such mission profile has been duplicated on a Soyuz craft. It appears that the operation on 12 May was connected with the later docking of Cosmos 1267. According to Asif Siddiqi, in the second part of his history of Almaz: '[The crew] suited up in the airlock, and opened the forward hatch to attach a device into the forward docking port to allow [Cosmos 1267] to dock with the DOS-type vehicle. The Salyut 6 station, when it had been originally built in the mid-1970s, had not been designed to receive heavy vehicles such as the TKS'.[3] The source for this is an article in *Zemlya I Vselennaya*, although it is not certain whether this operation involved a depressurisation of the airlock. Neither is it included in the IVA/EVA lists in the NK book. Moreover, it appears that Salyut 6's forward port was compatible with Cosmos 1267, and this had nothing to do with determining whether Soyuz could receive other spacecraft.

Soyuz T-4 was vehicle 10L, while Soyuz T-3 had been vehicle 8L. Vehicle 9L had been on stand-by at Baikonur to fly to Salyut 6 should problems develop with 8L in orbit. (The practice of having a back-up vehicle on stand-by for such emergencies was introduced with Salyut 6, because it was the first station with two docking ports and was hence capable of receiving a 'rescue Soyuz'). Because 9L was in the three-man configuration, it could not be used for T-4, and in June 1982 it was used for the next three-man mission, Soyuz T-6.

When the last Interkosmos crew departed on 22 May, having completed their programme of scientific experiments and observations, the final Salyut 6 resident crew left the station for the return to Earth on 26 May, after a 74-day mission. Again according to some Western accounts, the OM was left attached to the Salyut as the two cosmonauts undocked, and was then jettisoned on 31 May, but Russian sources do not confirm this. Soviet news reports also stated that the older Soyuz ferry had been retired following the Soyuz 40 flight, and that from the next mission Soyuz T would become the operational spacecraft. It was also confirmed that no further crews would be sent to Salyut 6.

SOYUZ T FLIGHT OPERATIONS, 1982

Salyut 7 (DOS 5-2) was launched on 19 April 1982, and though expected to include more than two docking ports, it was in fact the back-up station for Salyut 6. It was a similar design, with ports fore and aft, but it had improved onboard systems, including improved rendezvous and navigation equipment for smoother docking with Soyuz T craft. The Delta automated navigation system allowed the Salyut to manoeuvre without any input from the resident crew, while the Kaskad attitude control system had the capacity to orientate the station to an accuracy of less than 1°. To assist Soyuz T in finding Salyut, a long-range transponder (Mera) would assist in the final rendezvous phases, providing data to the spacecraft from the closest point in the transfer orbit to the point where the Igla system could lock on to Salyut from just a few kilometres away.

Western analysts monitoring Salyut 7 operations were soon able to determine whether a two- or a three-person crew would be launched to the station, because the station's orbit was lowered by some 35–45 km for a three-person crew. Fuel reserves for a lower orbit were determined as being 110 kg (three persons), while for a higher orbit (two people) it was 175 kg – the difference being the approximate mass of the third crew-member. To compensate for the third crew-member, the propellant load was reduced, thus restricting the capability to reach the higher orbit. Prior to launch, therefore, Salyut's orbital parameters would be reduced.

Soyuz T-5: the first operational mission

The first spacecraft to evaluate this improved system was Soyuz T-5, launched on 13 May 1982. Once in orbit, the crew took off their Sokol suits and 'hung them out to dry', because of the perspiration produced during the several hours that they were wearing them. Commander Berezovoi found it cold sleeping in the small spacecraft, and he therefore put his suit back on and dropped off to sleep. Flight Engineer Lebedev tried sleeping in the DM, but had difficulty in maintaining a comfortable position, first above the couch and then in it; and he even positioned himself spread-eagled in mid-air. The suits filled up the already-cramped OM. Sleep was essential in Soyuz T (even with the improvements) because of the pressure on the crew to achieve the docking and board the station.

Data from the computer recorded a separation distance of 457 km, but Mera did not activate until the final transfer orbit at 250 km. The system failed to lock on first time, and so the cosmonauts tested it. Mera did not 'find' Salyut until the distance between the craft was only 30 km, closing at 45 metres per sec. But once achieved, Igla soon locked on to the Salyut, and the crew saw their future home on the display screen. This allowed the Argon computer to control the final approach and docking. At 200 metres the crew was given permission to dock, which they did without incident – much to the relief of the controllers on the ground, who reportedly had 'lived through the docking with the cosmonauts.' However, the new Salyut was not yet operational. When the time came to enter the Salyut, the inner transfer hatch refused to open. Lebedev therefore resorted to placing his feet 'upside down' on either side of the hatch rim, for traction and stability, so that he could prise it open. He consequently entered the station at an unusual angle, and did not immediately recognise the 'floor' – which was actually one of Salyut's 'walls'.

Soyuz T-6: a manual override

By 24 June 1982, the Salyut's new crew-members were awaiting their first visitors – including the first French cosmonaut, Jean-Loup Chrétein – onboard Soyuz T-6. Under the command of veteran cosmonaut Vladimir Dzhanibekov, assisted by Flight Engineer Alexandr Ivanchenkov, they had launched the day before, and were in the final stages of docking to Salyut. As programmed, at 900 metres, and outside the zone of normal ground tracking coverage, the Soyuz T-6 onboard computer

The Soviet/French Soyuz T-6 back-up crew during training in the Soyuz T simulator at Star City: V. Solovyov, Leonid Kizim and Patrick Baudry.

turned the spacecraft around and initiated a brief braking burn of the main engines against the direction of flight, in order to slow the approach rate. The computer then initiated the rotation back to allow the docking unit to face forward. As it did so, it sensed the gyros approaching gimbal lock and halted the manoeuvre, sending the Soyuz into an end-over-end spin as the approach was aborted.

Onboard systems, monitoring the spacecraft functions and data from onboard gyros, reported the error to Dzhanibekov, who immediately regained manual control and proceeded to dampen the sudden rotation. Unable to locate Salyut visually, he manoeuvred the Soyuz in all three axes in order to determine his position in relation to Salyut, and the orientation of his spacecraft in relation to the station. On Salyut, the cosmonauts were monitoring the approach, and looking out of the portholes to ensure that the approach would take the Soyuz past the Salyut and not on a collision course with it. As Soyuz closed in under the control of Dzhanibekov, the Salyut cosmonauts could see Soyuz T-6 apparently 'just hanging there' 200 metres away, and occasionally saw the wedge-shaped flashes of gas-jets from the thrusters as the Soyuz manoeuvred towards them.

By then the spacecraft had appeared within range of the surface tracking vessel located in the Straits of Gibraltar, and, after reporting on the situation, Dzhanibekov was given permission to dock manually with the aft port of Salyut, some 14 minutes earlier than planned. The possibility of engine problems preventing the docking (as with Soyuz 33 three years earlier) were soon forgotten as Dzhanibekov guided Soyuz T-6 into the mooring port and secured a hard docking. This completion of a fully manual docking in a contingency situation was a significant achievement for a Soyuz Commander, and the praise by his peers may have contributed to his nomination for the 1985 Salyut 7 rescue mission (see p.309).

After such an intense docking, the return to Earth on 2 July, after a week onboard the station, was relatively uneventful; although Chrétein (who was on his first return from space) later stated that the landing seemed more dramatic than the launch. No exchange of Soyuz was conducted on this flight, as Soyuz T-5 had been in space for only six weeks.

Soyuz T-7: delivering a fresh spacecraft

On 20 August, Soyuz T-7 docked to Salyut 7 without incident. The Salyut crew noted that the docking was light but slightly off centre, and created a yaw rotation of 0°.3 per sec. With the docking confirmed, the crews were eager to open the hatches, and tapped on them and shouted to each other. They also decided to equalise the pressure between the two craft more quickly in order to open the hatches as soon as possible. When the hatched were open, Soyuz T-7 Commander Leonid Popov floated through first, followed by Flight Engineer Alexandr Serebrov. Then, aware that events were being broadcast on national TV, the third crew-member, Svetlana Savitskaya, brushed her before floating into Salyut. She had become only the second woman to fly in space, nineteen years after Valentina Tereshkova. Now, on her first spaceflight, she became the first female crew-member of a space station. Offered the relative privacy of the Soyuz OM, she instead decided to sleep in the main compartment of Salyut, with the rest of the crew.

On 27 August, the Soyuz T-7 crew returned to Earth after a week onboard Salyut – but not in the vehicle in which they were launched. For the first time, a Soyuz T was returned by a different crew, as the T-7 crew had already switched their contoured seat liners to Soyuz T-5 early in their stay onboard the Salyut, and came home in that spacecraft 106 days after it had been launched. Two days later, on 29 August, the Soyuz T-5 resident crew undocked the Soyuz T-7 spacecraft from the aft port, and backed away as the unmanned Salyut was commanded to turn 180° to allow its forward port to face the awaiting Soyuz T-7 craft. When completed, the crew re-docked with Salyut and re-entered the station to complete their mission, as well as freeing the aft port for further Progress vehicles.

Landing in a snowstorm

On 14 September the mission of the first Salyut 7 residency crew was extended by two months, with recovery planned for the week of Christmas, resulting in a flight duration of about 225 days. When Berezovoi became ill in early November there was talk of an early return, but his condition improved and the flight programme continued. However, they had almost depleted the notepaper for recording observations, and as continuing problems with the Delta automatic navigation system finally resulted in a complete failure early in December, it was decided to bring the cosmonauts home onboard T-7 on 10 December, which would set a record of 211 days.

As the landing took place two weeks prior to the normal landing window, the crew would have to land in darkness and without the normal programme of

The Soyuz T-7 crew: Savitskaya, Popov and Serebrov.

medication, and without the increased exercise which returning crews undertake as part of their pre-adaptation to Earth's gravity. The weather forecast for the landing area included winds of 21 kmh, a temperature of ˜15° C, and 10-km visibility; but in reality the conditions were much worse, as the capsule descended into low cloud and fog during a snowstorm, and a temperature of –9° C. The re-entry sequence followed a nominal trajectory, but the worsening weather caused problems. High winds dragged the DM over the side of a small incline, after which it rolled down the hill, to the discomfort of the two cosmonauts. Despite being strapped into their seats, Lebedev found himself on top of Berezovoi when the capsule finally stopped.

Due to the bad weather, the recovery helicopters could not find the capsule, and were flying almost blind. The leading pilot was told to use his own judgment in attempting a landing as he reported sighting the recovery beacon on the Soyuz. However, he landed in a dry riverbed, and upon touch-down the left wheel and support strut snapped off. He therefore informed the rest of the recovery forces that it was too dangerous to land, and so they headed for other safer landing area as distant as 150 km.

Inside the Soyuz, Lebedev and Berezovoi were still in radio contact with mission control, and from there, Vladimir Shatalov told them to remain in their suits and to not open the hatch until the recovery crews arrived. Despite improvements in the Soyuz T it was still not a comfortable place to await rescue, although with the external temperature at –15° C, it was comparatively warm.

After forty minutes, the recovery and medical teams arrived in wheeled transports, and the cosmonauts' first night back on Earth was spent in the back of one of these special vehicles. After resting, the following day they were transferred by helicopter to Dzhezkazgan and then to the cosmodrome.

SOYUZ T FLIGHT OPERATIONS, 1983

The three manned flights of Soyuz T during 1982 were highly successful, despite some in-flight difficulties. The new design was fully qualified for operational use on long-duration and visiting missions, and had overcome docking and landing incidents. From 1983 it was time to expand Salyut 7 operations, to further extend the duration record, and to launch another of the Cosmos modules to the station. For Soyuz T, new hurdles had to be overcome – some of them predictable, but others unforeseen. One of these occurred on the first mission of the year.

Soyuz T-8: a cancelled docking

The Soyuz T-8 crew was to stay in orbit for three months (April–July 1983), and were to have unloaded Cosmos 1443 and performed two EVAs to install additional solar panels delivered by it. Irena Pronina (if she had flown) would have become the third woman in space, and would have set a new flight endurance record for women. Then, after a brief unmanned period, Soyuz T-9 would have flown in August–September 1983, with Lyakhov, Aleksandrov and a third crew-member. Before Pronina's replacement this was Serebrov, and after Pronina's replacement probably Savinykh (who replaced Serebrov on the Soyuz T-8 back-up crew). The Lyakhov crew would have stayed on board until some time in 1984, and would have received Kizim, Solovyov and Volk as visiting crew in late 1983.

Vladimir Titov learned from his fellow cosmonauts that the beginning of a cosmonaut's first flight into space is perhaps the most fascinating and emotional experience imaginable, and that the success of attaining orbit results in joy and celebration. He could not wait to experience it. Selected to the cosmonaut detachment in August 1976, Titov had trained for more than six years to savour the 530-second ride to orbit and that first sensation of weightlessness. His moment finally arrived on 20 April 1983, as he flew as Commander of Soyuz T-8, together with Gennedi Strekalov and Alexandr Serebrov, who had both flown into space before.

At first, the flight plan proceeded as scheduled, with a pressure check on the first orbit, the unfurling of the solar arrays, and checking of the flight displays. However, problems arose during the second orbit. During the check-out of the Igla automated rendezvous system, its parabolic antenna – which normally folded under the launch shroud – appeared not to be providing any data on the crew display panel. At first the three cosmonauts thought that it had simply not unfolded, and recycled the switches to try again. Telemetry confirmed that the boom of the antenna had deployed, but not to the full extent. Strekalov suggested that the boom had caught up in something, and proposed using the RCS thrusters to try to shake the antenna loose. Mission control permitted this manoeuvre, but (not wishing to reveal the true nature of the problem) indicated that it was a test of the attitude control motors.

The result was disappointing, with no new data recorded, and so mission control told the crew to proceed with their flight programme while the situation was evaluated on the ground. After the expectation of flight, the thrill of his first rocket ride, and the onset of weightlessness, Titov's feelings onboard Soyuz T-8 over the next few hours were in sharp contrast, and he later wrote: 'Our mood was different. To tell you honestly, we did not feel like rejoicing. So we just got down to work, calmly and busily, as we had trained to do'.[4]

As the crew worked, time seemed to slip by. Suddenly, it seemed, the sixth orbit was completed and it was time for them to rest. While Serebrov stayed in the DM for the first shift, to monitor the onboard systems, Titov and Strekalov moved into the OM. They obtained very little rest, however, as the problem played on their minds. The crew (as all crews) had trained for both normal and contingency situations, but the current problem was not expected. It was also clear that without the Igla data they lacked the distance and closing speed measurements to ensure a safe and controlled docking approach in either automatic or manual mode. However, after discussions with the ground the next day, it was agreed that a fully manual approach would be attempted using visual clues to the distance and speed of approach to the Salyut. Ground simulations (carried out while the crew slept) indicated that such a docking was possible under certain conditions, but that even then the chances of success were slim.

As the spacecraft emerged from the darkness on the nineteenth orbit, the approach would begin after Soyuz T-8 had been positioned 1,000–1,500 metres from the rear port of Salyut (the front port was occupied by Cosmos 1443). The principle problem was the doubtful reliability of the spacecraft's drift indicator in providing an assessment of Salyut's distance and the speed of Soyuz. Titov later recalled that as they approached it was 'difficult to assess the speed. The size of the station on the screen does not change; it is visible as a dot.' As he attempted to control the distance and speed, mission control requested a progress report and offered ground-based distance computations. Titov realised that the ground had a clearer picture of the situation than they had in orbit. When he stated the Salyut was occupying just half a grid square on the drift indicator screen and did not appear to be moving, he was told that they were still too far out and should fire their engines for 50 seconds. As he looked, Salyut now appeared to begin filling one of the squares on his screen. Titov was reminded by mission control to 'watch the station and switch on the searchlight', and noted that the craft would soon again enter darkness. With mission control cautioning about the safety of flying too close too quickly, they passed out of radio range; and having lost closing-distance information, the crew were now on their own.

Titov later recalled that a fully manual docking was not part of his training programme for this mission, and that he was not sure of his depth perception while judging his closing velocity. With Salyut's acquisition lights on, and the Soyuz searchlight illuminating the station, he was fearful of collision: 'We hurtled on, and the distance to the station was 280 metres. We [maintained] the station's position on the screen, but we could feel that the rendezvous speed was high, so I fired the de-boosting engine. The distance was 160 metres.' From here he was trained to perform a manual docking, but he was not comfortable with the situation. 'Still the speed remained quite high. At night, when the distance and speed of rendezvous are difficult to assess visually, the danger of a collision is quite real. I fired the engine to make the spacecraft descend a little,' thus aborting the approach and flying past the station. 'After exiting from the darkness in which we had stayed for 35 minutes, the distance between Soyuz T-8 and Salyut 7 had increased up to 3–4 kilometres. So the docking operation had failed.'

Errors inherent in the solely visual determination of the distance of the station and the speed of approach were obviously too great, and eliminated any chance of another attempt before the spacecraft re-entered the darkness. After a further wait of 35 minutes, the inevitable news came, during the next communication session from mission control, that the crew were to return to Earth. Mission control had fully evaluated the situation while the spacecraft was out of communication range. With the absence of any reliable measurements of speed and distance between the two spacecraft, and the low levels of fuel on board (possibly due to excess use during the initial attempt), there was no opportunity to try again. Despite losing the long mission, events had provided everyone with valuable experience during an arduous 24 hours – although this was no comfort for the crew.

The official report from TASS stated: 'Because of deviations from the planned approach régime, the docking of the Soyuz T-8 craft with the Salyut 7 orbital station was cancelled. Cosmonauts Titov, Strekalov and Serebrov have begun preparations for their return to Earth.' However, there was no mention of the 15-hour period of 'extremely tense' work by FCC teams and the crew to stabilise the Soyuz and return the crew to Earth.

In order to save what little fuel was left for the de-orbit burn, the Soyuz was placed in a spin-stabilised mode, with the attitude control thrusters turned off to conserve fuel until the next landing attempt (a contingency not used since Soyuz 9 in 1970). On 22 April, all three cosmonauts were safely back on Earth following a normal Soyuz T re-entry profile. None of the Soviet news reports mentioned the docking attempt, and relegated news of the flight to secondary bulletins. However, the tension of this flight was revealed when a note was passed to Dmitri Ustinov, the Soviet Defence Minister, during a televised reception. When informed of the safe recovery of the cosmonauts, he and several other Politburo members showed obvious signs of relief that the mission had ended successfully.

The problem had obviously to be resolved prior to the next launch, which would not take place before mid-June. Meanwhile, the Salyut 7/Comos 1443 complex continued, unmanned, in orbit. Reports later suggested that early in the mission, the antenna had been torn off by the separation of the Soyuz launch shroud, unbeknown to the crew until they tried to deploy it.

After the Soyuz T-8 docking failure, Lyakhov and Aleksandrov were reassigned to unload Cosmos 1443 on the Soyuz T-9 mission. Titov and Strekalov were recycled to Soyuz T-10 to carry out the EVAs, and would take over from Lyakhov and Aleksandrov, to perform the first ever in-orbit crew exchange (which had not been planned prior to the Soyuz T-8 docking failure). Their mission was to last about three months. They were supposed to have launched in August, shortly after the return of Cosmos 1443, but the launch was delayed for about a month. This may have been because the vehicle to be used for Soyuz T-10 was not 15L (as originally planned), but 16L. The 15L – the stand-by rescue vehicle for the Lyakhov/Aleksandrov mission – had to be returned from Baikonur to NPO Energiya after the problem with the 14L/Soyuz T-9 solar panel, which failed to deploy in orbit. The 15L required additional checks and would not fly until February 1984, while 16L was prepared for the Titov/Strekalov mission. Titov and Strekalov were to have performed their EVAs during the crew change-out period, while Lyakhov and Aleksandrov (who would have provided support) were still on board; but these plans had to be changed again when Soyuz T-10 failed to reach orbit.

Soyuz T-9: the failure of the solar array

On 28 June 1983, Soyuz T-9 achieved what T-8 had failed to do, by docking with the Salyut/Cosmos combination. The mission was originally intended to carry three crew-members, but because of the additional propellant loaded into Soyuz T-9, only two cosmonauts could fly the new mission – although they still expected to complete the programme that the three-person Soyuz T-8 was assigned to achieve. For almost twenty years the 'official accounts' of the Soyuz T-9 launch and docking to Salyut had indicated no serious problems. However, an article published in *Novosti Kosmonavtiki* in May 2002 revealed the little-known fact that Soyuz T-9 had failed to deploy one of its two solar arrays shortly after entering orbit, which had not occurred since the ill-fated Soyuz 1 mission in April 1967. Despite this failure, it apparently did not prevent Lyakhov from docking his spacecraft with Salyut to complete a successful mission.

It would be a long and busy mission for Lyakhov and Alexandrov, in more ways than they could have imagined. They would work on the station through to September, when Titov and Strekalov would be launched to Salyut to perform the EVAs for which they had trained on Soyuz T-8, and had taken over from the Soyuz T-9 crew. But it was not to be. On 9 September, during refuelling operations from Progress 17, the Salyut suffered a ruptured line used to fed oxidiser from the Progress to the Salyut ODU tanks. With only half of its thirty-two thrusters available for use, it seemed probable that Salyut would have to be abandoned; but MCC decided to work around the problem and to let the mission continue.

Soyuz T-10-1: a very short long-duration flight

All Soyuz manned spacecraft utilise a launch escape tower for emergency separation of the vehicle from the booster, either on the pad or in the first 46 km of the ascent past max Q. In training, all crews simulate launch abort profiles from the pad or in flight, using the escape system, but none of them ever expect to use it in a real situation. Since 1961, emergency escape from spacecraft of the ballistic type had been available for all crews (except the two Voskhod crews) by either ejection seats (Vostok and Gemini) or escape towers (Mercury, Apollo and Soyuz). In eighty-three manned launches of ballistic spacecraft between April 1961 and June 1983 (USSR, fifty-two; USA, thirty-one), no in-flight use of the launch escape systems was necessary, although during the 5 April 1975 launch of what was intended to be Soyuz 18, the crew had to perform an emergency ballistic re-entry due to a failure on the launch vehicle *after* normal separation of the launch tower. In addition, despite the pad aborts of Soyuz 4 and Soyuz 10, no crew had resorted to using the escape system to escape from their vehicle.

On 26 September 1983, Titov and Strekalov were once more in the final stages of the count-down for launch onboard a new Soyuz T, which was expected to become Soyuz T-10 once it reached orbit. As the prospective ninety-fourth (?) manned space launch to orbit, the ascent was not expected be an outstanding event in space history. However, what transpired was a 'first' which both cosmonauts would no doubt have gladly foregone.

At about 9.00 pm that evening, the two cosmonauts had climbed on board their Soyuz T, with approximately 3.5 hours before the planned launch at around 00.38 local (Baikonur) time. The plan was for Titov and Strekalov to spend two weeks with Lyakhov and Alexandrov (call-sign, Proton), after which they would take over occupancy of the Salyut.

All pre-launch activities appear to have progressed according to plan until the final two minutes before launch. The temperature at the pad area was recorded at 27° C during the day, but had fallen to 10° C during the night, with winds gusting to 40–43 kmh. Inside the capsule, the cosmonauts listened to music piped in over the radio, as did many crews before them. In the blockhouse, monitoring the launch and acting as CapComs, were the back-up crew of Kizim and Vladimir Solovyov.

At 90 seconds before launch, a fuel value, supplying fuel in one of the first-stage strap-ons, failed to close, and leaked propellant around the base of the booster as it was about to ignite. Within a minute a fire had started, and flames began licking up the side of the R-7 with its 270 tonnes of highly explosive propellant. Watching the pad through a periscope from the safety of a nearby bunker was the launch director, Aleksei Shumilin, who noted a fire at the base of the rocket. Inside Soyuz, Titov and Strekalov had no windows in the shroud and could not see the fire, although they could certainly hear the tense radio commentary from the blockhouse.

The LES is designed to instantly separate the launch tower and shroud and to take the OM and DM with it, severing it from the EP (which is left on the top of the launch vehicle) and the rest of the launch vehicle by explosive charges. But in this case the fire had spread so rapidly that it had burned the wiring that would have initiated the escape system. It became apparent to the launch controllers that they had seconds to initiate the launch escape system and save the cosmonauts. They had a back-up radio system, but the command would have to be carried out manually by two separate controllers (A. Mochalov and M. Shevchenko), in different rooms but in the same building (the 'Saturn' station in Area 23 of the cosmodrome, some 30 km from the launch pad), initiating the abort command simultaneously – a safety precaution to prevent accidental initiation under normal conditions. Together in the bunker were Shumilin and Soldatenkov (the launch vehicle technical leader), and each of them individually had to send the code word (in this case 'Dnestr') to the Saturn station. There, one of the controllers was in touch with Shumilin, and the other with Soldatenkov. For the command to work, the two controllers had to press their buttons within five seconds of each another; but all this took ten seconds, by which time the entire rocket was engulfed in flame and was about to explode. Once the command was given, the duty personnel in the Measuring Complex for the Emergency Rescue System initiated the sequence to evacuate the cosmonauts from the pad by the LES.

Seconds had elapsed since the fire was first detected, but within a second of the command, pyrotechnic separation of Soyuz T was followed by the ignition of the 80,000-kgf-thrust solid-rockets clustered at the top of the launch tower, which tugged the Soyuz away from the rest of the stack at high velocity, and at an angle that imparted a high g-force acceleration on the vehicle and the crew inside. Inside the crew compartment, Titov felt the booster swaying in the wind and two waves of vibration as the explosive charges separated the spacecraft and the upper and lower parts of the launch shroud. The sharp jerk which pinned them in their seats (there was no engine noise) indicated to Titov that the escape rocket had fired, and that after the disappointment of Soyuz T-8 five months earlier, he would again not be visiting Salyut 7. Observers noticed a cloud of yellow, red, black and orange smoke surrounding the top of the vehicle, and an object that suddenly shot upwards with sparks shooting from it, indicating that the LES was working. After three seconds the cosmonauts reached Mach 1 – vertically. The acceleration lasted for only five seconds, but took the Soyuz to an altitude of 950 metres, with the crew having to endure a couple of seconds at 14–17 g. A second later, and six seconds after the capsule was pulled from the booster, the rocket exploded in a ball of flame on the pad.

According to the official Energiya history (in which the account is probably based on documentation, and not on individual memories) 11.2 seconds elapsed between the moment that the fire was first noticed and the activation of the emergency escape system (six seconds for Shumilov and Soldatenkov to realise what was happening and to send the code word, four seconds for the two controllers in the Saturn station to send the command, and 1.2 seconds for the command to be executed by the spacecraft's onboard automatic systems). The same source states that an initial explosion took place one second *before* the activation of the LES, and that at that point the rocket had already begun to lean over. The rocket collapsed some 3–4 seconds after the activation of the LES. In an interview later, Strekalov said that they pulled a maximum of 10 g.

After burn-out of the solid booster, the four aerodynamic panels at the base of the shroud opened like petals and slowed the ascent. Then, a second set of pyrotechnics fired (and were heard by the crew) to sever the connections between the DM and the OM inside the launch shroud. Now free, the DM dropped out of the base of the still-ascending shroud at 1 km. The reserve parachute then deployed almost immediately, and the heat shield was detached to expose the soft-landing engines at the base of the

DM, which fired at a height of 1.5 metres above the ground to soften the landing, which was considerably harder than a normal landing from orbit. Titov later recalled that he attempted to record his observations of the performance of the spacecraft from data on the instrument panel during the fully automated abort, but that the flight was so short, and imparted such high g, that he was unable to do so.

The DM landed 4 km from the pad,which by then was well ablaze. Immediately after landing, radio communications were established and rescue crews dispatched to the cosmonauts. As launch Capcom, Kizim informed the crew that the abort system had been initiated – but having just experienced the ride of their lives, and seeing the pad burning in the distance, they need hardly have been told. Titov later recalled how the behaviour of this launch vehicle was not as smooth as Soyuz T-8, and that he felt ripples of unusual vibration in the seconds leading up to the abort. There was no fear in the crew, he said, but rather the mixed emotions of relief at surviving the abort, and bitter disappointment due to all the wasted preparations for this and his previous flight.

In a later interview, Strekalov said that the rescue teams reached them about half an hour after landing, and that he and his colleagues' first request was for cigarettes. (Ironically, he had also telephoned his mother earlier in the day, and she had begged him not to fly.) The rescue teams found both cosmonauts unhurt by their ordeal, and after a brief medical check it was concluded that neither of them would require hospital attention, although both were given a glass of vodka. The pad – the same pad which was used to launch Sputnik and Gagarin (Pad 1) – burned for twenty hours after the explosion. The mission, although aborted, was at least a successful manned demonstration of the Soyuz abort capabilities.

The abort was not acknowledged by the Soviets until a month after the event, at an IAF congress in Budapest. Radio Moscow finally indicated 'another accident' in the Soyuz programme and a successful operation of the safety system, with the two-man crew surviving the incident. This flight became known as Soyuz T-10-1 (A in the West). It was not until the return of the Soyuz T-9 crew in November 1983 that the Soviet public learned of the incident. After the US Space Shuttle *Challenger* accident in January 1986, the Soviets revealed the details of their own space accidents and mishaps in a series of articles and speeches. These included the Soyuz pad abort, which, according to one Soviet official, was 'a very serious accident... just six seconds from a Soviet *Challenger*'.

Issues arising from the abort of Soyuz T-10-1

In all the excitement and drama of the launch abort, several factors had to be addressed almost immediately. First, although the crew did not require hospitalisation, they were rested from flight status for a while. The second problem was the loss of an operational pad, and although a second pad was available this would strain the preparation and sequencing of subsequent Soyuz T and Progress launches for some time. The US media reported that the cost to the Soviets of refurbishment to the damaged pad utility, propellant lines and support structures would be an estimated $250-500 million.

The Soyuz T-10 spacecraft (no.16L) was intended to be swapped with the Soyuz

T-9 capsule. Previously, Soyuz exchanges had taken place after no more than a hundred days, and at the time of the abort, Soyuz T-9 had already passed this point. The Western press began to indicate that the Salyut crew were 'stranded in orbit' – marooned because of the pad abort – and that the Soviets had reportedly asked for a rescue Shuttle flight by NASA. More experienced Soviet space-watchers attributed this to scaremongering, observing that the 100-day 'expiry' date of Soyuz could rather be likened to a 'best before' date on a can of food. There was nothing to prevent its use beyond that date – within reason. Soyuz T was designed to stay in orbit for at least 180 days; and Soyuz T-9 remained well within that limit – even with the mission extension.

Onboard the station, the deployment of the additional solar arrays – intended to be carried out by Titov and Strekalov – would have to instead be performed by the current crew, and the task was completed on 1 and 3 November. Meanwhile, systems on the Salyut caused further concern, with a malfunction of the environmental control system causing an acrid odour (reminiscent of the problem on Salyut 5 in 1976), and an internal temperature drop to 18° C due to the reduced solar array power system. The humidity also increased to 100%. Clearly, the next Salyut resident crew would be required to perform major work on the station's systems.

One unrevealed aspect of leaving the Soyuz T-9 capsule attached to Salyut was the loss of one of the solar arrays and its impact on exceeding the desired operational lifetime. On 14 November, the Soyuz T-9 engines were used to lower the station's orbit from 324 × 340 km to 322 × 337 km, and also to test their reliability for the de-orbit burn. The end of the Soyuz T-9 mission was indicated shortly afterwards. The OM was filled with rubbish, to be destroyed during re-entry, and the DM landed uneventfully, rolling on its side, on 23 November after a flight of 149 days, demonstrating the value of Soyuz T and its improved system for the extended support of a crew.

SOYUZ T FLIGHT OPERATIONS, 1984

At the end of the Soyuz T-9 mission, the Soviets indicated that there would be a lull in manned Salyut operations while the results of the flight were evaluated. This included the condition of the recovered DM after such a long stay in orbit, the findings of the T-10-1 abort investigation, the status of Salyut 7, the availability of the next Soyuz spacecraft and launch vehicle, and additional training for the next crew to effect repairs to the station systems. The vehicle (15L) to be used for the launch, in February 1984, had been ready to fly in June 1983, but was returned to NPO Energiya in connection with the Soyuz T-9 solar array failure. It would not have taken too long to fix whatever was required, and return it to Baikonur.

Soyuz T-10: repairs and records

On 8 February 1984, Soyuz T-10, with three cosmonauts onboard, successfully entered orbit. There were three main objectives to the mission; to repair the engine fuel leak, to set a new endurance record, and to conduct a programme of medical experiments and observations during the long flight. Priority was given to the series of EVAs to repair the Salyut, and Commander Kizim and Flight Engineer Vladimir Solovyov had conducted extensive EVA training at the hydrotank at TsPK. Once completed, their task would be to try to break the 211-day endurance record of Salyut 7's first crew by the IAF margin of 10% (22 days) to quality for a new record. Thus, a flight of 233 days or more was the target. The third member of the crew (call-sign, Mayak) was Dr Oleg Atkov, who would monitor their health and condition during the long flight. Although he would not participate in the EVAs, he was the first of a planned programme of medical specialists that would accompany every subsequent record-breaking crew (although this was plan was later abandoned).

The systems performed normally during the approach to Salyut, but as Kizim brought his Soyuz in to dock, intense sunlight masked the docking target markings, and so he pulled away. Shortly afterwards, as the spacecraft moved into shadow, he docked successfully. When they had checked the integrity of the docking, the cosmonauts removed the docking apparatus and entered Salyut; but as they did so, they noticed the smell of burnt metal from the drogue of the docking apparatus. Onboard the station, the main objective of the flight began with the first of a series of six EVAs to conduct repairs to the station and to erect a second set of panels on one of the main three solar arrays.

Visitors come and go, and Soyuz T-10 lands

On 4 April 1984, the Soviet–Indian visiting mission arrived at the station onboard Soyuz T-11, for a week of joint experiments with the resident crew. On 11 April the international crew returned to Earth in the Soyuz T-10 capsule, leaving the fresher Soyuz T-11 as the primary crew. The landing of Soyuz T-10 was an impressive 1 minute ahead in the flight plan, and was only 1 km from the target landing point. Furthermore, the vehicle landed the right way to rest on its base, instead of on its side. Two days later, Soyuz T-11 was moved from the aft port to the forward port of the Salyut.

Three months later, Soyuz T-12 was launched to the station – again with a three-person crew. In command was Dzhanibekov, with Savitskaya making her second flight as Flight Engineer and, Igor Volk as Cosmonaut–Researcher, who was undertaking the short twelve-day flight as part of his preparation to fly the Soviet shuttle Buran, which was then under development. (At this time, Buran was still a state secret, and this part of Volk's mission was therefore rather vague.) Dzhanibekov and Savitskaya would instruct the Salyut crew on techniques developed to repair and maintain the Salyut since their launch, and would conduct an EVA of their own. Docking occurred without incident on 18 July, but in an orbit 30 km higher than normal for three-person crews. Soyuz T-12 was the first manned vehicle to use the uprated Soyuz U-2 rocket, which in its core stage used a synthetic fuel called 'sintin' or 'tsiklin', with a higher specific impulse than kerosene. (Sintin was also used in some versions of the Proton's Blok D upper stage. By the mid-1990s it had become too expensive, and Soyuz U was again used. Western space analyst Phil Clark has suggested that the earlier Soyuz T flights to about 300 km were a result of limited capabilities of the launcher vehicle, or a safety measure to be used until its full capabilities were proven.) On 29 July – after completing their programme onboard the Salyut – all three Soyuz T-12 cosmonauts returned to Earth, without incident. Soyuz T-11 remained attached to Salyut for the remainder of the record-breaking flight, which ended without incident on 2 October after a flight of 236 days 22 hrs 50 min. The disappointments of 1983 were forgotten, as the Mayak crew had successfully repaired the fuel leak in a dramatic series of EVAs, and had set a new endurance record in the process. The prospects for flight operations with Soyuz T to the Salyut looked promising for 1985.

SOYUZ T FLIGHT OPERATIONS, 1985–86

The original flight plan for 1985–86 (prior to the problems in February) was as follows:

April	Launch of TKS-4 (which eventually became Cosmos 1686).
May 15	Launch of Soyuz T-13 (vehicle 19L); Vasyutin, Savinykh and Volkov; mission to last about six months.
November	Launch of Soyuz T-14 (vehicle 20L); Savitskaya, Ivanova and Dobrokvashina; a two-week visiting flight, timed to coincide with the anniversary of the October Revolution; to be launched after the undocking of TKS-4, and to take place shortly before the return of the Soyuz T-13 resident crew.
1986	Launch of Soyuz T-15 (vehicle 21L); Viktorenko, Aleksandrov and Salei; the final mission to Salyut 7.

At the end of 1984, Salyut 7 seemed to be in good shape; and at the beginning of 1985, status reports indicated that the onboard systems were functioning normally in autonomous flight, with internal conditions suitable for its next crew. On 11 February, however, this was to change. There was loss of telemetry from the station, and it entered free-drift. With the Igla transponder now unable to pick up an oncoming spacecraft in the automated mode, a freighter could not dock with the station to provide the independent stabilisation required to allow a crew to board and once more try to repair the Salyut. For the present, Salyut was essentially 'dead in the water'. By 1 March, reports from the Soviets indicated that the primary mission of Salyut 7 was at and end; but after so much effort to repair the Salyut during the previous year, they were not about to abandon it, and decided to once more attempt a rescue to prevent a situation in which Salyut 7 would make an uncontrolled re-entry (as with Skylab). In order to achieve docking with an uncontrollable station, an extraordinary crew would be needed. Dzhanibekov – the veteran mission commander with four missions to his credit – had already demonstrated his skills at manual override on T-6. He was joined by veteran Salyut 6 Flight Engineer Viktor Savinykh, who was part of the next residency crew and was already trained for EVA.

Soyuz T-13: a rescue mission

Soyuz T-13 was launched on 6 June 1985, on a space station rescue mission that was compared by some to the 1973 Skylab OWS rescue by the Skylab 2 crew. In order to reach the station, it was decided that Dzhanibekov and Savinykh would fly a two-day rendezvous to save onboard propellants. The final approach was made without the benefit of the Igla radar system, but an unplanned bonus of the previous year's Soyuz T-8 mission was that Titov had demonstrated that it was possible to move the Soyuz T close to the station without using radar. The problem with the Soyuz T-8 approach was that ground radar had been enough to place the spacecraft within 1 km from Salyut, but it relied on Titov's judgement of speed and distance to effect the last stages of rendezvous, which proved difficult. The Soyuz T, however, was not equipped for prolonged station-keeping, and Titov did not have sufficient propellant left to maintain position and conduct repeated docking attempts. Neither had he trained for such an exercise.

With this known, the fuel conservation approach of Soyuz T-13 also provided ground trackers with additional time to more accurately pinpoint the orbital parameters of the spacecraft. The crew's training was focused on this docking approach, and included flying approaches from as distant as 30 km, so that their final choice would depend upon the accuracy with which Soyuz T-13 had been placed in orbit. Additional equipment was fitted to the Soyuz T-13 to assist in their approach. On 8 June they sighted Salyut from a distance of 10 km. To carefully examine the Salyut, Dzhanibekov looked through the portholes while controlling the orientation of the Soyuz. It was in a slow roll of about 0°.3 per second, which he could easily match with the Soyuz T attitude control system that had been modified to include control levers for proximity operations and station-keeping. Fortunately, the station was not tumbling, although the solar arrays had lost solar lock and were 75° apart, indicating a power supply failure rather than the feared communications failure. The station was obviously dead, however, as it did not respond to Igla signals, and did not turn itself to face the Soyuz. With no power, conditions inside would at best be cold, dark and potentially hazardous. But first, the crew had to dock with the Salyut.

Using a low-light image intensifier, the crew could see the Salyut against the Earth's shadow, revealing its position and orientation more clearly than Titov had been able to see. The crew also had a second instrument available, to determine their exact distance from the Salyut. A laser-ranging device would be used to gather information on the distance and closing rate once they began the final approach. With Dzhanibekov controlling the spacecraft alignment and approach speed with help of the optical sight, Savinykh (known in the team as the 'human computer', due to his calculation skills) called out closing range and rates, and keyed data into the onboard computer. They then approached to 200 metres and performed a station-keeping exercise, assessing the situation and consulting with the ground controllers by describing the visual condition of the Salyut. Receiving permission to continue, the Commander brought Soyuz round to the front of the station and aligned the nose of Soyuz T-13 with the front port of Salyut. He then initiated a slow roll to match the station, and slowly brought the two spacecraft together as they flew into the shadow of the Earth. The skills of the two cosmonauts were demonstrated when they achieved the docking in darkness at the first attempt – much to the relief of the jubilant flight controllers, who were still in contact with them.

Although the vehicles were safely docked, the work of the two cosmonauts had only just begun. Following tests of the integrity of the docking, they had to determine whether there was any air in the station. With no power, there was no way to check other than by opening a plug valve in the docking assembly to equalise the pressure between the two spacecraft. There was air in Salyut, but to test whether it could support them, they had to open the hatch and take a sample. Fortunately, no harmful or toxic gases were recorded, and the two cosmonauts could therefore enter the station. They reported that Salyut was dark, the air was musty, and microgravity icicles seemed to be 'growing' from everywhere, in an estimated internal temperature of –10° C.

The work to restore Salyut would take the pair of cosmonauts several days, but system by system it was brought back to operational status, ready to resume its scientific work.

Konstantin Feoktistov described the mission of Soyuz T-13 and the efforts of its crew as 'a major technological accomplishment that holds great promise for the further development of manned flights. It will become possible to approach satellites for examination, or the execution of repair and maintenance operations. This technique assumes even greater importance in the light of the potential need to rescue crews unable to return to Earth because of a crippling malfunction on board.'

Glushko recommend that Dzhanibekov should receive a third Gold Star, as he believed the bravery involved in this mission far exceeded anything previously exhibited in space. Dzhanibekov had already been awarded two Gold Stars, and Savinykh had received one after his first mission. However, Glushko's recommendation was rejected. He was told that Korolyov had only two awards, and that this was sufficient for a cosmonaut.

Soyuz T-13 had proved that the improvements incorporated into the basic Soyuz design were advantageous in 'off-normal situations', and with new improvements already in progress for a further upgrade, its role in Soviet manned spaceflight seemed assured. In 1985, Feoktistov predicted a future role for it as a rescue craft for a space station crew, and this prediction has recently been fulfilled with the ISS; but what he could not have foreseen is how soon the role would be demonstrated.

Soyuz T-14: the first partial crew exchange

The planned Salyut science programme was to continue, with Soyuz T-14 carrying three cosmonauts to the station on 17 September 1985. The subsequent docking with the aft port, on the following day, allowed for the first partial crew exchange in a space station programme. Previously, crews had either exchanged Soyuz T spacecraft, or vacated a station for a period of automated flight between residency crews. Soyuz T-14 demonstrated crew exchange capabilities, indicating that the Soviets were ready to begin permanent occupation of the station. Commander Vasyutin and Cosmonaut–Researcher Alexandr Volkov remained on board the station with Savinykh (who had been part of the original long-duration crew before reassignment to the Salyut rescue mission), while Dzhanibekov, his role completed, undocked Soyuz T-13 on 25 September, alongside Soyuz T-14 Flight Engineer Gregori Grechko, who was already a veteran of three missions to different space stations.

Unlike on previous missions, they did not land later that day, and the Soyuz T instead remained in space to complete 30 hours of manoeuvring tests. First, they withdrew to a point where, with all Salyut docking aids temporarily deactivated, they could evaluate a rendezvous with Salyut to the 1,000-foot {1,000-metre?} point using only optical sighting devices. This experience and information would be used in training and mission planning for future crews in the transition from the final transfer orbit to the point of final approach using optical devices only. Essentially, it was a re-run of the docking approach that Dzhanibekov had completed on 8 June, providing supplementary information to compare with the initial data. No docking was planned in this experiment, although they used the new ability of Soyuz T to station-keep for a short while before pulling away and flying around the station, photographing the docked Soyuz T/Salyut 7 combination. In all, Dzhanibekov brought Soyuz T-13 to within 5 km of Salyut three times, then closed to within 1 km and positioned for station-keeping. The exercise over – and deemed a complete success – the crew prepared for final separation, to land on 26 September in a new zone located north-east of Dzhezkazgan.

The problems with Salyut 7 in 1985 had necessitated two Soyuz T vehicles being required to fulfil the objectives of the originally planned 1985 resident mission (Soyuz T-13 for the repairs, and Soyuz T-14 to complete the experiment programme). Therefore, after the launch of Soyuz T-14 in September 1985 only one Soyuz T vehicle (21L) remained, and this was to be used for the all-female mission, with no further long-duration flights planned. The idea was that the all-female crew would launch on Soyuz T-15 in early March 1986 (timed to coincide with International Women's Day on 8 March). After the departure of the women, the Vasyutin crew would return onboard Soyuz T-14 later in March, leaving Salyut 7 behind unmanned, and with Savinykh setting a new space endurance record.

During November, however, Vasyutin fell ill, and in discussions with the ground it was decided to terminate the mission early and to return all three cosmonauts to Earth on 21 November – the most favourable recovery conditions in the next landing window. Savinykh was given the command for returning the Soyuz T to Earth. The wisdom of having a Soyuz docked to the station at all times, as an emergency evaluation spacecraft, was clearly demonstrated in this situation. However, several objectives would remain, although the Salyut would be mothballed prior to leaving the station. The crew would also be unable to complete their pre-landing conditioning and exercise programme to the best advantage.

Soyuz T-15: the end of an era, the beginning of another

Due to the early evacuation of Salyut, some of the experiment results had not been returned; and the Soviets were also running low on Soyuz T hardware. Vehicle 21L (Soyuz T-15 – eventually flown by Kizim and V. Solovyov) used the refurbished Descent Module of vehicle 16L, which had been involved in the Soyuz T-10A pad abort. With a Soyuz T available for a flight to the station and the replacement Soyuz TM still not ready for flight, the Soviets took the opportunity to send a crew to the new Mir core module and revisit Salyut 7 to complete the work left undone by the Soyuz T-14 crew. Instead of continuing with the Salyut after the problems with the station and after Vasyutin's illness, it had been decided to press on with a new station, even though only the core module was ready for launch. The add-on modules were behind schedule.

On 15 March, Kizim and V. Solovyov were on the second day of their second mission together on Soyuz T-15, and were closing in on the new station – Mir. The Mir core module featured an aft docking port that incorporated the older Igla rendezvous and docking system, which was compatible with the Progress and Soyuz T spacecraft. However, the forward port was a new design with a multiple docking adapter, featuring one lateral forward facing port and four radial ports. The radial ports were designed to accept larger scientific modules, while the forward port featured the new Kurs docking system for the forthcoming Soyuz TM spacecraft, which was the upgraded Soyuz T crew transfer vehicle.

As Kizim and Solovyov were using the older Soyuz T spacecraft, automatic docking at the front port was impossible, and so they performed an automated approach from a distance of 200 metres, where Igla was shut off. Kizim then manually controlled the Soyuz to the forward port by using ranging devices similar to those used by Dzhanibekov on Soyuz T-13. It was a complicated manoeuvre, but because of his experience in Soyuz T development and training, and in command of Soyuz T-3 and Soyuz T-10, plus data from Soyuz T-8 and Soyuz T-13, he achieved docking at the first attempt, after Mir had been turned slightly to provide the best lighting conditions. This kept the aft port free for the arrival of the first Progress freighter (Progress 26) at Mir on 27 April. History shows that Soyuz T-15 – the last of the series – became the first and only Igla-equipped Soyuz to dock with Mir. All further craft would incorporate the new Kurs system.

With the activation of Mir proceeding as scheduled, flight controllers also monitored the unoccupied Salyut 7 which, some 2,500 km from Mir, was a minute ahead and in a similar orbit. By 4 May the crew had loaded 500 kg of equipment into the OM of Soyuz T-15 for the short trip to Salyut. They also powered up the systems, charged the batteries, and tested the PM engines. Despite the fuel-saving two-day approach to Mir, the Soyuz did not have enough propellant for the flight to and from Salyut as well as the de-orbit burn at the end of the mission, and it was impossible to refuel the Soyuz in orbit. Planners had allowed for this by using the engines of Progress 26 to adjust the orbit of Mir to help in the transfer to the Salyut.

The crew undocked Soyuz T-15 from Mir on 5 May, but did not immediately burn the engines for fear of damaging the brand new station. Once a safe distance had been reached, the first inter-space-station flight began, monitored by three control rooms at Kaliningrad: one for Salyut 7/Cosmos 1686, one for Mir, and a third for Soyuz T-15. The crew used a propellant-efficient approach to Salyut that took 29 hours to complete. Using the Igla system, they paused 200 metres from Salyut, the orientation of which was controlled from the ground via the docked Cosmos module. Kizim then assumed manual control, and docked Soyuz T-15 with the aft port of Salyut. After fifty days onboard Salyut 7 (where they had spent eight months in 1984) – completing two EVAs originally planned to be completed by the Vasyutin crew, and deactivating the station – the crew reloaded 400 kg of equipment and experiment results into the OM of Soyuz T-15 for the 29-hour flight back to Mir. They undocked on 25 June, and arrived at Mir – using the same approach as with the first docking with the station – for the final twenty days of their mission. With this additional time in space, on 3 July Kizim surpassed Ryumin's 363-day record for total time spent in space; and two days later he became the first (of several cosmonauts) to spend a career total of 1 year in orbit.

With no additional module available to supplement the few experiments located in the core module, the crew finally departed Mir on 16 July after spending fifty days onboard Mir, fifty days onboard Salyut 7, and a further twenty days onboard Mir in a 124-day mission. It was a remarkable achievement which demonstrated the maturity of the Soviet programme – a maturity that was about to be evolve even further with new missions to Mir, with a new Soyuz.

SUMMARY

The flight of Soyuz T-15 marked the final flight of the series. In total, the twenty-one Soyuz T (or T-related) spacecraft (five Cosmos, one unmanned and fifteen manned) accumulated more than 1,195 days of flight time (including orbital storage or powered-down status docked to space stations). They demonstrated the ability to ferry crews to and from the station, both in automatic and manual modes, to be powered down and brought back on line without difficulty, to act as a crew rescue and recovery vehicle and, in the case of the last mission, to ferry crews to and from a second space station.

REFERENCES

1 Severin, G.I., Abrahamov, I.P. and V.I. Svertshek, 'Crewman Rescue Equipment in Manned Space Missions: Aspects of Application', presented at the 38th Congress of the IAF, 10-17 October 1987, Brighton, UK.
2 *Soviet and Russian Cosmonauts, 1960–2000*, Novosti Kosmonavtika, 2001.
3 *Journal of the British Interplanetary Society*, January–February 2002, 55.
4 Titov, V., 'The Docking is Cancelled', in *Orbits of Peace and Progress*, reprinted from *Red Star*, 1984.

Soyuz variants: Soyuz T missions and hardware

Spacecraft (serial number) name	Design designation	Launch date	International designation	Docking date	Undocking date	Target spacecraft (port)	Landing/ decay date	Flight duration dd:hh:mm:ss
Key: (U), unmanned								
11F732 (Soyuz T improved transport manned spacecraft for DOS								
(01L) Cosmos 670	7K-S (U)	1974 Aug 6	1974-061A				1974 Aug 8	003:01:28:00
(02L) Cosmos 772	7K-S (U)	1975 Sep 29	1975-093A				1975 Oct 2	003:10:00:46
(03L) Cosmos 869	7K-S (U)	1976 Nov 29	1976-114A				1976 Dec 17	017:18:34:30
(04L) Cosmos 1001	7K-ST (U)	1978 Apr 4	1978-036A				1978 Apr 15	010:20:05:42
(05L) Cosmos 1074	7K-ST (U)	1979 Jan 31	1979-008A				1979 Apr 1	060:01:11:00
(06L) Soyuz T	7K-ST (U)	1979 Dec 16	1979-103A	1979 Dec 19	1980 Mar 23	Salyut 6 (Fwd)	1980 Mar 26	100:09:31:40
(07L) Soyuz T-2	7K-ST	1980 Jun 5	1980-045A	1980 Jun 6	1980 Jun 9	Salyut 6 (Aft)	1980 Jun 9	003:22:19:30
(08L) Soyuz T-3	7K-ST	1980 Nov 27	1980-094A	1980 Nov 28	1980 Dec 10	Salyut 6 (Fwd)	1980 Dec 10	012:19:07:42
(09L) Soyuz T-6	7K-ST	1982 Jun 24	1982-063A	1982 Jun 25	1982 Jul 2	Salyut 7 (Aft)	1982 Jul 2	007:21:50:53
(10L) Soyuz T-4	7K-ST	1981 Mar 12	1981-023A	1981 Mar 13	1981 May 26	Salyut 6 (Fwd)	1981 May 26	074:17:37:23
(11L) Soyuz T-5	7K-ST	1982 May 13	1982-042A	1982 May 14	1982 Aug 27	Salyut 7 (Fwd)	1982 Aug 27	106:05:06:12
(12L) Soyuz T-7	7K-ST	1982 Aug 19	1982-080A	1982 Aug 20	1982 Aug 29	Salyut 7 (Aft)		
				1982 Aug 29	1982 Dec 10	Salyut 7 (Aft)	1982 Dec 10	113:01:50:44
(13L) Soyuz T-8	7K-ST	1983 Apr 20	1983-035A	*Unable to dock with target spacecraft*		Salyut 7 (Aft)	1983 Apr 22	002:00:17:48
(14L) Soyuz T-9	7K-ST	1983 Jun 27	1983-062A	1983 Jun 28	1983 Aug 16	Salyut 7 (Aft)		
				1983 Aug 16	1983 Nov 23	Salyut 7 (Fwd)	1983 Nov 23	149:10:46:01
(15L) Soyuz T-10	7K-ST	1984 Feb 8	1984-014A	1984 Feb 9	1984 Apr 11	Salyut 7 (Fwd)	1984 Apr 11	062:22:41:22
(16L) Soyuz T-10-1	7K-ST	1983 Sep 26	*Launch pad abort; not classified as a spaceflight*			Salyut 7 (Aft)	1983 Sep 26	000:00:05:13
(17L) Soyuz T-11	7K-ST	1984 Apr 3	1984-032A	1984 Apr 4	1984 Apr 13	Salyut 7 (Aft)		
				1984 Apr 13	1984 Oct 2	Salyut 7 (Fwd)	1984 Oct 2	181:21:47:52
(18L) Soyuz T-12	7K-ST	1984 Jul 17	1984-073A	1984 Jul 18	1984 Jul 29	Salyut 7 (Aft)	1984 Jul 29	011:19:14:36
(19L) Soyuz T-13	7K-ST	1985 Jun 6	1985-043A	1985 Jun 8	1985 Sep 25	Salyut 7 (Fwd)	1985 Sep 26?	112:03:12:06

Spacecraft (board number) name	Design designation	Launch date	International designation	Docking date	Undocking date	Target spacecraft (port)	Landing/ decay date	Flight duration dd:hh:mm:ss
(20L) Soyuz T-14	7K-ST	1985 Sep 17	1985-081A	1985 Sep 18	1985 Nov 21	Salyut 7 (Aft)	1985 Nov 21	064:21:52:08
(21L) Soyuz T-15	7K-ST	1986 Mar 13	1986-022A	1986 Mar 15	1986 May 5	Mir (Fwd)		
				1986 May 6	1986 Jun 25	Salyut 7 (Aft)		
				1986 Jun 26	1986 Jul 16	Mir (Fwd)	1986 Jul 16	125:00:00:56

Soyuz TM, 1986–2002

On 19 February 1986 the Soviets launched the base block of the new Mir (Peace) space station. Featuring multiple docking ports and improved sub-systems and onboard facilities, it could also accept large permanent scientific modules, besides resupply and mission support from smaller ferry and freighter craft of the Soyuz and Progress classes as well as the much larger Buran shuttle orbiter. Many of Mir's onboard systems were improvements and upgrades that required parallel development in the Soyuz T, with a new version called the TM (Transport, Modification). When Soyuz T-15 was launched to Mir, reports from the Soviets indicated that it was the last of that type of Soyuz to fly. *Pravda* stated that because of the limited capability of the Soyuz T, Soyuz T-15 was 'the last Salyut-type ship'. Mir required its own generation of transporters.

THE ORIGINS OF SOYUZ TM

In February 1976 the Soviet government approved the development of the next generation of space stations, featuring multiple docking ports and several upgrades to the earlier DOS designs (Salyuts 1, 4, 6 and 7). These included improved sub-systems to extend the operational lifetime of the station, the capacity to expand the 'core module' with several scientific modules, and improved rendezvous and docking facilities for more reliable and extended visits by teams of cosmonauts using the Soyuz spacecraft as a ferry.

An essential element of this transition to the next generation of space stations involved the development of the Soyuz T (Transport) spacecraft. With the increased capability of the next generation of multi-module stations, it was soon recognised that considerable amount of propellant would be required to move the station into alignment with the smaller ferry. It would be much more practical if the Soyuz T itself could be further improved to allow free movement around the station in order to dock at any of the available ports.

During planning for these new multi-modular space stations (designated DOS 7 (Mir) and DOS 8 (Mir 2)) in the early 1980s, the intention was to place them at 65°

A Soyuz TM ferry docked to the ISS. Soyuz has been used as a space station crew transport vehicle for more than thirty years.

inclination instead of the traditional 51°.6 flown by earlier Salyut and Almaz stations, in order to provide a better ground coverage for planned remote sensing of the Soviet Union. Sending crews to them would require a Soyuz lighter than the Soyuz T, and a slightly improved performance of the launch vehicle (the 11A511U2, also known as Soyuz U-2. By November 1984 however, it became clear that the engineers designing DOS at NPO Energiya had seriously miscalculated the mass concentration of the core module. Despite some weight-saving measures and an increase in the thrust performance of the DOS Proton launch vehicle, it was obvious that this would still be insufficient, and in January 1985 iy was decided to revert to the 51.6-degree inclination for DOS 7. For almost a decade, however, planning for DOS 8 (Mir 2) would still call for it to remain at the higher inclination.

The development of Soyuz TM began with preliminary designs in 1980, followed by the signing of the draft plan in April 1981. The new vehicle was designated Soyuz TM, although it retained the article number 11F732 and manufacturer (NPO Energiya) designation of 7K-ST (the same as Soyuz T). The leading designer for the Soyuz TM modifications was Vladimir Guzenko, an engineer with NPO Energiya.[1]

By early 1982, most of the engineering blueprints had been completed, and work could begin in fabricating the first Soyuz TM spacecraft in time to support the launch of the DOS 7 station in 1986. However, it soon became evident that the delivery of the core module was behind schedule, and that if the normal processing flow was followed, the core would not be launched until late 1986. It was therefore decided to transport DOS 7 from the Khrunichev factory – bypassing the normal electrical systems testing at the NPO Energiya ZEM factory – directly to Baikonur, where tests would be completed. It was also decided to launch the core as early as possibly, as it was easier (and cheaper) to store it in orbit until the additional modules were ready, than it would be to leave it at Baikonur. And again, politics entered the arena, with the Soviet leadership calling for a new space 'achievement'.

The launch of DOS 7 (Mir) in February 1986 was timed to coincide with the twenty-seventh congress of the Communist Party in Moscow – but it would be the final politically-influenced launch time in the Soviet space programme. Although Mir was in orbit, the additional modules were behind in their production, and with Soyuz TM production timed for mid- to late 1986, and the first manned flight planned for the end of 1986 or early 1987, it was decided to use the last Soyuz T (T-15) to send a crew to Mir to begin the long process of changing the core module from launch to orbital configuration (see p.312). (This mission would also include a short close-out flight to Salyut 7.) The Soyuz T-15 crew would complete a series of 'get-ahead tasks' before the arrival of the first Mir crew in the new TM vehicle some months later.

Unlike the previous versions of Soyuz, no unmanned technology test flights were planned under the Cosmos label, as most of the vehicle testing and system qualification could be man-rated during the Soyuz T flights. But there would be one inaugural unmanned flight to the new station to qualify the systems before the crews took over operational missions.

SOYUZ TM MODIFICATIONS

Outwardly, Soyuz TM resembled Soyuz T, with its independent flight life of fourteen days and orbital storage duration of 180 days. The launch vehicle used to orbit the Soyuz TM was the Soyuz 11A511U2 (see p.90), with improved performance to match the design improvements of the spacecraft. The launch escape tower was fitted with lighter escape motors, enabling an increase in delivered payload weight of 200–250 kg into an orbit at 51°.6, and an increased payload return capability of 70–90 kg. Additional payload capacity was achieved by jettisoning the escape tower at T + 115 sec rather than at T + 123 sec. The Soyuz TM/Soyuz U-2 improvements allowed access by a three-person crew to Mir, operating in an orbit at 325–350 km, without the complex and fuel-expensive requirement to lower the station. However, in the original planning for Soyuz TM missions to the Mir station at 65°, it seems evident that even with the upgrade to the launch vehicle and the weight-saving programmes, Soyuz TM would have been unable to carry three cosmonauts to the higher inclination. All of these changes also allowed the Soyuz TM to be more flexible in executing its assigned mission, should the real-time situation require a deviation from the flight plan.[2]

Rendezvous and docking system

One of the most significant changes incorporated in the new spacecraft was in the rendezvous and docking system. The new Kurs (Course) replaced the older Igla (Needle) system flown on Soyuz T, and allowed the spacecraft to lock on to the station without the space station having to manoeuvre to face it, which was a further saving of the space station's propellant. The Soyuz TM made radar contact and measurements at a distance of 200 km (rather than the 30 km of Igla), and locked on to the target station at 20–30 km. A feature of Kurs was that the spacecraft antenna was omni-directional, and did not require line-of-sight orientation to lock on to the station. The onboard computer guided the spacecraft towards the station and, depending on the assigned docking port, manoeuvred the Soyuz TM around the station to approach that port. The Kurs docking facilities were initially located on the forward lateral docking port of Mir, with the older Igla system at the rear port; but when the Kvant astrophysical module docked at the aft port in 1987 it incorporated both Igla and Kurs rendezvous systems, allowing the Soyuz TM to

At Star City, astronaut Frank Culbertson exits a Soyuz TM simulator via a hatch in the side of the mock-up.

Controllers and engineers at TsPK monitor docking operations in support of mission control, Moscow.

dock at either forward or aft locations, thus increasing the flexibility for visiting vehicles. Kurs rendezvous equipment was also installed on the Kristall module that docked with the complex in June 1990 (see Soyuz TM-16, p.345). The reason for installing the older Igla system on the aft port was that the Kvant module was intended for launch to Salyut 7, and was constructed with the Igla system. When it was delayed and then reassigned to Mir, it was easier to install Igla on the aft port of Mir to allow Kvant to dock, and add the Kurs to the rear port of the module rather than completely refit Kurs throughout Kvant.

The Orbital Module

In addition to the installation of the Kurs rendezvous hardware on the exterior of the OM, the OM itself was modified. One of the most significant changes was the installation of a new window in order to improve visibility during docking. The window was installed in the OM next to the docking unit, to provide a direct view – rather than a view through only the DM periscope system – during docking approach. Normally, all crew-members were in the DM for docking, with only the Commander having a periscope view of the approach; but with the installation of the new forward-looking window in the OM, one of the crew could move to the OM to observe the approach directly. According to Yuri Semyonov: 'Control of the spacecraft can be made either from [the DM] or from new control situated in the OM. This will make docking manoeuvres easier in the future, and offer greater direct crew visibility, providing redundancy with a back-up docking control station.'[3]

The landing system

The second major innovation on Soyuz TM (resulting from an initiative that had begun in 1973) focused on the landing and recovery systems – notably the primary and reserve parachutes used at the end of the mission. The overall weight of the parachute canopies was reduced, and further significant weight-saving measures were incorporated by using new synthetic fibres woven into the reinforcement structure and shroud lines. This reduced the volume of the parachute containers inside the DM, and allowed more volume for other improvements in the vehicle. In total, 140 kg was saved in the design of the new parachute system.

In conjunction with the new parachute system, landings were rendered safer and more comfortable with improvements in the design of the landing altimeter for firing the soft-landing engines, and in the design of the crew seats. (Most Soyuz capsules, after landing in a high wind, roll on their side.)

The propulsion system

According to statements by V. Pouchkayev, Head of the Flight Control Centre's ballistic unit at the time of the Soyuz TM-2 landing in July 1987, firing the new spacecraft's braking engine for precisely 208 seconds at the end of the mission can reduce the velocity by 115.2 metres per sec. An error of 5 metres per sec would result in a landing 'within a few hundred kilometres of the desired landing point', but in extreme circumstances a landing with this burn error was possible into the landing area using 'a special system', although at entry this would increase the loads on the crew from 3 g to 11 g.

Improvements in the PM included a new 'base block' consisting of fuel tanks, a fuel feed system and a sustainer engine, including the use of metal tank membranes to separate the oxidiser from the fuel. The tanks had previously had organic membranes, which had a tendency to perish over time and contaminate the fuel or affect performance. The revisions included helium as the expulsion gas (to pressure-feed the supply of fuel to the engines), sectionalised fuel reserves, a network of back-up fuel lines, and a non-cooled nozzle in the new sustainer engine. The average propellant load for a Soyuz TM was approximately 800 kg, with about 150 kg for the rendezvous and docking operation, 200 kg for the de-orbit burn at the end of the mission, and a further 200 kg contingency.

The Soyuz TM main engine was designated KTDU-80 (or S5.80) and although similar to the Soyuz T engine (the 11D426), with a thrust of 300 kg, it had different dimensions and was slightly heavier. The combined engine installation also included fourteen 13.5-kg high-thrust approach and orientation engines (DPO-B), and a set of 2.5-kg low-thrust approach and orientation engines (DPO-M). The DPO-B thrusters were used for final orbit manoeuvres just before docking and also for orientation, and could be used as a back-up de-orbit system in case the main KTDU fails. The DPO-M thrusters could be used only for orientation.

The onboard sub-system

Throughout the vehicle, the main electrical circuits featured triple redundancy and any reserve lines had both pneumatic and hydraulic circuitry. The life support system included an additional reserve for the Crew Module that could sustain the crew from orbit to the ground at the next available (emergency) landing window in the event of a full failure of the main system. Modernised power supply units were also installed, although the solar arrays remained the same, with a span of 10.6 metres and an area of 10 square metres, generating 0.60 kW, and linked to rechargeable batteries.

Crew provisions

The standard crew of three was retained, as the Soviets stated that there was no operational reason to have more than three crew-men onboard Mir on long-duration missions.[3] The 10-kg Sokol pressure garments were again improved, and retained the capacity for emergency EVA operations. Each cosmonaut also had improved communication gear, with separate voice channels for each member of the crew (a maximum of three), and improved onboard reception quality. The new communication equipment (Rassvet) could relay through the Luch network of satellites (similar to Mir) while out of direct contact with Soviet tracking stations. The Soviets had previously relied on a number of tracking ships deployed around the world to support key areas of operations, docking and re-entry; but this was expensive, and required a large number of personnel. The cosmonauts referred to it as 'talking to the dead', as most of the ships were named after deceased cosmonauts and Korolyov.

Control of the spacecraft was improved with new platform-less inertial motion control systems, based on onboard computers, allowing commands to be input automatically or by the crew. New and highly precise angular velocity meters and linear acceleration transducers were specially developed for the TM module, to provide far more precise navigational computations that on previous vehicles. The upgraded computer programme was supported by redundancy throughout the spacecraft.

The additional spare volume produced by the smaller parachute housing and improved seats also allowed more interior space for personal equipment stowage.

THE FIRST OCCUPATION OF MIR: SOYUZ TM-1–TM-7

During the Soyuz T-15 mission, while the two cosmonauts were onboard the Salyut station (5 May-26 June), the Mir/Progress 26 complex continued in autonomous flight. Taking advantage of the lull in activity on Mir, the Soviets prepared to launch the first of the new TM spacecraft in order to man-rate its systems and upgrades prior to the first crew beginning permanent occupation of Mir from 1987.

Soyuz TM crewing

The initial crews – formed in September 1985 – followed the pattern established on Salyut 7 in which crews were named about four months apart. The first crew was V. Titov and Serebrov; the second crew was Romanenko and Manarov (who due to a medical problem was replaced by Laveikin); and the third crew, who began training in March 1986, was A. Volkov and Yemelyanov. This continued the established model of having one flight-experienced crew-member. Their training was extended when the Soyuz T-15 crew was launched to Mir, due to the delays in the construction of the modules; and another change was introduced because Serebrov became ill, and so the second crew crew replaced the first crew, as they were fully prepared. The first two crews to Mir were due to conduct extended record-breaking missions. Serebrov was replaced by Manarov (restored to flight status), and paired with V. Titov to form a new number 2 crew.

Three Soyuz TM craft were also assigned to taxi-style missions. The third seat was occupied by a cosmonaut–researcher on an Interkosmos style mission, which allowed a Syrian to fly on TM-3, a Bulgarian to fly on TM-5, and an Afghani on TM-6. Slovakia signed up for a mission by using the money owed to it by Russia, although some of the bill was paid in Bohemian glass. Bella – the Slovak – flew in a third seat on a resident crew in 1999.

Soyuz TM-2 – carrying Romanenko and Laveikin – was launched in February 1987 for a 300-day-plus mission, and were due to receive one visiting crew: Viktorenko and Aleksandrov, with A. Solovyov and Savinykh as back-up. They were joined by two Syrian Air Force officers, who both spoke fluent Russian. Laveikin later developed a heart problem during his residency, and it was therefore decided that Aleksandrov would replace him on the second half of the expedition mission.

The next expedition crew – who were due to remain in orbit for one year – therefore consisted of V. Titov and Manarov (who was paired with him in March 1987), with A. Volkov and Yemelyanov as back-up. In May 1987, Yemelyanov was replaced (for medical reasons) by Kaleri, who in March 1988 was in turn replaced by Krikalev, having backed up Soyuz TM-4. The Soviets decided to fly the second Buran pilot in the third seat of Soyuz TM-4, and Levchenko (who had supported Volk on Soyuz T-12) was selected, with Shchukin as his back-up. Because of the duration of the planned mission, Titov and Manarov would receive two visiting crews – the first of which consisted of A. Solovyov and Savinykh, backed up by Lyakhov and initially Zaitsev, until he was replaced by Serebrov as the engineer in March 1988. They were to fly with a Bulgarian cosmonaut (the second Bulgarian mission), as the original flight (Soyuz 33) had been cut short due to technical problems in 1979 (see p.228).

In September 1985 the Soviets had selected a group of pilots (Lyakhov, Berezovoi and Malyshev) to train to fly Soyuz alone, in order to provide a rescue capability. This programme was initiated mainly because of the problems on Salyut 7 in 1985. In 1988, it was decided to test this capability by launching a mission with a cosmonaut–doctor in the engineer's seat. Polyakov and Arzamazov trained for the doctor's seat, and Lyakhov was named as Commander, with Berezovoi as his back-up. The third seat was occupied by an Afghani cosmonaut. Polyakov would stay on board the station to monitor the medical condition of the core crew.

After the record-breaking mission, the crews were rotated as normal, with the fourth expedition (Volkov and Krikalev) being launched on Soyuz TM-7. Their back-ups were Viktorenko and Serebrov, and the third seat was occupied by French Air Force pilot Chrétien, following an agreement between the Russians and CNES. There were no plans for a visiting crew, as they were moving to a normal six-month mission pattern with crew replacing crew. This allowed for permanent use of the third seat, and also reduced the required number of Soyuz vehicles. The planned

Soyuz TM-8 mission would have consisted of Viktorenko and Serebrov as the prime crew, with A. Solovyov and Balandin as back-up. Another Buran shuttle pilot, Stankyavichus, would have been in the third seat. His back-up was Zabolotsky.

Viktorenko and Serebrov were scheduled to launch in April 1989 – one of their main tasks being to fly the SPK (the Soviet MMU) from Kvant 2. However, in February 1989 it became clear that Kvant 2 would be delayed, and it was decided to replace Serebrov (who had specialised in the SPK) with Balandin. Viktorenko and Balandin were expected to launch on 19 April, onboard Soyuz TM-8 (vehicle no.58); but in March 1989, Soyuz TM no. 59 – which was supposed to act as the back-up/rescue vehicle for the Viktorenko/Balandin mission – was damaged during a test in an altitude chamber at Baikonur (the Propulsion Module was partially destroyed due to over pressurisation), and had to be shipped back to Energiya in Moscow. Thc State Commission therefore decided to delay the launch of vehicle no.58 until another back-up vehicle (no.60) was ready for shipment to Baikonur. This resulted in Mir flying unmanned for several months. The damaged Soyuz was repaired and fitted with a new Propulsion Module, and was eventually launched as Soyuz TM-10 (with a new serial number: 61A).

The maiden flight of Soyuz TM

The initial launch of a Soyuz TM design took place on 21 May 1986. The objective of the unmanned Soyuz TM flight was stated as being a 'test of the overall performance of the spacecraft in autonomous flight, and jointly with the orbital station Mir'. Over the next two days the spacecraft approached to the unoccupied station and, using the Kurs rendezvous system, docked with the front port on 23 May. During its approach, the spacecraft's onboard systems, equipment and structural elements were evaluated prior to placing the spacecraft on its final approach to Mir.

During the next six days, further tests were conducted on the spacecraft's onboard systems and structural integrity, and on the performance of its onboard engine. On 29 May the spacecraft was undocked and moved away from the station. The next day, after further systems checks, the OM was separated, and the onboard braking engine fired at a predetermined point. This was followed by the separation of the DM from the PM to follow an automated descent profile and parachute landing in the Soviet Union, to end a successful test of the new spacecraft, and qualifying it for manned use.

Permanent occupation begins

The first phase of manned operation with the Soyuz TM was preceded by the launch of Progress 27 to the Mir complex on 16 January (it departed on 26 March for the arrival of Kvant). Three weeks later, Yuri Romanenko (Commander) and Alexandr Laveikin (Flight Engineer) became the first cosmonauts to ride a TM spacecraft (TM-2) into orbit on an intended ten-month mission (call-sign, Tamyr) to initiate permanent occupation of the station. Unusually, on 28 January the names of the crew were announced in advance, and in another departure from normal Soviet practice, the launch on 6 February was carried live on TV, after the launch time had

At the completion of the training programme, the Soyuz flight crew-members (in this case, TM-2) report their readiness to the Chiefs of Cosmonaut Training. (*Left–right*) Leonov, Rozhdestvensky, Shatalov, Romanenko and Laveikin.

been announced earlier that day. This was only the third time that a manned launch had been broadcats on live TV, and was another example of 'space glasnost'. Two days later, after a slow 50-hour approach to the station, the crew docked with the front port of Mir and soon began reactivating the station for a long stay. By April, with the crew working on the station, the first of several modules intended for Mir (the Kvant astrophysical module) was ready for launch.

Kvant arrived near the Mir complex on 5 April, and as a precaution, both cosmonauts retreated to relative safety in the Soyuz TM-2 DM, in case anything untoward should happen during docking. To some Western observers, this was a time-consuming process that ought to have been discontinued, considering the Soviets' experience of docking Soyuz and Progress spacecraft to the Salyut 6 and Salyut 7 space stations. However, at 11 tonnes, Kvant was much larger than either of the transport craft, and as events transpired, the precautionary move to the TM was prudent.

During the approach to Mir, the Igla approach system locked on to the aft port; but at 200 metres the lock on Mir was lost, and the Kvant passed by the side of Mir within 10 metres – much to the surprise of the cosmonauts. The Soviets played down the mishap, and soon stated that the problem was known and that there would be a second attempt. Inside Mir – ninety minutes after the near-collision, and as the spacecraft came back into range of ground communication – Romanenko came on line; and he was certainly not a happy cosmonaut!

On 9 April, Kvant was finally brought to the docking port of Mir – only to be prevented from hard docking due to a rubbish bag which had been left in the docking mechanism by the departing Progress freighter. On 11 April, Romanenko and Laveikin performed a contingency EVA of 3 hrs 40 min to remove the obstruction and to monitor the automated docking of Kvant, its propulsion unit undocking on 13 April. Kvant became the first permanent addition to Mir. Ten days later, on 23 April (the twentieth anniversary of the launch of Soyuz 1), Progress 29 docked to the Kvant aft port in its first operational use. Over the next few days, the crew tested the integrity of the four-spacecraft docking (Soyuz TM-2/Mir/Kvant/Progress 29) in tests similar to those which Romanenko had conducted onboard Soyuz 26/Salyut 6/Soyuz 27 in January 1978. This was part of the qualification for leaving Soyuz TM (and Progress and the later modules) attached to the station for extended periods, to ensure the safety of the habitation modules and the integrity of the complex.

The Tamyr crew continued their science programme onboard Mir, and conducted two EVAs over the next three months. During this time, recordings of Laveikin's heart-beat revealed slight irregularities, which caused some concern among the medical team on the ground. He had trained hard for his first space mission, and the stress of that training programme, the excitement of his first launch and flight into space, the desire to perform well, and the burden of the contingency EVA during the Kvant docking, all contributed to the problem. Although he was not in immediate danger, the management team decided to replace him with Alexandr Alexandrov, who was assigned to the seven-day Soyuz TM-3 visiting mission, as his experience on Soyuz T-9 in 1983 apparently compensated for his lack of training for a five-month stay on Mir.

Launched on 22 July, Soyuz TM-3 docked two days later at the rear port of Kvant – the first craft to do so. Onboard were Commander Alexandr Viktorenko, Flight Engineer Alexandrov, and Syrian Cosmonaut–Researcher Mohammed Faris. The ascent and docking was performed without major incident, but the visiting crew experienced difficulty in opening the hermetic seal of the TM forward hatch, and had to resort to a lever to open it because it opened inwards against the internal pressure, which was sealing it. After a week onboard the station, a disappointed Laveikin handed over to Alexandrov, who would continue with Romanenko to finish the ten-month mission. Laveikin returned to Earth with Viktorenko and Faris in the Soyuz TM-2 spacecraft, leaving the fresher Soyuz TM-3 for Romanenko and Alexandrov. As part of this process, Romanenko's seat-liner was removed to be installed in the Soyuz TM-3 Descent Module, while those of Viktorenko and Faris were transferred to the Soyuz TM-2 Descent Module. Despite being launched with only two cosmonauts, Soyuz TM-2 was fully capable of supporting three cosmonauts for re-entry and landing. On 29 July, after sealing themselves inside TM-2, the three cosmonauts were informed of a two-orbit (three-hour) delay in undocking, due to rain-storms over the planned landing area. During the descent, the high winds blew the descending spacecraft far off course and almost into a small settlement, but the DM landed safely without incident. Laveikin's disappointment at his early return was evident in post-flight activities. His heart irregularity was apparently a symptom of his physiology, and he was soon restored to flight status and cleared for future missions. The day after the undocking of Soyuz TM-2, the Tamyrs moved Soyuz

TM-3 from the aft port, staton keeping for thirty minutes while Mir was turned so that the front port of the station faced them for re-docking.

A year in space

The next TM crew was preparing to fly to the station in December 1987. Commander V. Titov was to attempt to work on a station for the third time, while his Flight Engineer Musa Manarov was on his first mission. These two cosmonauts (the Okean crew) would replace Romanenko and Alexandrov and attempt a full year in space, to return home in December 1988. Launched with them on 21 December 1987, but returning with the Tamyr crew eight days later, was 'merited' test pilot Anatoli Levchenko, who was assigned to the flight as part of his preparation for planned missions on the Buran space shuttle.

After Soyuz TM-4 docked with Mir two days later, and when the crews began the exchange, it was the first time in history that a full resident crew exchange had taken place onboard a station. The Soviets hoped that this would continue at the end of the Okean residency and be repeated many times, to demonstrate a permanent presence in space which had long been the dream of Soviet planners and designers. According to reports, all future Mir resident crews would include full or partial exchanges to maintain a permanent presence on Mir. Soyuz TM-3 undocked from Mir, with Romanenko, Alexandrov and Levchenko onboard, on 29 December. All three had been launched to Mir on different Soyuz TMs, and were now returning on one vehicle. Romanenko had remained in space for 326 days, Alexandrov for 160 days, and Levchenko for almost eight days.

The landing was quite rough, with visibility of 4 km and a temperature of –12° C. The weather forecast warned of severe ice, which could ground the recovery aircraft, forcing ground-based vehicles to recover the cosmonauts (as with Soyuz T-5, five years earlier). Part of the descent was filmed, showing the capsule descending into a bank of fog; and it was also later filmed positioned on its side, with the commentator explaining that winds of 25–30 mph had blown it over. The strong winds also prevented the construction of a medical tent at the landing site, and so all three cosmonauts were transferred into helicopters. Romanenko and Alexandrov were carried from the capsule, but two men supported Levchenko as he walked to a helicopter for a flight to an airport. Soon after his spaceflight he would go to Baikonur to fly aircraft in simulated landings of Buran vehicles.

On 30 December the Okean crew repositioned Soyuz TM-4 from Kvant to the forward docking location. During their long spaceflight, they had completed three EVAs, and were visited by two international crews. The first (Soyuz TM-5) was crewed by Commander A. Solovyov, Flight Engineer Viktor Savinykh, and Bulgarian Commander–Researcher Alexandr Alexandrov (not the Soviet Alexandr Alexandrov who had just returned to Earth). The launch on 7 June was the first to take place from the newly-named Yuri Gagarin Launching Complex – the same pad from which Gagarin had opened the manned space programme and, according to the Soviets, the 290th launch (manned and unmanned combined) from that site. Fifty hours later, Soyuz TM-5 docked to the aft port of Kvant – one orbit later than planned, due to faulty readings from the Kurs system onboard the Soyuz. It had sent an erroneous signal, causing the spacecraft to deviate from the straight docking approach, and forcing Solovyov to assume manual control and terminate the docking, after which he withdrew to a safe distance and performed station-keeping while the data were evaluated on the ground. The data (which had suggested that there was a fault in the approach) were checked, and it was found that the rendezvous was proceeding without any problems. The readings were therefore overruled and bypassed, allowing the docking to go ahead ninety minutes later. Had the Soyuz TM-5 then been unable to dock, with the docked TM-4 spacecraft approaching its prescribed storage safety limit, the Okeans may have had to evacuate the station and return to Earth, leaving Mir unoccupied. Fortunately everything went well, and Soyuz TM-5 docked. But the gremlins in the Soyuz TM-5 onboard systems would return three months later, in one of the most dramatic end-of-mission scenarios in the history of manned spaceflight.

Soyuz TM-4 had been in space for nearly six months, and the Soyuz TM-5 crew would therefore return in the older capsule, leaving the fresher craft for the resident crew in case of emergency evacuation. There would be no crew exchange this time. In order to provide a Soyuz TM with a 'good shelf life' for the return of the main crew, a second international mission, planned for three months ahead, would swap out the Soyuz TM-5. As part of the preparations for the return of the Okean crew after a record breaking twelve-month mission, a medical doctor would work with the crew to help condition them during the final months of their flight, as well as perform his own medical experiments on a long spaceflight. The doctor was supposed to stay in space for more than a year, to work with three resident crews (EO-3, EO-4, and EO-5). The Soyuz TM-4 craft, carrying the three returning cosmonauts, undocked from Mir on 10 June. After photodocumentation exercises while flying around the Mir complex, they landed three hours later, without incident.

'A combination of circumstances'

Soyuz TM-6 arrived as scheduled on 31 August after a standard two-day approach for a one-week visiting mission. On board was Commander Vladimir Lyakhov, who became the first cosmonaut to command the three different generations of Soyuz (Soyuz 32, Soyuz T-9, and now Soyuz TM-6) to three different stations (Salyut 6, Salyut 7 and Mir). His most recent training was in the small group trained to fly TM solo for rescue scenarios; and his assignment to this flight was to prove fortuitous. Because of Lyakhov's training, a Flight Engineer was not required, and so that place was filled by Dr Valeri Polyakov, who would remain on Mir to monitor the final weeks of the Okean mission and then remain onboard to oversee the next resident crew's adaptation to spaceflight. Polyakov was hoping to stay on Mir for more than a year, and to return with Soyuz TM-8. The third member of the crew was Afghani cosmonaut Abdul Mohmand, who would return with Lyakhov in Soyuz TM-5 six days later.

Undocking occurred without incident on 5 September, and at the specified time, the OM was jettisoned. The crew then proceeded through the sequence of events leading to retro-fire; but as they did so – just 30 seconds before the planned burn time – the horizon sensors on both the primary and back-up systems were confused by the glare as Soyuz TM-5 passed over the terminator into sunlight. The navigation computer then signalled itself unable to confirm the spacecraft's correct attitude for alignment for the burn. Despite all the upgrades, the Soyuz TM still relied on the old flight requirement to pass into sunlight at least ten minutes prior to retro-fire, to allow the onboard sensors sufficient time to orientate and stabilise the spacecraft; but because Soyuz TM-5 passed across the terminator at dawn, the system apparently became confused. The decision to bring down Soyuz TM-5 in the early morning instead of late afternoon had had a serious impact on the computer programming.

The computer thus issued a command to cancel the burn, but allowed the engine to remain armed. As Soyuz TM-5 passed beyond the range of ground communications, the system apparently cleared the confusion and signalled a 'Go' for the burn. However, the vehicle was by then seven minutes past the original burn-time and a further 3,500 km along its orbital path, which would have resulted in a landing far beyond the planned landing area, and a difficult task for recovery teams to reach the crew quickly. (In March 1965 a similar incident at the end of the Voskhod 2 mission resulted in a landing in the snow-covered forests of Siberia.[4]) After only six seconds, Lyakhov manually terminated the burn, knowing that they would have landed apparently north of Manchuria. Soon afterwards, following release on Moscow morning TV, news of the postponed return of Soyuz TM-5 was flashed around the world. However, as the sensor had worked and the engine had fired, there was no immediate concern over the safety of the cosmonauts.

Two orbits later there was a second attempt at re-entry, and this time the IR sensors were disengaged and the inertial measurement unit was used to control the burn. At the moment of ignition, the spacecraft's autopilot signalled the back-up computer to complete the manoeuvre that it had previously completed during the rendezvous of the spacecraft with Mir in June (a result of the crew forgetting to clear the memory or reset the program). After only six seconds of a planned 230-second burn, the prime computer – realising that it was out of alignment with the back-up computer – promptly shut down the engine again. It was a situation foreseen by the computer programmers, but it was not really expected to arise on a real mission. As soon as it became apparent to the Commander that again his engine was not behaving as planned and that his ground track was fast passing beyond that of the normal recovery footprint, he manually started the engine by pressing a nearby override button on his control panel in front of him, hoping that a landing on the next orbit would ensure a fairly prompt recovery near the original landing zone. However, after a burn of 50 seconds, the onboard navigation systems of Soyuz TM-5 yet again detected incorrect orientation, and automatically shut down the system for the third time. (Another version of these events was later reported by Fyodorov in *Novosti Kosomnavtiki*.[5])

The commands sent up by TsUP for the second attempt did not take into account that the engine had already fired for six seconds. This somehow had the effect of erasing the required 115.2 metres per sec de-orbit impulse from the computer's memory, and replacing it by a 3 metres per sec impulse which was left in the computer's memory from the approach phase. This time the infrared sensor worked normally, but the engine fired for only seven seconds (at 7.35.34). Lyakhov then manually restarted the engine at 7.35.50, but it shut down again after only fourteen seconds. At 7.36.06 he fired it yet again, and this time it shut down after 33 seconds. Overall, therefore, on the second attempt the engine had fired three times with a total burn time of 54 seconds.

Even worse, however (according to the report in *Novosti Kosomnavtiki*[5]), was that ten seconds after the final shut-down, there was activated a timer that would command the Service Module to separate in 20 min 58 sec. Out of range of mission control, Lyakhov tried to shut down the timer, but without success. At 7.57.00 the Soyuz came back within range, and as Lyakhov explained the situation, a 2 min 14 sec sequence to separate the Service Module was initiated at 7.57.47. Lyakhov asked permission to override the timer by giving the command 'ODR' (the meaning of which has not been explained). But there was no reply from the ground, and Lyakhov therefore decided to give the command without permission. If he had not done so, the Service Module would have separated about a minute later, and the crew would have been stranded.

Western analysts later estimated that the second burn had been terminated just seconds before disaster struck. Had the burn continued, the crew would have been sent into a lower orbit, and atmospheric drag would have begun their re-entry sequence. If they had been unable to return to a suitable altitude to attempt a further, correct, burn – in darkness, and with an already confused navigation system – they might have been forced to separate the Descent Module and continue the descent. Their orbit might not have been low enough to decay immediately, and thus they could have re-entered over any part of their orbital ground track. Their re-entry angle might also have been too shallow, which would have subjected the capsule to heating and descent stresses beyond designed limits, thereby burning up the capsule and the two cosmonauts inside.

The onboard computer sub-system that cancelled the burn undoubtedly averted a potentially life-threatening situation, and allowed a 24-hour wave-off, with the craft remaining in a stable orbit. Overnight, the computer data was analysed, and the error with the old rendezvous programme was revealed. This indicated a software problem, and not a problem with the engines – which relieved the tension at mission control, if not among the media. However, despite having sufficient propellant for an engine burn to take them back to Mir, having jettisoned the OM (and with it the Kurs equipment and docking hardware), such a flight would make manual rendezvous with the station difficult, docking impossible, and a contingency EVA transfer using the Sokol suits between two undocked spacecraft extremely hazardous. Monitoring the situation from onboard Mir, the Okean crew showed concern but could not contribute anything. In theory, Soyuz TM-6 could have flown over to Soyuz TM-5, but without adequate docking equipment or suitable EVA transfer equipment, a crew transfer would have been highly improbable. Again, suggestions were offered about creating unified docking equipment, emergency EVA suits, and contingency plans with the Americans in the event of future stranded crews – although none of these would have helped the crew of Soyuz TM-5. (During the early 1990s these ideas would form part of the discussions leading to Russia's involvement in the ISS programme.)

All that the Soyuz TM-5 crew could do was wait until the next day. As they continued their unplanned circuits around the Earth, the media began to highlight their situation in space – and even counted down the number of hours left until their supplies ran out. The two cosmonauts were flying the Soyuz vehicle that had taken the Bulgarian cosmonaut to Mir in June, and this ensured that links were made to the 1979 Soyuz 33 Bulgarian flight to Salyut 6, which had to be aborted during the docking approach due to a malfunction in the Soyuz PM engine. Coincidentally, Lyakhov was onboard Salyut during that attempt, and four years later he was on Salyut 7 when a Soyuz visiting crew (T-10-1) had to abort on the pad as their launch vehicle exploded. And now he was in the middle of his own 'off-nominal situation'. This was a remarkable series of incidents linked with one cosmonaut.

Inside the cramped Descent Module, the crew settled down for an extra night in space. With Polyakov remaining onboard Mir they had a little extra room in the DM, but at 10° C, the inside of the module was decidedly chilly – andthey could detect it, despite wearing their Sokol suits. The crew had been told to remove the suits, but they wanted to leave them on to save having to don them again in the confines of the Soyuz. It would in any case have been difficult to take of their suits in the DM, and would also have consumed more precious oxygen, and they therefore remained suited. The only manoeuvre which they attempted was to orientate their craft – to allowing the solar panels attached to the Propulsion Module to face the Sun and recharge the batteries – and then to go into a slow roll for stability and to maintain an even temperatur in the onboard systems. The configuration of their Soyuz (DM/PM only) had never been man-rated for orbital flight, as it was intended for use only during the final orbits prior to descent and landing.

Having discarded the Orbital Module carrying the supplies for independent orbital flight, the crew was forced to remain in the confines of the DM. Lyakhov elected not to touch the emergency rations (three days of basic survival food per man), and so they had no warm food. They also regretted the loss of the discarded OM's toilet facilities, and as makeshift toilet bags they used the plastic sleeves of Mohmand's sleeping bag, which he had brought with him from Mir.

After the engine burns, the orbit of Soyuz TM-5 had dropped from 364 × 338 km. Orbital mechanics decreed that they would re-enter, but with no engine burn this would take a few weeks, by which time the oxygen would have long been depleted. The crew requested mission control to play some music while Radio Moscow reassured listeners that contact with the crew was being maintained, that mission control was carefully evaluating all options, and that a new attempt would take place the next morning. There were emergency supplies onboard for two days, and contingency landing sites existed for every orbit, although most of these were outside the Soviet Union or in an ocean, which was not ideal for the Soviets.

Next morning, back in communications range, the revised computer programme was read up to Lyakhov, who then entered it manually into the computer (a procedure that the Americans had used several times during the aborted lunar mission of Apollo 13 in 1970). The orbit had decayed slightly as a result of the various aborted manoeuvres, but by this time using the back-up engine as a precaution, the alignment was perfect. Retro-fire took place without incident, and the crew landed in the primary recovery site a day and a few minutes after the planned time.

During the post-flight press conferences, Lyakhov explained that when the engine cut off for the second time, he was tempted to command manual descent. He therefore started the engine himself, although he was fully aware of the possible consequences. When the computer shut down the command automatically, he freely admitted that he was in error in trying to manually override the computer signal. But at the time – with split-second decisions required – the desire to land was at the forefront of his mind, especially after two cancellations of the engine burn. Despite the Western media's sensationalist headlines, the Soviets calmly explained away the events by stating that Lyakhov had taken timely actions that had prevented further serious trouble that might have occurred in orbit. 'Lyakhov shut down the programme, and thus averted possibly very serious trouble in which the crew might have found themselves in orbit.' During interviews, several former cosmonauts stated that there were several hours of worry on the ground, but they knew that their colleagues were safe in orbit for at least a day, giving the ground time to assess the situation. Lyakhov also suggested that in future the OM not be discarded until after retro-fire, even though this would be costly on propellant. Other comments indicated a lapse in attention to detail, in what Deputy Flight Director Viktor Blagov described as a 'combination of coincidences' that all contributed to the first incident in which a crew could have been stranded in orbit.

Mohmand later explained that although it was a dangerous situation, they were more tense than afraid. The events of the Soyuz TM-5 mission once again demonstrated the hazardous nature of spaceflight, and showed that any mission is far from routine.

Mir temporarily vacated

With the conclusion of the drama of returning the Soyuz TM-5 crew to Earth, routine activities onboard Mir continued. On 8 September, all three Okean crew-members entered Soyuz TM-6 to reposition it from the aft to the forward port on Mir. During the EVA by Titov and Manarov on 20 October, Polyakov remained in the Soyuz so that he was secure in the event of being unable to repressurise the docking adapter during post-EVA activities. The EVA crew would have had to use the Soyuz OM as an airlock, and then vacate the station. This became a standard operating procedure employed by a three-person crew conducting a two-person EVA.

On 28 November 1988, Soyuz TM-7 docked with Mir, bringing with it the EO-4 crew of Alexandr Volkov and Sergei Krikalev (call-sign, Donbass), who would take over from Titov and Manarov and, together with Polyakov, continue the occupation of Mir for the next five months. Also on board was the French cosmonaut Jean-Loup Chrétien, on his second Soviet flight. He would perform an EVA with Volkov, and then accompany the Okean crew back to Earth a month later.

On 21 December 1988 – exactly one year after their launch – Titov, Manarov and visitor Chrétien undocked Soyuz TM-6 from Mir. Using revised flight plans as a result of the Soyuz TM-5 entry, they retained the OM until after retro-fire. Although the onboard reserves of propellant were sufficient for the additional 10% consumption, the crew had to update the computer programme in flight, which proved more complex than first thought. The new programme was not compatible with the software already loaded on to the computer, and shortly prior to the planned ignition of the de-orbit engine, the computer detected an overload and cancelled the operation. Using the back-up procedure, an alternate programme was initiated for the next landing opportunity three hours later, and this time all went well, with a 270-second burn. On, landing the capsule remained upright. On 22 December the Donbass crew relocated Soyuz TM-7 to the front port of Mir. When this crew was in training, they had expected to receive the first of the larger modules at Mir during their residency, but in February 1989 it was announced that the D-module (Kvant 2) would be delayed because the next module (T, or Kristall) was slipping in production. After the accident with Soyuz TM vehicle no.59, and as well as not wishing to operate Mir in an asymmetrical configuration with only one of the lateral ports occupied, it was decided that the Donbass crew would return to Earth, and that after the end of their stay in April 1989, Mir would be mothballed for several months.

On 26 April 1989, Soyuz TM-7 left Mir unoccupied for the first extended period since February 1987. During the landing, Krikalev's knee struck one of the display panels; and although it did not cause a serious injury, he still needed minor attention after landing.

THE END OF THE SOVIET UNION: SOYUZ TM-8–TM-13

During the three years between the return of the Donbass crew and the spring of 1992, a further six Soyuz TMs – each of them featuring a new resident crew, and three of them carrying international visiting cosmonauts – were launched to Mir,. It was also a period during which the Soviet Union was in a state of change, ultimately leading to its collapse and the formation of a number of independent states. The tremendous upheavals across the former Soviet Union would also have significant influences on the structure, direction and funding of future space exploration programmes. During this period, funds that had been allocated to the Mir programme were not forthcoming, leading to delays and cancellations. With the new commercial approach of selling mission seats, the shortage of cash was obvious.

In other parts of the space programme, launches were being slowed as production lines declined and eventually closed. These changes were most obvious in the flight of Krikalev. He was launched on a Soviet Soyuz from Baikonur in the Soviet Union, and when he ended his mission – which had been extended – he landed in Kazakhstan, onboard a spacecraft of the Russian Federation. The Western press afterwards described him as 'the last citizen of the Soviet Union.

The effects of this period would have a far-reaching impact on long-term Russian space operations.

Soyuz TM crewing, 1989–91

In 1989 – when Mir operations were due to be restarted – four long-duration crews were in training for the next series of missions. They were, in order of flight, Viktorenko and Serebrov, backed up by A. Solovyov and Balandin, who would fly next; and the two support crews of Manakov and Strekalov, and Afanasyev and Sevastyanov. Their mission profiles demanded a crew rotation, with one crew following another. No third seat was to be used, and so Soyuz TM-8, TM-9 and TM-10 flew with a core team of two. The only change in planning occurred when problems with Soyuz TM-9 arose and Berezovoi began training to fly a rescue Soyuz in case the problem continued.

In June 1990, Sevastyanov failed a medical examination and was replaced by Manarov. The next three expedition crews were therefore Afanasyev and Manarov on Soyuz TM-11, Artsebarsky and Krikalev on Soyuz TM-12, and A. Volkov and Kaleri on Soyuz TM-13. Due to the new commercial reality, and to raise much-needed funds, the Soviets had planned to use the third seat on these missions by selling them to a Japanese TV company and two private companies based in Britain and Austria.

Soyuz TM-13 was due to be launched in 1991, and mission planners decided to fly a visiting mission (Soyuz TM-14) to Mir, with the intention of flying a Kazakh citizen, as no ethnic Kazakh had flown in space. The crews were formed, with the prime crew being Korzun, Aleksandrov and Aubakirov, backed up by Tsibliyev, Laveikin and Musabayev. Aubakirov and Musabayev were ethnic Kazakhs and pilots. The reality of economics then hit the programme, and the mission was merged with the expedition crew due to fly on Soyuz TM-13. A. Volkov remained the Commander, and would fly the mission in Soyuz rescue mode. His crew-men would be the Kazakh Aubakirov and the Austrian Viehbock, who was originally intended to fly with the Volkov/Kaleri long-duration crew on Soyuz TM-13. Kaleri was moved to the next expedition crew. It was decided that Krikalev would stay on Mir and be joined by Volkov, while Artsebarsky would return with the other members of the Soyuz TM-13 crew. Aubakirov would be Pilot-Cosmonaut of the USSR – the last and seventy-second person to receive this award. The next Soyuz to fly would be a Russian spacecraft containing the first Russian cosmonaut, Aleksandr Kaleri.

A NEW ERA DAWNS

The reoccupation of Mir

After five months with no manned operations, the Soviets ended this period of inactivity by launching Soyuz TM-8 on 6 September 1989, to re-occupy the Mir complex to receive the first of four large scientific modules (Kvant 2). These would eventually increase the mass of Mir to approximately 140 tonnes, making it difficult to manoeuvre, even with full power. The new operational parameters of Soyuz TM – with its ability to launch three-person crews to higher orbits than could its predecessors – became a valuable tool in Mir operations. During this time a problem appeared in the microchips supplied by PO Elektronika in Voronezh, and used in

Soyuz TM and Progress M. In ground tests, four failures were attributed to corrosion in chips from the same 1985 batch as used on Kvant 2. This would account for the delays in spacecraft testing during 1986–87, and probably also early ground testing with Soyuz TM-8.

The Commander of Soyuz TM-8 was Alexandr Viktorenko, with Alexandr Serebrov as his Flight Engineer (call-sign, Vityaz). On 7 September they were approaching the Mir complex when, at a distance of only 4 metres from the rear port (the new Progress M1 freighter had been docked to the forward port since 25 August 1989), they noticed the complex suddenly oscillate. This caused the Kurs system to abort the approach, but Viktorenko – who was carefully monitoring the sequence of events – quickly assumed manual control and pulled Soyuz TM-8 back to the 20-metre point to assess the situation. After a visual inspection he closed in for a second attempt and docked without incident – much to the relief of both the cosmonauts and the ground controllers. One hour later the hatches were opened, and for the fourth time, cosmonauts occupied the space station. Although permanent occupation was required, in 1991 Mir had an expected lifetime of only about five years. No-one could have predicted that successive teams of cosmonauts would remain onboard the space station until August 1999.

On 1 December 1989, Progress M1 undocked, freeing the forward port for the arrival of the Kvant 2 expansion module (launched on 26 November) on 6 December. Two days later it was manoeuvred from the forward port to the upper lateral docking port by the Lappa manipulator arm located on the module. Kvant 2 included a larger airlock for EVAs, and the SPK (Sredstvo Peredvizheniya Kosmonavta – Cosmonaut Manoeuvring Unit), to be tested by the Soyuz TM-8 cosmonauts. On 12 September the Soyuz was relocated to the front port.

'Like petals of a flower'

Soyuz TM-9 was launched on 11 February 1990, with the EO-6 two-man crew (Commander, A. Solovyov; Flight Engineer, A. Balandin; call-sign, Rodnik) intended to take over the operation of Mir from the Vityaz crew. As the Soyuz TM-9 cosmonauts configured their spacecraft for orbital flight shortly after separating from the upper stage of their booster, they realised that something was attached to the outside of their DM, as the view out of the Vzor optical sight was partially obscured. To the crew inside the DM, with their restricted view out of the portholes, it appeared that some of the thermal blankets covering the DM had worked loose. However, this did not prevent the rendezvous with Mir, and two days later, as the spacecraft approached the station at the aft docking port, Kvant 1's camera picked up several loose objects on one side of the Soyuz as it closed in. As the exact nature of the problem could not be determined, it was decided to allow the departing Soyuz TM-8 cosmonauts to visually inspect and photograph the newly-arrived TM in a fly-around, prior to their departure and return to Earth.

On 19 February the Vityaz crew undocked and moved around to the aft port, reporting that three of the six thermal blankets were indeed loose at the heat shield and were waving about as the thruster plumes of Soyuz TM-8 'blew' against them. One projected at 90°, and the others at about 60°, and to the Soyuz TM-8 crew they looked 'like petals of a flower'. The damage probably occurred during the ascent as the aerodynamic launch shroud separated at 165 seconds into the flight at a height of 80 km. With the fly-around completed, the Vityaz crew separated from Mir to begin the preparations for return to Earth. As they did so, they were ordered to postpone the descent for 24 hours due to high winds and a temperature of –30° C at the landing site. But fortunately, before they began powering down the Soyuz they were told that there was a small break in the weather that would allow them to land on the next pass; and they did so, without further incident.

From the photographs of the Soyuz TM-9 thermal blankets taken by the Soyuz TM-8 crew, and from their visual observations, it was determined that the Rodnik crew might be able to reattach them during an EVA, so that they did not block the infrared horizon sensors during orientation for the return to Earth. If this was not possible, the crew would cut them off. For this task, specialist tools would need to be developed to achieve the unplanned EVA towards the end of the mission, and these would arrive on the Kristall technological module planned for launch in mid-April. Meanwhile, in conjunction with TsUP, the cosmonauts had determined procedures in which the temperature onboard Soyuz TM-9 would be monitored and the Mir complex manoeuvred, in order to place the Soyuz in or out of sunlight to maintain the correct temperature for the internal systems. Viktor Blagov, Deputy Flight Director for Mir stationed at Mission Control, explained that as well as keeping the surfaces of the Soyuz at an even temperature so that it did not roast in the sunlight (130° C) sunlight or freeze in the shadows (–130° C), the inside of the DM needed to be kept warm so that condensation would not form inside and threaten the integrity of the circuitry (again, an incident reminiscent of Apollo 13).

On 21 February the Rodniks successfully flew Soyuz TM-9 from the rear to the front port without incurring any further damage. Meanwhile, at TsPK, cosmonauts entered the hydrotank to practice that EVA techniques which they were developing for the Rodnik crew to follow in space. At the same time, consideration was being given to launching a solo rescue cosmonaut (probably Berezovoi) to Mir to return the resident crew. On 20 April the Soviets revealed that the Kristall module would not be launched until early June (although it had been scheduled for docking on 7 April). By 28 May – after two further Progress spacecraft had resupplied Mir – Soyuz TM-9 was moved back to the rear port (again without further damage) in order to free the forward port for Kristall, which was launched three days later. Kristall docked with the forward port on 10 June (after an aborted attempt on 6 June), and the following day was moved to the nadir lateral port by its Lappa robotic arm, to create a T-shaped station with the Kvant 2 and Kristall, the base block/Kvant, and Soyuz TM-9. With the delay of Kristall's arrival and the impending EVA, the crew requested (and were granted) a ten-day extension to their mission.

The EVA to repair the Soyuz TM

Onboard the new module was the special equipment needed to support the repair on the Soyuz TM. Since the end of May, Western media reports had again grossly exaggerated the plight of the crew by once more suggesting that the cosmonauts were stranded in space, with little hope of rescue. The Russians, of course, quickly rejected these reports, and stated that Soyuz could be manually orientated if required, that a second Soyuz could be launched to replace the Soyuz TM-9, and that the blankets would normally burn off during re-entry.

Onboard Mir, the crew prepared for their EVA repair by first moving Soyuz TM-9 back to the front port on 4 July – not to free the aft port for a Progress docking, but to allow them to work through the problem more conveniently. Once again, no additional problems were encountered in this short flight. In planning for the EVA, two options were open for the crew to access the Soyuz. With two of the four radial ports occupied by Kvant and Kristall, it would have been possible to exit Mir via one of the two unoccupied hatches, and to progress the short distance across to the Soyuz. However, in this option, though much shorter in duration, the EVA path would take them across the Soyuz OM to the area in which they needed to work. Here there were no hand or footholds, and it was feared that the crew could inadvertently damage some of the exterior appendages on the Soyuz. The second option was therefore chosen: to exit via the Kvant 2 airlock. After relocating back down to the base of the module, the crew would erect an extension ladder at an angle that would take them over the OM directly to the DM, and from there detach or secure the loose blankets.

The crew had not trained to perform this specific EVA on this mission, but they would have received general EVA training as part of their mission preparations. Neither of them had ever conducted an EVA, and so they watched video uplinks of the EVA simulations in the hydro lab. On 17 July they were ready to exit, but, partly as a result of their lack of training and experience, they opened the outer hatch of Kvant too early. The remaining 0.05% of atmosphere imparted a force of 400 kgf on the door, which forced it to fly open and back against its hinges, startling the two cosmonauts for a moment. The implications of this error were not immediately apparent.

The traverse down the 13.73-m length of Kvant 2, with the folded ladders and tool-set, took much longer than expected (about 90 minutes), and it took another three hours to install the 7-metre straight ladder to bypass the OM and the smaller curved ladder to reach the area between the heat shield and explosive bolts where the thermal blankets had separated. Because of the distance that had to be traversed, using umbilicals to feed consumables from inside Kvant was impractical, and relyiance on internal suit supplies placed a safety limit on the length of the EVA. In addition, TV cables would not stretch as far as the work station, and ground control had to rely on verbal commentary only (although the cosmonauts videotaped the area in close-up, for later playback to TsUP).

In their inspection, the cosmonauts found no obvious damage to either the explosive bolts used to separate the PM from the DM or to the heat shield, nor any other obvious damage as a result of launch-shroud separation or prolonged exposure to the harsh environment in orbit. The mission plan was to have the two cosmonauts re-attach the blankets with spare pins, but this failed when they found that they had shrunk and that they were unable to secure them to the locking ring. The alternative was to fold up the blankets and secure them to the OM, at the same time keeping them away from the sensors' field of view during orientation for descent. This they managed to do with two of the three blankets, but the third (the most severely torn) proved to be 'disobedient', and they therefore tied it back as much as possible. Fortunately, it was the furthest from the sensors, and was not expected to cause an obstruction.

At the end of this task, the EVA had taken about six hours – approaching the safety limit of their suits. Unlike the excursion to the work site, which had included rests during night passes, this time they had to return to the airlock as quickly as possible. Leaving their tools and ladders *in situ*, and traversing during the night passes by means of helmet lamps, they made the trip back to the Kvant 2 hatch. It was here, as they reached six hours of suit time, that they discovered that the unexpected force of opening the hatch had buckled it, and it refused to close the last few millimetres to create the seal needed in order to depressurise the compartment. Leaving it ajar, they transferred into the middle compartment of Kvant 2, to use it as an emergency airlock to regain entry into the main part of the station. At the point at which they could open their visors and breathe station air, they had remained reliant on the suit systems for seven hours – which was very close to the design limit.

Bringing home Soyuz TM-9

The video playback convinced controllers that Soyuz TM-9 would be safe enough for the cosmonauts' return; but before they could undock, the ladders and tools had to be removed, and the problem with the Kvant 2 hatch had to be investigated. On 25 July they performed a second EVA – lasting 3 hrs 31 min – during which they removed the items from the TM and stowed them on the hull of the Kvant 2 module. But the hatch proved stubborn, and had to be forced shut rather than repaired (a task completed by a later crew).

With the repair to Soyuz TM-9 completed, the work site cleared, and the Kvant hatch secured, the crew could finally begin winding up their mission and prepare to hand over station operations to the Soyuz TM-10 crew (Gennedi Manakov and Gennedi Strekalov; call-sign, Vulkan), who, 1 August 1990, were launched together with four live Japanese quails for the Inkubator 2 experiment on Mir. On 3 August they docked at the rear port, without incident. *En route* to the station, one of the quails had laid an egg, and this was returned to Earth on the Soyuz TM-9 spacecraft. Soyuz TM-9, with the Rodniks onboard, was undocked from Mir on 9 August. Normally, the re-entry sequence is preceded by the jettisoning of the Orbital Module and then the PM, but on this occasion it was decided to simultaneously jettison the OM and the PM to prevent the loose blankets from fouling the Descent Module during orientation for re-entry. Re-entry and landing was accomplished without incident, but because of the added EVAs at the end of their residency, the crew had interrupted their regular exercise programme, which caused slight discomfort for both cosmonauts after they had landed on Earth.

International visitors

The seventh principle expedition had a relatively quiet occupancy of Mir, but there was a day of emergency situation training (as with all station missions) on 30 August, simulating situations that might arise on the growing complex – such as a collision, a fire, or a loss of pressure that would require emergency evacuation of the station using the Soyuz TM. In 1997 the value of these drills became evident with the fire and collision onboard Mir (see p.355) and in the decision to select the Soyuz as a crew rescue vehicle on the ISS. The residency also featured the first use of the Progress-delivered Raduga return capsule (see p.251).

The next residency crew – Viktor Afanasyev and Musa Manarov; call-sign Derbent – was launched on Soyuz TM-11 on 2 December, and two days later docked to Mir to take over from the Vulkan crew. Also onboard Soyuz TM-11 was Japanese TV journalist Toyohiro Akiyama, who had spent most of the two-day approach to the Mir strapped into his couch enduring a prolonged bout of space adaptation syndrome (space sickness). Throughout his week onboard he never overcame the effects (which normally takes two or three days), and returned to Earth with Manakov and Strekalov onboard Soyuz TM-10 on 10 December. They landed safely – much to the relief of Akiyama who, as perhaps the most unwilling space explorer, was glad to be back on Earth.

The following year a further potential problem with Progress M equipment surfaced. On 21 March 1991, Progress M-7 broke off its approach to the station just 500 m from the aft docking port. A second approach on 23 March was also terminated, this time just 20 metres from the rear port. The Progress passed only 5–7 metres from the station – just missing the left solar array and antenna as it glided by. Later diagnosis of the Kurs system failed to reveal any problem, but something was clearly not operating correctly – although whether it was the Progress equipment or that on Mir was not evident. The only way to find out was to test the Soyuz TM-11 docking system at the rear port.

On 26 March the crew undocked the TM ferry from the forward port, transferred it to the rear, and used the Kurs system to draw it towards the Kvant docking port. The spacecraft duplicated the profile of Progress M7 a few days earlier, revealing that it was the Kvant equipment that was at fault. Realising that a manual docking was called for, the crew safely brought the TM to the docking port; and two day slater, Progress M7 was brought to an automated-approach docking at the front port. It was suspected that the rear Kurs antenna had been inadvertently knocked or damaged during one of the earlier EVAs, and on 25 April this was confirmed when the two cosmonauts performed their fourth EVA and revealed that one of the system dishes was missing. A repair was added to the next residency crew's tasks, but in the meantime, resupply craft would continue to use the forward port of the station.

On 18 May, the next residency crew (Artsebarsky and Krikalev; call-sign, Ozon) was launched onboard Soyuz TM-12, and docked with Mir two days later. This time, the Soviets were joined by British cosmonaut–researcher Helen Sharman, flying under the UK–USSR Juno programme. At 200 metres from Mir, the crew found that one of the exterior antennae on their spacecraft was producing erroneous readings and was not providing the crew with relative motion data. Artsebarsky therefore assumed manual control, and docked with Mir's forward port. During the approach, Sharman handled the operation of the TV cameras (wide- and narrow-angle lenses), viewing the approaching Mir from her location in the DM. Krikalev was looking out of the forward window in the OM.

After a week on the station, Sharman returned to Earth with the Derbent crew, to undock from Mir onboard Soyuz TM-11 on 26 May. As Afanasyev backed away from Mir, Manarov tried to photograph the separation and the condition of Mir. Wedged above his seat near the small right-hand window in the DM, he struggled to use the camera while wearing his Sokol suit with an open visor and unpressurised gloves – a difficult operation that took longer than planned.

Re-entry and the early parachute descent took place without incident, but as it was a windy day the capsule moved laterally towards the ground. On 'dust-down' it tipped over and then rolled several more times, and the occupants, who were spun with it in their seats, lost all sense of direction. Though the arms, legs and body are restrained quite well, the head in the helmet is not, and Sharman received some minor bruising to her face from the helmet microphone. Soyuz TM-12 demonstrated once again that the so-called 'soft landing' at the end of the mission is anything *but* that!

With Soyuz TM-12 docked at the front port, and the Kurs system inoperable at the rear, the TM had to be relocated to free the front port for Progress resupply missions until a replacement antenna could be installed by EVA. On 28 May, Soyuz TM-12 was relocated from the front port and manually docked to the rear. The EVA to install a new dish to the Kurs, and to restore the aft docking port for freighters, was performed on 24 June.

Soyuz TM-13 was the first mission to be launched after the failed coup in the Soviet Union on 21 August. The mission was commanded by Alexandr Volkov, who would replace Artsebarsky on Mir and form the Donbass crew with Krikalev. Volkov flew the Soyuz in the solo rescue cosmonaut mode, without a Flight Engineer. Flying to Mir with Volkov, but returning with Artsebarsky, were cosmonaut–researchers Toktar Aubakirov from Kazakhstan, and Franz Viehbock from Austria – a combination of two planned commercial flights and one week-long mission. On 10 October the trio undocked Soyuz TM-12 from Mir, for an uneventful recovery. Five days later, the resident crew relocated Soyuz TM-13 from the front port to the rear port, in a final test of Kurs. The crew performed several successful approaches before finally docking to the Kvant port. The second docking location on Mir was back in operation.

RUSSIA IN SPACE

Soyuz TM-13 would be the final Soviet manned space mission before the dissolution of the USSR at the end of the year, and the last of a series that had begun with Yuri Gagarin 30 years and 6 months earlier. Even with two cosmonauts still on board Mir, the future of the Russian space programme – both operationally and financially – remained in doubt as the Commonwealth of Independent States (CIS) replaced the USSR. As a new era dawned, the next few years would be pivotal in both the Russian and global human exploration of space.

Soyuz TM crewing, 1990–94

The original idea was for a Soyuz to be on stand-by on the pad during the manned test flights of the Buran shuttle, and, if necessary, to rescue the Buran crew. The Soyuz would be launched with a single 'rescue pilot' on board, dock with Buran's APAS-89 docking port, and bring the two Buran pilots back to Earth. In 1989, Vladimir Titov and Anatoly Berezovoi began preparing for such a rescue mission. Berezovoi had been part of the Soyuz rescue team since 1985, and Titov had replaced Malyshev in July 1988 (when Malyshev left the cosmonaut team). Now Titov and Berezovoi would specialise in rescue missions to Buran, while Lyakhov remained available for rescue missions to Mir.

However, all these plans changed as the first manned flights of Buran continued to slip. In 1990, Berezovoi and Titov rejoined Lyakhov for Mir rescue training, and Afanasyev joined the Mir rescue team in 1991, but left in 1992 (as did Berezovoi and Titov). From 1992–94, Volkov and Korzun also performed Mir rescue training (alongside Lyakhov), but in 1994 the team was disbanded.

Because of the delays in the manned Buran flights, it was decided in 1990 to fly a Soyuz to dock with Buran in Earth orbit. The flight plan was for Buran 2 (fitted with a docking module in its cargo hold) to launch unmanned and dock with Mir's Kristall module. The Mir crew would board Buran 2 to perform some tests of its remote manipulator arm, and then, after one day at Mir, Buran 2 would undock (unmanned). Next, the rescue Soyuz would be launched to dock with the unmanned Buran, and the Soyuz crew would perform one day of tests. After undocking, Buran 2 would return to Earth unmanned, and the Soyuz would fly on to dock with Mir's Kristall module.

A total of three Soyuz ships with APAS-89 docking ports were initially ordered for rescue missions to Buran. These had serial numbers 101, 102 and 103. No.101 eventually flew as Soyuz TM-16, and elements of the other two vehicles were incorporated into subsequent Soyuz TM vehicles (with different serial numbers). The training group was established in November 1990, with pilots (Bachurin, Borodai and Kadenyuk) who were all Air Force test pilots attached to the Buran flight test

A group photograph of the specialists working with the Soyuz TM-15 crew to prepare them for the mission.

A rare distance photograph of docking operations at Mir, taken from Soyuz TM-17 in July 1993. It shows Progress M-18 undocking from the forward port of Mir, allowing the TM-17 craft to dock at the station. At the lower docking port on Kristal is the Soyuz TM-16 spacecraft, and at the aft port, Progress M-17 docked to the Kvant astrophysics module. (Courtesy ESA.)

programme, and engineers (Stepanov, Illarionov and Fefelov) who were all Air Force cosmonauts based at TsPK. Some civilian engineers – including Ivanchenkov and Laveikin – also worked on the programme. When it became clear that the Buran would not fly, due to the cost of launching it, the group was disbanded in 1992. In 1993 the Soyuz flew as TM-16, with an expedition crew to Mir.

By 1992, three crews were in training for Mir expedition flights: Viktorenko and Kaleri (Soyuz TM-14), A. Solovyov and Avdeyev (Soyuz TM-15), Manakov and Poleshchuk (Soyuz TM-16), and lastly Tsibliyev and Balandin, who were paired in May 1992. Immediately prior to the Soyuz TM-16 mission, a number of experienced cosmonauts failed their medical examinations, which led to some expedition crew changes. Kaleri became the first person to be awarded the title Pilot Cosmonaut of the Russian Federation.

Major changes were also incurred on the ground. Shatalov was removed as Head of the Cosmonaut Training Centre, to be replaced by Major General Klimuk, another space veteran. The Soviet authorities, and then the Russians, signed commercial agreements with the national space agencies of Germany and France for some short-duration missions to Mir, and the European Space Agency also signed up for two missions, one of which would be of long duration. These agencies would also provide a range of experiments for the station, and this led to a number of Europeans training at Star City. The Germans flew a short mission on Soyuz TM-14; French spationauts flew on Soyuz TM-15 and Soyuz TM-17; and the European Space Agency flew a short mission on Soyuz TM-20.

The Russians flew the next four expedition crews announced following the changes of October 1992, although Soyuz TM-17 was crewed by Tsibliyev and Serebrov. Soyuz TM-18 was crewed by Afanasyev and Usachev, and Soyuz TM-19 by Malenchenko and Musabayev – the first Russian crew, since 1977, that consisted of rookies. The flight of Soyuz TM-20 marked another major step with the long-duration flight of Yelena Kondakova – the third flight by a Russian woman. Her commander was Viktorenko.

Onboard Soyuz TM-18 was the doctor–researcher Valeri Polyakov, making his second flight. This time the aim was for him to break the duration record with a flight that would last 437 days. He returned onboard Soyuz TM-20, with Kondakova, who had also just set a duration record for a woman. Polyakov began training for this mission in January 1993, and was supported by Arzamazov and Morukov; but when the actual crews were formed in the summer of 1993, Morukov was not included. Polyakov was named prime, with Arzamazov as his back-up. Then, in December 1993, Arzamazov was also removed from the back-up crew in a disciplinary move (he had written a letter to the State Commission claiming that he was better prepared than Polyakov).

The next flights would mark the beginning of American missions to and on Mir.

Russian international missions

The first manned missions under the new Russian space programme began on 17 March 1992 with the launch of Soyuz TM-14, commanded by Aleksandr Viktorenko, with Alexandr Kaleri (the eleventh primary crew; call-sign, Vityaz) and the German cosmonaut Klaus Flade (who would spend a week on board Mir and return with the Donbass crew). The Soyuz TM-13 spacecraft was moved from the aft port to the forward port to allow Soyuz TM-14 to dock at the rear of the station. The 19 March approach and docking of TM-14 was achieved automatically to further qualify the repaired Kurs antenna at that location, especially from long range. A week later, on 25 March, Volkov, Krikalev and Flade undocked Soyuz TM-13 from the front of Mir, and began their preparations for descent. Krikalev had launched on Soyuz TM-12 as a Soviet citizen, but returned on Soyuz TM-13 as a Russian. He was the second cosmonaut to land with colleagues with whom he had not trained (the first being Romanenko in 1987), demonstrating the flexibility of crewing within the core training group.

The next mission – Soyuz TM-15 – was launched on 27 July, with the twelfth resident crew (A. Solovyov and Sergei Avdeyev; call-sign, Rodnik) and French cosmonaut Michel Tognini, who would fly a twelve-day (Antares) mission to the station before returning with the eleventh resident crew. They approached the station from the front, but once again the Kurs system failed during final approach, forcing Solovyov to assume manual control for docking. In the reports, nothing further was mentioned about the cause of the failure. Tognini and the Vityaz crew undocked the

Soyuz TM-14 from the Kvant aft docking port, and landed on 10 August, without incident.

By 1992 a regular pattern of Soyuz TM flights was emerging under the new Russian space programme. Two-man rotational resident crews would fly to the station, with the third seat occupied by an international representative (on a short commercial mission) who would return with the former resident crew – usually in the older Soyuz TM vehicle. It seemed that this would continue for the foreseeable future of Mir operations, but with the very next mission the Russians demonstrated that they were still preparing to expand the Mir programme and use Soyuz TM as more than just a crew ferry vehicle.

A new port of call

On 17 June 1992 an historic agreement was reached between US President George Bush Senior and Russian President Boris Yeltsin, initiating cooperative programmes in manned spaceflight. This would lead to flights of Soviet cosmonauts on American Shuttle missions, and residential visits by American astronauts to Mir. There would also be a series of Shuttle/Mir dockings (see p.349) by means of the Androgynous Peripheral Docking Assembly (APAS-89) attached to Kristall. This apparatus – positioned at the aft end of the Kristall module – had been installed for dockings by the Soviet Buran shuttle.

Although the idea of a Buran docking with Mir was now abandoned, the agreement with the Americans gave new life to the APAS docking port, and Soyuz TM-16 was assigned to test this facility prior to the first docking of an American Shuttle in 1995. The next stage of preparations for this came on 15 September 1992, during the fourth EVA of the twelfth resident crew. The cosmonauts relocated the Kurs antenna from the front of Kristall (where it had been used to dock with the Mir in June 1990) to the rear of the module near the androgynous docking unit, to facilitate approach and docking by Soyuz TM-16.

When Soyuz TM-16 was launched on 26 January 1993, the spacecraft carried androgynous docking apparatus for the first time on a Soyuz since the ASTP (Soyuz 19) in July 1975. Onboard the spacecraft was the thirteenth resident crew: Gennedi Manakov and Alexandr Poleshchuk (call-sign, Elbrus). This was the first two-man crew since August 1990, with the planned Israeli commercial mission being cancelled, and only the fifth two-man TM launch in the fifteen launches. Manakov and Poleshchuk approached to within 150 metres of Mir, and then disengaged the automated rendezvous system. Soyuz TM-16 then closed to within 70 metres of the Kristall docking port and, following a final test of systems, the alignment and a visual check of the port, the cosmonauts were given the go-ahead for docking. Soyuz TM-16 slipped into the new mooring on the Kristall module without problems. The successful demonstration of the APAS facility on Kristall paved the way in the planning of the American Shuttle missions, and also provided additional data on docking a spacecraft off the longitudinal axis of a target spacecraft.

This procedure had also been planned for Skylab CSM rescue flights in the 1970s (although it had not been executed), during which a two-man CSM would dock at the radial second port on the multiple docking adapter, to rescue a stranded Skylab crew in the Orbital Workshop if a disabled CSM on the longitudinal port could not be undocked. The Skylab scenario would have seen three spacecraft link up; but with the Soyuz TM-16 docking, seven spacecraft – Soyuz TM-16, Soyuz TM-15, the Mir core module, Kvant, Kvant 2, Kristall and Progress M-15 – were linked for the first time, creating a space complex with a mass of 100 tonnes. Part of the test of the structural integrity of this combination was carried out on 28 January when a Rezonans (Resonance) experiment evaluated the dynamics and structural characteristics across the complex. The first docking in this configuration was offset by the different masses of the Soyuz TM and the much larger Mir complex, and so there was little danger of the new TM spacecraft sending the Mir into an unplanned rotation. Despite demonstrating a new capability of the Soyuz at a station, the Russians were not planning to capitalise on this success at the time by sending APAS-configured Soyuz spacecraft to Mir. This was the only flight by a Soyuz to the Kristall location, which would be the primary location for docking the American Shuttle. However, discussions were already in progress about supplying the TM vehicle with a number of upgrades as a crew rescue vehicle for the American space station Freedom (see p.381).

On 1 February the outgoing Rodnik resident crew handed over operations of Mir to the Elbrus crew, and undocked Soyuz TM-15 from the front of Mir for their return to Earth. In the latter stages of their descent, low clouds hid the landing area, and the DM touched down on a small hillock, causing the spacecraft to roll part-way down and come to rest just 150 metres from a frozen marsh.

Soyuz TM-17 strikes Mir

Despite the appearance of normality in the Russian space programme, there were occasional reminders of the difficulties facing the changing Russia. Immediately before the launch of Soyuz TM-17 on 1 July, with the crew already on board, there was a short power black-out at the pad, and just an hour after launch, the electricity supply to the nearby city of Leninsk failed completely.

When Soyuz TM-17 arrived at Mir on 3 July 1993, all docking ports were occupied. The Soyuz TM-16 spacecraft, with the APAS docking system, was still attached to the Kristall module, and could not dock anywhere else. Progress M-17 was at the rear Kvant 1 port, and Progress M18 was at the forward port. Onboard Soyuz TM-17 was the fourteenth resident crew (Vasily Tsibliyev and Alexandr Serebrov; call-sign, Sirius), accompanied by French cosmonaut Jean-Pierre Haignere, who was on a three-week visit to the station (designated the Altair mission). The Russians took the opportunity for a unique photo-shoot from Soyuz TM-17 from its station-keeping position 200 metres from the complex, as Progress M-18 was undocked from the front port of Kvant and pulled away. Twenty-six minutes later – with the Progress clear of the complex – Tsibliyev brought Soyuz TM-17 in, to dock with the vacant front port.

For three weeks, five cosmonauts (four Russian and one French) performed joint experiments onboard the Mir complex, stretching the capacity of the station in supporting such a large crew over an expanded time-frame. On 22 July, Soyuz TM-16 undocked from Kristall for the return to Earth with the Elbrus crew and

Haignere, leaving control of Mir to the new crew. They were resident there during the annual Perseid meteor shower (which peaks on 12 August), and an unusually high rate was forecast for 1993. NASA delayed the launch of STS-51 for a month, but with the Russian crew already onboard the station, other precautions were adopted. A number of search-and-rescue aircraft were on standby should an emergency evacuation of Mir be required, and at the height of the shower the cosmonauts retreated to the Soyuz TM-17 DM and closed the hatches, in case the station should be damaged. Although the number of hits on Mir was higher than normal, no serious impacts were recorded, although holes were created in the solar arrays, and ten impacts pitted the windows panes – as a result of which an EVA was arranged to inspect the exterior of the station on 28 September.

The replacement crew for Mir (the fifteenth; call-sign, Derbent) was launched on 8 January 1994, and docked with Mir two days later. Onboard Soyuz TM-18 were cosmonauts Viktor Afanasyev, Yuri Usachev and Dr Valeri Polyakov. For the first time since the Soyuz T-14 mission to Salyut 7, a three-person long-duration crew was flown; and for Polyakov it was to be the beginning of a record-breaking 15-month stay onboard Mir. With the hand-over completed by 14 January, Tsibliyev and Serebrov loaded up Soyuz TM-17 for the return to Earth. But this time it would not be a standard withdrawal from the station; instead, Tsibliyev took Soyuz TM-17 towards the Kristall APAS port to obtain photographs (for NASA) of the docking apparatus (including a NASA docking target which had been installed during one of the EVAs) and Mir configuration at that angle, to assist in the training of Shuttle pilots for the forthcoming docking missions.

However, instead of pulling back a reasonable distance before moving toward the docking unit end of Kristall, the plan was for Tsibliyev to move the Soyuz TM back only 45 metres, and to then take the TM along the length of the Kristall module. He began the approach, with Serebrov in the OM ready to take photographs, but found that the spacecraft was not responding as expected. Despite the attempt to avoid a collision, the Soyuz struck two glancing blows on the Mir spherical docking adapter; but fortunately it hit an area where no external appendages were located. Although it seemed to be a large collision it was actually only slight, with the crew inside Mir not feeling the impact (unlike in 1997). The station attitude control system immediately compensated for the unplanned movement as it drifted temporarily, losing the communication link with the Luch satellite and adding to the tension at TsUP in Moscow. It has been reported that Tsibliyev's piloting skills in avoiding the solar arrays on Mir, successfully prevented more serious damage.

After the incident the two cosmonauts checked their spacecraft while the crew onboard Mir inspected the inside of Kristall, reporting no apparent damage. Soyuz TM-17 was also apparently in good order, and the crew landed safely less than four hours later. The incident generated an inquiry, which determined that a switch for an aft thruster was in the standby mode instead of the active mode. It was also concluded that the crew was not at fault. Back on Mir, the resident crew repositioned Soyuz TM-18 from the aft port to the front port on 24 January, flying some 150 metres from the station. In the process they flew past Kristall, and reported only a few scratches on the hull where Soyuz TM-17 had struck.

Shortage of hardware

The problems of supplying hardware to support Mir operations in cash-starved Russia were already becoming evident when, in October 1993, the Russians announced that the launch of Soyuz TM-18 had been delayed until the following January due to the unavailability of the more powerful Soyuz U-2 launcher required to lift the three-man crew. The flight manifest for Soyuz TM-19 originally included a launch on 20 June 1994, but a shortage of payload fairings and delays in accepting the payload fairing for flight on the Soyuz TM-19 mission booster pushed this back to 1 July. The growing cooperation with the Americans was a fortuitous intervention at a time of cash crises in the Russian programme, and the need for Americans to gain much-needed space station experience prior to the ISS.

The launch of Soyuz TM-19 on 1 July 1994 was uneventful, as was the docking with Mir two days later. The two cosmonauts on board – Yuri Malenchenko and Talgat Musabayev (call-sign, Agat) – took over as the sixteenth resident crew, along with Polyakov, while the two members of Derbent crew returned to Earth in Soyuz TM-18 on 9 July. This was an all-rookie crew, which reflected confidence in the Kurs system and in the ability of the crew to perform a manual docking if called upon to do so. It was to be a short mission of four months, to line up the crew rotation with the American Shuttle schedule for returning the first American visitor from Mir.

Soyuz TM-20 was launched on 4 October 1994 – the thirty-seventh anniversary of the flight of Sputnik – and for the second time, Polyakov would greet a new team of resident cosmonauts (Alexander Viktorenko and Yelena Kondakova; call-sign, Vityaz). Arriving at Mir with them was German astronaut Ulf Merbold – representing ESA, and already a veteran of two Shuttle/Spacelab missions – who would spend a month onboard Mir, conducting experiments as part of the Euro Mir 94 science programme. He would return with the Agat crew in early November 1994.

It was planned that the docking with the front port of Mir on 6 October would be automated, but 150 metres from the docking port, Soyuz TM-20's Kurs system initiated a yaw and diverted the spacecraft from its intended flight path. Viktorenko quickly assumed manual control and, with Kondakova's assistance from the OM, completed a safe docking just six minutes late. The incident occurred only a few weeks after problems with Progress M24 at the same port – indicating that after eight years of faultless operation, the front port seemed to have developed an error in its Kurs rendezvous system.

This could have been caused by a drain on Mir's resources due to six cosmonauts being onboard for a month; but it was actually a short circuit onboard the core module on 11 October (as the crew recharged the Soyuz TM batteries) that disabled the computer which guided Mir's main solar arrays for power, thus causing the station storage batteries to drain. The crew used the Soyuz TM RCS thrusters to move the station for solar lock-on, and they then switched to a back-up computer. By 15 October, normal operating conditions had been restored to Mir.

A month later, on 2 November – after completing the Euro Mir '94 programme, and packing the results in the Soyuz TM-19 DM – Malenchenko, Musabayev and Merbold undocked Soyuz TM-19 and backed out about 200 metres from the rear port. They then commanded the Kurs system to complete an automated redocking from 190 metres, and within 35 minutes of the initial undocking. This proved that it was only the front port system that seemed to be at fault, and not the system as a whole. With Soyuz TM-19 already packed with experiment results, the crew could have returned to Earth had the docking failed. As the experiment worked perfectly however, they transferred back into Mir for a further night, leaving Soyuz TM-19 packed for the return the next day. The landing on 4 November took place in strong winds, which caused the Soyuz TM-19 capsule to travel 9 km from the planned landing site. It landed roughly, and bounced once before coming to rest; but despite this 'soft landing', all three cosmonauts were unhurt.

Reports of the problems with Kurs during dockings with the front port began after the docking of Kvant 2 and Kristall. Apparently, reflections of radio signals by these two modules confused Kurs measurements of the spacecraft's roll axis, which had caused Progress M-24 to be misaligned during its docking attempts. After the Progress M-24 problems it was decided to test Kurs in a back-up mode (which involved the use of software originally developed for Buran shuttle dockings with Mir). Soyuz TM-20 was supposed to test it, but ran into problems before it could be activated and thus forcing a manual docking. However, Soyuz TM-19 successfully tested the back-up mode during the redocking, and it became the prime mode for subsequent automatic Progress and Soyuz dockings.

The next phase of TM operations began in February 1995 with the cooperative Shuttle/Mir programme, during which ten more TM spacecraft would dock with Mir over the ensuing five years.

Soyuz TM crewing, 1994–2000

Under the agreement between the USA and Russia, there would be a series of flights by Russian cosmonauts on the Shuttle, and an American on a Soyuz TM to Mir. The first astronauts to be assigned were Space Shuttle veterans Norman Thagard and Bonnie Dunbar, who soon began training in Star City. The mission profile was for one American and two Russian compatriots to be launched on a Soyuz and land on the Shuttle, with the replacement expedition crew to Mir being launched on the same Shuttle. Thagard was paired with Dezhurov and Strekalov, and launched on Soyuz TM-21, while their back-up Russian crew (A. Solovyov and Budarin) was later launched on STS-71 to Mir as the replacement expedition crew. Thagard landed safely in the USA, on the Shuttle, having set a duration record for an American, and breaking the record set by the Skylab 4 mission in the early 1970s.

The next mission – Soyuz TM-22, crewed by Gidzenko, Avdeyev and Reiter – set a duration record for an ESA astronaut. Reiter flew under an agreement signed with the European Space Agency for a long-duration mission, and this pattern was established for the next series of expedition crews. Again, some of the core crews were accompanied by French and German cosmonauts under new commercial agreements. The French flew two missions – Soyuz TM-24 and Soyuz TM-27 – and the Germans flew on Soyuz TM-25. Future agreements – jointly signed by CNES and ESA – included a long-duration flight by a European, and Haignere flew on Soyuz TM-29. Slovakia also made their first spaceflight on the same mission. During the period beginning with the launch of STS-76 to Mir in 1996, when the Soyuz TM-

23 crew were in occupation, six Americans worked on Mir, setting a number of duration records while working with Russian crews.

Soyuz TM-29 landed on 28 August 1999, leaving Mir empty for the first time in nearly ten years. Also in 1999, there was a project to fly an actor to Mir as part of a feature-film project; but although the actor, Steklov, trained for the mission and was even included in an official crew photograph, the money was not forthcoming and the mission never flew. The last mission to Mir was Soyuz TM-30, carrying Zaletin and Kaleri – the twenty-eighth crew – on 4 April 2000.

THE BEGINNING OF PHASE 1 AT MIR

While discussions continued about the utilisation of Soyuz TM spacecraft at Freedom and then the ISS, operations at Mir proceeded with ISS Phase 1. (Phase 2 would mark the beginning of ISS operations.) In February 1994, Sergei Krikalev became the first Russian to be launched on an American spacecraft when he flew as Mission Specialist on the STS-60 *Discovery* mission. Then. in February 1995 Vladimir Titov was Mission Specialist onboard *Discovery*, flying the STS-63 mission that included a rendezvous with Mir, crewed at the time by the EO-17 crew of Viktorenko, Kondakova and Polyakov. STS-63 was also called the 'near-Mir' mission, as no docking was planned, with the Shuttle evaluating the approach and proximity operations (rendezvous) at Mir. Docking would be attempted on the STS-71 mission during the summer of that year.

America's first cosmonaut

At 09.11 hrs on 14 March 1995, American astronaut Norman Thagard – veteran of four Space Shuttle missions – made history by becoming the first American to enter orbit by means of a Russian launch vehicle. Thagard was Cosmonaut–Researcher on Soyuz TM-21, alongside the EO-18 crew – Commander Vladimir Dezhurov and Flight Engineer Gennedi Strekalov (call-sign, Uragan). With this launch, a new record was set for the total number of people (thirteen) in space at the same time. Onboard Mir were the three EO-17 cosmonauts, and there were also seven astronauts onboard the Shuttle *Endeavour*, flying the STS-67 Astro 2 astronomical research mission. Soyuz TM-21 was the first of a series of flights by American astronauts to the Mir station, but only Thagard would ride to Mir on a Soyuz launch. All the others would arrive via the Shuttle, and all (including Thagard) would return to America on the Shuttle. Each American astronaut (and back-up) was trained in the emergency landing and survival techniques of the Soyuz, but were not expected to have to use their new skills.

It had been a cold and windy day at the cosmodrome as the Soyuz left the pad, and shortly after launch, a small fire broke out at the pad. The gale-force winds directed the rocket exhaust across the pad – not into the flame trench – and the fire damaged some ground support equipment and cabling, although no injures were reported. Such fires are commonplace after a launch, but in this case additional damage to the pad was caused by a strong north-easterly wind which ripped up lumps of the melted asphalt around the pad. Thagard remembered the launch as being similar to a Shuttle launch, but with less noise and vibration. Once the launch shroud separated, all he could was one or two clouds out the side (right-hand) window, but he mostly remembered the difference in power from his Shuttle flights, and that the flight was not as smooth as the Shuttle in its second stage after separation of the solid rocket boosters. Thagard also recalled that, unlike the Shuttle, the Soyuz main engine shut-off is accompanied by a 'clang and a crank', as the engines are at full power when cut off, and are not throttled back as during a Shuttle launch.

Soyuz TM-21 docked with Mir, and without incident, two days later. Thagard later recalled his activities during the docking part of this mission. 'My responsibility during the rendezvous and docking was control of the radios and the television cameras, and to help... through the systems checks.' He took a brief look through Dezhurov's periscope, and found the view no different to that seen through the TV camera, so he could view without interfering with the Commander's field of vision. The docking, he recalled, 'was not violent. In fact, the way I describe it is if you've ever backed a car into a loading dock and they have those rubber cushions... Kind of a little bump, but nothing awesome, nothing scary.'

On 22 March, with the hand-over complete, the EO-17 crew successfuly undocked their Soyuz TM-20 spacecraft from Mir, and completed a successful landing. Viktorenko and Kondakova had spent 169 days in space, but Polyakov, launched in January 1994, had logged an impressive 437 days 17 hrs 58 min 31 sec on one spaceflight.

Thagard and his colleagues would spend nearly four months onboard the space station before their replacement crew arrived. During their residence, a new module – Spektr – was launched on 20 May, and docked to Mir at its first attempt on 1 June. The next day it was moved to the port directly opposite Kvant 2. The next Russian resident crew arrived onboard an American vehicle, as STS-71 docked with Mir at the Kristall module docking port on 29 June. For the first time since July 1975, American and Russian spacecraft were linked in orbit. During the five days that the Shuttle was docked to Mir, the crews transferred supplies and logistics between the two craft. Using the Spacelab module carried in the Shuttle payload bay, extensive medical tests and experiments were conducted on Thagard and the Mir 18 crew by STS-71 astronaut Ellen Baker (a medical doctor) and Bonnie Dunbar, who had trained as back-up to Thagard. In the original planning, Dunbar should have stayed onboard Mir until the next Shuttle docked later that year, but this conflicted with plans for a European astronaut to remain on the station through to the early part of 1996, and so Dunbar's extended mission had been cancelled. The official hand-over time of the resident crews was marked as the time that the crews exchanged the personal seat-liners in the Soyuz. From then on, the Mir 18 crew became part of the Shuttle crew, and the Mir 19 pair – Anatoly Solovyov and Nikolai Budarin (call-sign, Rodnik) – became the latest resident crew on Mir.

On 2 July, the Rodniks put on their Sokol suits for a suit leak test in the Soyuz TM-21 spacecraft in which they would return to Earth. Later that day they also completed a communications test in preparation for relocating in the Soyuz for station-keeping. This allowed them to photodocument the first Shuttle undocking from Mir on 4 July, after they had undocked station keep TM-21 from the Kvant docking port and backed away to about 100 metres from the station. About sixteen minutes later the *Atlantis* crew undocked from the Kristall module. The procedure was filmed by the Soyuz TM crew, and then the astronauts onboard the Shuttle filmed the re-docking of Soyuz TM-21 before flying around the station and heading off to prepare for their landing in Florida on 7 July. The EO-18 crew lay prone for landing on recumbent seat assembles specially developed for the return of long-duration crew-members on the Shuttle.

The flight plan called for Soyuz TM-21 to re-dock with the forward port of Mir about thirty minutes later, but the re-docking took place several minutes early, as onboard the unoccupied space station the attitude control computer (the Salyut 5B) shut itself down. With the Shuttle docked, the onboard system could not cope with the asymmetric configuration, and so the station was left in free-drift. When *Atlantis* undocked and was clear of the station, the computer automatically reactivated; but then, as it initiated the pre-programmed pitch, it became confused and promptly shut down. Fortunately the station remained stable, enabling Soyuz TM-21 to re-dock without incident. Transferring back into the station, the cosmonauts' first task was to check the computer to discover why it had failed. It was discovered to be a small problem with the software, which was easily corrected.

A short stay and a longer stay

With Soyuz TM-21 having been in space since March it needed to be exchanged, and so this residency mission would be a short one of only 75 days before a new crew was launched to resume the longer-duration missions. The Rodnik's time on Mir was taken up with maintenance and repair tasks until the launch of Soyuz TM-22, on 3 September 1995, carrying a three-man crew: Commander Yuri Gidzenko and Flight Engineer Sergei Avdeyev (the EO-20 crew; call-sign, Uran), and German cosmonaut Thomas Reiter (representing ESA on the Euro Mir 95 mission). They docked to Mir on 5 September, and following the hand-over period, the Rodnik crew departed in Soyuz TM-21 on 11 September. Planned at 135 days, the new mission was extended – while the crew was in orbit – by an extra 44 days, for additional experiment time; but because os this, Soyuz TM-22 would not be exchanged until it neared the end of its optimum design life in February 1996.

In November 1995, STS-74 delivered a Docking Module (for the American Shuttles), to be permanently attached to the Kristall module docking facility, providing greater clearance for docking the orbiter with Mir. This was the only Shuttle docking mission with no onboard American who was beginning or ending a residence on Mir.

Phase 1B operations

On 20 February 1996 the base block of Mir had been in space for exactly ten years, and on the following day, Soyuz TM-23 was launched with the twenty-first resident crew: Yuri Onufrienko and Yuri Usachev (call-sign, Skif; the press referred to them as 'the two Yuris'). They docked without incident at the rear port two days later, and after a six-day hand-over period. on 29 February the Uran crew re-entered their Soyuz TM-22 spacecraft and undocked from the front port of Mir, touching down safely. It was bitterly cold at the landing site, and after they were removed from the DM, the crew-members were wrapped in warm coats and carried from the capsule.

The next event was the arrival of the second resident American astronaut – this time onboard the Shuttle. Thagard's visit to Mir was termed Phase 1A, but all the remaining Americans to reside on Mir would fly under the Phase 1B programme. The next astronaut was Shannon Lucid, who arrived on STS-76 on 24 March. Once her Soyuz couch-liner had been moved from *Atlantis* into the Soyuz, she became an official member of the Mir EO-21 crew. *Atlantis* undocked on 29 March, leaving Lucid onboard Mir with the 'two Yuris' for a planned 4.5-month visit. It was a visit that would be extended for both the Russians and Lucid, due to delays caused by funding difficulties in the construction of Soyuz rockets at the Samara factory, and a NASA decision to replace the solid-rocket boosters on the *Atlantis* STS-79 mission that was scheduled to return Lucid and deliver her replacement, John Blaha, in September. During Lucid's tour, the final Mir science module (Priroda) was launched to the station on 23 April, docking with the Mir front port on 26 April. The following day its small Lyappa robotic arm transferred it to the remaining vacant radial port opposite Kristall. The Mir configuration had been considerably delayed over many years, but was now complete – a decade after the core module had been launched. The next three years at Mir would be quite busy and eventful for the Russians, NASA and ESA, as they prepared for the start of ISS operations.

During the weekend of 11–12 August, the prime crew of EO-22 – Gennedi Manakov and Pavel Vinogradov (call-sign, Vulkan) – were stood down from the flight less than a week before launch, due to unspecified heart problems suffered by Manakov. Since his Commander could not fly, Vinogradov also lost his seat on the mission, and the back-up crew – Commander Valeri Korzun and Flight Engineer Aleksandr Kaleri (call-sign, Freget) – replaced them. The third member of the crew – French researcher Claudie Andre-Deshays – was not part of the long-duration crew and would return after only two weeks, and so she remained assigned to the flight.

Preparations for Soyuz TM-24 were subject to far more technical problems than were previous missions. Delays in supplying the Soyuz U-2 launcher and its replacement by the less powerful Soyuz U vehicle had slipped the launch several times from June into August. The Soyuz was consequently 275 kg too heavy for that launcher, and to help reduce the launch weight the cosmonauts had to endure a harsh diet to shed 6 kg, and were not allowed to take their 4 kg of personal effects on the mission. In addition, the TM was reportedly stripped of certain back-up systems, and the amount of fuel loaded into the spacecraft was less than normal. To assist in reaching the station, Mir's orbit was lowered using the docked Progress M-31 freighter. Soyuz TM-24 launched on 17 August, and the smooth docking two days later rewarded all the efforts to place the crew on Mir. On 2 September – after bidding farewell to Lucid, who now became part of the EO-22 crew – the two Yuris and Deshays boarded the older Soyuz TM-23 spacecraft, and subsequently landed without incident.

On 19 September, *Atlantis* returned to Mir for the fourth time to pick up Lucid at the end of a record-breaking American residency (188 days from launch to landing), and to deliver her replacement, John Blaha. The official transfer of Blaha (the only NASA pilot–astronaut to reside on the station) to the EO-22 crew came when he and Lucid exchanged the seat liners in the Soyuz TM-24 DM. Lucid landed with the STS-79 crew on 26 September, while Blaha would remain onboard with the EO-22 crew until early 1997. In December, reports from Russia indicated that unless additional funds were forthcoming, Mir operations would have to cease and Russian preparations for the ISS might fall behind schedule. Stocks of carrier rockets were also close to exhaustion, casting doubt about how much longer Mir could be maintained without draining resources from the ISS programme.

'An afternoon spin in a spaceship'

The new year on Mir began with the EO-22 crew receiving *Atlantis* once again. Flying the STS-81 mission, the Shuttle docked on 15 January 1997 to deliver Blaha's replacement, Jerry Linenger. On 20 January, Blaha (now part of the STS-81 crew) departed Mir, and landed two days later in Florida. With three long-duration missions completed, and one rendezvous and five dockings, the Shuttle/Mir programme was approximately half-completed, and was proceeding relatively smoothly. It was providing both the Americans and the Russians with valuable experience that had direct application for the ISS programme scheduled to begin the following year.

On 7 February, Linenger became the first American to ride in a Soyuz without having previously launched in it. The Soyuz TM-24 had to be relocated from the front port to the rear to help protect it against constant radiation and bombardment by micrometeoroids, and to free up the front port for Soyuz TM-25 three days later. Linenger had the opportunity to participate in what he termed 'an afternoon spin in a spaceship', and later wrote of the experience, revealing an insight into routine operations not generally covered in detail in the press reports.

First, all non-essential equipment on Mir was powered down in case they could not return. Then the three cosmonauts helped each other into the Sokol suits and took their places in the DM. Linenger was tall, but was still able to fit into the seat couch, although his knees were forced almost up to his chin. After the relatively large volume of Mir, the inside of the Soyuz seemed 'like being stuffed in the capsule like sardines in a can.' There then followed the methodical power-up of Soyuz, with the requirement to pre-flight check all systems before undocking. The occupant of the right-hand seat (cosmonaut–researcher) was responsible for the communications system, the forward-looking TV camera control, the life support system, and the pressure valves; while in the left-hand seat, Flight Engineer Kaleri activated the thrusters, monitored the guidance system, and monitored electrical power usage. In the central couch, Commander Korzun would control the computer and actually 'fly' the spacecraft.

The hooks holding the Soyuz OM to the docking ring of Mir were automatically released, and springs built into the docking ring provided a smooth but firm push-off backwards. Korzun, peering through the periscope between his knees, could indirectly view where they were heading, and initiated the smaller thrusters to back

Soyuz away from Mir. There was little for Linenger and Kaleri to see out of the side windows until the Earth came into view. The thrusters then flashed, and the view of the receding docking port was slowly replaced by the sight of Priroda out of the astronauts' windows; and then all of the station came into view as Soyuz moved around it. As Linenger later recalled: 'On station, you fly around inside, but you don't feel like the station is flying – especially if you don't look outside. But in the spaceship, it feels like a car or aeroplane or jet – sitting in the cockpit and flying... The space station changes position outside the window, and you feel the gentle thrust.'

After fifteen minutes the Kvant port came into view, and Korzun initiated the thrusters to head towards a firm docking. '[You] feel and hear a thud,' recalled Linenger, '... feel your spaceship being yanked around a bit, but not frightfully so.' There was relief when the pressure built up to allow them to re-enter the space station – 'A feeling like coming home after a vacation; all the lights are out [and it feels] familiar, but not overly familiar – a kind of a good feeling inside.'

A fire and a collision

On 10 February, three cosmonauts left the launch pad at Baikonur onboard Soyuz TM-25. The two Russian cosmonauts – Vasily Tsibliyev and Aleksandr Lazutkin (call-sign, Sirius) – were to join Linenger as the EO-23 crew onboard Mir, while German researcher Reinhold Ewald was to spend eighteen days onboard the station and then return with Korzun and Kaleri. During the approach for docking, the Kurs system once again failed in the final seconds, forcing Tsibliyev to override it and assume manual control to attempt a second docking.

Inside Mir, Linenger was in Kvant 2 monitoring the approach to the front port of the station recently vacated by Soyuz TM-24. He watched – holding his breath as he saw the Soyuz approach and then disappear, only to reappear again, but this time backing away. Nothing seemed wrong on the Soyuz as it stopped and began a second approach. Korzun – who was watching via a TV camera with the same view as the approaching crew – floated into Kvant 2 with a worried look on his face, and ask if Linenger could determine why they were backing way. Linenger said that he could see nothing obvious as to why the TM had not docked, but they then both felt the impact as Soyuz TM-25 hard-docked to the station, the NASA accelerometers recording the slight impact. Korzun, now much happier, rushed back into the main module to welcome his replacements.

In April 1998, Lazutkin expanded on the accounts of the docking attempts during a lecture in Korolyov.[6] Apparently, Kurs failed during the first approach when the vehicles were only 2.5 metres apart, and due to momentum, Soyuz TM-25 moved as close as 50–60 cm from Mir before it began to draw back. The crew took ten seconds to decide to try again, and with just 12 metres left, Tsibliyev was ordered to assume manual control – although when training he had not practised so close to the target. With less than a minute left before the vehicle was due to enter Earth's shadow, Soyuz TM-25 hard-docked to Mir.

In a further effort to reduce overheads in the Russian programme, it had already been decided to terminate the use of the now increasingly expensive Kurs system (built in the Ukraine), despite its having accomplishing more than seventy consecutive and successful dockings of Soyuz and Progress spacecraft. It would be replaced with the Russian TORU system on both Soyuz and Progress, and the dockings would be controlled from a console inside the station. Tests of this new system would have a more dramatic effect later on this mission.

In 1996, studies were performed to remove Kurs from Soyuz TM (but not from Progress) in order to save mass after the switch from the Soyuz U-2 to the Soyuz U; but this proposal was not accepted. However, Kurs hardware began to be reused by returning it to Earth onboard the Shuttle. Beginning with Soyuz TM-25, the gyro-stabilised Kurs antenna was also replaced with a fixed antenna that had less mass, although some changes to the on-board software were required.

The two crews joined each other inside Mir for the hand-over activities and the completion of Ewald's scientific research. On 23 February that programme had been dramatically interrupted with a flash fire erupting in Kvant, caused by one of the oxygen candles inside a small oxygen generator used to periodically clean the air over a twenty-minute period. A blowtorch-like flame erupted, billowing thick black smoke into Mir's compartments in less than a minute. Having donned facemasks and goggles, the crew spent the next few minutes trying to extinguish the flames. After 90 seconds the fire burned itself out – but it had seemed a much longer time for the six cosmonauts onboard Mir.

Fortunately the smoke was non-toxic, as it had seeped into both Soyuz spacecraft, so that use of the spacecraft in an emergency evacuation was more problematic than impossible. News of the fire (initially not picked up by NASA) soon spread around the media as a major incident threatening the lives of the crew. The Americans were upset at not being informed immediately, but the Russians – with years of past experience in station operations – knew that they could handle the incident. With the smoke clearing onboard Mir, Linenger, a medical doctor, performed preliminary examinations of himself and his colleagues to ensure they had not been injured and would not suffer lasting effects. The fire – on an already ageing station – was worrying for the Americans, but with two Soyuz vehicles attached, the crew could theoretically have easily escaped the station had the fire not been brought under control... or could they?

The permanently docked Soyuz at Mir was for just such an emergency – which was also a selling point for the other partners in the ISS. Fortunately, the crew were not called upon to put theory into practise, and they contained the fire. However, it was revealed that in this case, with the crew in the main module and in Kvant, the flames briefly blocked the path to Soyuz TM-24. Shortly after docking, the crew had exchanged seat-liners. At the front port, Soyuz TM-25 contained the seat-liners for Tsibliyev, Lazutkin and Linenger, while Soyuz TM-24 held the liners for Korzun, Kaleri and Ewald – but it was on the other side of the fire source in Kvant!

Secondly, the Soyuz retro-burn schedule – which was constantly updated throughout the mission as the ground track changed – was up-linked to an onboard printer for the cosmonauts to down-load and install in the DM, so that the most recent schedule could be used to manually orientate the Soyuz by entering the programme into the re-entry computer for the retro-fire burn at the end of the mission. Under normal circumstances, the information was stored in the next Soyuz to depart. However, during the fire the cosmonauts realised that two such reports were needed immediately, and Kaleri rushed to print off the second as the spacecraft filled with smoke.

What was not immediately realised was that with both Soyuz following identical schedules, if they had undocked at the same time (which in all probability they would have done in an emergency) they would have followed approximately the same descent trajectory. No-one had thought to programme two separate sets of data for two different Soyuz vehicles entering at the same time. There had never previously been a need to provide such data, and it could have theoretically heightened the risk of collision as the spacecraft orientated and the modules separated for re-entry.

With the fire extinguished, and control of the station restored, Korzun, Kaleri and Ewald undocked Soyuz TM-24 (with air samples taken after the fires for analysis) and returned to Earth on 2 March 1997. The Soyuz TM-24 descent capsule re-entered with the hatch facing forward after separation of the compartments, and did not assume its normal position until 160 seconds after separation. Apparently, this had something to do with the separation of the thermal blankets from the descent capsule, and Soyuz TM-23 and TM-25 experienced similar problems. Also during the Soyuz TM-24 re-entry, the Propulsion Module approached dangerously close to the Descent Module after separation, due to the use of the high-thrust approach and docking engines (DPO-B), rather than the low-thrust approach and docking engines (DPO-M), for stabilisation after separation. The crew reported a bumpy re-entry, and being thrown from side to side (which apparently is normal during for a Soyuz).

This was followed by the replacement of Linenger with Mike Foale, who arrived onboard *Atlantis* (STS-84) on 17 May. After the change-over procedures, *Atlantis* undocked again on 22 May, with a very happy and relieved Jerry Linenger onboard.

If the fire had been considered an emergency, then the events of 25 June and the collision of Progress M-34 would be an even greater threat to the whole future of Mir operations (see p.268).[7] When it became evident to Lazutkin (who was monitoring the approach of Progress, controlled by Tsibliyev via the TORU system) that Progress would hit the station, he shouted to Foale, coming from Kvant 1: ‘Michael, go to the Soyuz, *NOW!*’ Foale reacted instinctively to the instruction and dived through the hatches in the node and in Soyuz, as Progress struck Spektr and the solar arrays. The crew immediately felt the change of pressure as air seeped out of the punctured module’s hull. Lazutkin and Foale quickly severed the interconnecting cables, and sealed off the hatches into Spektr to secure their safety, for the time being, in the rest of the station. Luckily, their quick reactions saved Mir and their lives, but it would take some time to bring Mir back to full power and orientation. Fortunately the Progress had not hit the node, as a rupture would have been more serious, and would probably have prevented the crew from accessing the sole Soyuz docked on the frontal port of the station. If this had happened, they would have been marooned in the dying core module, with little hope of escape or escape. Once again, having a Soyuz docked to station with an independent supply of power and control proved vital, as the crew used its engines to help re-orientate the stricken Mir and regain solar lock to recharge the batteries. Tsibliyev controlled the firings from inside the Soyuz, while Foale aligned several stars to direct the thrust to re-orientate the station. Mindful of excess usage of valuable onboard fuel, they used just 10 kg of propellant for the realignment.

The last Americans on Mir

After an eventful six months, on 7 August the replacement crew of EO-24 launched onboard Soyuz TM-26 docked at the aft port of Mir. A week later, and after an emotional farewell, Tsibliyev and Lazutkin pulled Soyuz TM-25 away from the station and landed safely – but not without further incident – on 14 August. In April 1998, Lazutkin expanded upon events during the landing of Soyuz TM-26. Initially, all systems worked satisfactorily during the pre-undocking check, but when they undocked the Soyuz they tried to activate the on-orbit recorder, which is used to record onboard conversations and problem analysis in the event of in-flight failures. However, it did not work.

The re-entry burn was initiated on time; but although this was Tsibliyev's second descent, it was Lazutkin's first, and despite all the training and simulations, and the previous proven flight history of the Soyuz, he was still very tense. Having endured a mission full of failures, he expected a few more mishaps during landing. When the indicator on the panel continued to display a higher than 1-g overload, even after parachute deployment, Lazutkin became worried, although they soon realised that it was the display that was at fault. Communications with flight control also failed, with the crew only able to talk to the patrol aircraft, which reported that all seemed nominal. Even so, the cosmonauts asked where they were heading, and were told: 'Falling in the Kazakh Desert... Going to a salt area... Going right in.' Waiting for the impact, the crew became tense; and when the impact came it felt as though they

After a long mission, three cosmonauts relax during post-landing activities at the Kazakhstan landing site.

The effects of a fiery re-entry are very evident on this recently landed Descent Module with open parachute hatch.

had been kicked in the back and the stomach. Once the recovery crew had opened the hatch, they were asked if they were aware that the soft-landing retro-rockets had failed to work, and that the seat shock absorbers had saved them. It was obvious that something had not worked, but in a flight full of malfunctions, this was just one more to add to the list. Originally, French cosmonaut Leopold Eyharts was scheduled to be in the third seat, but his mission had been delayed due to the collision on Mir. If he had been onboard he might have suffered serious injury, as his seat position bore the brunt of the landing impact. It was later established that the soft-landing engines had been fired at an altitude of 5.8 km, and that the touchdown speed was 7.5 metres per sec.

The day following the landing of Soyuz TM-25, Foale took the second Soyuz relocation ride by an American, as the three crew-members moved Soyuz TM-26 from the aft port to the forward port. During the flyaround, Foale took photographs of the station, and the damage to the Spektr module and solar arrays. Back on Mir, the EO-24 crew – Anatoly Solovyov and Pavel Vinogradov (the Rodniks) – would attempt to restore Mir back to operational service. and completed an EVA programme with Foale to permanently seal the Spektr and try to find the leak in the hull. During the IVA of 22 August, Foale stayed in the Soyuz DM while his colleagues performed the IVA, with the OM depressurised and internal hatches sealed between the DM and OM in case they needed to use it for quick evacuation from the station.

The Rodniks also played host to the next two Shuttle flights, which exchanged Foale for Wolf (STS-86 in September–October 1997), and then brought the last American to live onboard Mir – Andy Thomas (STS-89 in January 1998). On 20 October, Solovyov and Vinogradov performed a second IVA, and once again the American crew representative (this time Wolf) remained in the sealed DM of Soyuz TM-26. Life on Mir, though at times difficult, was returning to normal, so when Andy Thomas arrived on Mir (courtesy of STS-89) on 24 January, the day-to-day media releases of activities had dropped considerably. During the change-over it was feared that Thomas might not after all complete the last American mission on Mir, as his Sokol suit had been improperly sized on the ground – and now that he was in space, it would not fit him. If the suit could not be adjusted then he could not use it for emergency evacuation in Soyuz, and he could therefore not remain on the station. Fortunately, with a little help from Solovyov the suit was adjusted, and Thomas was allowed to remain onboard as planned. After the seven-day visit, *Endeavour* undocked from Mir, with Wolf onboard, just 23 minutes after the next Mir resident crew left the pad at Baikonur and were in orbit heading for the station.

Soyuz TM-27 – carrying EO-25 Commander Talgat Musabayev and Flight Engineer Nikolai Budarin (call-sign, Kristal), together with French researcher Leopold Eyharts – docked with Mir's rear port on 31 January – the same day that *Endeavour* landed back on Earth. Eyharts spent twenty days onboard the station and then returned with the Rodniks, departing Mir onboard Soyuz TM-26 on 19 February. To prevent possible collision with the descending capsule during a blizzard, only one Mil-8 helicopter was dispatched to recover the crew. In case the crew landed off target and out of the landing zone, TASS reported, they carried 'Makarov handguns with which to meet wild animals or ill-willing bipeds.' The deteriorating weather prevented air traffic control from locking radar on the capsule, but the pilot of the lead helicopter, Anatoli Mikhalishev, tracked them, from the onboard radar, from an altitude of 4,500 metres. Soyuz TM-26 landed in snow, with the helicopter close by with its engines still running to prevent the snow freezing over the capsule. The three cosmonauts were lifted from the Soyuz and placed on the recovery stretcher/couches, and then into the helicopter, which took them away from the harsh conditions at the landing site.

Back on Mir, Thomas joined his new colleagues in the third Soyuz relocation involving an American astronaut. On 20 February, Soyuz TM-27 was undocked from the aft port, and the crew then waited for Mir to swing around to allow re-docking at the front port. During the latter half of the 45-minute flight, radio contract was lost with Soyuz TM-27 when communications through the Altair communications relay satellite went off line. Without the support of mission control, the crew had a difficult time re-docking.

On 4 June 1998, the last Shuttle (STS-91, *Discovery*) docked with Mir to pick up Thomas, and ended the Shuttle/Mir programme. Among the Shuttle crew was the head of the Russian part of the joint programme – former Salyut 6 cosmonaut Valeri Ryumin, who was returning to space after an absence of eighteen years. His official capacity was to evaluate the condition of Mir for future use after the last Americans had left. The first elements of the ISS were to be prepared for launch later that year, and all resources would soon be required to sustain the effort in creating the new station.

The last occupants?

Towards the end of 1998, the Russians – whose commitment was meant to be focused on the ISS – were being pressured by NASA and the US Government to cease operations with Mir as soon as possible. The Russians acceeded, and brought forward the planned de-orbit date from December 1999 to June 1999. At least one or two manned flights were scheduled before that date, but funding to support them was significantly reduced. Food and hygiene stocks were both in good supply, but useful research was the final casualty.

On 13 August, Soyuz TM-28 was launched with Commander Gennedy Padalka, Flight Engineer Sergei Avdeyev (EO-26, call-sign Altair) and Russian researcher and former presidential aide Yuri Baturin, who would spend twelve days in space before returning with the EO-25 crew. Two days later, just 12 metres from the docking port, Padalka was instructed to assume manual control to dock Soyuz TM-28, as further failures on Kurs were recorded. An earlier test of two sets of Kurs-P hardware onboard Mir showed that both sets were faulty. It is not exactly clear what role this hardware played in the overall Kurs system, but they were apparently electronics boxes inside Mir and connected to the Kurs antennae. They were, however, very important, because the problem threatened manned operations on Mir. A back-up Kurs-P set therefore had to be urgently delivered to Mir by Soyuz TM-28; but since this weighed 63 kg, other equipment (including scientific equipment, food rations, water, and even a back-up control stick) had to be removed from the Soyuz, whilst the combined engine installation was fuelled to the minimum possible level. Meanwhile, it was decided to activate a Kurs-P set onboard the Kvant module that had not been used for eleven years. It was successfully tested during the undocking of Progress M-39 on 12 August, but there were more problems during the approach of Soyuz TM-28, forcing the crew to switch to manual control at a range of about 50 metres (although this was not entirely unexpected). After their arrival the crew connected the newly delivered Kurs-P set to the station's Kurs antennae, and the system worked properly during the redocking of Soyuz TM-28 two days later. Soyuz TM-28 also tested a new satellite navigation system, using Glonass and GPS data as well as carrying a modified set of high-thrust approach and orientation engines (DPO-B) which spent less propellant.

Following the hand-over, the EO-25 crew departed with Baturin in Soyuz TM-27 on 25 August, landing without incident later that day. Two days later, the Altair crew relocated their Soyuz TM-28 craft from the rear Kvant port to the forward port – an operation that took 23 minutes, but which was carried out without incident. As the crew resumed orbital activity onboard the station, rumours emerged on 25 September that the life of Mir might be extended beyond June 1999, if sufficient funds were to be available. Five days later it was announced that the launch of the Russian Service Module, Zvezda, to the ISS – which was required for independent manning of the station without a docked American Shuttle – had slipped into 2000. During the night of 17–18 November the crew, as a normal precaution, sheltered in the Soyuz DM to await the predicted Leonid meteor storm, but when they returned to the main core module they reported that they had not seen or heard anything unusual.

Two days after the Leonid meteor storm, a new era of space station operations began with the launch of the first element of the ISS. The Zarya control module was placed in orbit, and as the first Shuttle mission to the ISS was launched in December, so Mir dropped out of the news. In the background, discussions continued with potential financial investors to secure a fully commercial flight to the station, hoping to 'sell' the third seat on TM flights to 'tourists' for about $20 million each.

On 22 February 1999, the twenty-seventh main crew docked to Mir onboard Soyuz TM-29, having been launched two days before. Onboard were Commander Viktor Afanasyev, French cosmonaut and Soyuz Flight Engineer-qualified J.-P. Haignere, and Slovakian citizen Ivan Bella. Bella was flying a short scientific mission and would return with Padalka, but Avdeyev (Flight Engineer 1) would continue with Afanasyev and Haignere (Flight Engineer 2) as the Derbent crew, to complete what was expected to be the final crewed mission on Mir.

On 28 February 1999, Soyuz TM-28 – carrying Padalka and Bella – landed without incident just 3 hrs 22 min after undocking from Mir, and on 1 June 1999,

Recovery of the Soyuz TM-32 crew: Tito, Musabayev and Baturin.

news was released in Moscow that the current crew would return home on 23 August, leaving Mir unmanned for six months before a crew was launched to prepare to de-orbit the complex. It was 'with grief in our souls' that the crew vacated Mir – closing the hatches with a wave to the onboard TV camera, and leaving it without a crew for the first time in almost a decade,. After undocking from the forward port, the crew completed two orbits of Earth before retro-fire. After the hot capsule touched down, it set fire to the surrounding grass as the cabin rolled over on its side; and for the next fifteen minutes, the crew hung from their seat harnesses.

MirCorp, and the final TM mission to Mir

By January 2000, efforts to secure funding for one or more missions to Mir had succeeded, with the possibility that Mir could fly a further five years if the funding and the condition of the station could support it. The EO-28 crew – Commander Sergei Zaletin and Flight Engineer Aleksandr Kaleri (call-sign, Yenisey) – were tasked to evaluate whether Mir could indeed sustain further missions. Initially, there was to be a third person on this crew (Steklov, the Russian actor, who was involved in making a feature film about a stranded cosmonaut), but his backers did not pay his flight fee, and he was dropped from the flight.

On 4 April 2000, Soyuz TM-30 was launched from Baikonur, and approached the empty Mir after the normal two-day flight. The onboard Kurs system recorded a deviation of 1–2° at a distance of 9 metres from Mir, due, it was reported, to a fault in the Mir attitude control system. Zaletin therefore assumed control, and manually docked his Soyuz at the front port of Mir without further difficulty. As the crew worked onboard the station, there were indications that further missions were planned, but not until after the EO-28 crew returned to Earth. After their return, in the early hours of 16 June, Kaleri said that it was the roughest of his three landings, as the capsule had bounced several times before stopping.

Mir was cleared for future use – but it was not to be. Finances were not immediately forthcoming, and the Russians came under pressure to divert the Soyuz TM and Progress M1 vehicles to the ISS, on which construction had begun. So, no more crews ventured to Mir. It was left to unmanned Progress spacecraft to perform the de-orbit burn, and 23 March 2001 – after fifteen years of service – the unoccupied Mir re-entered Earth's atmosphere, and was destroyed.

Another crew was named to fly to Mir if there was a problem with the re-entry plans. Padalka and Budarin were backed up by Sharipov and Vinogradov. But they were not needed, and were stood down.

SOYUZ TM AT THE ISS: SOYUZ TM-31–TM-34

When the new design for the ISS emerged towards the end of 1993, the change or orbit to a 51.6-degree inclination enabled the Soyuz TM to serve as a crew ferry and rescue vehicle while a three-person crew occupied the station. After 2002, when the US Habitation Module was scheduled for delivery, a permanent crew of six (rotated in threes by the Space Shuttle or Soyuz) would require two Soyuz TMs permanently docked to the facility for crew rescue until the more advanced ACRV became available. This was a role that Soyuz had provided throught the previous three decades, and one for which it was perfectly suited. In addition, becuase of the revised inclination, Progress vehicles could be used to supply logistics to the station, if the funding to produce the supply of vehicles could be maintained.

A lifeboat for the ISS

Despite the immediate availability of Soyuz TM for the ISS, there remained the problem of providing the whole six-person crew with a reliable escape option on one vehicle. Studies into a single-crew return vehicle continued in Europe and America, as well as in Russia. The Russian concept – which emerged in 1995 – was a joint Energiya–Rockwell–Khrunichev design which returned to the Zarya-type concept, although (indicated by drawings in the RKK Energiya history) it had a standard parachute system like that of Soyuz, rather than the multitude of soft-landing engines on Zarya.

It featured an 8-tonne descent capsule, 7.2 metres long and 3.7 metres in diameter, that could hold up to eight crew-members for a short flight back to Earth. Launched in the payload bay of the Shuttle, it could be attached to the ISS for up to five years (1,825 days), and included a solid-propellant retro-fire motor within a protective cover during orbital storage, and cold gas thrusters for orientation, with power supplied by onboard batteries.

On 1 June 1996, after further consideration of all escape options and designs, the Soyuz-derived large lifeboat option at the ISS was dropped in favour of a NASA-led lifting-body-type vehicle called X-38. By 2001, delays in funding X-38, and requirements for further flexibility in the design of the Soyuz TM, resulted in the development of the Soyuz TMA to provide crew rescue capability at the ISS for the foreseeable future.

In 1996, RKK Energiya began production of Soyuz TM vehicles for the ISS. These were called the '200 series', because they would be assigned serial numbers beginning with 201. Initially, five vehicles were ordered – 201 to 205. Vehicles 201–203 would have hybrid docking ports (a cross between APAS and probe/drogue) so that they would be compatible with the hybrid nadir port of Zvezda, and vehicles 204–205 would have probe/drogue systems. The two vehicles on which work began in 1996 were 201 and 204. In 1997 it was decided to turn the Zarya nadir port into a standard probe/drogue type (rather than a hybrid), so that a Soyuz could be permanently attached there without the requirement for the hybrid port. During the second half of 1997, therefore, it was decided not to build vehicles 202 and 203, and to instead refit vehicle 201 with a probe/drogue docking port (after which it was redesignated 206). Vehicle 204 was intended to carry the first ISS crew, but when it became clear, in 1999, that Mir would support another mission, it was reoriented to Mir and launched as Soyuz TM-30. The first vehicle to fly to the ISS was 205 (Soyuz TM-31). Vehicle 206 (the former 201) was on stand-by for a rescue mission to help de-orbit Mir, but after that became unnecessary, it flew to the ISS as Soyuz TM-32.

Soyuz TM docked to the ISS in its crew rescue role. Note the thermal insulation on the Orbital Module and Descent Module. During the Soyuz TM-9 in 1990, some of this insulation on the Descent Module became unfastened, and it required an EVA to reattach it.

Soyuz TM/ISS crewing

The Soyuz TM craft was to be used as a rescue craft for the main ISS crew of three. It was decided that the first mission crew would fly on the Soyuz, for it to be on station immediately on initial occupation. The original mission crew consisted of A. Solovyov, Krikalev, and William Shepherd, and after they had been training for some weeks it was announced that Shepherd would be the first station Commander. Solovyov – who was named Soyuz Commander – objected to the station command being given to someone with so little flight experience, and his objections led to his being removed from the crew (he was replaced by Gidzenko) and assigned to Soyuz TM-26 to go to Mir. The back-ups for Shepherd's crew were Dezhurov, Tyurin and Bowersox.

The key to the success of the ISS rested on the launch of the Russian Service Module, Zvezda, in July 2000. Zvezda would perform an automated docking with Zarya, and allow a resident crew to remain onboard the station without a Shuttle to support them. With no back-up module readily available should the Zvezda fail to dock, a Russian ISS 'Zero' crew of Padalka and Budarin was trained to fly a Soyuz TM mission to Zvezda and use the TORU system to manually dock the two station elements. They would have been in Zvezda to bring in the FGB (which would have been the active vehicle). In the event, all worked as planned, and the 'Zero' crew was stood down.

The need to replace the Soyuz has resulted in a short-visit crew flying to the ISS every six months. The crewing has been entirely influenced by the need for the Russians to raise money for their cash-strapped programme, and so the Russian Space Agency and Energiya signed agreements with the European Space Agency to fly an astronaut as a Flight Engineer on the first six taxi missions. The money would come from the astronaut's home country, with ESA also contributing. Claudie Haignere made her second flight on Soyuz TM-33, and Roberto Vittori, an Italian pilot, flew on Soyuz TM-34. The Russians also devised the idea of 'tourist' flights. In return for $20 million, the third seat would be sold to a Spaceflight Participant (SFP). Two multi-millionaires bought their seats at the asking price: Dennis Tito – an American businessman – flew on Soyuz TM-32; and Mark Shuttleworth – a South African 'dot com' millionaire, flew on Soyuz TM-34. The Air Force cosmonaut group provided the Commanders and Flight Engineers, all of whom had a good command of English and had completed previous spaceflights.

'Off we go!'

The first resident crew for the ISS began training in 1995, but due to several delays in launching the first elements – most notably, the Russian Service Module (Zvezda) – it was 2000 before the station was ready to support a crew without a Shuttle orbiter docked to it. The Zvezda module contained the hardware and systems to support a three-person crew, resupplied by Progress freighters and supplemented by Shuttle assembly flights that both expanded the components and facilities on the station and provided larger logistics movement to and from the station. With a resident crew onboard the station, at least one Soyuz would need to be permanently docked, and it was therefore logical that the first permanent crew should launch to the station from Baikonur to deliver the initial rescue craft.

The rotation of a new Soyuz vehicle each six months did not, at that time, fit with the planned 4–5-month crew rotation of main crews. (Russian medical staff believe

The first expedition crew for the ISS pose for a farewell wave prior to ascending to their Soyuz TM-31 spacecraft. The American Expedition 1 Commander is in the centre, with Soyuz Commander Yuri Gidzenko in front of him and Flight Engineer Sergei Krikalev behind. (Courtesy NASA.)

that the optimum stay is around 4–4½ months on a long-duration flight.) Because of financial realities, however, many crews on Mir completed 180–190 days, matching the design life of Soyuz TM. On the ISS, 4–5-month residencies were planned, but mission delays have resulted in some missions being extended. It was also more reasonable to use the Shuttle logistics flights to change out the main crews, and to introduce shorter visiting missions to exchange the Soyuz 'taxi' spacecraft twice a year as their design life was reached.

After months of delay, on 31 October 2000 the first crew (call-sign, Uran) was launched to the ISS onboard Soyuz TM-31. In command of the Soyuz was Yuri Gidzenko, who was already a veteran of a residency flight to Mir. The Flight Engineer, Sergei Krikalev, had flown two long-duration missions to Mir, in 1994 was the first Russian to fly on a Shuttle (STS-60) in the inaugural mission in the Phase 1 programme, and had flown on the first ISS assembly mission (STS-88) two years earlier. Once on board the station, command of the mission would pass to American astronaut William Shepherd – a veteran of three previous missions on the Shuttle. He was the second American to be launched in a Soyuz spacecraft.

During preparations for launch, Shepherd and his colleagues observed the numerous 'cosmonaut traditions' begun by Gagarin almost forty years earlier. At the moment of ignition of the Soyuz launch vehicle, Shepherd borrowed the famous words of Gagarin at the start of his mission: '*Poyekhali!*' – '*Off we go!*' Nine minutes later, the Soyuz was in orbit and heading for the ISS. A standard two-day approach was made, with deployment of the solar arrays, aerials and docking probe in the first two orbits, and phasing burns on the third and fourth orbits. Although called Soyuz TM-31 by the Russians, the Americans added confusion to the process by referring to the mission as Soyuz '1' (as the first Soyuz to fly to the ISS, and apparently forgetting that the official 'Soyuz 1' flew in April 1967). This pattern that has continued with each Soyuz flight to the station.

After a night's sleep in Soyuz, the crew continued with a full systems check of their spacecraft in preparation for docking the next day. At the ISS, meanwhile, the Progress M1-3 that had been docked to the station since 8 August was undocked from the rear port of Zvezda to clear the port for the Soyuz. A small third burn of the engines on Soyuz on the second day in space was followed by a large manoeuvre which resulted in the trio of cosmonauts trailing the station by about 8,500 km, to close in at about 1,148 km per 90-minute orbit.

On the morning of 2 November, the crew woke early to complete nine hours of final approach manoeuvres with the ISS. The automated Kurs rendezvous sequence began 2 hrs 15 min before docking, and ended 35 minutes later. Gidzenko completed a nine-minute partial flyaround of the station, to place Soyuz 500 feet {500 metres?} from the Zvezda port before initiating the final approach. As the combination flew over the borders of south-east Russia, Kazakhstan, Mongolia and China, Soyuz TM-31 arrived at the ISS with a perfect docking. Ninety minutes later the crew entered Zvezda to begin the permanent habitation of the space station.

The Soyuz remained at the Zvezda aft port for the next fifteen weeks, with the single Progress M1-4 resupply craft using the nadir port on Zarya. On 24 February 2001, the Soyuz would be moved to free up the port for regular docking by Progress vehicles. The previous experience of the two Russians on Mir assisted in speeding up mothballing of some of the onboard systems for the flyaround, and during the long wait for pressure checks the crew filled their spare time by watching a movie on a new portable DVD player. After the undocking, Gidzenko backed away the Soyuz to a distance of about 30 metres, and then flew the spacecraft around the complex, during which communications were intermittent. After docking with Zarya's nadir port 31 minutes after undocking, the crew were told to remain in the Soyuz until necessary latching and pressure checks had been completed. Not wishing to waste time for ninety minutes, the crew enjoyed a second film on their DVD. The rest of that day was spent in again powering up the ISS, which raised the temperature (when they had re-entered the complex, it was cool).

On 10 March, the Space Shuttle *Discovery* (flying the STS-102 mission) docked with the ISS at pressurised mating adapter 2 (PMA-2), on the American side of the complex. Over the next nine days, the ISS-1 crew worked with the Shuttle crew in delivering supplies and equipment to the station, and with the ISS-2 crew – Commander Yuri Usachev (call-sign, Flagman), Flight Engineer 1 Jim Voss, and Flight Engineer 2 Susan Helms, who had arrived on STS-102 – in handing over command of the station.

It was a staggered change of shift, beginning almost immediately after the hatches opened on 10 March, and with Usachev and Gidzenko exchanging Soyuz seat-liners so that the new station commander could gain experience by working with Shepherd for a week onboard the station. Had the spacecraft had separated early, Gidzenko would have returned on the Shuttle, while Usachev would have remained onboard the ISS. During the morning of 12 March, Voss and Krikalev exchanged seats and crews, followed by Helms and Shepherd two days later. Although officially part of the Shuttle crew, Shepherd would remain responsible for the ISS until the moment the hatches were closed between the two vehicles for the final time. This took place on 19 March, just two hours prior to *Discovery* undocking and departing the station for a landing on 21 March. This began the pattern of resident crew exchanges that has become a standard mode of operations on the ISS.

Taxis and tourists

Although the main station crew had changed, the Soyuz TM-31 that had taken them to the ISS was still docked to the nadir port of Zarya. By the end of April, Soyuz TM-31 needed to be exchanged with a fresher vehicle to support crew rescue capabilities for another six months. The Soyuz TM-32 crew were preparing for a short flight to the ISS specifically to change out the older Soyuz, which was being prepared for relocation again on the ISS.

On 12 April 2001, the fortieth anniversary of the world's first spaceflight, by Yuri Gagarin, and the twentieth anniversary of the first Shuttle flight, were observed around the world and on ISS, whose crew passed on congratulations to the teams of controllers in Moscow and Houston. Usachev also noted that Gagarin's pioneering flight had lasted for just one orbit. Four decades later, the crew take that long to eat breakfast.

During 14–16 April the crew conducted systems checks on the Soyuz TM-31 craft

The Soyuz TM-32 crew, on the first Soyuz taxi mission to the ISS. (*Left–right*) American businessman Dennis Tito, Soyuz Commander Talgat Musabayev, and Flight Engineer Yuri Baturin, during a break in the pressure integrity tests of their Sokol suits.

in preparation for the relocation, which took place on 18 April. This time the flight was a reverse of the earlier relocation, moving from the nadir port on Zarya to the aft port on Zvezda. Already, the movement of ISS elements was becoming like a large game of celestial chess, with the flexibility of the Zarya and Zvezda docking locations, and Soyuz and Progress, helping to provide greater opportunities in scheduling orbital operations. The ISS-2 crew entered the Soyuz and spent 41 minutes flying around the station to dock at the Zvezda aft port. The removal of Soyuz TM-31 from the Zarya nadir port to the Zvezda aft port provided clearance for docking the MPLM on STS-100, and by undocking Soyuz TM-31 from the Zvezda aft port at the end of the taxi mission, the port was vacated for Progress.

On 28 April the Soyuz TM-32 mission (or Soyuz '2') was launched, with Commander Talgat Musabayev, (call-sign, Kristal), Flight Engineer Yuri Baturin, and Spaceflight Participant US millionaire Denis Tito. During filming in the crew compartment during ascent, Tito enthusiastically gestured a 'thumbs up' to the camera – but too enthusiastically for Musabayev, who prompted Tito to lower his arms away from the control panel.

Onboard the ISS, the crew were experiencing computer problems with the recently installed Canadarm 2, and with STS-100 still docked it was decided to extend the Shuttle's mission an extra two days to help support the effort to bring the computers on line. NASA requested the Russians to delay the launch of Soyuz TM-32 for 24 hours. The Russians announced that they would launch on schedule on 28 April, but would hold the Soyuz TM in a parking orbit for an extra day if this helped NASA. On 28 April the computer problem was overcome, and it was announced that the orbiter would be undocked the next day – and it did so, without any further incident. On 30 April, Soyuz TM-32 docked with the Zarya nadir port on the ISS. According to reports, Tito had difficulty in adapting to the space environment, and was ill during most of the two-day flight to the ISS. Onboard the station he contradicted this assertion by stating the he 'loved space'. Tito was confined to the Russian segment, as NASA had banned him from their part of the station due to his lack of training.

The taxi crew did not have much to do during the week onboard the station, and were mainly involved in preparing Soyuz TM-31 for the return to Earth. On 1 May it was reported that Usachev had spent time modifying the toilets onboard both Soyuz craft – presumably to accommodate the exchange of crew-members. After a week of joint 'commemorative activities by the three cosmonaut researchers' (as the axi crew was designated), they transferred their seat-liners in preparation for the return to Earth on the older Soyuz TM-31. Musabayev – assisted by Baturin – had conducted Earth observations and crystal experiments onboard the Russian elements, and Tito had conducted 'symbolic activity, amateur photography, and videotaping'.

On 6 May, Soyuz TM-31 undocked from the rear port of Zvezda and landed just over three hours later, although the vehicle overshot the planned landing zone by 56 km – reportedly because of the improper functioning of the infrared vertical sensor. It was, according to reports, a rough and hard landing – so hard, that Commander Musabayev checked whether his $20-million passenger was still alive! Whatever the controversy of paying millions of dollars for a short flight into space, the mission certainly generated interest in other future 'seats for sale' opportunities. Moreover, the flight proved the concept of a short-term specialist crew flying to exchange the Soyuz periodically and performing specific experiments that did not interrupt the main operations onboard the station.

On 12 August, STS-105 *Discovery* docked to the ISS. On board was the third ISS crew, who the following day transferred their Soyuz seat-liners out of the Shuttle and into the Soyuz TM-32. The ISS-2 liners were taken back to the Shuttle, marking the official change of main crews. Discovery undocked on 20 August, and landed two days later.

Andromède: a French taxi mission

On 17 September, Progress M-SO1 (see p.270) docked with the ISS at the nadir port on Zvezda, and delivered the Pirs (Pier) docking and EVA module. A month later, on 19 October, the ISS-3 crew transferred Soyuz TM-32 from the nadir port on Zarya to the new docking port at the rear of Pirs, in a sixteen-minute flight. The Russians wanted the Soyuz attached to Pirs rather than to Zarya, so that if anything went wrong during an EVA from Pirs, the third crew-member, remaining behind on the ISS, still had access to the TM rescue vehicle.

Onboard Soyuz TM-33 (Soyuz '3') was a three-person crew: Commander Viktor Afanasyev (call-sign, Derbent), on his fourth TM mission; Flight Engineer 2

Konstantin Kozeev – a rookie test engineer from Energiya; and, as Flight Engineer 1 – flying her second mission, and representing CNES – Claudie Haignere (formerly Andre-Deshays), who would conduct a programme of experiments under the Andromède programme for the French space agency. Two days later, on 23 October, Soyuz TM-33 docked with the nadir docking unit of Zarya.

Unlike Tito, Haignere was a fully qualified cosmonaut (having previously spent two weeks onboard Mir in 1996), and so this crew was fully integrated into the main ISS crew's work schedule, albeit following separate research programmes. Dezhurov worked with Afanasyev in the hand-over of Soyuz TM ferries. By 31 October – their work completed, and the personal couches transferred to Soyuz TM-32 – the crew undocked from the ISS at the Pirs port, for re-entry and landing later that day.

The first South African, and the last Soyuz TM

The ISS-3 crew was rotated with the trio of ISS-4 cosmonauts during the STS-108 logistics mission flown in December 2001. The new year would also see the final flight in the Soyuz TM series. But first, on 20 April, the Soyuz had to be moved.

After closing deactivated systems – including pressure sensors, C&W powered panels, the Vozdukh carbon dioxide absorber, and the toilet in Zvezda – in the US and Russian segments, the three ISS-4 crew-members moved into the Soyuz. Four hours and 16 minutes after closing the hatches on Soyuz – during which time they configured the spacecraft for flight, and conducted leak checks and pressure checks – they undocked. After backing away a short distance from the FGB port, Onufrienko translated the Soyuz down the short distance to re-dock with Pirs, twenty-one minutes later. The crew then spent a further ninety minutes in Soyuz – re-checking the pressure integrity of the connections, and powering down the vehicle – before returning to the ISS and reversing the earlier work, enabling them to continue living and working in the station.

The thirty-fourth and final launch of a TM craft – 208 (ISS designation Soyuz '4') – took place at Baikonur on 25 April with a crew of three onboard. In command was Yuri Gidzenko (call-sign, Uran), who had been a member of the first ISS residency crew and was a former Mir resident. His Flight Engineer – representing ESA – was Roberto Vittori from Italy, who (like Claudie Haignere before him) was scheduled to carry out a programme of scientific research (in his case designated Marco Polo) during his brief stay onboard the ISS. The third member of the crew was South African businessman Mark Shuttleworth – a private-paying participant, and the first representative of his country to fly in space.

On 27 April, Soyuz TM-34 docked to the nadir port of Zvezda after a planned flyaround of the complex. During their stay onboard ISS, the taxi crew relocated the seat-liners between Soyuz TM-34 and TM-33, transferring the science equipment into the station for Vittori and Shuttleworth to conduct their assigned programmes of experiments. Unlike Tito's, Shuttleworth's daily programme on the ISS was far more scientifically orientated. Following Tito's flight, an agreement was reached between ISS partners so that all personnel flying to the ISS on Soyuz have to meet NASA safety standards as well as Russian safety standards. Because of this, Shuttleworth integrated more easily into the main crew's activities. On 4 May, Soyuz

TM-33 undocked from Pirs, and landed some three hours later, leaving the last TM attached to the station for return by the TMA-1 taxi crew in October 2002.

The return of the last TM

Soyuz TM-34 undocked from the Zvezda module at 23:44 Moscow Time on 9 November 2002, and landed on the Kazakh steppes, some 81 km north-west of Arkalyk, at 3:04 Moscow Time the following morning. The three cosmonauts who had launched in TMA-1 to return Soyuz TM-34 were Commander Zaletin, Flight Engineer 1 De Winne – a Belgian working for ESA – and Flight Engineer 2 Lonchakov. This landing brought the Soyuz TM era to a close.

SUMMARY

Between May 1986 and November 2002, the TM series built upon the experience of the Soyuz T series, provided a reliable crew ferry craft to Mir, and supported initial ISS operations. But more remarkable was that despite difficulties encountered with the Kurs system, no TM spacecraft failed to dock with its intended target station; which was a rewarding achievement for Soyuz designers and mission planners, considering the difficulties with docking during the earlier part of the programme. With one unmanned launch and thirty-three manned launches to its credit, Soyuz TM logged more than 5,550 days in space – more than fifteen years, during a flight programme lasting 16 years and 6 months.

REFERENCES

1 *History of Energiya*, 1946–1996, pp.220–221.
2 Soviet Space Programs, 1981–1987, US Library of Congress, May 1988, pp.100-101; *Moscow News*, No.23, 8 June 1986; *Aviation Week and Space Technology*, 5 December 1988, p.32; *Ibid.*, 23 October 1989, p.49.
3 *Aviation Week and Space Technology*, 23 October 1989, p.49.
4 Hall, Rex D. and Shayler, David J., *The Rocket Men*, Springer–Praxis (2001), pp.248–251.
5 *Novosti Kosomnavtiki*, Nos.19/20, 1998.
6 Salmon, Andy, 'Off-Nominal Situations', *Spaceflight*, **40**, (September 1998), 346–348.
7 Shayler, David J., *Disasters and Accidents in Manned Spaceflight*, Springer–Praxis (2000), pp.320–340.

Soyuz variants: Soyuz TM missions and hardware

Spacecraft (serial number) name	Design designation	Launch date	International designation	Docking date	Undocking date	Target spacecraft (port)	Landing/ decay date	Flight duration dd:hh:mm:ss
Key: (U), unmanned								
11F732 (Soyuz TM improved transport manned spacecraft for DOS Zarya; ISS)								
(051) Soyuz TM	7K-M (U)	1986 May 21	1986-035A	1986 May 23	1986 May 29	Mir (Fwd)	1986 May 30	008:20:56:17
(052) Soyuz TM-2	7K-M	1987 Feb 5	1987-013A	1987 Feb 7	1987 Jul 29	Mir (Fwd)	1987 Jul 30	174:03:25:56
1987 Apr 9: the Kvant Astrophysical Module was permanently docked to the Mir core module aft port; all future dockings were at the Kvant (Aft-K) port on Mir								
(053) Soyuz TM-3	7K-M	1987 Jul 22	1987-063A	1987 Jul 24	1987 Jul 30	Mir (Aft-K)		
				1987 Jul 30	1987 Dec 29	Mir (Fwd)	1987 Dec 29	160:07:16:58
(054) Soyuz TM-4	7K-M	1987 Dec 21	1987-104A	1987 Dec 23	1987 Dec 30	Mir (Aft-K)		
				1987 Dec 30	1988 Jun 17	Mir (Fwd)	1988 Jun 17	178:22:54:30
(055) Soyuz TM-5	7K-M	1988 Jun 7	1988-048A	1988 Jun 9	1988 Jun 18	Mir (Aft-K)		
				1988 Jun 18	1988 Sep 5	Mir (Fwd)	1988 Sep 7	091:10:46:25
(056) Soyuz TM-6	7K-M	1988 Aug 29	1988-075A	1988 Aug 31	1988 Sep 8	Mir (Aft-K)		
				1988 Sep 8	1988 Dec 21	Mir (Fwd)	1988 Dec 21	114:05:33:49
(057) Soyuz TM-7	7K-M	1988 Nov 26	1988-104A	1988 Nov 28	1988 Dec 22	Mir (Aft-K)		
				1988 Dec 22	1989 Apr 26	Mir (Fwd)	1989 Apr 27	151:11:08:24
(058) Soyuz TM-8	7K-M	1989 Sep 5	1989-071A	1988 Sep 7	1989 Dec 12	Mir (Aft-K)		
				1989 Dec 12	1990 Feb 19	Mir (Fwd)	1990 Feb 19	166:06:58:16
(059) Soyuz TM	7K-M	*Service Module was damaged during ground tests; landing capsule and Orbital Module were reassigned to fly as 61A (Soyuz TM-10)*						
(060) Soyuz TM-9	7K-M	1990 Feb 11	1990-014A	1990 Feb 13	1990 Feb 21	Mir (Aft-K)		
				1990 Feb 21	1990 May 28	Mir (Fwd)		
				1990 May 28	1990 Jul 3	Mir (Aft-K)		
				1990 Jul 3	1990 Aug 9	Mir (Fwd)	1990 Aug 9	179:01:17:57
(061A) Soyuz TM-10	7K-M	1990 Aug 1	1990-067A	1990 Aug 3	1990 Dec 10	Mir (Aft-K)	1990 Dec 10	130:20:35:51

Spacecraft (serial number) name	Design designation	Launch date	International designation	Docking date	Undocking date	Target spacecraft (port)	Landing/ decay date	Flight duration dd:hh:mm:ss
(061) Soyuz TM-11	7K-M	1990 Dec 2	1990-107A	1990 Dec 4	1991 Mar 26	Mir (Fwd)		
				1991 Mar 26	1991 May 26	Mir (Aft-K)	1991 May 26	175:01:50:41
(062) Soyuz TM-12	7K-M	1991 May 18	1991-034A	1991 May 20	1991 May 28	Mir (Fwd)		
				1991 May 28	1991 Oct 10	Mir (Aft-K)	1991 Oct 10	144:15:21:50
(063) Soyuz TM-13	7K-M	1991 Oct 2	1991-069A	1991 Oct 4	1991 Oct 15	Mir (Fwd)		
				1991 Oct 15	1992 Mar 14	Mir (Aft-K)		
				1992 Mar 14	1992 Mar 25	Mir (Fwd)	1992 Mar 25	175:02:51:44
(064) Soyuz TM-14	7K-M	1992 Mar 17	1992-014A	1992 Mar 19	1992 Aug 9	Mir (Aft-K)	1992 Aug 10	145:14:10:32
(065) Soyuz TM-15	7K-M	1992 Jul 27	1992-046A	1992 Jul 29	1993 Feb 1	Mir (Fwd)	1993 Feb 1	188:21:41:15
(066) Soyuz TM-17	7K-M	1993 Jul 1	1993-043A	1993 Jul 3	1994 Jan 14	Mir (Fwd)	1994 Jan 14	196:17:45:22
(067) Soyuz TM-18	7K-M	1994 Jan 8	1994-001A	1994 Jan 10	1994 Jan 24	Mir (Aft-K)		
				1994 Jan 24	1994 Jul 9	Mir (Fwd)	1994 Jul 9	182:00:27:02
(068) Soyuz TM-19	7K-M	1994 Jul 1	1994-036A	1994 Jul 3	1994 Nov 2	Mir (Aft-K)		
				1994 Nov 2	1994 Nov 4	Mir (Aft-K)	1994 Nov 4	125:22:53:36
(069) Soyuz TM-20	7K-M	1994 Oct 3	1994-063A	1994 Oct 6	1995 Jan 11	Mir (Fwd)		
				1995 Jan 11	1995 Mar 22	Mir (Aft-K)	1995 Mar 22	169:05:21:35
(070) Soyuz TM-21	7K-M	1995 Mar 14	1995-010A	1995 Mar 16	1995 Sep 11	Mir (Aft-K)	1995 Sep 11	181:00:41:06
(071) Soyuz TM-22	7K-M	1995 Sep 3	1995-047A	1995 Sep 5	1996 Feb 29	Mir (Fwd)	1996 Feb 29	179:01:41:45
(072) Soyuz TM-23	7K-M	1996 Feb 21	1996-011A	1996 Feb 23	1996 Sep 2	Mir (Aft-K)	1996 Sep 2	193:19:07:35
(073) Soyuz TM-24	7K-M	1996 Aug 17	1996-047A	1996 Aug 19	1997 Feb 7	Mir (Fwd)		
				1997 Feb 7	1997 Mar 2	Mir (Aft-K)	1997 Mar 2	196:17:26:13
(074) Soyuz TM-25	7K-M	1997 Feb 10	1997-003A	1997 Feb 12	1997 Aug 14	Mir (Fwd)	1997 Aug 14	184:22:07:41
(075) Soyuz TM-26	7K-M	1997 Aug 5	1997-083A	1997 Aug 7	1997 Aug 15	Mir (Aft-K)		
				1997 Aug 15	1998 Feb 19	Mir (Fwd)	1998 Feb 19	197:17:34:36
(076) Soyuz TM-27	7K-M	1998 Jan 29	1998-004A	1998 Jan 31	1998 Feb 20	Mir (Aft-K)		
				1998 Feb 20	1998 Aug 25	Mir (Fwd)	1998 Aug 25	207:12:51:02
(077) Soyuz TM-28	7K-M	1998 Aug 13	1998-047A	1998 Aug 15	1998 Aug 27	Mir (Aft)		
				1998 Aug 27	1999 Feb 8	Mir (Fwd)		

				1999 Feb 8	1999 Feb 27	Mir (Aft)	1999 Feb 28	198:16:31:20
(078)	Soyuz TM-29 7K-M	1999 Feb 20	1999-007A	1999 Feb 22	1999 Aug 28	Mir (Fwd)	1999 Aug 28	188:20:16:19
(101)	Soyuz TM-16 7K-M	1993 Jan 24	1993-005A	1993 Jan 26	1993 Jul 22	Mir (Kristall)	1993 Jul 22	179:00:43:46
(204)	Soyuz TM-30 7K-M	2000 Apr 4	2000-018A	2000 Apr 6	2000 Jun 15	Mir (Fwd)	2000 Jun 16	072:19:42:16
(205)	Soyuz TM-31 (1S) 7K-M	2000 Oct 31	2001-070A	2000 Nov 2	2001 Feb 24	ISS Aft Zvezda		
				2001 Feb 24	2001 Apr 18	ISS Nadir Zarya		
				2001 Apr 18	2001 May 6	ISS Aft Zvezda	2001 May 6	186:21:48:41
(206)	Soyuz TM-32 (2S) 7K-M	2001 Apr 28	2001-017A	2001 Apr 30	2001 Oct 19	ISS Nadir Zarya		
				2001 Oct 19	2001 Oct 31	ISS Pirs	2001 Oct 31	185:21:22:40
(207)	Soyuz TM-33 (3S) 7K-M	2001 Oct 21	2001-048A	2001 Oct 23	2002 Apr 20	ISS Nadir Zarya		
				2002 Apr 20	2002 May 5	ISS Pirs	2002 May 5	195:05:07:42
(208)	Soyuz TM-34 (4S) 7K-M	2002 Apr 25	2001-020A	2002 Apr 27	2002 Nov 9	ISS Nadir Zvezda	2002 Nov 10	193:06:21:15

Soyuz TMA, 2002–

In October 2002 the latest Soyuz variant to support a human crew made its orbital debut. It was a vehicle designed to provided expanded crew recovery options over the standard Soyuz TM – a lifeboat for the ISS. Designated TMA (Transport, Modification, Anthropometric), the new vehicle resulted, in part, from crew-sizing issues encountered during the Shuttle/Mir programme in the mid-1990s. The occasion of the flight of the first TMA spacecraft was also the first time a new Soyuz variant had been manned without first test-flying it unmanned. Several Cosmos missions had preceded the original Soyuz manned flights, and there were also unmanned missions for Soyuz T to Salyut 6, and Soyuz TM-1 to Mir.

SOYUZ TM AND MIR 2

Throughout most of the 1980s, the hardware that was fabricated as the space station core module DOS 7K no.8 (serial number 17KS no.12801) was initially assigned as back-up to DOS 7K no.7, which in 1986 became Mir. Once Mir had reached orbit, the back-up hardware was scheduled to be used as a follow-on station that resembled the Mir design – essentially, Mir 2. In 1984, plans were changed to incorporate core module no.8 into a much larger orbital complex, designated Orbital Assembly and Operations Centre (OSET). This space station would feature the central core module, solar array truss structures, various scientific modules, a fuel storage depot and a satellite repair facility, and would support a variety of specialised space tugs. The logisitics and servicing of this huge facility would be undertaken by the Buran shuttle, larger unmanned supply ships derived from the Progress vehicle but launched by the Zenit launch vehicle, and a Zenit-launched reusable crew ferry designated Zarya. This featured an enlarged Soyuz DM covered with Space Shuttle-type heat-resistant tiles as a thermal protection system. In reserve were the standard Soyuz TM spacecraft being used at Mir, or an advanced TM to be launched by Zenit. Indications of improvements to the basic TM design also began to be disclosed in the late 1980s.

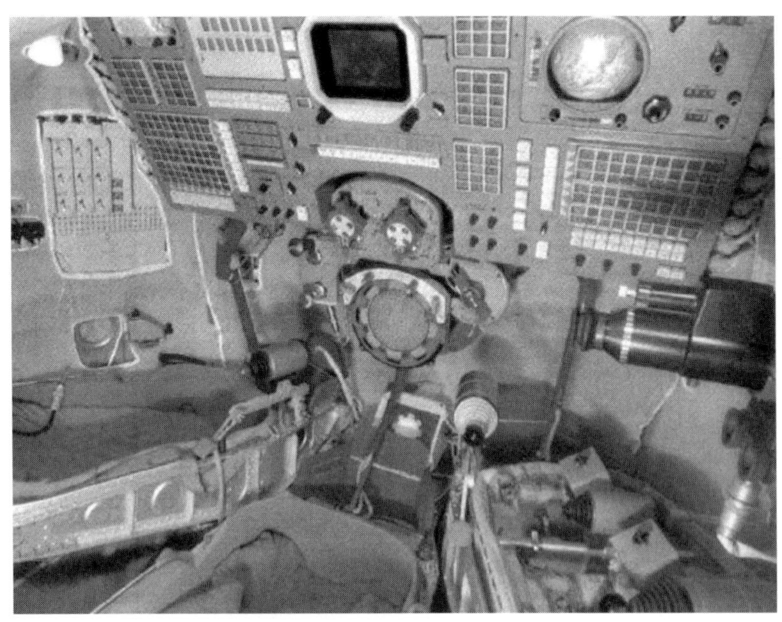

Inside the Descent Module of the Soyuz TM.

Further upgrades to Soyuz TM?

In December 1988, Vladimir A. Shatalov, Head of the Cosmonaut Training Centre, indicated that further improvements to the TM design were not only planned, but were in the process of becoming flight-ready: 'The vehicles exist not only on paper, but they also are taking shape in hardware form. The new versions are improved from existing Soyuz [and Progress], but they do not represent a major departure from the spacecraft we have been using for many years.'[1] In a subsequent interview in October 1989,[2] Yuri Semyonov's comments indicated that further modifications might be incorporated in an updated Soyuz TM to appear in 1990; but he could have been referring to the multitude of small improvements that followed almost every flight after crew debriefing and mission evaluation. Or it could simply have been a misquote by the Western media. In any event, there were plans for a larger crew ferry vehicle to support a larger space station after Mir.

Zarya: an enlarged Soyuz

During the late 1980s, Soviet designers evaluated a proposal for a larger logistics vehicle, which had evolved from the Soyuz design and was intended to supplement future Soyuz and Progress operations at the giant Mir 2 space station. This 'super-Soyuz' – Zarya – was planned as a reusable spacecraft concept; but because of financial problems it did not progress far, and work on the design ceased in January 1989.

On 27 January 1985 a government decree (Article 14F70) was issued to evaluate the concept. Preliminary design work began at the Energiya design bureau, and on 22 December 1986 the Military Industrial Commission approved development of the

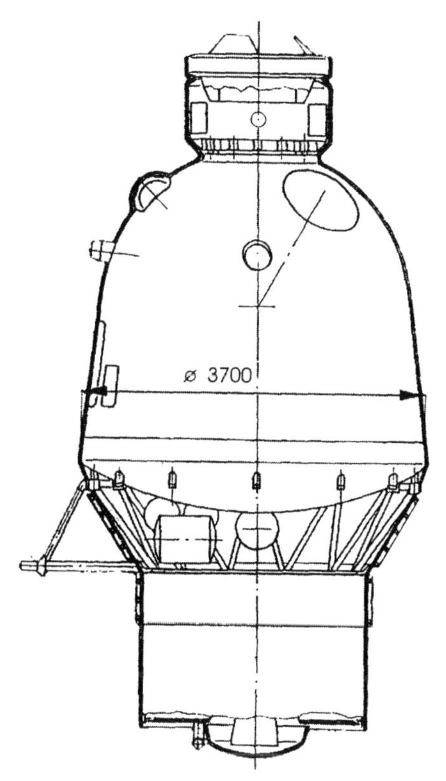

Zarya – the enlarged Soyuz proposal. (Courtesy Bart Hendrickx and the British Interplanetary Society.)

draft plan. The programme was to be developed in two-stages. Initially, the basic manned transport vehicle would be developed; then, in the second phase, certain modifications would be initiated that would allow Zarya to 'carry out special tasks during autonomous flights and joint flights with other spacecraft in a wide range of orbits and inclinations up to 97°.' The draft plan was completed during the first quarter of 1987, and was then submitted to the Ministry of General Machine Building for evaluation. Numerous changes were recommended, and due to their implementation, the modified draft plan was not ready until May 1988. When this draft plan was presented to a joint meeting of the Ministry of General Machine Building and TsUKOS (the forerunner of the Military Space Forces), approval was given for the further development of Zarya. By January 1989, however, financial problems forced the termination of the Zarya programme. This was the month in which Glushko died. Together with former cosmonaut Konstantin Feoktistov, Glushko was one of the main supporters of the Zarya programme; so perhaps within the bureau, opposition to the vehicle was a factor in its demise.

The original plan was to have a fully reusable vehicle without having to jettison any element prior to re-entry. The design featured an enlarged Soyuz DM configuration and a larger crew compartment, with the added capacity to deliver and return a far heavier payload in addition to the crew. A docking port was located in the front of the vehicle, with crew couches for five or six cosmonauts in the centre, and a cargo compartment in the aft. The whole vehicle would have been covered with heat-resistant tiles, with all the support sub-systems sandwiched between the TPS and the pressure shell of the crew compartment. The deployable antenna would have emerged through hatches after orbital insertion, and be stowed for entry. However, all this compactness resulted in very little volume inside the crew compartment. Zarya would therefore have to use an expendable Propulsion Module, and consequently would not be as fully reusable as the original 1987 draft plans had indicated.

Zarya would have measured 4.1 metres in diameter and 5 metres in length, with a maximum mass of 15,000 kg to a 190-km orbit at 51°.6 (near the limit of the proposed Zenit launch vehicle), and a payload capability (unmanned) of 3,000 kg. It would normally be crewed by two, three or four cosmonauts, but in later development plans it was foreseen that the vehicle could support up to eight cosmonauts. However, this was proposed only for a later version of Zarya (which was also the capacity of the later ISS Assured Crew Return Vehicle). Problems were encountered in convincing many (including the Air Force) that recovery would be safe. It appears that Zarya would have employed a small single stabilising parachute, rather than the larger canopy as on Soyuz. Descent to Earth after re-entry would therefore take place at high speed, and would rely totally on the twenty-four 1,500-kg-thrust hydrogen peroxide/kerosene engines (combined engine installation – ODU) to reduce the landing velocity to almost zero.

In the proposed Zarya design, braking engines replaced the main parachute. This system could reduce both vertical and horizontal landing speed to almost zero, which would prevent the capsule from toppling over, from being dragged, or rolling, as it landed. The resulting damage reduction would have a significant impact on reusability, with the potential for landing on planets or moons without atmosphere (where parachutes are useless). Initial flights of Zarya would have included a back-up system for the braking rockets as well as ejection seats, limiting the maximum crew size to four. No increase beyond that number would be possible until the landing system had been man-rated.

The docking system could be either probe/drogue or androgynous – depending on the intended mission and the vehicle with which it was to dock – and a crew would be transferred via an internal transfer tunnel. The vehicle would also feature sixteen 62-kgf mono-propellant ODU engines of the same combined engine installation as the twenty-four soft-landing engines. These would probably have used hydrogen peroxide (similar to the thrusters on the Soyuz DM), and would have been used for stabilisation during re-entry. All propellants stored in the crew compartment had to be non-toxic to prevent accidental contamination inside the habitation area. The exterior of the Zarya would have been covered with thermal 'bricks' similar to those used on the Buran shuttle, to ensure the reusability of each capsule between thirty and fifty times.

The unpressurised Propulsion Module was a far smaller unit than the Soyuz Propulsion Module. It featured two 300-kgf engines (multifunctional engine unit –

MDU) using N204/UDMH, and offering a total thrust of 600 kgf. It would have been used for orbital manoeuvring in conjunction with the on-orbit approach and orientation thrusters and retro-fire. Being the only non-reusable element, it would be jettisoned prior to re-entry, to be burned up. The planned orbital storage time for Zarya was 195 days, later leading to 270 days. Four mission profiles were planned:

A space station ferry to Mir-class stations, launched into a 190 km × 51°.6 orbit. A crew of two to four cosmonauts, with a 2-tonne cargo and logistics, and a 1.5-tonne return capability. For a crew of two: a payload delivery of 1.5 tonnes, and 2 tonnes return. In the unmanned mode: 3 tonnes up and 2.5 tonnes down.

A construction or repair spacecraft in low Earth orbit, with a crew of two or three, and perhaps equipped with a type of remote manipulator system (RMS). Autonomous flights of Zarya would have been conducted in the interests of the Academy of Sciences and the Ministry of Defence. There was no indication of an airlock or EVA capacity.

A crew rescue or space station lifeboat, after confirmation of the landing system and removal of ejection seats. Zarya would be launched with one or two cosmonauts, or unmanned, with the capacity to return up to eight cosmonauts, intended as a rescue vehicle for Buran crews.

An unmanned resupply vehicle. Zarya could have resupplied a space station; but it had no refuelling capability, and so Progress would still be required.

The obvious benefits over Soyuz or Progress would be the increased crew and cargo capability. In addition, Zarya's reusability was of benefit to mission planners in that fewer vehicles could orbit more hardware or crews, thus saving on launch and production costs. But there would be disadvantages. Orders would be reduced at factory level, and without a Soyuz OM, all crew provisions (such as the waste management system) would have to be incorporated in the crew compartment. Unwanted docking equipment and de-orbit engines would have to be returned, adding to the landing weight but making them available for reuse.

There were also concerns about the acoustic levels of the soft-landing engines (which were to be situated adjacent to the crew cabin), and about the reliability of soft-landing engines without a back-up system, which was a high risk to crew safety. Combined with the cost of production, and the requirement to still use Progress to resupply fuel to the stations, all this led to the demise of Zarya, although it briefly resurfaces as an ACRV for Freedom.

SOYUZ TM AND FREEDOM

During 1992 and 1993, new plans that emerged for Mir 2 reflected the growing international cooperation, from both Europe and Japan, in the new Russian space programme. During these cooperative studies, plans emerged to modernise the Soyuz and Progress vehicles. A further important development was also emerging from talks held in October 1991, between Yuri Semyonov, Energiya General Designer, and R. Grant, Vice President of Boeing (one of the leading primary contractors for Space Station Freedom), at the Congress of the International Astronautical Federation in Montreal. At this time, the American lead space station (under development with Canada, Europe and Japan) was vastly over budget, seriously delayed, and close to Presidential cancellation. In an inspired and timely proposal to help reduce costs with Freedom, and to secure much needed foreign investment in the Russian programme, Semyonov suggested that Russia could provide the Soyuz TM as an assured crew return vehicle for Freedom while the costly and protracted development of a more advanced ACRV continued. Over the next few months, in addition to the flight-proven and much cheaper Soyuz TM, the old Zarya 'big Soyuz' proposal was dusted off and also proposed as a Freedom rescue vehicle.

Inclinations and durations

On 18 June 1992, a $1-million contract – including a joint study in using TM as the interim lifeboat for Freedom – was signed by NASA and NPO Energiya. The main disadvantage in using a TM at Freedom was that the station was then planned to operate at an orbital inclincation of 28°, which was not accessible by a Soyuz launched from Baikonur. Alternative plans for launching manned Soyuz TM on Zenit or Proton rockets with additional Blok-D upper stages, would allow the spacecraft to reach Freedom's operating orbit, but with a reduced overall mass. The Russians requested that NASA consider raising the inclination of Freedom to 33°.5, to render it more accessible from the Baikonur cosmodrome; but this severely hampered Shuttle payload delivery capacity from Florida, and so NASA declined.

Other evaluations included the use of the Titan or Atlas launch vehicle from Cape Canaveral, or ESA's Ariane rocket from the Kourou range in French Guyana. Another plan was to place the unmanned rescue TM in the payload bay of Shuttle, and to ransfer it to the ISS by using the RMS system. This would require minimum modification to the basic TM design, and would dispense with rendezvous equipment requirements. Another problem in operating the TM at the proposed Freedom inclination of 28° concerned its intended role as a crew recovery vehicle. From such an inclination, Soyuz TM would not be able to land on the steppes of Kazakhstan, and so alternative sites – including three very promising landing zones in Australia – were evaluated.

The Soyuz TM/ACRV feasibility team

Since 1992, several American aerospace companies have been competing for the contract for the American ACRV. When plans emerged for adopting the Soyuz TM vehicle as a stop-gap, two of these companies (Rockwell and Lockheed) approached NPO Energiya to undertake joint contract work on evaluating the Soyuz TM ACRV proposals – possibly with the view to obtaining a licence from Energiya to adapt the Russian-built TMs in the United States.

During August 1992, an ACRV feasibility team travelled to Moscow to evaluate the feasibility of using Soyuz.[3] The team was charged with evaluating the capabilities of the current design of the TM, rather than a straight comparison of the Soyuz with the design specification of the proposed long term ACRV vehicle. As well as NASA management and industrial representatives, the NASA CB (astronaut office) assigned astronaut Ron Grabe as the crew representative for this evaluation. In their studies, the team found that if the interim or pre-Permanently Manned Configuration (PMC) period lasted five years, then Soyuz would be suitable until the point when the PMC reached or exceeded four persons (although they recognised that, at least on Freedom, there was a very real possibility that this milestone would never be reached – mainly due to budget restrictions). The main stumbling block identified by the team was that although Soyuz was not something the station crew would want to endorse for the duration of the SSF programme, it was a readily available vehicle. The team also identified the problem that the complex design of the ACRV would continue to impose high operating costs and operational delays, pushing its availability back several years. Therefore Soyuz – as the only viable crew rescue vehicle – would severely restrict the number of permanent crew onboard Freedom.

A four-man crew would require two Soyuz craft to be docked either in a zenith or nadir port to ensure clean separation from the station. At the time of the study, the assembly sequence of Freedom indicated that prior to PMC, only two nodes would be available. Both nadir ports were reserved for Payload Logistics Module (PLM) berthing, and one of the zenith ports would be occupied by an airlock. The task team recommended moving the airlock to an axial port if the assembly sequence and clearance allowed for such. Alternatively, a different means of conducting PLM change-out had to be devised, changing only one berthing port. 'If neither of these is possible, the fundamental concept of Soyuz/ACRV operations [two vehicles for a four-man Freedom crew at PMC] is flawed,' the team concluded.

Soyuz escape trajectory analysis

The analysis reviewed a variety of candidate locations, based on various accommodation issues, and focused on two Soyuz vehicles docked at one time to support the PMA (four crew). The location of Soyuz for escape trajectory narrowed down to three options: +Z (primary nadir), –Z (primary zenith), and –X (primary minus velocity). The analysis was also influenced by six factors: the magnitude of the ΔV; the direction of departure; the configuration of the solar arrays while tracking the Sun during the orbit; the density of the atmosphere; the attitude control of the station; and the size of the docking adapter connecting the Soyuz to Freedom. A set of assumptions was also included in the analysis as to the operating orbit, a modelled docking adapter and a 'canted' docking adapter, two atmospheric densities, two configurations of solar array positioning and three attitude control modes, and a nominal Soyuz departure rate of 12 cm per sec.[4]

When all of these factors had been defined, each of the three departures were computed, resulting in a combined total of eleven case studies. The conclusion of the report stated that the aerodynamic density or the solar array configuration had no significant effect on departing from any of the three locations where the ΔV required was less than 1 cm per sec. However, the contingency attitude rates of Freedom were significant in the increases of ΔV to depart safely from the station, depending on the pitch or roll rate of Freedom. It was determined that either +Z or –Z departure directions were equally desirable for a clear departure path, with canted docking adapters reducing the ΔV a little, while the –X departure suffered significantly in the mode of excess station pitch rate. The analysis – which was carried over for ISS studies – provided the mathematical data required to determine that Soyuz could be used at Freedom as a crew escape vehicle, but that the location had to be clearly determined to ensure efficient and safe departure. For that, +Z nadir or –Z zenith would be the optimum.

Crew accommodation

A further primary concern with regard to the Soyuz at Freedom was that the crew module excluded members of the current astronaut corps because of the height and weight constraints of the Soyuz couches. According to information presented to the team, although the limits quoted a seating height of 99 cm, a total height of 187 cm, and a weight of 90 kg, applicable to each position, the Russians conceded that the centre seat could accommodate a crew-member of slightly greater total height.

The seating height was a firm limit governed by the physical size of the metal shell that supported the individual couch inserts. As this insert provides head and spine protection during re-entry and landing, the occupant's body 'cannot spill over the couch frame.' The weight limit was defined by the ability of the couch shock

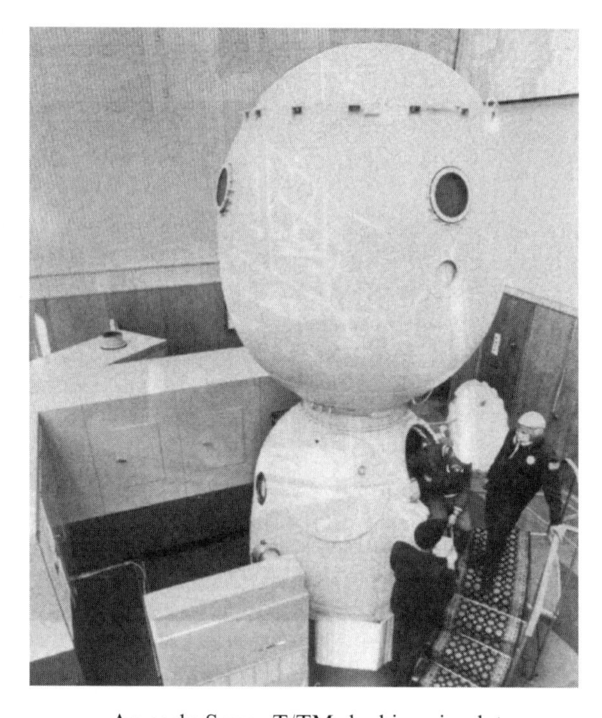

An early Soyuz T/TM docking simulator.

mechanism to absorb crash loads if the retro-rockets did not fire. The task team indicated that since Soyuz was an emergency-only return vehicle, crew weight could be waived to some degree, while accepting the risk of degraded crash landing capability.

However, the team was concerned that these limits excluded some members of the current astronaut team from assignment to resident crews on Freedom. They would be restricted to pre-Permanently Manned Configuration assembly flights and Shuttle crew assignments, and stricter weight and height limits might have to be applied to future NASA astronaut selections. The anthropometric limits quoted above were relaxed from current TM operations by 3 cm and 10 kg, 'under the assumption that the Soyuz, when used as an ACRV, would *not* require crew-members to wear pressure suits.' The CB endorsed this assumption, and was impressed by the measures demonstrated by NPO Energiya to ensure pressure integrity. It appears that, as Soyuz was to be used only in a contingency situation, there was sufficient crew safety for shirt-sleeved operations. Energiya was requested to provide further hard data to support this claim, to allow formal acceptance by NASA. If it was a case of proving pressure garments, then this would have a greater impact on crew assignments.

Medical and emergency evacuation

A capacity for medical evacuation was also required. Soyuz could return a sick or injured crew-member only if he/she was strapped into the couch, approximating a foetal position (as Vasyutin had to endure at the end of the Soyuz T-14 mission in 1985; see p.312). The team identified that leg or spinal injuries could not be accommodated, which led to the compromise that such injuries remained a remote possibility.

With a crew of three, there is no room onboard Soyuz DM for medical support equipment, so in a med-evac role, Soyuz's third seat would be required for stowage of medical support hardware. This limited the return of Soyuz to a two-person rather than a three-person crew. The task team suggested that the array of medical support equipment could be stowed in the SSF or OM of Soyuz until required, to retain the flexibility for a three-person escape option. Other concerns focused on the limitation of TM to handle open-loop oxygen resuscitators. As the DM carried no nitrogen for cabin atmosphere replenishment, it was impossible to purge the oxygen-rich atmosphere resulting from the existing resuscitators. To overcome this, the task team recommended (to NPO Energiya) that an independently-produced closed-loop resuscitator be developed. Furthermore, during the discussions it was found that NPO Energiya accepted an oxygen concentration of 40%, allowing Soyuz to operate with only oxygen replenishment in the DM. The task team therefore asked the question: 'Would NASA really accept a 40% oxygen concentration limit, given our sensitivity to flammability?' Even twenty-five years after the loss of three astronauts in the Apollo 1 pad fire (27 January 1967), caused by a spark in an oxygen-rich atmosphere inside the sealed Command Module, the spectre of a fire in a spacecraft still haunted NASA.

One operational problem foreseen was that in nominal Soyuz TM operations, a lengthy (1-hour) leak-check period preceded separation. In an emergency evacuation from Freedom this was not practical, with only hatch closure preceding separation, and strapping into the seats completed only after safely undocking. The team also found that there was no time-delay requirement to activate dormant batteries to ensure adequate battery temperature at separation. They warned that an increase in Soyuz's orbital lifetime to a year must be without adding a waiting period for powering up batteries, which would render Soyuz unacceptable as an ACRV.

A final disadvantage in using Soyuz as a rescue vehicle was in the assigned seating dictated by the custom-built couch-liners. The NASA team rightly observed that it was impossible to predict which crew-members would arrive at which Soyuz ACRV in event of a catastrophic emergency on station. If both vehicles could be located on the same node then this was not a major problem, but in practical terms the vehicles would be at different locations. This would entail finding a way around having a crew-member near the wrong Soyuz at the wrong time! It was suggested that for a four-person permanent crew, both Soyuz vehicles be maintained in three-person configuration, with a fourth seat stowed in each OM. This would allow for any crew configuration after separation. The attachment of a new seat-liner in the Soyuz DM was a relatively simple task, frequently demonstrated on Salyut and Mir missions since 1978. The impact was one of logistics, in that for every four-person main crew exchange, eight seat liners (two per person) would have to be flown to and from the station, taking up valuable stowage in the OM or the Shuttle. A more serious problem would occur if all four crew-members were isolated near one of the three-seat Soyuz ACRVs. This scenario was not mentioned in the NASA report.

Training for pilots and passengers

In concluding their study, the team identified training requirements for all Freedom crew-members to qualify for Soyuz certification: one who could operate or fly the Soyuz, and those who were passengers with no piloting skills. Energiya estimated that it took 200 hours to train a crew-member as a Soyuz Operator (enough to operate Soyuz from separation through de-orbit and landing) – and this included the skills to handle systems malfunctions and to back up automated systems via manual control. Ideally, piloting skills would be advantageous in selecting Soyuz Operators, but not necessarily a pilot astronaut or cosmonaut (thus engineer cosmonauts or NASA Mission Specialist astronauts could qualify for this role). A number of astronauts from ESA have already undertaken the additional training necessary to gain this qualification, and have flown as Flight Engineer 1 on a number of missions.

Soyuz passengers (in the Soyuz ACRV role) would be required to close hatches, activate the separation sequence, work the communication systems, have knowledge of the automated control systems, and assist the Soyuz Operator. In total (according to NPO Energiya), 20–30 hours of training would be sufficient. In a four-person main crew, two people would be trained as Soyuz Operators, and two as Soyuz Passengers. In the unforeseen situation in which only a Passenger was able to return a Soyuz in an emergency, ground would up-link all non-automated commands for de-orbit and re-entry. In the event of a radio failure, the crew would resort to onboard documentation.

The Energiya training figures were equally split between classroom time and simulator time, and did not include water survival training or remote terrain training sessions. The American team considered that this time was over-estimated, and that a great deal of 'Soyuz familiarisation' could be accomplished during NASA astronaut candidate (Ascan) training, with refresher courses during mission-specific training time. The team also indicated that training would be better accomplished in the US (at Houston), using English as the primary language, and that NASA 'could not afford the inefficiency of putting crews sequentially through a course originally developed to suit the Russian systems, taught by instructors whose native language is not English.'

A Soyuz TM lifeboat for Freedom

The team conclude that the difficulties of incorporating the Soyuz as a short-term ACRV involved the location of the two vehicles in the current Freedom design, and that 'there are numerous areas in which the capacity provided by Soyuz falls short of what is desired. The net sum of these limitations is a capability which would make the Soyuz acceptable as a short term solution only.' In September 1992 – after two weeks of joint discussions and evaluation – it was publicly announced that there were no major hurdles in adapting the Soyuz TM for Freedom lifeboat assignment, but that more work was required before the Soyuz officially became part of Freedom.

In March 1993, new studies for extending the on-orbit 'shelf life' of the Soyuz TM (from the six months it remained docked at Mir, to the 1–3 years that NASA required it to operate at Freedom) were begun between NASA and NPO Energiya. NASA officials were on hand to witness the TM landing in July that year as a dress rehearsal of a lifeboat recovery scenario. By December 1993, there were good prospects for using Soyuz in conjunction with Freedom; but that same month, the contract with NOP Energiya was cancelled in favour of accepting Russia as a full partner in the new International Space Station, and changing the orbital inclination to 51°.6 – within the range of the Russian launch vehicles and the American Shuttle. With that agreement, a new role emerged for Soyuz TM as a space station ferry and rescue craft, able to land in Kazakhstan.

SOYUZ ASTRONAUTS: CERTIFICATION ISSUES

In October 1992, NASA signed contracts with the Russian Space Agency to create what would become known as the Shuttle/Mir programme. This involved flying a cosmonaut on a week-long Shuttle mission, and then a NASA astronaut on a 90-day Mir mission. The training called for the astronaut to qualify as a Soyuz crew-member, leading to a launch on a Soyuz TM vehicle, but returning on an American Shuttle after it had docked to the Mir station. One of the criteria for selecting any American astronauts for Mir training was that they had to fit into the cramped Soyuz DM – both for launch and, if necessary, for emergency return prior to the Shuttle docking to the station.

Shuttle/Mir, Phase 1A

On 3 February 1994, NASA announced the prime and back-up astronauts for what became known as the Phase 1A programme:

NASA	*Prime*	*Back-up*
Mir 1	Norman Thagard	Bonnie Dunbar

Dunbar would also be assigned to the Shuttle flight (STS-71) manifested to return Thagard to Earth. Although the Russian cosmonaut training programme and cultural adjustments were an additional challenge for the Americans, they would have no difficulty in fitting in the Soyuz couches and Descent Module, due to their height and build (Thagard was 1.75 metres and 70.76 kg; Dunbar was 1.66 metres and 53 kg).

Shuttle/Mir, Phase 1B

By December 1994, the agreement had expanded to include further flights by American astronauts to Mir under the Phase 1B programme. In each case, the astronaut would be launched and returned on NASA Shuttles, reducing Soyuz training to only include emergency evacuation and recovery training. The team of six astronauts (announced on 30 March 1995) that began 'cosmonaut' resident crew training for long flights on Mir from 1996 were:

NASA	*Prime*	*Back-up*
Mir 2	Shannon Lucid	John Blaha
Mir 3	Jerry Linenger	Scott Parazynski
Mir 4	John Blaha	Wendy Lawrence

It was expected that Parazynski and Lawrence would rotate to fly the fifth and sixth missions in the programme, with their back-ups announced closer to their missions. However, as the Shuttle programme had fewer physical limitations on crew-members than did Soyuz, several problems arose as training progressed. On 14 October 1995, NASA announced that Parazynski would discontinue his training because he was too tall (at 1.87 metres) to safely fit inside the Soyuz Descent Module. When Parazynski was assigned to the back-up role, with the probability that he would rotate to a prime position on the Shuttle/Mir 5 mission, both NASA and the RSA understood that he was slightly beyond the normal height to fly on a Soyuz capsule (after an extended period in space, the body 'grows' 2 cm); but as a contingency return was a remote possibility, it was agreed that further discussions between the agencies would take place.

The preliminary evaluation cleared Parazynski for back-up training, but further detailed discussions between the two agencies over seated height and deceleration loads on emergency landing indicated that the safety margins against injury were unacceptably reduced. The Russians indicated that the Soyuz seating system could not be specially modified in time for Parazynski's mission, and so regrettably – and despite his scoring one of the highest levels of performance in training, and gaining the respect of his Russian trainers – he returned to NASA JSC for assignment to a Shuttle mission. Just ten days later, on 24 October, there was a second blow to the programme. Lawrence – about to begin a year's training as back-up on Shuttle/Mir 4 – did not meet the minimum height limit (she stood at 1.60 metres) to fit safely inside the Soyuz descent craft. She was also to remain in JSC, to be replaced in Mir training; but shortly afterwards she was (surprisingly) cleared for Soyuz flights, and was reinstated into the NASA training team based in Moscow. On 16 January 1996, NASA announced adjustments to the crewing, bringing in Mike Foale and Jim Voss:

NASA	*Prime*	*Back-up*
Mir 2	Shannon Lucid	John Blaha
Mir 3	John Blaha	Jerry Linenger
Mir 4	Jerry Linenger	Michael Foale
Mir 5	Michael Foale	James Voss

Then, on 16 August 1996, NASA announced the revised Shuttle/Mir 6 and Shuttle/7 crews:

Mir 6	Wendy Lawrence	David Wolf
Mir 7	David Wolf	Andrew Thomas

During the summer of 1997, however, the damage caused by the Progress M-34 collision with the Mir Spektr module necessitated a series of EVAs. Russia asked NASA to fly a Russian EVA-qualified astronaut to assist in any repairs that might be necessary during the later NASA/Mir residency. The problem was that Lawrence could not fit inside the Orlan suit, and had not trained for EVA. She was therefore replaced by Wolf (who did fit); but she received a 'consolation prize', flying the STS-86 mission (with Parazynski also assigned to that Shuttle crew) that would have delivered her to Mir; and again on STS-91 – the final visit to the station. On 10 October 1997, NASA named the final Shuttle/Mir crews, having found quite a problem convincing any other astronaut to serve on a dead-end assignment without a flight. But Jim Voss was willing, having already trained as back-up to Foale, and in line for assign ment to the second ISS crew. The names of Scott 'Too Tall' Parazynski, Wendy 'Too Short' Lawrence, and Dave 'Just Right' Wolf, circulated for some time after the reshuffling of assignments.

NASA	*Prime*	*Back-up*
Mir 6	David Wolf	Andrew Thomas
Mir 7	Andrew Thomas	James Voss

The safety issues of continuing to fly Americans on the ageing and damaged Mir were examined by a NASA review board and an independent task force chaired by former NASA astronaut and ASTP commander Tom Stafford. After further discussions with Foale (who had been onboard Mir during the collision) and with Wolf and Thomas (who were still to fly), NASA management decided to allow the programme to reach its conclusion, providing safety was not further compromised. But NASA was not happy, as this placed serious limitations on future crewing for the ISS, where Soyuz craft would be used for crew rotation and as the short-term resident crew lifeboats until the planned X-38 became available.

The initiation of the TMA series

With the problems encountered by Lawrence and Parazynski during Shuttle/Mir programme, NASA evaluated the seating arrangements of Soyuz more closely, and found that 40% of its current cadre of astronauts would not fit in the TM. This would have a serious effect on selecting NASA astronauts for long-duration mission training assignments for ISS, and so NASA asked the Russians whether slight modifications could be incorporated into the internal arrangements of the Descent Module to accommodate more of the US astronauts currently in training. In January 1996, a preliminary agreement was reached to incorporate suitable modifications at US expense.

The Soyuz TM crew-member size limits were set as follows:

Measurement	*Unmodified Soyuz TM (before June 1999)*	*Modified Soyuz TM (after June 1999)*
Maximum standing height	182 cm	190 cm
Minimum standing height	164 cm	150 cm
Maximum sitting height in Soyuz seat	94 cm	99 cm
Minimum sitting height in Soyuz seat	80 cm	80 cm
Maximum crew mass	85 kg	95 kg
Minimum crew mass	56 kg	50 kg
Maximum foot length	not defined	29.5 cm
Maximum bideltoid breadth	not defined	52 cm
Maximum interscye breadth	not defined	45 cm
Maximum hip breadth, sitting	not defined	41 cm
Maximum thigh-to-thigh breadth	not defined	41 cm
Maximum chest circumference	112 cm	not defined
Minimum chest circumference	96 cm	not defined

The new modified Soyuz design would replace the Zarya re-entry vehicle (the lifeboat system originally planned for the now abandoned Freedom/Alpha space station). The new, modified Soyuz TM would thus serve as the short-term crew rescue vehicle for the ISS until the purpose-built X-38 was available. On 19 September 1996, NASA signed the agreement with RSA (Article 11F732, designated 7K-STMA) for adapting the Soyuz TM into the TMA. Unfortunately, initiation of the TMA production programme was delayed during the late 1990s due to non-payment by the Russian Government to RKK Energiya, the prime contractor for the vehicles. As a result, NASA refused to pay for the upgrades until Russia ensured the production and delivery of the flight articles. Initially, four vehicles were planned for construction and flights through 2004, with no firm commitment or funding for any more.

SOYUZ TMA DESIGN MODIFICATIONS

Outwardly, the TMA does not differ greatly from the TM. Its overall length is 6.98 metres, and the maximum diameter remains at 2.72 metres. The diameter of the habitable modules has been stated as being 2.2 metres, with a habitable volume of 9 cubic metres. The span of the solar array, when extended, is 10.7 metres, producing a total area of 10 square metres and providing, on average, 0.60 kW. The TMA carries 900 kg of propellant (N204/UDMH) for orbital manoeuvring and de-orbit burns, with the primary engine thrust of 400 kgf producing a total burn-time of 305 seconds and spacecraft ΔV of 390 metres per sec. The crew size is three, with an operational design life of up to fourteen days and and orbital storage time of up to 180 days (or six months). Additional flexibility would probably extend this up to 200 days. Module masses are similar to Soyuz TM: 2.9 tonnes for the DM, 1.3 tonnes for the OM, and 2.6 tonnes for the PM.

The major design adjustment lies in the arrangement of the seats and their support structures. First, the individual frameworks had to be enlarged to carry tailored seat-liners for crew-members with a standing height of 150–190 cm and a sitting height of 50–95 cm. To accommodate these new seats, changes to the foot-well area in the hull of the Soyuz had to be included on the left and right crew positions.

The crew display and control panel has also been modified to incorporate new computer displays and to relocate crew equipment under the three seats in the base of the module. At the same time, to incorporate the larger crew-members, the seats' shock-absorbers had to be upgraded to accommodate higher weight limits, from 56–85 kg on Soyuz TM, to 50–95 kg on the TMA. No two crew-members, of course, would weigh exactly the same, and the difference in the weight between the occupants of the two outer couches can vary as much as 45 kg, which can have a serious effect on the descent craft's centre of mass as it enters the atmosphere. Therefore, the spacecraft's automated landing system has also been adjusted to cope with such variations during future flights.

A problem also occurred with the heavier vehicle at soft-landing, during which the six landing rockets are fired 1.5 metres above the ground to cushion the landing. In the TMA, two of the six engines have been modified to operate at two different levels of thrust, as determined by the landing mass of the DM. This ranges from approximately 2,980 kg to 3,100 kg. In all, the TMA will have a maximum mass of approximately 7,200 kg – about 200 kg more than the TM. This increase requires the TMA to be launched by an uprated R-7 with improved first- and second- stage performance – designated the Soyuz FG – and then eventually a further uprated version of the R-7, designated Soyuz 2 (see p.{?})

Soyuz TMA modifications

The modifications to the TMA required the expansion of the seat liners, an increase in the overall length, and modification of the shock-absorber struts to allow for both lighter and heavier crew-members. The internal structure of the spacecraft had to be modified for the taller crews. Panels were also moved to allow clearance for the knees, and hardware was relocated to support both reach and visibility. These changes included:

- A reduction in the physical size of the Neptune display panel, leading to a change in display layout and display drivers.

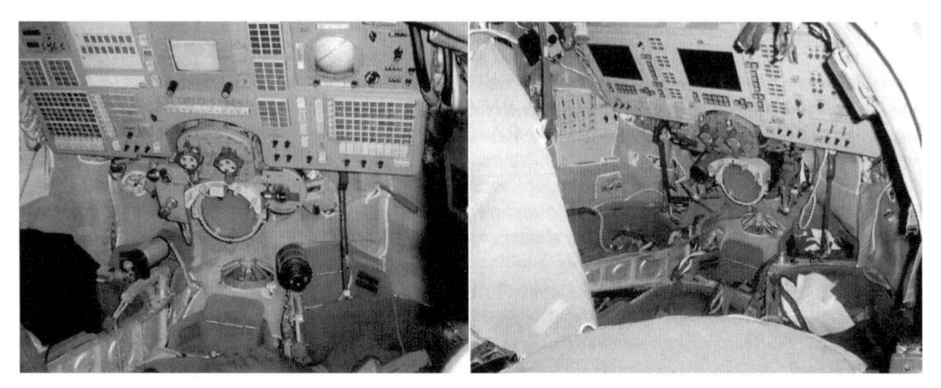

The cockpits of Soyuz TM and Soyuz TMA.

- Replacement of the mechanical recorder with a solid-state device, to reduce size and mass.
- A reduction in the cooler/dryer unit to provde extra room for the additional required crew-size.
- A change of the computer and software controlling the Descent Control System, to compensate for changes in the centre of gravity of the vehicle and overall mass of the crew.
- Modification and reduction in size of the sequencer used for landing actions. This was part of the overall landing system complex (the soft-landing engines, the altimeter, and the landing system sequencer).
- An upgrade and reduction in mass of the TV system, from black-and-white (as on Soyuz TM) to colour.
- An upgrade and reduction in mass of the voice communications system.
- Modification of the onboard complex control system used to interpret control commands, to be compatible with the other modified systems
- Modifications to be certified for 1 year.

A revision of the sub-system design proposals was published on 30 January 1997, and the preliminary design report on the modifications was issued on 5 February 1997. An evaluation report for the mock-up TMA was published on 18 March 1997, and the final installation documentation was published on 8 December 1997.

Over the next few months, a series of test programmes was initiated to evaluate the changes incorporated into the TMA design. This included a series of development tests of new and modified equipment such as the instruments, seats, shock-absorbers, soft-landing engines and associated components, and tests on modified systems such as displays, descent control actuators, and parachute systems. A series of fit tests, dynamic structural tests, ground-based drop tests and airborne drop tests were initiated on a modified Soyuz.

Ground tests A structural mock-up of TMA was created to demonstrate the additional anthropometric measures. This included the seat modifications and the relocation of equipment. In the mock-up, solutions to the problem of relocating the hardware were confirmed, and tests of the ingress and egress of crew-members demonstrated the ease of future crew operations in the vehicle. There was also a programme of vibration testing of a structural mock-up, during which the dynamic loads on the character of the structure were determined. The structure, components and attachments were subjected to vibration loads. The vehicle had been assembled as per flight vehicle design, which included plumbing and electrical cabling. A final report on the DM mock-up (3D) dynamic strength tests was published on 30 April 1998.

Ground-based drop-tests A series of three drop tests verified strength characteristics. This was followed by eight further drops with nominal landing loads, designed to verify the loads that affected the crew after the modification of the DM. This programme verified maximum operational landing velocities on the modified seats and structure of the DM, and determined the g-loads in the crew areas and the level of shock loads under a variety of landing conditions. The final report on the DM mock-up drop-tests was published on 30 March 1999.

Airborne drop-tests Four drop-tests were conducted, evaluating normal and reserve modes for parachute recovery. These tests verified the modified systems during parachute operations, and the velocities and shock loads that could affect the crew during landing. Also tested was the interaction of the modified systems, together with verification of their operational use during parachute recovery. The final DM mock-up airborne test report was published on 30 April 2000.

A series of reports listing failures and their corrective actions was published between 20 March 1997 and 27 September 1999. A landing impact assessment report was issued on 26 October 2000, together with additional documentation updating and approving all hazards testing of Soyuz TMA, and a series of updated Soyuz TMA training manuals.

Following the test and modification programme, the NASA team charged with evaluating and monitoring the TMA changes recommended that ISS technical assessments on the seats, crew displays and controls, and command and data handling issues, should be performed during phases of rendezvous, docking, on-orbit operations, and undocking. The team did not assess the entry guidance and Earth landing system, although additional information was accrued on contracted modifications to these sub-systems that were not part of the NASA programme.

In addition, software modifications to the motion control system were evaluated, and three-axis accelerometer testing was performed on the Progress '7' (M1-8) mission. The NASA/Soyuz crew training team was also briefed on TMA systems and crew procedures and the TMA simulator. The Russian vehicle team held technical discussions on the TMA modifications, including visual aids from the airborne drop-tests using the new descent control and landing systems.

Soyuz TMA-1 crewing

The inaugural flight of Soyuz TMA (TMA-1) was the first mission in which a new flight model carried a crew. This was because the modifications were mainly internal, and did not affect key operating systems. However, the crewing reflected the changing times and economic realities facing Russian manned spaceflight.

The prime crew Commander was Lieutenant-Colonel Sergei Zaletin of the Air Force, who was selected as a cosmonaut in 1991, and was the Commander of the last expedition to Mir in 2000. His Flight Engineer was ESA astronaut Lieutenant-Colonel Frank De Winne, a test pilot with the Belgian Air Force before joining the ESA Astronaut Team in January 2000. The Russian Space Agency (Rosaviakosmos) and the Energiya Production Association have entered into an agreement with ESA for a series of flights by European astronauts onboard Soyuz. De Winne's flight represents the third of an agreed series of six such flights under the initial contract. The cost of the mission is met by the astronaut's home country and ESA. As well as paying for the seat, they are also providing a series of experiments that will be taken up in the Soyuz and in one of the Progress craft. ESA astronauts assigned to these missions have to undergo training and qualification as Soyuz Flight Engineers.

The crew-member for the third seat on TMA-1 was expected to be another tourist candidate. (There is normally no numerical designation on an initial identified flight of a Russian spacecraft, but there was on this occasion). Many people have attempted to raise the funds to take part in a flight, which costs around $20 million (£12 million). The price covers initial medical screening, cosmonaut training at TsPK, medical support, a Sokol pressure suit from the Zvezda Production Association, launch, nine days on the ISS, mission recovery, and post-flight medical support. It usually takes six months for all the training to be accomplished, and this includes a visit to NASA's Johnson Space Flight Center near Houston, Texas, for ISS safety briefings and training under the Multilateral Crew Operations Panel agreement on principles regarding expeditions and visiting crews to ISS.[5]

A leading candidate for this flight was Lance Bass, a 23-year old American singer with the US rock band N'Sync, whose backing came from a consortium of record and film companies based in Hollywood. Previous participants, Dennis Tito and Mark Shuttleworth, are multi-millionaires who found the money out of their own funds. Unfortunately, by the end of September, Bass's sponsors were unable to raise the required funds, and he was dropped from the flight. Had he flown, he would have become the youngest person to enter orbit, surpassing the record of Gherman Titov in August 1961. Titov was 25 years 11 months.

The back-up crew-members were all Russian. The Commander was Lieutenant-Colonel Yuri Lonchakov, who flew to the ISS as a Mission Specialist on STS-100 in 2001. He was in line to fly as Commander of TMA-2, until a GMVK meeting in Moscow on 1 October 2002 tentatively named him for the third vacant seat on TMA. The back-up Flight Engineer was Alexandr Lazuktin, who became a cosmonaut in 1992 and flew a long-duration mission to Mir in 1997. There was no third-seat back-up. In late September it was decided that rather than fly a cargo container, the Commander of the back-up crew (Lonchakov) would become Flight Engineer 2, and De Winne would be Flight Engineer 1.

The maiden flight of Soyuz TMA

Soyuz TMA-1 was launched, with a crew of three (Zaletin, De Winne and Lonchakov) on 30 October 2002. The mission was called Odissea, and after two days the craft docked to the ISS without any problems. At the end of their short mission, the Odissea crew returned to Earth in TM-34. In its role as a rescue craft, TMA-1 is due to return to Earth with members of the expedition 6 crew, following the loss of *Columbia*.

The crewing of TMA missions in 2003

In late 2002, the State Commission named the crews for TMA-2 and TMA-3. TMA-2 would be commanded by Colonel Gennedy Padalka, with the ESA astronaut Pedro Duque (a Spanish engineer); while TMA-3 would be commanded by an Energiya engineer Pavel Vinogradov, who would be joined by André Kuipers (a Dutch doctor). It would be Kuipers' first spaceflight, but Duque flew on STS-95 (the John Glenn mission) in 1998, and has already completed training as back-up for a 1994 Soyuz – the ESA's Euro Mir '94 mission. The third seat was to be available for potential Tourist flight participants. A number of people have undergone initial medical assessments at the Institute of Biological Medical Problems (IBMP), but no disclosed agreements are in place. A 'Big Brother'-style contest (with people training for a mission and being voted off, carried on American TV) was a project that could have filled one of the seats on a flight in autumn 2003. The back-up crews would all be Russian.

Within days of the loss of *Columbia* on 1 February 2003, all named crews were stood down, and it was decided that the Soyuz TMA would now be used for ISS expedition crew rotation. The next Soyuz TMA will be launched to the ISS in May 2003, and will be crewed by one Russian and one American, who will maintain the station until normal operations resume or another Soyuz is required in late autumn. Assigned three-person resident crews will therefore be temporarily replaced until the Shuttle returns to flight, further restricting the seats available for the other ISS partners.

FUTURE OPTIONS

Although TMA has only recently flown as the latest manned version of Soyuz, there are already plans underway for a 'next-generation' Soyuz that would incorporate a wider range of upgrades. The order to begin studies on upgrading the Soyuz (and Progress) was issued by RKK Energiya President Yuri Semyonov in November 1995, prior to the authorisation to develop the TMA version.[6]

The primary reason was to ensure that the new Soyuz spacecraft would feature lighter and 'state-of-the-art' sub-systems developed solely in the Russian Federation, to avoid the inflated prices charged by former Soviet Republics which had previously supplied the components under the USSR. There has also been a design study to determine which of the Descent Module's components could be refurbished and re-used on new vehicles, and for increasing the maximum orbital storage time to year. This would avoid using two Soyuz craft per year to support the ISS, thus reducing craft requirements and not exceeding Soyuz lifetime safety guidelines, which currently stands at six months.

Soyuz TMA in flight.

Soyuz TMM improvements

Because of the shortage of funds available to the Russian programme, which was already struggling to fulfil the commitment to contracted spacecraft, contracts for such improvements were not signed between Energiya and RSA until June 1997. Under the subsequent draft plan approved in August 1998, the improved model of Soyuz, called TMM (double modification), was planned to include:

- Replacement of the computer previously installed in the propulsion compartment on Soyuz TM, together with the computer in the DM, with one single upgraded unit in the DM.
- The installation of an improved telemetry system throughout the spacecraft.
- The installation of a satellite data relay system, designated Regul.
- The installation of an autonomous satellite navigation system and an improved Kurs rendezvous system, designated Kurs-MM.
- The installation of a pair of additional braking engines in the approach and docking engine sub-system. This was apparently incorporated as a result of experiences of near collisions and collisions during Mir missions.
- Modifications to the automatic landing control system, and a decision to deploy the recovery parachutes at a lower altitude in order to improve landing accuracy.
- An increase in the maximum flight duration margin to 380 days (or 12.6 months) by the introduction of a thermoelectric cooling system for the

hydrogen peroxide thrusters on the DM, the installation of improved storage batteries, and the fabrication of the oxidiser tanks in the propulsion compartment out of steel instead of the aluminium alloy previously used.

As the Russians evaluated the cost of all these changes, it became apparent that to initiate all of them at the same time would be very expensive. The TMM design therefore remains very much on the drawing board for a possibly future variant. However, by the summer of 1999, alternative plans were devised for an intermediate upgrade, designated Soyuz TMS (the meaning of which is unclear), with Semyonov approving these designs in December 1999.

Soyuz TMS: an intermediate answer

This version of Soyuz is expected to incorporate only the two extra braking engines and the improvements in landing accuracy that will eventually see the traditional Russian manned spacecraft recovery zones shifted from the steppes of Kazakhstan to within Russian territory as a cost saving measure. The hydrogen peroxide cooling system will be incorporated in the design to allow increased orbital duration, but to only 180–210 days (6–7 months), and not the full year capacity hoped for, thus still requiring double the amount of vehicles to support the ISS than with the full TMM plans.

The two-computer design in the descent and propulsion compartment will remain, as on the Soyuz TM, but there are plans for significant upgrades and improved software. In addition, updates to the flight control system will increase confidence in safe and reliable dockings and a 'reliable' rescue capacity (an important point in selling the TMS series to NASA). Plans to introduce the TMS to flight operations are dependent upon adequate and sustained funding and regular supply. The TMS would be launched by the R-7 Soyuz 2 launch vehicle, and could see its first flight by 2006 or 2007 – which would coincide with the fortieth anniversary of the first unmanned and manned Soyuz missions in 1966 and 1967.

Supply and demand

There currently exist four Soyuz TMA vehicles in different stages of construction, and one vehicle that has flown. The first one was launched in October 2002, and the others will fly in 2003/04. This meets Russian contractual obligations to supply an escape craft for the three-person resident crew currently operating on the ISS. Russia is no longer not building Soyuz craft on speculation. If a craft has a launch or docking problem, there would be no immediate replacement available – which would throw the permanent occupancy of the ISS into doubt.

There is an acceptance that for the ISS to be really operational, a permanent crew of six is necessary to carry out the mix of maintenance and experiments, and to begin to meet the potential of the station and the resources poured into it. NASA has stated that it does not expect the size of crew, in the foreseeable future, to exceed three. This has placed the Americans in conflict with its international partners, who are developing hardware for the ISS, and in exchange have agreed flight opportunities for up to 180 days a year when their modules are launched. The

Original Soyuz Orbital Modules lined up during production flow.

long-duration opportunities would seem to be severely curtailed unless the current crew-size is reviewed. One option being studied by United Space Alliance (the Space Shuttle processing contractor) is to fly extended-duration orbiters to the the ISS to allow expanded Shuttle science crews to perform additional scientific research for periods of up to 28–45 days. But this would still fall short of the international agreements, and would require substantial additional funding from the already restricted NASA budget. Meanwhile, the Russians have offered to increase construction of Soyuz to up to four per year, although funding would have to be found to pay for the extra craft. So far, no-one has produced the additional resources to support this option. The permanent docking of two Soyuz vehicles would allow the crew-size to double from three to six. Currently, there is no long-term resolution to the issue of crew rescue and return. Comments from Russian officials towards the end of 2002 indicate that without additional funding, permanent occupation of the ISS may end in May 2004 when the optimum design life of the third TMA is reached (the fourth vehicle being reserved as a rescue craft).

But no-one would be surprised if the need for additional craft were to become a reality. In 2003, Soyuz is available, cheap, and flight-proven – and it is entirely possible that another Soyuz variant will emerge to serve manned spaceflight operations through the next two decades.

REFERENCES

1 *Aviation Week and Space Technology*, 5 December 1988.

2 *Aviation Week and Space Technology*, 23 October 1989.

3 Soyuz feasibility as an ACRV. CB memo (CB-92-250) dated 28 August 1992, from Ron Grabe to the Manager of ACRV Projects Office, NASA JSC History Archive.

4 NASA Contractor Report 4508, May 1993, by Michael L. Heck, Contract NAS1-18935, BIS archives, London.

5 Principles Regarding Processes and Criteria for Selection, Assignment, Training and Certification of ISS (Expedition and Visiting) Crewmembers, ISS Multi-lateral Crew Operations Panel, Revision A, November 2001.

6 Hendrickx, Bart, 'From Mir 2 to the ISS Russian Segment', in, *ISS: the First four Years, 1998–2002*, BIS, London, 2002, pp.36-40.

Soyuz variants: Soyuz TMA missions and hardware

Spacecraft (serial number) name	Design designation	Launch date	International designation	Docking date	Undocking date	Target spacecraft (port)	Landing/ decay date	Flight duration dd:hh:mm:ss
11F732 (Soyuz TMA (Transport Modernised Anthropometric) improved transport manned spacecraft for the International Space Station)								
(211) Soyuz TMA-1	(5S)	2002 Oct 30	2002-050A	2002 Nov 1	In orbit	ISS Pirs	2003 May	
(212) Soyuz TMA-2	(6S)	2003 May				ISS	2003 Nov?	
(213) Soyuz TMA-3	(7S)	2003 Nov				ISS	2004 May?	
(214) Soyuz TMA-4	(8S)	2004				ISS		
(215) Soyuz TMA-5	(9A)	2005				ISS		

Conclusion

The statistics of Soyuz-related missions for the period 1966–2002 reveal a staggering record of in-flight activity. During this period, the manned Soyuz in all its variants – including unmanned test flights and Zond lunar missions – accumulated more than 8,050 days (193,200 hours – 22.05 years) in space; and even though from 1971 the majority of this time was accrued whilst docked to the Salyuts, Mir and the ISS, and not in free flight, this still remains a remarkable achievement.

Between 1967 and 2002, eighty-six Soyuz spacecraft took cosmonauts to orbit: thirty-eight original Soyuz, fourteen Soyuz T, thirty-three Soyuz TM, and one Soyuz TMA (including the Soyuz 18-1 launch abort, but not the Soyuz T-10-1 pad abort).

Flown Soyuz Descent Modules stored in the musuem at the Energiya design bureau near Korolyov, Moscow.

Of the 127 individuals (including five females) who have flown onboard a Soyuz, fifty-five flew twice, twenty-eight flew three times, seven flew four times, and three flew five separate missions. Eight-seven Russian cosmonauts have flown onboard a Soyuz variant in Earth orbit, and forty international cosmonauts have occupied a crew position in orbital flight (although not necessarily during launch or landing): ten from the USA, five from France, five from Germany, and one each from Afghanistan, Austria, Belgium, Bulgaria, Cuba, Czechoslovakia, Hungary, India, Italy, Japan, Kazakhstan, Mongolia, Poland, Romania, Slovakia, South Africa, Syria, the UK, and Vietnam. The unmanned Progress series – for most of the time docked to Salyut 6, Salyut 7, Mir or the ISS – logged more than 5,800 days (139,200 hrs) – 15.89 years of unmanned flight time. Between 1966 and the end of 2002, the Soyuz family – in all its manned and unmanned variants (including Progress) – accumulated more than 13,850 days (332,400 hrs) – almost 37.94 years of flight time (some of it simultaneous).

However, as the authors were writing about the impending maiden flight of the latest Soyuz variant (TMA) to the International Space Station, there appeared disheartening news from Russia. Towards the end of 2002, senior figures within RKK Energiya announced that due to insufficient funds, construction of the TMA craft – planned for launch in October 2003 – had been suspended, and that only three Progress craft – rather than the planned six – will be built for launch in 2003. In a letter to NASA, Valeri Ryumin (a Deputy Director of Energiya) has stated that the Russian government already owes Energiya $25 million, and that it has offered only $38 million dollars towards the $133 million dollars that the programme will cost. The company has consequently eaten into its profits, and has had to borrow money from banks to cover the deficit.

Energiya has calculated that a Soyuz takes two years to build, and costs $65 million dollars to fly. This is twice as much as NASA has calculated or considers that it should pay for additional craft to expand ISS occupancy; and the situation has been further aggravated by the wasteage of a budgeted $20 million dollars due to the withdrawal of Lance Bass from the flight of the recently-launched Soyuz TMA.

The authors understand that Energiya's budget for 2003 assumes the sale of four Soyuz seats, although by December 2002 only two had been contracted to ESA, for flights by a Spanish and a Dutch astronaut. There are numerous rumours concerning potential projects which have not been finalised. Many Western observers consider that this may be indicative of Russian tactics to persuade NASA to buy Soyuz craft and to push the Russian government and the Russian space agency into providing more money; but it more obviously reveals a major internal problem in the funding of space projects.

The days of the open cheque book are over. National pride has a price – but within the Russian economy, its value is not clear. In its many different forms and with a variety of mission profiles, Soyuz has flown for more than thirty-five years, and it would be very sad if this astonishing record of exploration were to be forced to an end due to the economics of international co-operation.

This situation changed dramatically on 1 February 2003. The loss of another Shuttle orbiter, and the temporary grounding of the fleet, left ISS operations entirely

The landing of a Chinese Shen Zhou Descent Module after an unmanned test flight in Earth orbit. The Chinese are scheduled to launch their first crew into orbit during 2003. (Courtesy Mark Wade.)

in the control of Soyuz and Progress. As the primary concern is the provision of an adequate supply of fresh water on the station, at least one extra Progress will be launched in 2003 to help alleviate this problem. Meanwhile, the tourist programme is suspended and announced flight crews are stood down while the Columbia accident is investigated and various profiles for the continuing manning of the ISS – including the supply of additional Soyuz and Progress hardware – are studied.

Soyuz is a simple, proven design – ideal for its current role of taking crews to and from a space station, and providing a recovery capability for long-duration crews. Its inadequacies are balanced by the bigger American Space Shuttle, which delivers the heavy cargo and the consumables. It is noteworthy that due to competition between the superpowers throughout the first four decades of human spaceflight, two complementary craft have been created. It is also said that imitation is the greatest form of flattery. When the Chinese decided to commit to manned flight, they were clearly impressed by the durability of Soyuz. The Shen Zhou is designed to suit the Chinese programme and planned missions, but is clearly based on the Soyuz.

The title of this book reflects our feelings about Soyuz, and we hope that the obstacles will be overcome and that it will continue to fly. It is the hard work, skill, commitment and imagination of the designers, engineers, military personnel, trainers and cosmonauts that has made Soyuz a universal spacecraft.

Soyuz variants, 1961–2000

This table summarises the primary known variants of Soviet developed over the past four decades, and includes the distinction between public names and system names. The public name is the name announced after launch, while the system name applies to projects which never left the ground. (It is not known what the latter would have been called.) Stated masses apply only to flown vehicles, and are typical and variable (especially for the Progress series), dependent on the amount of cargo carried. (Courtesy Bart Hendrickx.)

Public name	KB name	Production index	System	Development period	First launch	Final launch	Comments
	5K (?)	?	Sever	1961–62	–	–	Early space station ferry; diameter, 2.2 metres
	7K (?)	?	Vostok 7*	1961–62	–	–	Assembly ship in first version of the Soyuz Complex, 7 metres long, 2 metres diameter, mass 5,500–5,800 kg
	1L	?		1961–62	–	–	Circumlunar vehicle in first version of the Soyuz Complex
	7K	?	Soyuz A**	1962–64	–	–	Circumlunar vehicle in second version of the Soyuz Complex; mass 5,500–5,800 kg, 7.7 metres long
	9K	?	Soyuz B**	1962–64	–	–	Rocket stage in second version of the Soyuz Complex; 7.8 metres long, 2 metres in diameter, 5700 kg
	11K	?	Soyuz V**	1962–64	–	–	Tanker in second version of the Soyuz Complex; 4.2 metres long, 2 metres in diameter, 6,100 kg fuelled
Cosmos	7K-L1P	11F91		1965–66	10.03.67	08.04.67	Prototype L1 to test Blok D upper stage for circumlunar missions
Zond	7K-L1	11F91		1965–67	27.09.67	20.10.70	Circumlunar vehicle
	7K-L1S	11F92		1966–69	21.02.69	03.07.69	N1-launched unmanned lunar orbiter; it is not known what the L1 orbiter would have been called had it been manned.

	7K-L1E	?		?–1969	28.11.69	02.12.70	Prototype L1 to test Blok D for lunar landing missions
	T1K	?		1967–70	–	–	Lunar orbiter for Earth-orbital test flights
	7K-LOK	11F93		1964–72	23.11.72	23.11.72	Lunar orbiter of the N1/L3 manned lunar landing complex
	7K-PPK	?	Soyuz P	1963–64	–	–	ASAT version of Soyuz
	7K-R	11F71***	Soyuz R	1963–66	–	–	Military space station
	7K-TK	11F72***		1964–67	–	–	Ferry for Soyuz R, initially also planned as ferry for Almaz
	7K-VI	11F73	Zvezda	1965–68	–	–	Military research vehicle
	OB-VI	11F731		1967–70	–	–	Orbital block of the Soyuz VI/OIK military space station
Cosmos	7K-S	11F732		1967–74	06.08.74	29.11.76	Ferry for the Soyuz VI/OIK military space station; only flown solo
	7K-SI	11F733		1967–?	–	–	7K-S for short-duration military solo flights
	7K-SII	11F734		1967-?	–	–	7K-S for long-duration military solo flights
Soyuz	7K-OK	11F615		1964–66	28.11.66	01.06.70	Soyuz for docking, research flights; mass about 6,600 kg; length about 8 metres (including the docking collar); maximum diameter, 2.3 metrs
Soyuz	7K-T	11F615A8		1970–71	23.04.71	26.06.72	Salyut (1) ferry (three-man, with solar panels); mass about 6,800 kg; length about 7.5 metres; maximum diameter, 2.72 metres
Soyuz	7K-T	11F615A8		1971–73	15.06.73	14.05.81	Salyut 4/6 ferry (two-man, no solar panels); 6,700 kg; length about 7.5 metres; maximum diameter, 2.72 metress

Public name	KB name	Production index	System	Development period	First launch	Final launch	Comments
Soyuz	7K-T	11F615A9		1971–74	27.05.74	07.02.77	Salyut 3/5 (Almaz) ferry; 6,700 kg; length about 7.5 metres, maximum diameter, 2.72 m
Soyuz	7K-TM	11F615A12		1972–74	03.04.74	15.09.76	EPAS (ASTP) vehicle; mass about 6,550 kg; length about 7.5 metre, maximum diameter, 2.72 metres
Soyuz T	7K-ST	11F732		1974–78	04.04.78	13.03.86	Salyut 6/7 ferry; mass, 6,900 kg; length 6.98 metres; maximum diameter, 2.72 metres; habitable volume, 6.5 cubic metres
Soyuz TM	?	11F732		1981–86	21.05.86	25.04.02	Mir/ISS ferry; mass, 7,000 kg; length, 6.98 metres; maximum diameter, 2.72 metres
Soyuz TMA	?	11F732		1996–2002	30.10.02		ISS ferry; 6.98 metres long; 2.72 metres maximum diameter; habitation modules, 2.2 metres diameter; mass, 7,200 kg
	?	?	Soyuz TMM	1997–99	–	–	ISS ferry
	?	?	Soyuz TMS	1999–			ISS ferry
	?	14F70	Zarya	1985–89	–	–	Mir 2 ferry
	7K-SG	11F735		1967–70	–	–	Cargo ship for Soyuz VI space station
Progress	7K-TG	11F615A15		1973–78	20.01.78	05.05.90	Salyut 6/7/Mir cargo ship; mass, ˜7,020 kg; length, 7 metrs; max imum diameter, 2.72 metres
Progress M	7K-TGM	11F615A55		1986–89	23.08.89		Mir/ISS cargo ship; dimensions similar 7K-TG
	?	11F615A75	Progress MT	Second half of 1980s			Mir 2 cargo ship
	?	11F615A77	Progress M2	Early 1990s–96		–	Mir 2/ISS cargo ship

Progress M1	?	11F615A55		1996–2000	01.02.00		Mir/ISS cargo ship with higher refuelling capacity; dimensions simliiar to 7K-TG
	?	?	Progress M3	1999	–	–	ISS cargo ship
			Progress MM	1997–99	–	–	ISS cargo ship
			Progress MS	1999–			ISS cargo ship
Progress M-SO1	?	11F615A55		1990–2001	14.09.01	14.09.01	Pirs airlock module plus Progress M Service Module. Originally designed for Mir 2, launched to ISS; mass (with cargo), 3,676 kg; length, 4.9 m; maximum diameter, 2.55 m
Gamma	19KA30	?			1974–90	11.07.90	Progress-based gamma-ray astrophysics satellite
	?	?	Aelita				Progress-based cryogenically cooled infrared telescope satellite
	?	?			Early 1990s	?	Progress-based Earth observation and remote sensing spacecraft; mass, 10,000 kg
	?	?		1976 – ?	5		Progress-based tests of anti-satellite missiles; following cancelleation they were rebuilt as ordinary ressuply craft with different serial numbers

* Vostok 7 variants strictly do not belong to the Soyuz/Progress family.

** 7K was definitely used for the circumlunar vehicle of the Soyuz Complex, as it appeared on the drawing boards during the latter half of 1963. Soyuz A is a 'cover designator' devised in 1980 when Korolyov's 'Creative Legacy' book was published. Timothy Varfolomeyev *speculates* in one of his articles that it was also the designator for the Vostok 7 assembly craft, but this is not confirmed by Russian sources. Soyuz A, B and V are cover names first published in 1980.

*** After the cancellation of Soyuz R and its transport vehicle, the designators 11F71 and 11F72 were 'recycled' for the Almaz space station and the TKS ferries respectively

Soyuz crew assignments (as flown)

Soyuz flight	Crew position	Cosmonaut (mission)	Call-sign	Back-up cosmonaut	Launch date	Crew duration dd:hh:mm:ss	Landing date	Notes
Key: CDR, Commander; FE, Flight Engineer; RE, Research Engineer, CR, Cosmonaut Researcher; DR, Doctor Researcher (Polyakov, TM-6); EO, Resident crew sequence; SFP, Spaceflight Participant								
11F615 series (Soyuz basic manned spacecraft for test flights)								
1	CDR	Komarov (2)	Rubin (Ruby)	Gagarin	1967 Apr 23	001:02:47:52	1967 Apr 24	Died during landing accident; Soyuz 2 cancelled
2	–	Unmanned	–	–	1968 Oct 25	–	1968 Oct 28	Docking target for manned Soyuz 3
3	CDR	Beregovoi (1)	Argon (Argon)	Shatalov	1968 Oct 26	003:22:50:45	1968 Oct 30	Failed to dock with unmanned Soyuz 2
4	CDR	Shatalov (1)	Amur (Amur)	Shonin	1969 Jan 14	002:23:20:47	1969 Jan 17	Docked with Soyuz 5
5	CDR	Volynov (1)	Baikal (Baikal)	Filipchenko	1969 Jan 15	003:00:54:15	1969 Jan 18	Docked with Soyuz 4; Yeliseyev and Khrunov performed a 1-hour EVA transfer (Jan 16) and returned in Soyuz 4 with Shatalov; Volynov returned alone in Soyuz 5
	FE	Yeliseyev (1)		Kubasov		001:23:45:50	1969 Jan 17	
	RE	Khrunov (1)		Gorbatko		001:23:45:50	1969 Jan 17	
6	CDR	Shonin (1)	Antey (Antaeus)	Shatalov	1969 Oct 11	004:22:42:47	1969 Oct 16	Flown in troika group flight with Soyuz 7 and 8; originally a solo Soyuz flight to test space welding experiments
	FE	Kubasov (1)		Yeliseyev				
7	CDR	Filipchenko (1)	Buran (Snow-storm)	Shatalov	1969 Oct 12	004:22:40:23	1969 Oct 17	Intended to dock with Soyuz 8; group flight with Soyuz 6 and 7
	FE	Gorbatko (1)						
	RE	Volkov V. (1)						
8	CDR	Shatalov (2)	Granit (Granite)	Nikolayev	1969 Oct 13	004:22:50:49	1969 Oct 18	Intended to dock with Soyuz 7; group flight with Soyuz 6 and 7; Soyuz 8 was the lead spacecraft for the group
	FE	Yeliseyev (2)		Sevastyanov				
9	CDR	Nikolayev (2)	Sokol (Falcon)	Filipchenko	1970 Jun 1	017:16:58:55	1970 Jun 19	Set world endurance record for longest spaceflight, breaking the American Gemini 7 record set in December 1965, and remains the longest non-space-station manned spaceflight; biomedical Soyuz solo flight
		Sevastyanov (1)		Grechko				
11F615 A8 series (Soyuz transport manned spacecraft for DOS (Salyut space stations)								
10	CDR	Shatalov (3)	Granit (Granite)	Leonov	1971 Apr 22	001:23:45:54	1971 Apr 24	Docked to Salyut 1, but failed to enter station; intended to be a 30-day residency mission
	FE	Yeliseyev (3)		Kubasov				
	RE	Rukavishnikov (1)						
11	CDR	Dobrovolsky (1)	Yantar (Amber)	Shatalov	1971 Jun 6	023:18:21:43	1971 Jun 30	Docked to Salyut 1 (DOS-1); spent 24 days in space, setting new endurance record; first space space station crew; crew died after a pressure seal failed, venting atmosphere during automated recovery phase
	FE	Volkov V. (2)		Yeliseyev				
	RE	Patsayev (1)		Rukavishnikov				
12	CDR	Lazarev (1)	Ural (Urals)	Gubarev	1973 Sep 27	001:23:15:32	1973 Sep 29	Two-day test of two-man Soyuz space station ferry

	FE	Makarov (1)		Grechko				
13	CDR FE	Klimuk (1) Lebedev (1)	Kavkaz (Caucasus)	– –	1973 Dec 18	007:20:55:35	1973 Dec 26	Solo scientific Soyuz flight (with solar arrays) carrying astrophysical and biological experiments
17	CDR FE	Gubarev (1) Grechko (1)	Zenit (Zenith)	Lazarev Makarov	1975 Jan 10	029:13:19:45	1975 Feb 9	Docked to Salyut 4 (DOS-4) space station, setting new Soviet space space endurance record
18-1	CDR FE	Lazarev (2) Makarov (2)	Ural (Urals)	Klimuk Sevastyanov	1975 Apr 5	000:00:21:27	1975 Apr 5	Intended to spend 60 days on board Salyut 4; mission aborted during launch, due to erroneous R7 performance
18	CDR FE	Klimuk (2) Sevastyanov (2)	Kavkaz (Caucasus)	Kovalyonok Ponomarev	1975 May 24	062:23:20:08	1975 Jul 26	Docked to Salyut 4, setting new Soviet endurance record with 63 days in space; the joint ASTP was flown while the Soyuz 18 crew was on board Salyut 4
20	–	Unmanned	–	–	1975 Nov 17	–	1976 Feb 16	Automated docking to unmanned Salyut 4; carried biological experiments; tested long-duration orbital storage and profile of Progress resupply missions
25	CDR FE	Kovalenok (1) Ryumin (1)	Foton (Photon)	Romanenko Ivanchenkov	1977 Oct 09	002:00:44:45	1977 Oct 11	Failed to dock with Salyut 6 (DOS-5) due to docking system failure; intended 90-day residency
26	CDR FE	Romanenko (1) Grechko (2)	Tamyr (River)	Kovalyonok Ivanchenkov	1977 Dec 10	096:10:00:07	1978 Mar 16	Docked to Salyut 6, setting new world endurance record, surpassing US Skylab 4 (1974) 84-day record; recovered in Soyuz 27
27	CDR FE	Dzhanibekov (1) Makarov (2)	Pamir (Pamirs)	Kovalyonok Ivanchenkov	1978 Jan 10	005:22:58:58	1978 Jan 16	Docked to Salyut 6; first visiting crew and spacecraft swoop; returned in Soyuz 26
28	CDR CR	Gubarev (2) Remek (1) (Czechoslovakia)	Zenit (Zenith)	Rukavishnikov Pelczak (Czechoslovakia)	1978 Mar 2	007:22:16:00	1978 Mar 10	First Interkosmos mission (Czechoslovakian); Remek was also the first non-Soviet and non-US space-traveller
29	CDR FE	Kovalyonok (2) Ivanchenkov (1)	Foton (Photon)	Lyakhov Ryumin	1978 Jun 15	139:14:47:32	1978 Nov 2	Docked to Salyut 6, setting new endurance record; crew returned in Soyuz 31
30	CDR CR	Klimuk (3) Hermaszewski (1) (Poland)	Kavkas (Caucasus)	Kubasov Jankowski (Poland)	1978 Jun 29	007:22:02:59	1978 Jul 5	Docked to Salyut 6; second Interkosmos mission (Polish)
31	CDR CR	Bykovsky (3) Jahn (1) (GDR)	Yastreb (Hawk)	Gorbatko Koellner (GDR)	1978 Aug 26	007:20:49:04	1978 Sep 3	Docked to Salyut 6; third Interkosmos mission (East German); crew returned in Soyuz 29
32	CDR FE	Lyakhov (1) Ryumin (2)	Proton (Proton)	Popov Lebedev	1979 Feb 25	175:00:35:37	1978 Aug 19	Docked to Salyut 6; set new endurance record; crew returned in Soyuz 34; Soyuz 32 was recovered unmanned on 13 June 1979
33	CDR CR	Rukavishnikov (3) Ivanov (1) (Bulgaria)	Saturny (Saturn)	Romanenko Alexandrov (Bulgaria)	1979 Apr 10	001:23:01:06	1979 Apr 12	Intended to dock with Salyut 6; fourth Interkosmos mission (Bulgarian); onboard systems failure on Soyuz forced emergency return
34	–	Unmanned	–	–	1979 Jun 6	–	1979 Aug 19	Launched unmanned to Salyut 6 to replace ageing Soyuz 32 after failed Soyuz 33 docking; returned with Soyuz 32 launch crew
35	CDR FE	Popov (1) Ryumin (3)	Dneiper (Dnieper)	Zudov Andreyev	1980 Apr 9	184:20:11:35	1980 Oct 11	Docked to Salyut 6; set new endurance record; crew returned in Soyuz 37

Soyuz flight	Crew position	Cosmonaut (mission)	Call-sign	Back-up cosmonaut	Launch date	Crew duration dd:hh:mm:ss	Landing date	Notes
36	CDR	Kubasov (3)	Orion (Orion)	Dzhanibekov	1980 May 26	007:20:45:44	1980 Jun 3	Docked to Salyut 6; fifth Interkosmos mission (Hungarian); postponed from 1979; crew returned in Soyuz 35
	CR	Farkas (1) (Hungary)		Magyari (Hungary)				
37	CDR	Gorbatko (3)	Terek (Terek)	Bykovsky	1980 Jul 23	007:20:42:00	1980 Jul 31	Docked to Salyut 6; sixth Interkosmos mission (Vietnamese); crew returned in Soyuz 36
	CR	Tuan (1) (Vietnam)		Liem (Vietnam)				
38	CDR	Romanenko (2)	Tamyr (Tamyr)	Khrunov	1980 Sep 18	007:20:43:24	1980 Sep 28	Docked to Salyut 6; seventh Interkosmos mission (Cuban)
	CR	Tamayo-Mendez (1) (Cuba)		Lopez-Falcon (Cuba)				
39	CDR	Dzhanibekov (2)	Pamir (Pamirs)	Lyakhov	1981 Mar 22	07:20:42:03	1980 Mar 30	Docked to Salyut 6; eighth Interkosmos mission (Mongolian)
	CR	Gurraggcha (1) (Mongolia)		Ganzorig (Mongolia)				
40	CDR	Popov (2)	Dneiper (Dneiper)	Romanenko	1981 May 14	07:20:41:52	1980 May 22	Docked to Salyut 6; ninth and last Interkosmos mission (Romanian); last flight of original Soyuz
	CR	Prunariu (1) (Romania)		Dediu (Romania)				
11F615 A9 series (Soyuz transport manned spacecraft for OPS Almaz space stations)								
14	CDR	Popovich (2)	Berkut (Golden Eagle)	Sarafanov	1974 Jul 3	015:17:30:28	1974 Jul 19	First successful Salyut space station mission; docked to military Salyut 3 (Almaz 2)
	FE	Artyukhin (1)		Dyomin				
15	CDR	Sarafanov (1)	Dunay (Danube)	Volynov	1974 Aug 26	002:00:12:11	1974 Aug 28	Failed to dock with Salyut 3
	FE	Dyomin (1)		Zholobov				
21	CDR	Volynov (2)	Baikal (Baikal)	Zudov	1976 Jul 6	049:06:23:32	1976 Aug 24	Docked to military Salyut 5 (Almaz 3), spending 49 days in space; ECLSS system malfunctions and crew health issues terminated the flight
	FE	Zholobov (1)		Rozhdestvensky				
23	CDR	Zudov (1)	Radon (Radon)	Gorbatko	1976 Oct 14	002:00:06:35	1976 Oct 16	Failed to dock with Salyut 5; crew performed emergency re-entry and splashed down in a frozen lake during a blizzard, but were safely recovered
	FE	Rozhdestvensky (1)		Glazkov				
24	CDR	Gorbatko (2)	Terek (Terek)	Berezovoi	1977 Feb 7	017:17:25:58	1977 Feb 25	Docked to Salyut 5; final Almaz manned mission
	FE	Glazkov (1)		Lisun				
11F615 A12 series (Soyuz manned spacecraft for Apollo–Soyuz Test Project (ASTP) joint mission with the USA)								
16	CDR	Filipchenko (2)	Buran (Snow-storm)	Romanenko	1974 Dec 2	005:22:23:35	1974 Dec 8	Single spacecraft test of ASTP docking hardware
	FE	Rukavishnikov (2)		Ivanchenkov				
19	CDR	Leonov (2)	Soyuz (Union)	Dzhanibekov	1975 Jul 15	005:22:30:51	1975 Jul 21	Soviet ASTP spacecraft; docked (17–19 July) with American three-man Apollo 18 CSM; first international mission
	FE	Kubasov (2)		Andreyev				

22	CDR	Bykovsky (2)	Yastreb (Hawk)	Malyshev	1976 Sep 15	007:21:52:17	1976 Sep 23	Solo Earth observation/photography mission using back-up ASTP spacecraft
	FE	Aksenov (1)		Strekalov				
11F732 series (Soyuz T improved transport craft for DOS second generation)								
T-1	–	Unmanned	–	–	1979 Dec 16	–	1980 Mar 23	Test flight; docked with Salyut 6 on 19 December, following an overshoot the previous day
T-2	CDR	Malyshev (1)	Yupiter (Jupiter)	Kizim	1980 Jun 5	003:22:19:30	1980 Jun 9	First manned test flight of Soyuz T; docked to Salyut 6
	FE	Aksenov (2)		Makarov				
T-3	CDR	Kizim (1)	Mayak (Lighthouse)	Lazarev	1980 Nov 27	012:19:07:42	1980 Dec 10	First three-man Soyuz since 1971; docked to Salyut 6 for a refurbishment and evaluation mission
	FE	Makarov (4)		Savinykh				
	RE	Strekalov (1)		Polyakov				
T-4	CDR	Kovalyonok (3)	Foton (Photon)	Zudov	1981 Mar 12	074:17:37:23	1981 May 26	Docked to Salyut 6; last residency mission to the ageing station.
	FE	Savinykh (2)		Andreyev				
T-5	CDR	Berezovoi (1)	Elbrus (Elbrus)	Titov V.	1982 May 13	211:09:04:32	1982 Dec 10	Docked to Salyut 7 (DOS-6); set new endurance record
	FE	Lebedev (2)		Strekalov				
T-6	CDR	Dzhanibekov (3)	Pamir (Pamirs)	Kizim	1982 Jun 24	007:21:50:53	1982 Jul 2	Docked to Salyut 7; French international mission
	FE	Ivanchenkov (2)		Solovyov V.				
	CR	Chrétien (1) (France)		Baudry (France)				
T-7	CDR	Popov (3)	Dnieper (Dnieper)	Vasyutin	1982 Aug 19	007:21:52:24	1982 Aug 27	Docked to Salyut 7; visiting mission; Savitskaya became the second woman in space (nineteen years after Tereshkova) and the first to visit a space station
	FE	Serebrov (1)		Savinykh				
	CR	Savitskaya (1)		Pronina				
T-8	CDR	Titov V. (1)	Okean (Ocean)	Lyakhov	1983 Apr 20	002:00:17:48	1983 Apr 22	Failed to dock to Salyut 7; Serebrov was the first to fly back-to-back space missions
	FE	Strekalov (2)		Aleksandrov				
	CR	Serebrov (2)		Savinykh				
T-9	CDR	Lyakhov (2)	Proton (Proton)	Titov V.	1983 Jun 27	149:10:46:01	1983 Nov 23	Docked to Salyut 7; endured isolation due to no visiting missions
	FE	Alexandrov A. (1)		Strekalov				
T-10-1	CDR	Titov V (–)	Okean (Ocean)	Kizim	1983 Sep 26	000:00:05:13	1983 Sep 26	Launch pad abort; not a spaceflight; crew recovered by LES and endured 14–17 g
	FE	Strekalov (–)		Solovyov A.				
T-10	CDR	Kizim (2)	Mayak (Lighthouse)	Vasyutin	1984 Feb 8	236:22:49:04	1984 Oct 2	Docked to Salyut 7; set new endurance record; completed extensive EVA repairs to station propulsion system; Atkov first Soviet doctor on an extended mission
	FE	Solovyov V. (1)		Savinykh				
	CR	Atkov (1)		Polyakov				
T-11	CDR	Malyshev (2)	Yupiter (Jupiter)	Berezovoi	1984 Apr 3	007:21:40:06	1984 Apr 11	Docked to Salyut 7; Indian international mission
	FE	Strekalov (3)		Grechko				
	CR	Sharma (1) (India)		Malhotra (India)				
T-12	CDR	Dzhanibekov (4)	Pamir (Pamirs)	Vasyutin	1984 Jul 17	011:19:14:36	1984 Jul 29	Docked to Salyut 7; Volk , a Buran Shuttle Pilot, was to gain spaceflight experience; Savitskaya first woman to fly in space twice, and first woman to perform EVA
	FE	Savitskaya (2)		Savinykh				
	CR	Volk (1)		Ivanova				
T-13	CDR	Dzhanibekov (5)	Pamir (Pamirs)	Popov	1985 Jun 13	112:03:12:06	1985 Sep 26	Docked with Salyut 7; rescue mission; Dzhanibekov returned in T-13, but Savinykh joined and returned with the T-14 crew
	FE	Savinykh (2)		Alexandrov A.		168:03:51:00	1985 Nov 21	

Soyuz flight	Crew position	Cosmonaut (mission)	Call-sign	Back-up cosmonaut	Launch date	Crew duration dd:hh:mm:ss	Landing date	Notes
T-14	CDR	Vasyutin (1)	Cheget (Cheget)	Viktorenko	1985 Sep 17	064:21:52:08	1985 Nov 21	Docked with Salyut 7; intended residency of {???} days; Vasyutin
	FE	Grechko (3)		Strekalov		008:21:13:06	1985 Sep 26	became ill, forcing early termination of the mission; Grechko
	CR	Volkov A. (1)		Salei		064:21:52:08	1985 Nov 21	had returned on T-13, and Savinykh replaced him
T-15	CDR	Kizim (3)	Mayak (Lighthouse)	Viktorenko	1986 Mar 13	125:00:00:56	1986 Jul 16	Docked with Mir (staying 52 days); transferred to Salyut 7
	FE	Solovyov V. (2)		Alexandrov				(51 days), and then back to Mir (21 days)
11F732 series (Soyuz TM improved (transport modernised) spacecraft for DOS-7, Mir and early International Space Station								
TM-1	–	Unmanned	–	–	1986 May 21	–	1986 May 30	Docked with Mir in test of new spacecraft systems
TM-2	CDR	Romanenko (3)	Tamyr (Tamyr)	Titov V.	1987 Feb 5	326:11:37:59	1987 Dec 29	Docked with Mir; Romanenko set new endurance record;
EO-2	FE	Laveikin (1)		Serebrov		174:03:25:56	1987 Jul 30	Laveikin became medically unfit, and was replaced by Aleksandrov landing on TM-2
TM-3	CDR	Viktorenko (1)	Vityaz (Viking)	Solovyov A.	1987 Jul 22	007:23:04:55	1987 Jul 30	Docked with Mir; Syrian international mission;
	FE	Aleksandrov (2)		Savinykh		160:07:16:58	1987 Dec 29	Aleksandrov replaced Laveikin to complete the residency mission,
	CR	Faris (1) (Syria)		Habib (Syria)		007:23:04:55	1987 Jul 30	with Romanenko returning on TM-3
TM-4	CDR	Titov (2)	Okean (Ocean)	Volkov A.	1987 Dec 21	365:22:38:57	1988 Dec 21	Docked with Mir; Titov and Manarov completed the first 1-year
EO-3	FE	Manarov (1)		Kaleri		365:22:38:57	1988 Dec 21	space flight, setting a new record; Levchenko was the second
	CR	Levchenko (1)		Shchukin		007:21:58:12	1987 Dec 29	Buran pilot to fly a Soyuz mission; returned on TM-3
TM-5	CDR	Solovyov A. (1)	Rodnik (Spring)	Lyakhov	1988 Jun 7	009:20:09:19	1988 Jun 17	Docked with Mir; second Bulgarian international mission; crew
	FE	Savinykh (3)		Serebrov				returned on TM-4
	CR	Aleksandrov A.P (1) (Bulgaria)		(Bulgaria)	Stoyanov			
TM-6	CDR	Lyakhov (3)	Proton (Proton)	Berezovoi	1988 Aug 29	008:20:26:27	1988 Sep 8	Docked with Mir; Afghan international mission; Polyakov remained
	DR	Polyakov (1)		Arzamazov		240:22:34:47	1989 Apr 27	on Mir for eight months; Lyakhov and Mohmand experienced
	CR	Mohmand (1) (Afghanistan)		Dauran (Afghanistan)		008:20:26:27	1988 Sep 8	aborted/delayed re-entry sequence, but were safely recovered
TM-7	CDR	Volkov A. (2)	Donbass (Donbass)	Viktorenko	1988 Nov 26	151:11:08:23	1989 Apr 27	Docked with Mir; French international mission; Volkov and
EO-4	FE	Krikalev (1)		Serebrov		151:11:08:03	1989 Apr 27	Krikalev returned with Polyakov on TM-7; Chrétien returned
	CR	Chrétien (2)		Tognini		024:18:07:25	1988 Dec 21	with Titov and Manarov on TM-6
TM-8	CDR	Viktorenko (2)	Vityaz (Viking)	Solovyov A.	1989 Sep 5	166:06:58:16	1990 Feb 19	Docked with Mir; tested YMK (the Soviet MMU)
EO-5	FE	Serebrov (3)		Baladin				
TM-9	CDR	Solovyov A. (2)	Rodnik (Spring)	Manakov	1990 Feb 11	179:01:17:57	1990 Aug 9	Docked with Mir
EO-6	FE	Balandin (1)		Strekalov				
TM-10	CDR	Manakov (1)	Vulkan (Volcano)	Afanasyev	1990 Aug 1	130:20:35:51	1990 Dec 10	Docked with Mir
EO-7	FE	Strekalov (4)		Manarov				
TM-11	CDR	Afanasyev (1)	Derbent (Derbent)	Artsebarsky	1990 Dec 2	175:01:51:42	1991 May 26	Docked with Mir; Japanese international mission; Akiyama returned
EO-8	FE	Manarov (2)		Krikalev		175:01:51:42	1991 May 26	with Manakov and Strekalov on TM-10; Afanasyev and
	CR	Akiyama(1) (Japan)		Kikuchi		007:21:54:40	1990 Dec 10	Manarov returned on TM-12

TM-12 CDR	Artsebarsky (1)	Ozon (Ozone)	Volkov A.	1991 May 18	144:15:21:50	1991 Oct 10	Docked with Mir; UK international mission; crew returned on three
EO-9 FE	Krikalev (2)		Kaleri		311:20:01:54	1992 Mar 25	different spacecraft: Sharman on TM-11, Artsebarsky
CR	Sharman (1) (UK)		Mace (UK)		007:21:14:20	1991 May 26	on TM-12 and Krikalev on TM-13
TM-13 CDR	Volkov A. (3)	Donbass (Donbass)	Viktorenko	1991 Oct 2	175:02:52:43	1992 Mar 25	Docked with Mir; duel Austrian and Kazakh international mission;
EO-10 FE	Aubakirov (1) (Kazakhstan)		Musabayev		007:22:12:59	1991 Oct 10	Volkov remained on Mir, Artsebarsky returned with Aubakirov and Viehbock on TM-12
CR	Viehbock (1) (Austria)		Lothaller (Austria)		007:22:12:59	1991 Oct 10	
TM-14 CDR	Viktorenko (3)	Vityaz (Viking)	Solovyov A.	1992 Mar 17	145:14:10:32	1992 Aug 10	Docked with Mir; German international mission
EO-11 FE	Kaleri (1)		Avdeyev		145:14:10:32	1992 Aug 10	
CR	Flade (1) (Germany)		Ewald (Germany)		007:21:56:52	1992 Mar 25	
TM-15 CDR	Solovyov A. (3)	Rodnik (Spring)	Manakov	1992 Jul 27	188:21:41:15	1993 Feb 1	Docked with Mir; French international mission; Tognini returned
EO-12 FE	Avdeyev (1)		Poleshchuk		188:21:41:15	1993 Feb 1	with Viktorenko and Kaleri on TM-14
CR	Tognini (1) (France)		Haignere J.-P. (France)		013:18:56:14	1992 Aug 10	
TM-16 CDR	Manakov (2)	Vulkan (Volcano)	Tsibilyev	1993 Jan 24	179:00:43:46	1992 Jul 22	Docked with Mir
EO-13 FE	Poleshchuk (1)		Usachev				
TM-17 CDR	Tsibilyev (1)	Sirius (Sirius)	Afanasyev	1993 Jul 1	196:17:45:22	1994 Jan 14	Docked with Mir; French international mission; Haignere returned
EO-14 FE	Serebrov (4)		Usachev				with Manakov and Poleshchuk on TM-16
CR	Haignere J.-P. (1) (France)		André-Deshays (France)				
TM-18 CDR	Afanasyev (2)	Derbent (Derbent)	Malenchenko	1994 Jan 8	182:00:27:02	1994 Jul 9	Docked with Mir; Polyakov remained on Mir for 14 months, and
EO-15 FE	Usachev (1)		Musabayev (Kazakhstan)		182:00:27:02	1994 Jul 9	returned on TM-20
CR	Polyakov (2)		Arzamazov		437:17:58:31	1995 Mar 22	
TM-19 CDR	Malenchenko (1)	Agat (Agate)	Viktorenko	1994 Jul 1	125:22:53:36	1994 Nov 4	Docked with Mir; Musabayev flew as a resident crew-member;
EO-16 FE	Musabayev (1)		Kondakova				returned with Merbold
TM-20 CDR	Viktorenko (4)	Vityaz (Viking)	Gidzenko	1994 Oct 3	169:05:21:35	1995 Mar 22	Docked with Mir; ESA mission; Viktorenko and
EO-17 FE	Kondakova (1)		Avdeyev		169:05:21:35	1995 Mar 22	Kondakova returned with Polyakov; Merbold (who had flown on
CR	Merbold (3) (ESA, Germany)		Duque (ESA, Spain)		031:12:35:56	1994 Nov 04	two NASA Space Shuttle missions) returned with Malenchenko and Musabayev on TM-19; Kondakova was the third Russian female in space
TM-21 CDR	Dezhurov (1)	Uragan (Hurricane)	Solovyov A.	1995 Mar 14	115:08:43:02	1995 Jul 7	Docked with Mir; first NASA mission; crew landed on Shuttle
EO-18 FE	Strekalov (5)		Budarin				*Atlantis* (STS-71); Solovyov and Budarin arrived on STS-71 to
CR	Thagard (5) (USA)		Dunbar (USA)				take over residency. but landed in TM-21 on 11 September
EO-19 CDR	Solovyov A (4).	Rodnik (Spring)	Onufrienko	1995 Jun 27	075:11:20:21 (STS–Mir–Soyuz)	1995 Sep 11	Crew launched on NASA STS-71 *Atlantis*, but returned as Soyuz crew on TM-20
FE	Budarin (1)						
TM-22 CDR	Gidzenko (1)	Uran (Uranus)	Manakov	1995 Sep 3	179:01:41:46	1996 Feb 29	Docked with Mir; ESA mission; visited by STS-74

Soyuz flight	Crew position	Cosmonaut (mission)	Call-sign	Back-up cosmonaut	Launch date	Crew duration dd:hh:mm:ss	Landing date	Notes
EO-20	FE	Avdeyev (2)		Vinogradov				US Shuttle docking mission
	CR	Reiter (1) (ESA, Germany)		Fuglesang (ESA, Sweden)				
TM-23	CDR	Onufrienko (1)	Skif (Scythian)	Tsibilyev	1996 Feb 21	193:19:07:35	1996 Sep 2	Docked with Mir; worked with US astronauts Lucid and crew of STS-76
EO-21	FE	Usachev (2)		Lazutkin				
TM-24	CDR	Korzun (1)	Freget (Frigate)	–	1996 Aug 17	196:17:26:13	1997 Aug 14	Docked with Mir; ESA mission; André-Deshays returned with EO-21 crew on TM-23; EO-22 crew worked with NASA astronauts Lucid, Blaha and Linenger, and crews of STS-79 and STS-81
EO-22	FE	Kaleri (2)	–			196:17:26:13	1997 Aug 14	
	CR	André-Deshays (1) (France)		Eyharts (France)		015:18:23:37	1996 Sep 2	
STS-81								
TM-25	CDR	Tsibilyev (2)	Sirius (Sirius)	Musabayev	1997 Feb 10	184:22:07:41	1997 Aug 17	Docked with Mir; German international mission; crews experienced a fire and collision on Mir; EO-23 worked with NASA astronauts Linenger and Foale and crew of STS-84; Ewald returned with EO-22 crew on TM-24
EO-23	FE	Lazutkin (1)		Budarin		184:22:07:41	1997 Aug 17	
	CR	Ewald (1) (Germany)		Schlegel (Germany)		019:16:34:46	1997 Mar 2	
TM-26	CDR	Solovyov A. (5)	Rodnik (Spring)	Padalka	1997 Aug 5	197:17:34:36	1998 Feb 19	Docked with Mir; worked with Foale, Wolf and Thomas, and crews of STS-86 and STS-89
EO-24	FE	Vinogradov (1)		Avdeyev				
TM-27	CDR	Musabayev (2)	Kristall (Crystal)	Afanasyev	1998 Jan 29	207:12:51:02	1998 Aug 25	Docked with Mir; worked with Thomas and crew of STS-91; French international mission; Haignere landed on TM-26
EO-25	FE	Budarin (2)		Treshchev		207:12:51:02	1998 Aug 25	
	CR	Eyharts (1) (France)		Haignere J.-P. (France)		020:16:36:48	1998 Feb 19	
TM-28	CDR	Padalka (1)	Altair (Altair)	Zaletin	1998 Aug 13	198:16:31:20	1999 Feb 28	Docked with Mir; Avdeyev remained on Mir for more than a year; Baturin returned with EO-25 crew on TM-27; Paldaka returned on TM-28
EO-26	FE	Avdeyev (3)		Kaleri		379:14:51:10	1999 Aug 28	
	CR	Baturin (1)		Kotov		011:19:41:33	1998 Aug 25	
TM-29	CDR	Afanasyev (3)	Derbent (Derbent)	Sharipov	1999 Feb 20	188:20:16:19	1999 Aug 28	Docked with Mir; French and Slovakian international mission; Haignere remained on Mir for six months; Bella returned after eight days on TM-28
EO-27	FE	Haignere J.-P. (2) (France)		André-Deshays (France)		188:20:16:19	1999 Aug 28	
	CR	Bella (1) (Slovakia)		Fulier (Slovakia)		007:21:56:29	1999 Feb 28	
TM-30	CDR	Zaletin (1)	Yenisey (Yenisey)	Sharipov	2000 Apr 4	072:19:42:16	2000 Jun 16	Final manned mission to Mir funded under MirCorp
EO-28	FE	Kaleri (3)		Vinogradov				
TM-31	CDR	Gidzenko (2)	Uran (Uranus)	Dezhurov	2000 Oct 31	140:23:38:55	2001 Mar 21	First International Space Station resident crew; returned on STS-102; worked with STS-97 and STS-98 crews
ISS-1	FE	Krikalev (5)		Tyurin				
(3R-1S)		FE-2	Shepherd (4) (USA)	Bowersox (USA)				
TM-32	CDR	Musabayev (3)	Kristall (Crystal)	Afanasyev	2001Apr 21	007:22:04:08	2001 Apr 30	First Soyuz taxi mission to swop out older TM vehicle; Tito was the first 'space tourist'; crew return on TM-31
(2S)	FE	Baturin (2)		Kozeev				
	SFP	Tito (1) (USA)	–					

TM-33 (3S)	CDR	Afanasyev (4)	Derbent (Derbent)	Zaletin	2001 Oct 21	009:20:00:25	2001 Oct 31	Second taxi mission with a French cosmonaut; crew returned on on TM-32
	FE	Kozeev (1)		Kuzhelnaya				
	FE-2	Haignere C. (2) (France)	–					
TM-34 (4S)	CDR	Gidzenko (3)	Uran (Uranus)	Padalka	2002 Apr 25	009:21:25:18	2002 May 5	Third taxi mission; South African 'space tourist' flight, and ESA visiting mission; last Soyuz TM launch; crew returned on TM-33; TM-34 to be returned by the first TMA crew
	FE	Vittori (1) (ESA, Italy)		Kononenko				
	SFP	Shuttleworth (1) (South Africa)		–				

TKF11372 series (Soyuz TMA (transport modernised anthropometric) improved spacecraft for the International Space Station)

TMA-1 (5S)	CDR	Zaletin (2)	Yenisey (Yenisey)	Lonchakov	2002 Oct 30	10:20:53:09	2002 Nov 10	Fourth taxi mission; first TMA flight; ESA visiting mission; crew returned in TM-34
	FE	De Winne (1)		Lazutkin				
	FE-2	Lonchakov (2)						
TMA-2 (6S)	CDR	Malenchenko (3)	Agat (Agate)	Kaleri	2003 May		2003 Nov	ISS-7 crew
	FE	Lu (3) (USA)		Foale				
TMA-3 (7S)	CDR	Kaleri (4)		Tokarev	2003 Nov		2004 May(?)	ISS-8 crew; Duque on visiting mission only, and will return with the ISS-7 crew
	FE	Foale (6) (USA)		McArthur				
	FE-2	Duque (2) (ESA) (Spain)						

Soyuz flight sequence, 1967–2002

A listing in flight order (CDR, FE, RE/CR) of those who have flown on a Soyuz spacecraft, in sequence of first, second, third flight, and so on It does not distinguish between Soyuz types, and includes the occasions during which crew-members 'flew' in a Soyuz to change docking locations but did not launch or land in the craft

Pos.	Cosmonaut	Pos.	Cosmonaut	Pos.	Cosmonaut	Pos.	Cosmonaut	Pos.	Cosmonaut
First flight									
01	Komarov	31	Zudov	61	Atkov	91	Usachev	121	Kozeev
02	Beregovoi	32	Rozhdestvensky	62	Sharma (India)	92	Malenchenko	122	Walz (USA)[S]
03	Shatalov	33	Glazkov	63	Volk	93	Musabayev	123	Bursch (USA)[S]
04	Volynov	34	Kovolyonok	64	Vasyutin	94	Kondakova (third female)	124	Vittori (Italy)
05	Khrunov	35	Ryumin	65	Volkov, A.	95	Merbold (Germany)	125	Shuttleworth (SA)
06	Yeliseyev	36	Romanenko, Y.	66	Laveikin	96	Dezhurov[A]	126	De Winne
07	Shonin	37	Dzhanibekov	67	Viktorenko	97	Thagard (USA) [A]	127	Lanchenkov
08	Kubasov	38	Remek (Czechoslovakia)	68	Faris (Syria)	98	Budarin [D]	128	
09	Filipchenko	39	Ivanchenkov	69	Manarov	99	Gidzenko	129	
10	Volkov V.	40	Hermaszewski (Poland)	70	Levchenko	100	Reiter (Germany)	130	
11	Gorbatko	41	Jahn (GDR)	71	Solovyov, A.	101	Onufrienko	131	
12	Nikolayev	42	Lyakhov	72	Alexandrov, A. (Bulgaria)	102	Korzun	132	
13	Sevastyanov	43	Ivanov (Bulgaria)	73	Polyakov	103	Deshays (France; fourth female)	133	
14	Rukavishnikov	44	Popov	74	Mohmand (Afghanistan)	104	Lazutkin	134	
15	Dobrovolsky	45	Farkas (Hungary)	75	Krikalev	105	Ewald (Germany)	135	
16	Patsayev	46	Malyshev	76	Balandin	106	Linenger (USA) [S]	136	
17	Lazarev	47	Pham Tuan (Vietnam)	77	Manakov	107	Vinogradov	137	
18	Makarov	48	Mendez (Cuba)	78	Afanasyev	108	Foale (USA/UK)[S]	138	
19	Klimuk	49	Kizim	79	Akiyama (Japan)	109	Thomas (USA)[S]	139	
20	Lebedev	50	Strekalov	80	Artsebarsky	110	Eyharts (France)	140	

21	Popovich	51	Savinykh	81	Sharman (UK; second female)	111	Paldalka	141	
22	Artyukhin	52	Gurragcha (Mongolia)	82	Viehbock (Austria)	112	Baturin	142	
23	Sarafanov	53	Prunariu (Romania)	83	Aubakirov (Kazakhstan)	113	Bella (Slovakia)	143	
24	Dyomin	54	Berezovoi	84	Kaleri	114	Zaletin	144	
25	Gubarev	55	Chrétien (France)	85	Flade (Germany)	115	Shepherd, W. (USA)[A]	145	
26	Grechko	56	Serebrov	86	Avdeyev	116	Voss, J.S. (USA)[S]	146	
27	Leonov	57	Savitskaya (first female)	87	Tognini (France)	117	Helms (USA; fifth female)[S]	147	
28	Zholobov	58	Titov, V.	88	Poleshchuk	118	Tito (USA)	148	
29	Bykovsky	59	Alexandrov, A.	89	Tsibilyev	119	Tyurin[S]	149	
30	Aksenov	60	Solovyov, V.	90	Haignere, J.-P. (France)	120	Culbertson (USA)[S]	150	

Second flight

Includes 5 April 1975 Soyuz 18-1 launch abort ballistic flight by Lazarev and Makarov
Does not include 26 September 1983 Soyuz T-10-1 launch pad abort flight by V. Titov and Strekalov

01	Shatalov	16	Bykovsky	31	Solovyov, V.	46	Tsibilyev
02	Yeliseyev	17	Ryumin	32	Alexandrov, A.	47	Musabayev
03	Volkov, V.	18	Aksenov	33	Titov, V.	48	Budarin
04	Filipchenko	19	Romanenko	34	Volkov, A.	49	Haignere, J.-P. (France)
05	Rukavishnikov	20	Dzhanibekov	35	Chrétien (France)	50	Gidzenko (A)
06	Lazarev (ballistic trajectory)	21	Popov	36	Viktorenko	51	Baturin
07	Makarov (ballistic trajectory)	22	Lebedev	37	Solovyov, A.	52	Dezhurov (S)
08	Klimuk	23	Ivanchenkov	38	Manarov	53	Haignere, A. (France; second female)[F]
09	Sevastyanov	24	Strekalov	39	Krikalev	54	Onufrienko (S)
10	Kubasov	25	Serebrov	40	Manakov	55	Zaletin
11	Volynov	26	Lyakhov	41	Afanasyev	56	
12	Gorbatko	27	Kizim	42	Polyakov	57	
13	Grechko	28	Malyshev	43	Avdeyev	58	

Pos.	Cosmonaut	Pos.	Cosmonaut	Pos.	Cosmonaut	Pos.	Cosmonaut	Pos.	Cosmonaut
14	Gubarev	29	Savitskaya (first female)	44	Usachev	59			
15	Kovolyonok	30	Savinykh	45	Kaleri	60			
Third flight									
01	Shatalov	09	Kovalenok	17	Lyakhov	25	Krikalev^{A}		
02	Yeliseyev	10	Dzhanibekov	18	Serebrov	26	Usachev^{S}		
03	Makarov	11	Popov	19	Solovyov, A.	27	Musabayev		
04	Klimuk	12	Strekalov	20	Volkov, A.	28	Gidzenko		
05	Rukavishnikov	13	Grechko	21	Viktorenko	29			
06	Ryumin	14	Kizim	22	Avdeyev	30			
07	Kubasov	15	Romanenko	23	Afanasyev	31			
08	Gorbatko	16	Savinykh	24	Kaleri	32			
Fourth flight									
01	Makarov	04	Solovyov, A.^{D}	07	Afanasyev	10			
02	Dzhanibekov	05	Serebrov	08		11			
03	Strekalov	06	Viktorenko	09		12			
Fifth flight									
01	Dzhanibekov	03	Solovyov, A.	05		07			
02	Strekalov^{A}	04		06		08			
International cosmonauts (including ascent, descent by Shuttle and/or spacecraft swap-only 'flights')									
First flight									
01	Remek (Czechoslovakia)	11	Sharma (India)	21	Haignere, J.-P (France)	31	Shepherd, W. (USA)	41	
02	Hermaszewski (Poland)	12	Faris (Syria)	22	Merbold (German)	32	Voss, J.S. (USA)	42	
03	Jahn (East 13 Germany)		Alexandrov, A. (Bulgaria)	23	Thagard (USA)	33	Helms (USA)	43	

04	Ivanov (Bulgaria)	14	Mohammad (Afghanistan)	24	Reiter (Germany)	34	Tito (USA)	44
05	Farkas (Hungary)	15	Akiyama (Japan)	25	Deshays (France; second female)	35	Culbertson (USA)	45
06	Tuan (Vietnam)	16	Sharman (UK; first female)	26	Ewald (Germany)	36	Walz (USA)	46
07	Mendez (Cuba)	17	Viehbock (Austria)	27	Linenger (USA)	37	Bursch (USA)	47
08	Gurragcha (Mongolia)	18	Aubakirov (Kazakhstan)	28	Foale (USA/UK)	38	Vittori (Italy)	48
09	Prunariu (Romania)	19	Flade (Germany)	29	Thomas (USA)	39	Shuttleworth (South Africa)	49
10	Chrétien (France)	20	Tognini (France)	30	Bella (Slovakia)	40	De Winne	50
Second flight								
01	Chrétien (France)	04						
02	Haignere, J.-P. (France)	05						
03	Haignere, C. (France; first female)	06						

A Ascent only; descent by US Space Shuttle
D Descent only; ascent by US Space Shuttle
S Soyuz docking port swap only; ascent and descent by US Space Shuttle (US astronauts Lucid and Blaha did not swap Soyuz ports on their Mir missions)
F French cosmonaut C. Deshays married J.-P. Haignere prior to her second Soyuz flight.

Biographies

AIR FORCE COSMONAUTS

Afanasyev, Viktor Mikhailovich (Colonel) was born in Bryansk, Russia, on 31 December 1948. He graduated from the Kacha HAFPS in 1970, and then flew MiG-29s and MiG-31s in the Air Force. He went to the Chkalov Test Pilot School in 1976, and served as a test pilot for ten years before joining the Buran programme in 1985. Due to the delays in the programme, he transferred to the TsPK team in 1987, and over the next twelve years flew four Soyuz missions. Afanasyev is currently Deputy Commander of the cosmonaut team, and lives at Star City.

Beregovoi, Georgi Timofeyevich (Lieutenant-General) was selected as a cosmonaut in 1964, and was assigned to fly the first Soyuz craft after Soyuz 1 crashed into the ground, killing Komarov, in April 1967. He flew Soyuz 3 in October 1968, and tried, but failed, to dock the craft with the unmanned Soyuz 2. Beregovoi retired in 1969 and became the head of the Cosmonaut Training Centre (TsPK) from June 1972 until January 1987. He died on the 30 June 1995, and is buried in Novodevichi cemetery. (For further details, see *The Rocket Men*.)

Bykovsky, Valeri Fyodorovich (Colonel) was born on 2 August 1934. He joined the cosmonaut team in 1960, and flew his first mission on Vostok 5 in 1963. He was a member of the original Soyuz training group and was assigned to command the Soyuz 2 craft that would have docked with Soyuz 1. This mission was cancelled when Soyuz 1 developed problems. Bykovsky then joined the lunar programme, and his next assignment was as training manager for the Apollo–Soyuz Test Project. After that programme ended, Bykovsky was assigned to command Soyuz 22 and then Soyuz 31 in the Interkosmos programme. He retired in 1988, and now lives in Star City. (For further details, see *The Rocket Men*.)

Dobrovolsky, Georgi Timofeyevich (Lieutenant Colonel) was selected as a cosmonaut in 1963. He commanded the Soyuz 11 crew that carried out the first successful occupation of a space station, onboard Salyut 1, but on 30 June 1971, after completing 23 days in space, he was killed during re-entry. He is buried in the Kremlin Wall. (For further details, see *The Rocket Men.*)

Dzhanibekov, Vladimir Alexandrovich (Major-General) was born in Iskander, Uzbekistan, on 13 May 1942. He changed his name when he was married. In 1965 he graduated from Yeisk HAFPS, and remained at the school as an instructor. He joined the cosmonaut team in 1970, and was a back-up commander on the ASTP project before commanding 5 Soyuz flights between 1978 and 1985. This included the very dangerous mission to Salyut 7 on Soyuz T-13. He retired after that mission, and became a senior trainer at TsPK, finally leaving the centre in 1997. Dzhanibekov is twice Hero of the Soviet Union and a Pilot Cosmonaut of the USSR. He now lives in Moscow.

Gagarin, Yuri Alexeyevich (Colonel) was selected in 1960, and flew the first manned space mission on Vostok in April 1961. On the orders of the Soviet leadership he was then grounded due to his status, but he persuaded the commanders to reinstate him to flight status. Gagarin was a member of the original Soyuz training group, and in 1966 was selected to be the back-up to Komarov on Soyuz 1. He was involved in the investigation of the Soyuz accident, but was himself killed in an air crash on 27 March 1968. He is buried in the Kremlin Wall. (For further details, see *The Rocket Men.*)

Gorbatko, Viktor Vasilyevich (Major-General) was selected as a cosmonaut in 1960, and was a member of the first Soyuz training group. He backed-up a number of early Soyuz flights, and flew three Soyuz missions in 1969, 1977 and 1980 – on the last two as Commander. Gorbatko retired in 1982 and now lives in Moscow. (For further details, see *The Rocket Men.*)

Gubarev, Alexei Alexandrovich (Major-General) became a cosmonaut in 1963, and originally trained to fly the military Soyuz in 1966 and 1967. He rejoined the main Soyuz group in 1968, and trained for a number of missions (which were not flown) before eventually commanding the Soyuz 17 mission in 1975. He left the cosmonaut team in 1981 and held a number of Air force assignments. Gubarev is a Hero of the Soviet Union and a Pilot Cosmonaut of the USSR. He now lives in Moscow. (For further details, see *The Rocket Men.*)

Khrunov, Yevgeny Vasilyevich (Colonel) was selected as a cosmonaut in 1960, and was a member of the original Soyuz training group selected in 1966. In January 1969 he made his only spaceflight when he was the Research Engineer on Soyuz 5. This included an EVA transfer from one Soyuz to another. Khrunov backed-up a number of Soyuz missions, including the original mission cancelled in 1967. He retired from the cosmonaut team in 1980, and died on 19 May 2000. Khrunov was a Hero of the Soviet Union and a Pilot Cosmonaut of the USSR He is buried in Moscow. (For further details, see *The Rocket Men.*)

Klimuk, Pyotr Ilyich (Colonel-General) was born in the Brest Region of Byelorussia on 10 July 1942. He attended a pilot school and, after graduating, flew MiG-15s. He became a cosmonaut in 1965, and was immediately assigned to the lunar training group, while also working on the Kontakt missions. He commanded the Soyuz 13 Orion mission in 1973, and flew a second mission in 1976 before becoming the TsPK political chief, attending the Lenin Military Political Academy in 1983. In 1991, he succeeded Shatalov as director of the Gagarin Training Centre (TsPK) – a post he still retains. Klimuk is twice Hero of the Soviet Union and a Pilot Cosmonaut of the USSR. He lives near Star City. (For further details, see *The Rocket Men.*)

Kolesnikov, Gennedy Mikhailovich (Colonel) was born in the Chita Region of the USSR on 7 October 1936. He graduated from the Zhukovsky AF Engineering Academy in 1964, and joined the cosmonaut team a year later. He was a military scientist and was assigned as Flight Engineer to the Soyuz military project 7K-VI. This programme was cancelled in early 1968, but Kolesnikov had already left the cosmonaut team in late 1967 due to medical reasons. He continued to work at the training centre, and was a candidate for a Soyuz military mission between 1983 and 1986, but never flew. He currently works at the Air Force Academy named for Gagarin. He is a senior scientist and inventor, with more than a thousand parachute jumps to his credit. He still lives in Star City.

Kolodin, Pyotr Ivanovich (Colonel) is a military engineer, and was selected as a cosmonaut in 1963. He was a back-up to Voskhod crews, and became an original member of the Soyuz training group. He was assigned to some of the early Soyuz flights, and was a prime crew-member on at least two occasions, although he never flew. He retired in 1983, working for some time at mission control. Kolodin still lives in Star City. (For further details, see *The Rocket Men.*)

Komarov, Vladimir Mikhailovich (Colonel-Engineer) was selected as a cosmonaut in 1960. He made his first spaceflight in 1964 when he commanded Voskhod 1, and was an original member of the 1966 Soyuz group. He was selected to command the first Soyuz flight, which he flew in 1967, but was killed during the return to Earth after a serious problem developed on the Soyuz 1 spacecraft. He was the first Soviet cosmonaut to make two flights. Komarov died on 24 April 1967, and is buried in the Kremlin Wall. (For further details, see *The Rocket Men.*)

Kovolyonok, Vladimir Vasilyevich (Colonel-General) was born in Byelorussia on 3 March 1942. He graduated from the Balashov HAFPS in 1963, and flew in heavy aviation until he joined the cosmonaut team in 1967, where he became involved in the rescue and recovery squadron. He commanded three Soyuz missions in 1977, 1978 and 1981. Kovolyonok attended the General Staff Academy while also being a deputy director at TsPK. Over a number of years he has held several senior posts in research institutes, and from 1994 he was director of the Zhukovsky Air Force Engineering Academy. Kovolyonok is twice Hero of the Soviet Union and a Pilot Cosmonaut of the USSR. He lives in Moscow.

Kuklin, Anatoly Petrovich (Colonel) was born on 3 January 1932. He joined the cosmonaut team in 1963, and the Spiral training group in 1965. In 1968 he transferred to the lunar training programme, and then moved to training on the

Soyuz, for which he was the original back-up for Soyuz 7. He was medically disqualified from active training in 1970 and left the team in 1975. Kuklin is retired, and lives in Star City. (For further details, see *The Rocket Men.*)

Lazarev, Vasily Grigoryevich (Colonel) was selected as a cosmonaut in 1966, and was a pilot as well as a qualified doctor. He started training on Soyuz in 1968, and backed-up the long-duration Soyuz 9 mission before commanding the first flight after the death of the Soyuz 11 crew. Soyuz 12 was the first fully operational test of the Soyuz ferry. He was involved in a Soyuz abort in April 1975, and afterwards acted as back-up to at least one other mission. Lazarev retired in 1985, and died on 31 December 1990. He is buried near Star City. (For further details, see *The Rocket Men.*)

Leonov, Alexei Arkhipovich (Major-General) was selected as a cosmonaut in 1960. He made his first flight on Voskhod 2 in 1965, on which mission he made the first ever spacewalk. He was a member of the original Soyuz training group selected in 1966, and trained for a number of missions (which were cancelled). He would have commanded the ill-fated Soyuz 11 mission, but a colleague was medically disqualified and the whole crew was replaced. He commanded the ASTP Soyuz 19 mission in 1975. Leonov afterwards became a senior commander at TsPK, and retired in 1991. He is now a banker in Moscow, and still lives near Star City. (For further details, see *The Rocket Men.*)

Malyshev, Yuri Vasilyevich (Colonel) was born near Volgograd on 27 August 1942. He graduated from the Kharkov HAFPS in 1963, and went on to fly MiG-17s and Mig-21s. He enrolled in the cosmonaut group in April 1967, and also attended test pilot school, gaining the qualification of Test Pilot Third Class. He joined the Soyuz training group in 1974, and in 1977 he graduated from the Air Force Academy named for Gagarin. He served as back-up to Soyuz 22, and was then was paired with Aksenov, flying the first Soyuz T mission in 1980. After flying a second mission in 1984 he was considered for other missions, but was not selected. Malyshev retired from the team having served in a number of roles at TsPK. He was a twice Hero of the Soviet Union and a Pilot Cosmonaut of the USSR. He died on 8 September 1999, and is buried near Star City.

Musabayev, Talgat Amangeldyevich (Colonel) was born on 7 January 1951 in Kazakhstan. He is a native Kazakh, but made his three spaceflights as a Russian. He graduated from the Red Banner School for Civil Aeronautics Engineers in 1974, and then served as a pilot in a commercial airline service. InMay 1990 he joined the cosmonaut team to back-up Aubakirov on the first Kazakh mission. He then transferred to the Russian Air Force, and flew his first mission in 1994 as a Flight Engineer, before commanding two more Soyuz flights. Musabayev lives in Moscow and is still active. He is a Hero and a Pilot Cosmonaut of the Russian Federation.

Nikolayev, Andrian Grigoryevich (Major-General) was selected as a cosmonaut in 1960, and made his first flight in 1962 on Vostok 3. He was a member of the original Soyuz training group selected in 1966, and backed-up a number of early Soyuz missions before commanding the Soyuz 9 long-duration mission, setting a record of 18 days. He retired from active flight status in 1970, and worked at TsPK in a number of senior training and command positions. Nikolayev retired from the Air Force in 1992, and lives in Star City. (For further details, see *The Rocket Men.*)

Popov, Leonid Ivanovich (Major-General) was born in the Kirovogradsky region of the Ukraine on 31 August 1945. He graduated from the Chernigov Lenin Komsomol HASPS in 1968, and went on to fly MiG-19s. He enrolled in the cosmonaut team in 1970, joined a Soyuz training group in 1974, served as back-up Commander for two missions before eventually flying three Soyuz missions in 1980, 1981 and 1982, and left the cosmonaut team in 1987. In 1990 he graduated from the General Staff Academy, and was assigned to the directorate responsible for responding to natural disasters. Popov is twice Hero of the Soviet Union and a Pilot Cosmonaut of the USSR. He now lives in Moscow.

Popovich, Pavel Romanovich (Major-General) was enrolled in the cosmonaut team in 1960, and flew his first mission in 1962 on Vostok 4. He was assigned to command the first Soyuz as part of the 7K-VI military programme, but when that was cancelled he was assigned to the lunar programme. In 1972 he was given command of the first ferry Soyuz to go to a space station following the Soyuz 11 disaster, flying on Soyuz 14 to Almaz/Salyut 3 in 1974. He then undertook a number of senior positions at TsPK, finally leaving the centre in 1990. Popovich is twice Hero of the Soviet Union and Pilot Cosmonaut of the USSR. He now lives in Moscow. (For further details, see *The Rocket Men.*)

Shatalov, Vladimir Alexandrovich (Lieutenant-General) was selected as a cosmonaut in 1963, and became a member of the original Soyuz training group formed in 1966. He commanded three early Soyuz Flights in 1969 and 1971, and was the command cosmonaut on the 1969 group flights. Shatalov left the cosmonaut team in July 1971, becoming Director General of Cosmonaut Training in the Soviet Air Force (the post was abolished in 1987 when he left). In January 1987 he succeeded Beregovoi as Commander of TsPK, and held this position until his removal in 1991. He still lives at Star City. (For further details, see *The Rocket Men.*)

Shonin, Georgy Stepanovich (Lieutenant-General) was selected as a cosmonaut in 1960, and was a member of the original Soyuz training group formed in 1966. He was involved as back-up in a number of early missions before commanding the Soyuz 6 mission in 1969 – a mission involved the first welding experiment conducted in space. Shonin was intended to command a number of other missions, but was withdrawn due to medical problems. He became a senior trainer at TsPK before heading a number of Air Force Research Institutes until he retired in 1990. He died on 6 April 1997, and is buried near Star City. (For details, see *The Rocket Men.*)

Solovyov, Anatoly Yakovlevich (Colonel) was born in Riga on 16 January 1948. He graduated from Chernigov HAFPS in 1972, and then served in the Air Force before joining the cosmonaut team in August 1976. He trained as a test pilot in Akhtubinsk, gaining the qualification of Test Pilot Second Class, and served as a back-up on a number of occasions before making the first of his five space flights on Soyuz TM-5 in 1998. He afterwards commanded three more Soyuz flights, and was launched on STS-71 to Mir. Solovyov is a Hero of the Russian Federation and a Pilot Cosmonaut of the Russian Federation. He has now retired from the cosmonaut team, and lives near Star City.

Titov, Vladimir Georgyevich (Colonel) was born in the Chita region of Russia on 1 January 1947. He graduated from the Chernigov HAFPS in 1970 and stayed on as an instructor. He joined the cosmonaut team in August 1976 and also attended the test pilot school at Akhtubinsk, attaining the status of Test Pilot Third Class. He started working on Soyuz/Salyut systems training soon after and was a back up commander for Salyut 7 missions before his first flight. Titov commanded the Soyuz T8 mission in 1983 and was the commander of the Soyuz abort mission later that year. When he was launched on TM4, he became the commander of the first crew to be in space for more than a year (with Manarov). Titov also flew 2 missions as a Mission Specialist on board the Shuttle. He retired from the team in 1998 and lives in Star City. He is a Hero of the Soviet Union and a Pilot Cosmonaut of the USSR.

Tsibliyev, Vasily Vasilyevich (Major-General) was born in the Kirov district of Russia on 20 February 1954. He graduated from the Kharkov HAFPS in 1975, served as pilot in the Air Force before attending the Yuri Gagarin Air Force Academy in 1987, and then joined the cosmonaut team and qualified as a Soyuz TM Commander. He served on a number of back-up crews before commanding two missions – Soyuz TM-17 and Soyuz TM-25 to Mir. Tsibliyev left the cosmonaut team in 1997, and is currently the Deputy Commander of TsPK. He is a Hero and Pilot Cosmonaut of the Russian Federation, and lives in Star City.

Viktorenko, Alexandr Stepanovich (Colonel) was born in Kazakhstan on 29 March 1947, but is Russian by nationality. He graduated from the Polbin HAFPS in 1969, and served in the Black Sea area until his selection as a cosmonaut in 1978. After qualifying as a Test Pilot Third Class at Akhtubinsk, he was injured in an accident in an isolation chamber, which delayed his first assignment until 1982. Viktorenko commanded four Soyuz flights before retiring from the cosmonaut team in 1997. He is a Hero and Pilot Cosmonaut of the USSR, and lives in Star City.

Volkov, Alexandr Alexandrovich (Colonel) was born in the Donetsk region of the Ukraine on 27 April 1948. He graduated from the Kharkov HAFPS in 1970, and after working as an instructor at the school he joined the cosmonaut team in August 1976 and then went to the test pilot school at Akhtubinsk, where he gained the qualification of Test Pilot Second Class. He was a back-up Commander several times before flying three Soyuz missions – two of them as Commander – between 1985 and 1991, and was the Commander of the cosmonaut team for many years before his retirement in 1998. He is twice Hero of the Soviet Union and a Pilot Cosmonaut of the USSR, and now lives in Star City. His son, Sergei Alexandrovich Volkov, was selected as a cosmonaut in 1997.

Volynov, Boris Valentinovich (Colonel) was selected as a cosmonaut in 1960, and was involved as a back-up pilot in a number of Vostok and Voskhod missions. In 1996 he became a member of the original Soyuz training group, and commanded two Soyuz flights in 1969 and 1976. He then became a senior cosmonaut, commanding the cosmonaut team for more than a decade. Volynov is twice Hero of the Soviet Union and a Pilot Cosmonaut of the USSR. He retired in 1990, and still lives in Star City. (For further details, see *The Rocket Men.*)

Zaletin, Sergei Viktorovich (Lt. Colonel) was born in the Tula region of Russia on 21 April 1962. He graduated from the Borisoglebsk HAFPS in 1983, joined the cosmonaut team in 1990, and graduated from Moscow State University in 1994. He was assigned to a Mir training group in 1996, commanded the final Soyuz mission to Mir, and also commanded the first Soyuz TMA mission in 2002. Zaletin is a Hero and a Pilot Cosmonaut of the Russian Federation. He is still active, and lives in Star City.

NPO ENERGIYA COSMONAUT-ENGINEERS

All civilians.

Akyonov, Vladimir Viktorovich was born in Moscow on 1 February 1935. From 1950 he served in the military, and had a strong engineering background. In January 1957 he joined the Korolyov design bureau as an engineer, and in 1963 graduated from the All Union Polytechnic Institute. He then transferred to the flight test department, working on various systems including recovery and EVA techniques before joining the cosmonaut team in 1973. He then flew on Soyuz 22 in 1976, and again on the first manned flight of the Soyuz T (T-2) in 1982. Akyonov retired in 1988, and is currently a senior engineer in Energiya. He is twice Hero of the Soviet Union and a Pilot Cosmonaut of the USSR.

Alexandrov, Alexandr Pavlovich was born in Moscow on 20 February 1943. He initially enrolled in the Rocket Forces, but left to join Energiya in 1964. He also tried to join the cosmonaut team, but when he was medically tested in 1967, he failed. In 1969 he graduated from the Bauman Higher Technical School, and in December 1978 he finally joined the team. He flew two Soyuz missions in 1983 and 1987, involving Soyuz T and TM craft. In 1993 he retired, and became head of Department 291 at Energiya – the Manned Spaceflight Directorate with responsibility for all cosmonauts and training in the bureau. Alexandrov is twice Hero of the Soviet Union and a Pilot Cosmonaut of the USSR. He lives in Moscow.

Avdeyev, Sergei Vasilyevich was born in the Samarskaya region of Russia on 1 January 1956. He graduated from the Moscow Institute of Physical Engineering (MIFI) in 1979, and joined NPO Energiya. In March 1987 he joined the cosmonaut team – initially as a back-up engineer on a number of missions – and flew three Soyuz missions in 1994, 1997 and 1998. On these missions he spent a total of 749 days in space, placing him as the overall record-holder. Avdeyev is a Hero and a Pilot Cosmonaut of the Russian Federation. He is retired, and lives in Moscow.

Feoktistov, Konstantin Petrovich was a leading designer of manned spacecraft with the NPO Energiya design bureau, and was involved in the design of Vostok, Voskhod, Soyuz T and Progress. He was selected to be the Flight Engineer on Voskhod 1 in 1964. In 1987 he returned to the team and trained for a mission on a Soyuz T craft. Feoktistov retired due to medical problems, and is now a Professor at Bauman Higher Technical School in Moscow, where he lives. (For further details, see *The Rocket Men.*)

Grechko, Georgy Mikhailovich was born in Leningrad on 25 May 1931. In 1955 he graduated from the Leningrad Institute of Mechanics and went to work on the Sputnik project in the Korolyov design bureau. He also worked on the Luna craft. In 1964 he was considered for the cosmonaut team, and in May 1966 he joined the team to form part of the original Soyuz training group. In October 1966 he broke one of his legs, which delayed his participation in the crewing cycle, although he later resumed training and joined the lunar training group. In 1969 he rejoined the Soyuz group and served as a back-up on a number of occasions before flying on Soyuz 17 in 1975. He later flew on two other Soyuz missions, including a Soyuz T mission, before retiring in 1986 to become head of a laboratory for the Academy of Sciences. Grechko is twice Hero of the Soviet Union and a Pilot Cosmonaut of the USSR. He lives in Moscow.

Krikalev, Sergei Konstantinovich was born in Leningrad on 27 August 1958. He graduated from the Leningrad Institute of Mechanics in 1981, and earned a Master of USSR sports title for flying acrobatic aircraft. He then joined NPO Energiya, working in the flight documentation department. He joined the cosmonaut team in September 1985, and initially carried out work on a possible Buran flight. His first flight took place in 1988, and he has served as Flight Engineer on three Soyuz flights and as Mission Specialist on two Shuttle missions. Krikalev is a Hero of the Soviet Union and the Russian Federation and a Pilot Cosmonaut of the USSR. He lives in Moscow, and is currently training as an ISS expedition Commander.

Kubasov, Valeri Nikolayevich was born in Moscow on 7 January 1935. He graduated from the Moscow Aviation Institute in 1958, and went to work in the Korolyov design bureau on spacecraft trajectories. He was considered for the cosmonaut team in 1964, but formally joined in May 1966, to become a member of the original Soyuz training group. He was back-up to a number of early Soyuz flights before serving as Flight Engineer on Soyuz 6 in 1969. He was then involved in a number of other prime crews whose missions were cancelled, before completing two more Soyuz flights in 1975 and 1978, and commanding his third mission. In 1987 he took up a senior engineer post at Energiya, and was then appointed Deputy Director of a branch of Energiya concerned with life support systems. Kubasov is twice Hero of the Soviet Union and a Pilot Cosmonaut of the USSR. He lives in Moscow.

Laveikin, Alexandr Ivanovich was born in Moscow on 21 April 1951. He graduated from the Bauman Higher Technical School in 1974, and went to work at Energiya NPO on space structures. He enrolled in the cosmonaut team in December 1978, and was assigned to a training group in 1984. He flew as Flight Engineer on the first manned Soyuz TM flight to Mir, but returned to Earth early due a heart problem. This was his only flight in space. He left the team in 1994 and Energiya in 1995. Laveikin is a Hero of the Soviet Union and a Pilot Cosmonaut of the USSR. He now lives in Moscow.

Makarov, Oleg Grigoryevich was born in the Tver Region of the USSR on 6 January 1933. In 1957 he graduated from the Bauman Higher Technical School, and joined Korolyov's design bureau, where he was involved in the development of Vostok and the design of Soyuz. In May 1966 he joined the cosmonaut team, and was assigned to the prime crew of the first lunar mission with Leonov. He worked on this project until 1971, when it was cancelled. In 1973 he was Flight Engineer on Soyuz 12 – the first flight of the ferry craft. He was then involved in the launch abort in 1975, before flying on Soyuz 27 and then on Soyuz T-3 in 1980. He retired from the team in 1987, and now holds a senior position in NPO Energiya. Makarov is twice Hero of the Soviet Union and a Pilot Cosmonaut of the USSR. He lives in Moscow.

Manarov, Musa Khiramanovich was born in Baku, Azerbaijan, on 22 March 1951. In 1974 he graduated from the Moscow Aviation Institute and went to work at NPO Energiya as a tester. He joined the cosmonaut team in December 1978, but served as a mission controller at TsUO until beginning training in 1983. His first mission was as Flight Engineer on Soyuz TM-4, when he spent a year in orbit. He flew a second mission in 1990, and then retired from Energiya and the cosmonaut team in 1992. Manarov is a Hero of the Soviet Union and a Pilot Cosmonaut. He lives in Moscow.

Nikitsky, Vladimir Petrovich was born in the Ryzan region on the 8 March 1939. In 1961 he graduated from the Moscow Chemical–Technical Institute and joined the Korolyov design bureau, where he worked on manned spacecraft. He joined the cosmonaut team in August 1967 and was assigned to the lunar training group, but failed a medical and left the following year. Nikitsky lives in Korolyov, and continues to work as a departmental head at NPO Energiya.

Rukavishnikov, Nikolai Nikolayevich was born in Tomsk on 18 September 1932. In 1957 he graduated from the Moscow Institute of Physics and Engineering and went to work at the Korolyov design bureau, where he was involved in the development of automatic control systems. He was selected for the cosmonaut team in January 1967, and was assigned to the lunar programme. In 1970 he was assigned to the Soyuz 10 mission, which flew the following year. His next assignment was as Flight Engineer on the Soyuz 16 ASTP rehearsal mission, and he was afterwards Commander of the Soyuz 33 Interkosmos mission (during which engine problems forced an emergency landing). Following this he participated in the design of the Mir station. Rukavishnikov was twice Hero of the Soviet Union and a Pilot Cosmonaut of the USSR. He retired in 1987, and died on 19 October 2002.

Ryumin, Valeri Viktorovich was born in far-eastern Siberia on 16 August 1939. He enrolled in the Moscow Forestry Institute and graduated in 1966. He was a senior engineer during the development of the 7K-L1 vehicle and, in 1970, the Salyut station. He joined the cosmonaut team in 1973, and went on to make four flights in space – three as a Flight Engineer on Soyuz 25, Soyuz 32 and Soyuz 35, and the last as a Mission Specialist on STS-91 to the Mir station. In 1982 he became Lead Flight Director and Head of the Manned Spaceflight Department, and was involved in the design of Salyut 7. In 1987 he became a Deputy Chief Designer at NPO Energiya, responsible for Buran and Shuttle/Mir. Ryumin is twice Hero of the Soviet Union and Pilot Cosmonaut of the USSR. He is married to cosmonaut Yelena Kondakova, and lives in Moscow.

Savinykh, Viktor Petrovich was born in the Kirov Region on 7 March 1940. In 1969 he graduated from the Institute of Geodetic Engineering, Aerial Photography and Cartography, and joined the Korolyov design bureau the same year. In December 1978 he enrolled in the cosmonaut team, where he immediately began training on Soyuz T craft. He served as a back-up before making three flights in 1982, 1983, and on the Soyuz T-13 mission in 1985 (which has been described as the most dangerous ever flown by the Soviet Union).He left the cosmonaut team and Energiya in 1988. Savinykh is twice Hero of the Soviet Union and a Pilot Cosmonaut of USSR. He lives in Moscow, and is rector of his alma mater.

Savitskaya, Svetlana Yevgenyevna was born in Moscow on 8 August 1948. In 1972 she graduated from the Moscow Aviation Institute, and later became a very famous parachutist and pilot. In 1972 she attended the World Aerobatics Championship in Britain, returning as the world champion in all-round flying. She worked for Yakovlev's design bureau while setting many world records. Savitskaya joined the cosmonaut team in July 1980, and transferred to NPO Energiya in 1983. She flew twice into space flights, in 1983 and 1984, and on her second flight she became the first woman to carry out a space walk. Savitskaya was assigned to command an all-female Soyuz crew, due to fly in 1986, but the mission was cancelled. In 1993 she retired from the cosmonaut team but stayed at Energiya, and was later elected to become a member of the Duma (the Russian Parliament). Savitskaya is twice Hero of the Soviet Union and a Pilot Cosmonaut of the USSR. She lives in Moscow.

Serebrov, Alexandr Alexandrovich was born in Moscow on 15 February 1944. In1967 he graduated from the Moscow Physical–Technical Institute, and stayed there before joining TsKBEM in 1976. In 1978 he joined the cosmonaut team and began training on the Soyuz T spacecraft. He served as back-up on a number of missions before making the first of his four spaceflights: Soyuz T-7 in 1982, T-8 in 1983, TM-8 in 1989, and TM-17 in 1993. He was also the cosmonaut team's specialist on the MMU, which he flew for the first time on the TM-8 mission. He finally left the cosmonaut team in 1995, and is currently an independent consultant to President Putin. Serebrov and is a Hero of the Soviet Union and a Pilot Cosmonaut of the USSR. He lives in Moscow.

Sevastyanov, Vitali Ivanovich was born on 8 July 1935 near the city of Sverdlovsk. After graduating from the Moscow Aviation Institute in 1959 he joined the Korolyov design bureau, where he was part of the team that created Vostok. He lectured to the first groups of cosmonauts before joining the team himself in January 1967, when he was immediately assigned to work on a Zond mission as part of the lunar programme. When this was cancelled in late 1968, he was paired with Nikolayev to train for a number of missions before flying a solo Soyuz long-duration mission on Soyuz 9 in 1970. He then went on to work on Salyut stations, and in 1975 flew a second mission before heading his own department in Energiya and then training for a Soyuz TM mission in the late 1980s. He retired in 1993, and is now member of the Duma (the Russian Parliament). Sevastyanov is twice Hero of the Soviet Union and a Pilot Cosmonaut of the USSR. He lives in Moscow.

Solovyov, Vladimir Alexandrovich was born in Moscow on 11 November 1946. In 1970 he graduated from the Bauman Higher Technical School and went to work at the Korolyov design bureau, where he became an expert on space propulsion. In December 1978 he joined the cosmonaut team and worked on Soyuz T systems. He afterwards served on a number of back-up crews before flying two missions in 1984 and 1986 – the latter as part of the first crew to fly to Mir. He retired as a cosmonaut in 1994, and became the Flight Director at TsUP for Mir and currently ISS. Solovyov is twice Hero of the Soviet Union and Pilot Cosmonaut of the USSR. He lives in Moscow.

Strekalov, Gennedi Mikhailovich was born in Moscow on 28 October 1940, and worked as a coppersmith where Sputnik was constructed. In 1965 he graduated from the Bauman Higher Technical School and joined the Korolyov design bureau, where he was involved in the design and development of Soyuz. He joined the cosmonaut team in 1973 and went to work as a controller at TsUP prior to his first assignment, serving on a number of back-up crews before his first mission on Soyuz T-3 in 1980. He was then a Flight Engineer on four more Soyuz flights (including the launch abort in 1983), and retired from the cosmonaut team in 1995. Strekalov is twice Hero of the Soviet Union and a Pilot Cosmonaut of the USSR. He was head of NPO Energiya's cosmonaut team, and is now an Energiya senior engineer.

Volkov, Vladislav Nikolayevich was born in Moscow on 23 November 1935. In 1959 he graduated from the Moscow Aviation Institute and joined the Korolyov design bureau, where he worked closely with Feoktistov on Vostok and Soyuz. He finally joined the cosmonaut team in May 1966, having been considered for the Voskhod flights in 1964, and was a member of the original Soyuz training group. He was assigned as back-up to the first group flight of 1969, and flew on Soyuz 7 the same year. In 1971 he was Flight Engineer on the ill-fated Soyuz 11 crew, who died due to the depressurisation of the Soyuz capsule. Volkov was twice Hero of the Soviet Union and a Pilot Cosmonaut of the USSR. He is buried in the Kremlin Wall.

Yazdovsky, Valeri Alexandrovich was born on 8 July 1930 in the Ukraine. In 1954 he graduated from the Moscow Aviation Institute and joined the Korolyov design bureau. In 1957 he joined the department responsible for developing manned spacecraft, and helped design the Vostok, Voskhod and Soyuz spacecraft. In August 1967 he joined the cosmonaut team and was assigned to the Soyuz training group as back-up for Soyuz 9. He would have flown the Soyuz 13 mission in 1973, but was replaced only a few days before the mission due to disagreements with senior managers. He left the team in 1982, having never been assigned to another mission.

Yazdovsky continued to work at NPO Energiya until his retirement, and he now lives in Korolyov.

Yeliseyev, Alexei Stanislavovich was born in the Kaluga Region of Russia on 13 July 1934. In 1959 he graduated from the Bauman Higher Technical School and joined the Korolyov design bureau. (While at university he was twice USSR fencing champion.) In May 1966 he joined the cosmonaut team and was immediately assigned as Flight Engineer on the Soyuz 2 mission, which was later cancelled following problems with the Soyuz 1 craft in April 1967. However, he remained in training, and in 1969 and 1971 flew three missions as a Flight Engineer, including the first EVA vehicle transfer with Khrunov in 1969. Between 1973 and 1981 he was then in charge of all manned spaceflights at mission control, and was a Deputy General Designer at NPO Energiya until his resignation in 1986. He then became rector of his alma mater unti his retirement in 1991. Yeliseyev is twice Hero of the Soviet Union and a Pilot Cosmonaut of the USSR. He lives in Moscow.

OTHER RUSSIANS

Fartushny, Vladimir Grigoryevich was born in the Ukraine on 3 February 1938. He worked at the Paton Institute in Kiev before joining the cosmonaut team in 1968. He was the experimenter who would have operated the Vulkan welding experiment onboard Soyuz 6 in 1969, but did not fly due to weight constraints. Fartushny left the team in 1971 after being injured in a car accident. He still works for Paton, and lives in Kiev.

Commanders of the cosmonauts

Goreglyad, Leonid Ivanovich (Major-General) was born on 13 April 1916, and served in the Air Force during the Second World War. He was honoured as a Hero of the Soviet Union on 23 February 1948, and attended the General Staff Academy in 1950. He was deputy to Kamanin for more than ten years, with the official title of Inspector General for Spaceflight. He was present at the landing site when the Soyuz 11 crew returned to Earth.

Kamanin, Nikolai Petrovich (Colonel-General) was born – according to official records – on 18 October 1908, although in reality he was born on 18 October 1909. In 1934 he was one of the first to be honoured as Hero of the Soviet Union. In 1958 he was appointed Deputy Chief of the Soviet Air Force responsible for manned spaceflight. In 1960 this post was redesignated Director of Cosmonaut Training in 1960, and was later again redesignated Aide for Space Matters of the Air Force C-in-C. Kamanin remained in this post until July 1971, when he was replaced by Shatalov. He died on 11 March 1982, and is buried in Novodevichi cemetery, Moscow. *Full biography see Rocket Men p 309*

(See also Shatalov.)

Commanders of the Cosmonaut Training Centre (TsPK)

Kuznetsov, Nikolai Fedorovich (Major-General) was born on 26 December 1916. During the Second World War he served as a fighter pilot, and was honoured as a Hero of the Soviet Union on 1 May 1943. On 2 November 1963 he was appointed to command the training centre, and remained in command until May 1972. He then worked at NPO Energiya until his retirement. Kuznetsov died on 5 March 2000, and is buried in a cemetery on the outskirts of Moscow. (For further details, see *The Rocket Men.*)

(See also Beregovoi, Klimuk and Shatalov.)

Other

Seryogin, Vladimir Sergeyevich (Colonel Engineer) was born in Moscow in 1922, and is a graduate of the Tambov school for military pilots. During the Second World War he served at the front, and was afterwards honoured as a Hero of Soviet Union. After the war he went to the Zhukovsky AF Engineering Academy as a test pilot, and was Commander of the cosmonaut flight training squadron based at Chkalov Air Force base. One of his roles was to oversee the aircraft that the cosmonauts used to keep their flying skills up to par, and the aircraft used for weightlessness training. On 28 March 1968 he was overseeing such a flight with Gagarin when their MiG-15 crashed and they were both killed. Seryogin is buried in the Kremlin wall. The regiment – which is still based at Chkalov – is named after him.

FOREIGN COSMONAUTS

From 1976 up to the present day, a number of non-Soviet, Russian or American citizens have trained to be cosmonauts, and some of them have flown on Soyuz craft to different space stations. Details of those programmes are covered in the main text. The information below covers those foreign cosmonauts who have set a record by their participation in manned spaceflight.

First non-Soviet/Russian/American citizen in space

Remek, Vladimir was born in Czechoslovakia on 26 September 1948. On joining the Czech Air Force he was posted to Russia to attend the Yuri Gagarin Air Force Academy, from where he graduated in 1976. He was then immediately assigned to train as part of the Interkosmos programme for a mission to Salyut 6, which, with Commander Gubarev, he flew on Soyuz 28, spending eight days in space. After his flight he returned to the Czech Air Force. Remek is now a citizen of the Czech Republic, and is honoured as a Hero by both the USSR and his own country.

First non-Soviet/Russian/American female in space

Sharman, Helen Patricia was born in Sheffield, England, on 30 May 1963. She has a degree in chemistry, and has worked in the commercial sector. She applied for the flight opportunity following an advertisement in a newspaper, and in 1990 was selected to train for an 8-day mission to Mir. In 1991 she flew as Cosmonaut-Researcher on Soyuz TM-12. After her flight, she worked in public relations. Her mission created a number of new records, including her being the first citizen of the UK to fly in space.

First black person in space

Tamayo-Mendez, Arnaldo (General in the Cuban Air Force) was born in Cuba on 29 January 1942. He trained at a pilot school in the Soviet Union, and in 1978 was selected as a cosmonaut to fly an 8-day mission to the Salyut space station as part of the Interkosmos programme, flying Soyuz 38 with Romanenko as his Commander. After his flight he returned to Cuba to work for the government.

First professional astronaut on a commercial flight

Chrétien, Jean-Loup Jacques (General in the French Air Force) was born in La Rochelle, France, on 20 August 1938. He joined the French Air Force and became a test pilot, and in 1980 was one of two French pilots selected to train for an 8-day mission to Salyut 7. This was the first spaceflight paid for by a national space agency, and was deemed the first commercial mission. In1982 Chrétien flew on Soyuz T-6, and later flew twice to Mir – the first flight on Soyuz TM-7, and the second on STS-86. He is now retired, and lives in the USA.

First non-professional astronaut on a commercial flight on a Soviet spacecraft

Akiyama, Toyohiro was born in Tokyo on 22 July 1942. He was a professional journalist for TBS television, which signed a commercial agreement with the Soviets to fly one of their journalists to Mir on an 8-day mission. In 1989, Akiyama and a colleague were selected to train for this mission, and Akiyama was chosen – although he was a heavy smoker and very unfit. His flight as a cosmonaut-researcher on Soyuz TM-11 in 1990 cost the TV company $12 million. Akiyama afterwards returned to work for the station, but is now a farmer in a remote part of Japan.

First American astronaut to launch on a Russian (Soyuz) spacecraft

Thagard, Norman was born in Marianna, Florida on 3 July 1943. He gained advanced degrees in engineering science before serving in the USMC Reserves, flying 163 combat missions during the Vietnam War. He trained as a medical doctor before being selected by NASA in the first Space Shuttle astronaut class (1978). He served as a mission specialist on STS-7 (1983), STS 51-B/Spacelab 3 (1985), STS-30 (Magellan) (1989), and STS-42 (IML-1) (1992), and flew as the first NASA astronaut to Mir in 1995. Launched as NASA Board Engineer on Soyuz TM-21 on 14 March 1995, he became the first of seven American astronauts to complete long-duration stays onboard Mir. As part of EO-18 he logged 115 days 8 hrs in space – a new record for an American – and landed STS-71 *Atlantis* on 7 July 1995. After this flight, Thagard retired from NASA to become a Professor at Florida State University.

First 'tourist' spaceflight participant

Tito, Dennis Anthony was born in New York on 8 August 1940. He initially trained as an engineer, and went to work on the Mariner spacecraft at the Jet Propulsion Laboratory. In 1972 he founded an investment management company called Wilshire Associates, and, having always had a dream to fly into space, agreed to pay the Russians approximately $20 million for the opportunity to be the first Tourist in Space. His participation was agreed by all parties, although he was restricted to the Russian part of the International Space Station. In 2001, Tito flew with cosmonauts Musabayev and Baturin on the Soyuz TM-32 taxi mission, and spent eight days in space.

COMMUNIST PARTY SENIOR OFFICIALS

A number of these key party officials were given titles to reflect their position within the military, and are referred to as Civilian Generals.

Afanasyev, Sergei Aleksandrovich was born in the Moscow Region on 30 August 1918. He graduated from the Bauman Higher Technical School in 1946, and was afterwards involved in the military industries. In 1965 he was appointed Head of the Ministry of General Machine Building – the key production ministry for the space and missile industry. He was Chairman of the N1/L3 State Commission, and was called the Iron Man. He died on 13 May 2001, and is buried in Novodevichi cemetery. He was a twice Hero of Socialist Labour.

Kerimov, Kerim Aliyevich (Lieutenant-General) was born in Baku on 14 November 1917. In 1946 he joined the Ministry of Defence Main Artillery Directorate, and worked with Korolyov on a number of projects involving rockets. During the late 1940s he was in Germany with Korolyov. In 1965 he was appointed Chairman of the Soyuz State Commission on Flight Testing of Manned Space Complexes. This is 'the State Commission' whose Chairman is saluted by the cosmonauts when they leave for the launch pad. His identity was a state secret until 1986, and he retained his position until 1989. Kerimov is a Hero of Socialist Labour. He lives in Moscow.

Rudnev, Konstantin Nikolayevich was born in Tula in June 1911. In 1952 he became Deputy Minister for the Defence Industry, and in 1958 was appointed Chairman of the Council of Ministers State Committee for Defence and Technology (the State Commission) – a post he held until 1965. He was responsible for all major decisions relating to manned flight. In 1965 he became Minister for Instruments, Automation Equipment and Control Systems. Rudnev was honoured as a Hero of Socialist Labour and six times with the Order of Lenin. He died on 13 August 1980, and is buried in Novodevichi cemetery.

Serbin, Ivan Dmitriyevich was born in Krasnadoron 25 February 1910. In 1942 he was appointed to a post in the Communist Party's Central Committee, and by 1958 he had risen through the ranks to become Head of Defence Industries. He contributed considerably in solving the problems relating to the creation of space technology and its practical applications, and reported directly to the Minister of Defence, Dmitri Ustinov. Serbin was a member of the Supreme Soviet, and was five times awarded the Order of Lenin. He died on 15 February 1981.

Smirnov, Leonid was Deputy Chairman of the Council of Ministers and Head of the Military Industrial Commission (VPK) during the 1960s. This body oversaw the work of the eight ministries working in the Military Industrial Complex.

Tyulin, Georgi Aleksandrovich (Lieutenant-General) was born in Penze on 9 December 1914. He held a senior role in the Central Committee of the Communist Party, and was the deputy to Smirnov. He was Head of GKOT from 1961 to 1965, and Head of the State Commission for the L-1, as well as First Deputy Minister of General Machine Building from 1965 to 1976. Tyulin was honoured as a Hero of Socialist labour. He died on 22 April 1990.

Ustinov, Dmitri Fedorovich (Marshall of the Soviet Union) was born in Samara on 30 October 1908. In 1934 he graduated from the Bauman Higher Technical School, and afterwards worked in the military industries. In 1941 he was appointed People's Commissioner (Minister) for Armaments, and held this post until 1953, after which he was appointed Minister of the Defence Industry in charge of all military areas of production. In 1957 he became Deputy Prime Minister, and in 1965 a non-voting member of the Politburo, rising to full status in 1976. In the same year he became Minister of Defence, and thus had much influence over the space programme at all stages of its development. Ustinov was three times honoured as a Hero of Socialist Labour. He died in December 1984.

RKK ENERGIYA

This is the lead design bureau responsible for the construction of the Soyuz spacecraft and all its variants. It was originally part of another bureau, NII-88, based in Podlipki (which was renamed Kaliningrad and later renamed Korolyov). Over the years this bureau has had several names:

1956–66 OKB-1: the Korolyov design bureau.
1966–74 TsKBEM: Central Construction Bureau of Experimental Machine Building.
1974–94 NPO Energiya: Energiya Science and Production Association, named after Academician S.P. Korolyov.
1994– OAO S.P. Korolyov Rocket and Space Corporation Energiya (RKK Energiya).

Chief Designers

Korolyov, Sergei Pavlovich was born in the Ukraine on 30 December 1917. He was the creative genius behind the Soviet space programme from the 1940s until his death in 1966. He was the head of OKB-1 and the legendary 'Chief Designer of Carrier Rockets and Spaceships'. He designed the rocket and manned spacecraft Vostok, and headed the team that began the early Soyuz design studies, as well as creating many other types of spacecraft. Korolyov was a member of the State Commission, and was twice Hero of Socialist Labour. He died on 14 January 1966, and is buried in the Kremlin Wall. (For further details, see *The Rocket Men.*)

Mishin, Vasily Pavlovich was born on 5 January 1917. In 1946 he went to work in the Korolyov design bureau, and was deputy to Korolyov for many years, specialising in control processes. In 1966 he succeeded Korolyov as Chief Designer, and the bureau was renamed TsKBEM. He oversaw the development of the Soyuz, Salyut and lunar programmes, and the N-1 rocket. After he was dismissed in 1974 he went to work at the Moscow Aviation Institute, and tried to rebuild his reputation with a number of books and interviews over the years. Mishin was a Hero of Socialist Labour. He died on 10 October 2001, and is buried in Troekurovskoe cemetery, Moscow. (For further details, see *The Rocket Men.*)

General Designers

Glushko, Valentin Petrovich was born in Odessa on 2 August 1908. From 1929 he headed the Gas Dynamics laboratory responsible for the design of many of the engines used on rockets and missiles in the Soviet Union. He was Chief Designer of Rocket Engines, and was a member of the State Commission that approved the launch of Sputnik, Vostok and many other craft. In 1974 he became head of the reformed NPO Energiya, which incorporated Korolyov's design bureau. His title was General Designer of Manned Spaceships and Space Stations, which included Soyuz, Salyut, Mir, Buran and the rocket Energiya. Glushko was twice Hero of Socialist Labour. He died on the 10 January 1989, and is buried in Novodevichi cemetery. (For further details, see *The Rocket Men.*)

Semyonov, Yuri Pavlovich was born in the Ukraine on 20 April 1935. He graduated from the Dnepropetrovsk University and went to work at the Yangel design bureau, where he was involved in the development of the R-12 and R-14 missiles (which involved work at Baikonour and Kasputin Yar). In 1958 he went to work on the launch of the first Cosmos spy satellite. He first met Korolyov in 1960, but did not go to work at OKB-1 until 1964. There he became involved in design work on Vostok and Soyuz. In 1965 he was named Chief Designer for Zond/L-1, and in 1969 he was appointed Chief Designer for the Salyut orbital space station. In 1974 he was appointed Chief Designer of Manned Spacecraft and Orbital Stations, and in 1980 Deputy General Designer and Chief Designer of Buran. In 1989 he succeeded Glushko as General Designer for NPO Energiya. Semyanov is a Hero of Socialist Labour, and lives in Moscow.

Leading members of the Energiya design bureau

Bushayev, Konstantin Davidovich was born on 23 May 1914. He was Deputy Head of Construction at OKB-1 from 1954, and was the Soviet Director of the ASTP programme from 1972 until 1975. Bushayev was a Hero of Socialist Labour. He died on 26 October 1978, and is buried in Novodevichi cemetery. (For further details, see *The Rocket Men.*)

Chertok, Boris Yevseyevich was born on in Lodz (now part of Poland) on 1 March 1912. He has been a leading member of the Korolyov design bureau for more than half a century, and still works for NPO Energiya. He has been involved in the development of all manned craft produced by the bureau, and was a deputy to Chief Designer Mishin during the 1960s and 1970s. He later moved to the Ministry of General Machine Building to establish the Manned Spacecraft Experimental Construction Bureau. Chertok is a Hero of Socialist Labour, and lives in Moscow. (For further details, see *The Rocket Men.*)

Gorshkov, Leonid joined the Korolyov design bureau in 1960, and was involved in the design of Soyuz, civilian Salyuts, and Mir. He is head of the orbital station design department.

Kozlov, Dmitri Ivanovich was born on 1 October 1919, and has worked in the Korolyov design bureau since 1946. He was Deputy Chief Designer, reporting to Korolyov, and was a leading designer of the R-7 booster. After a reorganisation in 1967 he became head of the specialised design bureau in Samara, where he developed the concept of the reconnaissance Soyuz spacecraft to meet military requirements, thereby finding himself in competition with Mishin and Chelomei. He also developed the Soyuz R, Zvezda, and Soyuz VI, and had plans for the development of a photoreconnaissance satellite based on Soyuz R. He was also responsible for the serial production of the Vostok R-7 launch vehicle and various unmanned spacecraft. Kozlov is twice Hero of Socialist Labour, and now heads the design bureau called TsSKB Progress. (For further details, see *The Rocket Men.*)

Tikhonravov, Mikhail Klavdiyevich was born in Vladimir in 1900, and was with Korolyov when he went to Germany in 1945. He was Deputy Director of OKB-1, and headed a design team that developed Sputnik, Vostok, Voskhod and Soyuz. He also had many of the younger Korolyov cosmonauts working for him. Tikhonravov was a Hero of Socialist Labour. He died on 4 March 1974, and is buried in Novodevichi cemetery. (For further details, see *The Rocket Men.*)

Tregub, Yakov Isayevich was Deputy Chief Designer for the Flight Control of Piloted Spacecraft during the early years of Soyuz operations.

Voskresensky, Leonid Aleksandrovich was born on 14 July 1913 and was Deputy Chief Designer to Korolyov at OKB-1. He was in charge of the flight test department, including all Vostok test flights, and was also involved in the early development of Soyuz. Voskresensky was a Hero of Socialist Labour. He died on 15 December 1965, and is buried in Novodevichi cemetery.

(See also Feoktistov and Ryumin.)

Other leading Designers

Barmin, Vladimir Pavlovich was born in Moscow on 17 March 1909. He headed the design bureau responsible for the construction of most of the Soviet Union's launch facilities (KBOM), and was a member of the State Commission for many years. Barmin was a Hero of Socialist labour. He died on 17 July 1993, and is buried in Novodevichi cemetery. (For further details, see *The Rocket Men.*)

Bogomolov, Vladislav Nikolayevich was born on 14 September 1919. He was deputy to Isayev from 1946 until Isayev's death in 1971, and then headed the KB KhimMash (the old OKB-2) from June 1971 to 1985. He designed a number of liquid-propellant engines for the main engines and soft-landing engines for Soyuz T and Soyuz TM. Some of his other engines were also used on Progress, Progress M, Salyut, Mir, Kvant, TKS, FGB, and many other craft. After spending some years in retirement, he died in February 1997.

Chelomei, Vladimir Nikolayevich was born on 30 June 1914. He headed a design bureau – OKB-52, based on the outskirts of Moscow – responsible for military hardware (including cruise missiles). He also led the design work on the Proton booster and the Almaz military space stations that flew with the designations Salyut 2, Salyut 3 and Salyut 5. The bureau also developed the TKS manned spacecraft to support Almaz, and the modules used so successfully on Mir and the ISS. Chelomei was twice Hero of Socialist Labour. He died on 8 April 1984, and is buried in Novodevichi Cemetery. (For further details, see *The Rocket Men.*)

Isayev, Aleksei Mikhailovich was born on 24 October 1908. He was head of OKB-2, and was a leading designer of liquid propellant engines for Vostok, Voshkod and the Soyuz craft. Isayev was a Hero of Socialist Labour. He died on 18 June 1971.

Keldysh, Mstislav Vsevolodovich was born in Riga in 1911. He was Chief Theoretician of Cosmonautics, and was a member of the State Commission and the Head of the Academy of Sciences. Keldysh was three times Hero of Socialist Labour. He died on 24 June 1978. (For further details, see *The Rocket Men.*)

Kuznetsov, Viktor Ivanovich was born in Moscow on 27 April 1913. From the 1940s he worked with Korolyov, and headed the NII-94 design bureau, which produced the gyroscopes used on Sputnik, Vostok, Voskhod, Soyuz, Buran, Salyut and Mir. He was also a member of the State Commission. Kuznetsov was twice Hero of Socialist Labour. He died on 22 March 1991, and is buried in Novodevichi cemetery. (For further details, see *The Rocket Men.*)

Pilyugin, Nikolai Alekseyevich was born near St Petersburg on 18 May 1908. From the 1940s he worked with Korolyov, and was a specialist in automatic control systems. He headed the NII-885 design bureau, which produced the systems used on manned craft such as Vostok, Soyuz and space stations. He was also a member of the State Commission. Pilyugin was a Hero of Socialist Labour. He died on 2 August

1982, and is buried in Novodevichi cemetery. (For further details, see *The Rocket Men*.)

Polukhin, Dmitri Alekseyevich graduated from the Moscow Aviation Institute in 1950. He was involved in the development of the Proton booster, as well as other ballistic missiles under the control of Chelomei's OKB-52 bureau. From 1973 he was the Chief Designer and General Director of the Salyut design bureau, and was responsible for the design of the Mir modules Kvant, Kvant 2, Kristall, Spektr and Priroda. Polukhin was a Hero of Socialist Labour. He died on 7 September 1993.

Ryazanskiy, Mikhail Sergeyevich was born in St Petersburg on 5 April 1909. He was a specialist in radio control systems, and worked with Korolyov from the 1940s when he went with him to Germany. He headed the NII-845 design bureau, which produced technology used on Vostok, Voskhod, Soyuz and other space vehicles. He was also a member of the State Commission. Ryazanskiy was a Hero of Socialist Labour. He died on 5 August 1987. (For further details, see *The Rocket Men*.)

Severin, Gai I. was born in Leningrad in 1926. He has worked on aircraft ejector seats and pressure suits for many years. In 1964 he became General Designer for Zvezda. This bureau designed the space suits – including Sokol and Orlan, and those for Buran – for all Soviet and Russian cosmonauts, and also developed the airlock used on Voskhod and the MMU used on Mir. Severin is a Hero of Socialist Labour, and lives in central Moscow. (For further details, see *The Rocket Men*.)

Yangel, Mikhail Kuzmich was born on 25 October 1911. He worked closely with Korolyov from the end of the Second World War until 1954, after which he headed his own design bureau in Dnepropetrovsk. OKB-856 was entirely committed to the design and construction of missiles, but Yangel also developed the Zenit booster, which was used independently and as part of Energiya. He replaced Korolyov on the State Commission. Yangel was twice Hero of Socialist Labour. He died on 25 October 1971, and is buried in Novodevichi cemetery. (For further details, see *The Rocket Men*.)

STRATEGIC ROCKET FORCES

Ivanov, Vladimir Leontyevich (Colonel-General) was born in the Ukraine in 1936. He attended a military school before joining the Strategic Rocket Forces in 1958. In 1971 he graduated from the military academy named for Dzerzhinskiy, and in 1979 became head of the Plesetsk Cosmodrome, where he stayed for five years. In 1989 he became head of the Space Units of the USSR, and later succeeded Kerimov as the head of the State Commission for the Mir space station. He was retired by Yeltsin in October 1996.

Shumilin, Aleksei Aleksandrovich (Lieutenant-General) went to the cosmodrome in March 1959, began as a test engineer, and rose through the ranks until 1983, when he reached the position of head of the testing service. He was present at every manned launch until his retirement. He was a member of the Space Units of the USSR, and until 1991 he was Deputy Director of Baikonur Cosmodrome. When the Soviet Union came to an end in 1992, Baikonur passed into the control of the newly-formed Kazakhstan Republic, and in September 1992 Shumilin was appointed Director of the Cosmodrome on behalf of the Kazakhstan Republic. He is a Hero of Socialist Labour, and is now retired.

A note on burials

In the Soviet Union and the Russian Federation, the place of burial reflects status and position. The Kremlin Wall and the area around the Lenin Mausoleum was the primary place, but no-one has been buried there for many years. The next most important cemetery is situated in the grounds of Novodevichi convent in central Moscow, and it reflects the importance of the space programme that so many of the men who led the programme are buried there.

Organisations

The following are the leading organisations involved in the development and maintenance of Soyuz systems and operations. During the last few years there have been many changes in the names and status of Russian organisations and design bureaux, and this list is as accurate as available information permits.

AOOT RKK Energiya *Piloted spacecraft, including the design of variants of Soyuz T, TM, TMA and Progress, plus Salyut space stations, Mir and the ISS; launch vehicles.* Based at Korolyov (originally called Kaliningrad), near Moscow. Established as Department 3 of NII-88 in 1946. In 1956 the department was upgraded to Experimental Design Bureau 1, under the direction of the Chief Designer Sergei Korolyov. It was renamed TsKBEM when it came under the control of Mishin, and in 1974 was again reorganised under the control of Glushko. Currently, the General Designer is Yu. P. Semenov. It also runs the mission control centre (TsUP).

NPO MASH *Piloted manned spacecraft and military Almaz space stations; cruise missiles and military hardware.*
Based in Reutov (a suburb of Moscow). Established in 1955 as OKB-52, under the control of V.N. Chelomei. When Chelomei died he was replaced by G.A. Yefremov.

GNPRKTs TsSKB-Progress *Piloted military spacecraft, the 'military Soyuz', reconnaissance satellites; development, manufacture and operation of the Soyuz launch vehicle.*
Based in Kuybyshev and Samara. Established in 1959 as part of OKB-1, under the control of D.I. Kozlov, and became fully independent in 1974.

YKVZ *Ust-Katav Kirov Carriage-Building Works: construction of spacecraft.*
Based in the Chelyabinsk Region. Established in 1758 [*sic*].

NPO EnergoMash *Research and Production Association for Energy Engineering: engines for launch vehicles.*
Based at Khimki, near Moscow. Originally the Gas Dynamic Bureau, established in 1946. After being renamed as OKB-456, originally under the control of V.P. Glushko, in 1974 it was again reorganised (still under Glushko) to become part of

Energiya. In 1990 it became a separate organisation, under the control of Radovsky and then his successor B.I. Katorgin.

KB KhiMMash *Design Bureau for Chemical Engineering: liquid propellant engines for manned spacecraft, including all types of Soyuz and Progress craft; soft-landing engines for manned spacecraft.*
Based in Korolev. Originally established in 1943 under the control of Isayev, and in 1959 became OKB-2. Isayev was succeeded by Bogomolov and then by N.I. Leontyev.

KB Khimicheskoy Automatiki *Design Bureau for Chemical Automatics: liquid oxygen kerosene rocket engines used on launch vehicles; RD-0110 engine.*
Based In Voronezh. Established in 1941 under the control of S.A. Kosberg, and renamed OKB-154 in 1946. When Kosberg died in 1965 he was succeeded by Konopatov and later by V.S. Rachuk.

AO Krigenmash *Cryogenic and cryo-vacuum equipment.*
Based in Balashikha. Established in 1949.

AOOT NPO Nauka *Life support systems and scientific experiments.*
Based in Moscow.

AOOT OKTB Orbita *Simulators for the cosmonaut training centre (TsPK) since 1980.*
Based in Novocherkassk, near Rostov. Established in 1973.

NII Parachute Engineering *Parachutes for the recovery of manned and unmanned space craft.*
Based in Moscow. Established in 1946.

KB Obshchego Mashinostroyeniya (KBOM) *Design and development of launch complexes and facilities.*
Based in Moscow and at the Soviet/Russian launch complexes. Established in 1941 under the direction of V.P. Barmin. On Barmin's death in 1993 he was succeeded by his son, I.V. Barmin.

NPP Zvezda *High-altitude pressure suits; ejector seats; space and EVA suits (Sokol and Orlan); the MMU.*
Based in Tomilino, near Moscow. Established in 1952. The Director and General Designer is G.I. Severin.

GKNPTs Khrunichev *Khrunichev State Research and Production Space Centre: Proton booster; planning and manufacture of the modules used on Mir and the ISS.*
Based in central Moscow. Established as a branch of the Chelomei bureau in 1960. Originally responsible for the construction of the Salyut, and in 1973 was reorganised by being merged with KB Salyut. In 1975, A. Kiselev became Director of the plant; and when he left in 2001 he was replaced by A.A. Medvedev.

NPO AP *Scientific and Production Association for Automation and Instrument Engineering: guidance, navigation and control systems.*

Established in 1947 as NII-885 under the direction of N.A. Pilyugin. In 1982, Pilyugin was succeeded by V.L. Lapygin, who was the Director until 1998.

NII AO *Scientific Research Institute: aviation equipment; manual guidance equipment; information presentation equipment; instrument panels for Soyuz craft.*
Based in Zhukovsky. Established in 1982.

NPO Geofizika *Automatic and visual opto-electronic equipment; attitude measurement and guidance systems.*
Based in Moscow. Established in 1908 [*sic*].

GNPP Kvant *Solar cells and chemical battery accumulators.*
Based in Moscow. Established in 1987.

NII PM *Scientific Research Institute of Applied Mechanics: gyro instruments; inertial navigation equipment.*
Based in Moscow. Established in 1945 as NII-944 under the direction of V.I. Kuznetsov. Reorganised in 1956.

NII TP *Scientific Research Institute: precision instruments: radio systems for interactive measurement in connection with automatic tracking, docking and rendezvous; Igla and Kurs guidance systems.*
Based in the Ukraine, under the direction of A.S. Mnatsakanyan. The Kurs system was built by the Kiev Radio Factory, but production has now been transferred to Russia.

OKB MEI *Kontakt guidance system for lunar and Soyuz operation.*
Director, A.F. Bogomolov.

NII Argon *Argon computers used on Soyuz and space stations.*
Based in Moscow. Established in 1963 as the Research Institute of Electronic Machines, and renamed in 1986.

GNTs RF IMBP *Institute for Biomedical Problems State Research Centre: medical and research support for the manned space programme.*
Based in Moscow. Established in 1963 under the direction of V.V. Parin. He was succeeded by O.G. Gidzenko and the current Director is A.I. Grigoriev. Among the Deputy Directors is Y.V. Ilyin, and the Deputy Director for Medical Support of Human Space Missions is Prof. V.V. Polyakov. A mission control centre for medical staff is maintained to continually monitor cosmonauts in orbit.

SKTB BIOFIZPRIBOR *Equipment for medical monitoring of crews in orbit.*
Based in St Petersburg. Established in 1954.

NII PP and SPT *Scientific Research Institute for the Food Concentrate Industry and Specialised Food Technology.*
Based in Moscow. Established in 1936, and has provided food for space crews since 1961.

RGNIITs.P.K *The Cosmonaut Training Centre named for Yuri Gagarin.*

Based at Star City. Established in 1960. The training base for all cosmonauts, it includes the simulators, centrifuge and medical facilities. The current Commander is Colonel-General P.I. Klimuk. All Air Force cosmonauts are based at this facility.

FPSU *Federal Administration for Aerospace Search and Rescue.*
Based in Moscow, and organised regionally. Established in 1976. Headed by an Air Force General.

Bibliography

The authors used the following reports, articles, books and press kits as leading sources of information, in conjunction with reference material from their own archives collected from Interkosmos, ESA, NASA, and other Western, Soviet and Russian sources. The research and writings of James Oberg, Phil Clark, Gordon Hooper, Bill Barry, Michael Cassutt, Bart Hendrickx, Asif A. Siddiqi, Neville Kidger, Nick L. Johnson and Brian Harvey demonstrates what can be achieved by Western observers to reveal numerous aspects and details of the Soviet/Russian space programme.

Reports

1973 *From Spaceships to Orbital Stations*, A. Yu. Dmitriyev *et al.*, (1971), NASA Technical Translation, TT F-812, (December).

1974 *From Sputnik to Space Station*, Yu. I. Zaitsev, (1970), British Library translation.

1976 *Soviet Space Programs, 1971–1975, Part 1: Overview, Facilities and Hardware, Manned and Unmanned Flight Programmes*, US Congress, (August).

1981–90 *The Soviet Year in Space* (10 volumes), Nicholas L. Johnson, (Teledyne Brown Engineering).

1984 *Soviet Space Programs, 1976–1980, Part 2, Manned Space Programmes and Space Life Sciences*, US Senate, 98th Congress 2nd Session, (October).

1988 *Soviet Space Programs, 1981–1987, Part 1, Pilot Space Activities, Launch Vehicles, Launch Sites, and Tracking Support*, US Senate, 100th Congress 2nd Session, (May).

1993 *Europe and Asia in Space, 1991–1992*, Nicholas L. Johnson, USAF Phillips Laboratory, Kaman Sciences Corporation.

1993– Soviet Space History. Sale catalogues, Sotheby's, New York.

1995 *Europe and Asia in Space, 1993–1994*, Nicholas L. Johnson, USAF Phillips Laboratory, Kaman Sciences Corporation.

Mir Hardware Heritage, David S.F. Portree, NASA Reference Publication RF 1357, (March).

1996 *The Missile Design Bureaux and Soviet Manned Space Policy, 1953–1970*, William P. Barry, (Thesis for D.Phil. in Politics, University of Oxford).

1997 *Walking to Olympus: an EVA Chronology*, David S.F. Portree and Robert C. Trevino, NASA Monographs in Aerospace History, No.7, (October).

Articles

1967–2002 Publications of the British Interplanetary Society: *Spaceflight* and *Journal of the BIS*; in particular, the numerous articles on the development of the Soyuz programme by (among others) Phil Clark, Bart Hendrickx, Gordon Hooper and Neville Kidger, including 'The Kamanin Diaries' (1960–1971), by Bart Hendrickx, (*JBIS*, **50**, **51**, **53** and **55**), and 'Soviet Rocketry that Conquered Space', by Timothy Varfolomeyev (*Spaceflight*, **37**, **38** and **40–43**). Also articles in *Aviation Week and Space Technology* and *Flight International*.

Books

1971 *Russians in Space*, Evgeny Riabchikov, (Weidenfeld and Nicolson, UK edition).

1973 *Soviets in Space*, Peter Smolders, (Lutterworth, UK Edition).

1978 *The Partnership: a History of the Apollo–Soyuz Test Project*, Edward C. Ezell and Linda N. Ezell, NASA SP-4209.

1980 *Handbook of Soviet Manned Space Flight*, Nicholas L. Johnson, (AAS Publications, Vol.48, Science and Technology Series).

1981 *Red Star in Orbit*, James E. Oberg, (Harrap, London).

1988 *Diary of a Cosmonaut*, Valentin Lebedev, (GLOSS Co. USA).
Race into Space, Brian Harvey, (Ellis Horwood).
The Soviet Manned Space Programme, Phillip Clark, (Salamander).

1990 *Almanac of Soviet Manned Spaceflight*, Dennis Newkirk, (Gulf).
The Soviet Cosmonaut Team, Volume 1: Background Sections, Gordon R. Hooper, (GRH Publications).

1992 *Space Station Handbook*, Vladimir A.Pivnyuk, (Matson Press, USA).

1993 *Seize the Moment*, Helen Sharman and Christopher Priest, (Viktor Gollancz).

1996 *The New Russian Space Programme*, Brian Harvey, (Wiley–Praxis).

1997 *The Mir Space Station*, David M. Harland, (Wiley–Praxis).

1998 *Dragonfly*, Bryan Burrough, (Harper Collins).

1999 *Waystation to the Stars*, Colin Foale, (Headline).
Who's Who In Space: the ISS Edition, Michael Cassutt, (Macmillan).

2000 *Disasters and Accidents in Manned Spaceflight*, David J. Shayler (Springer–Praxis).
Challenge to Apollo, Asif A. Siddiqi, NASA SP-2000-4408.
Off the Planet, Jerry M. Linenger, (McGraw Hill).
The History of Mir, 1986–2000, Rex D. Hall (*ed.*), (British Interplanetary Society).

2001 *Mir: the Final Year*, Rex D. Hall (*ed.*), (British Interplanetary Society).
The Rocket Men, Rex D. Hall and David J. Shayler, (Springer–Praxis).
Russia in Space, Brian Harvey, (Springer–Praxis).

2002 *Star-Crossed Orbits*, James Oberg, (McGraw Hill).
ISS: Imagination to Reality, 1998–2002, Rex D. Hall (*ed.*), (British Interplanetary Society).
Heroes of the Cosmos, Rex D. Hall and David J. Shayler. (In preparation; detailing the selection and training of the cosmonaut team over a period of forty years.)

Books (Russian, in English)

1969 *Transfer in Orbit*, Novosti Press Agency (Moscow).

1982 *To the Stars*, V.A. Shatalov, M.F. Rebrov and E.A. Vaskevich, (Planeta Publishers, Moscow).

1988 *Orbits of Peace and Progress*, Pavel Popovich (*ed.*), (Mir Publishers, Moscow).

1994 *S.P. Korolev Space Corporation Energiya*, (Energiya, Moscow).

Books and articles (Russian language)

The authors used a number of Russian-language sources (too many to list individually), one of the primary sources being the magazine *Novosti Kosmonavtiki* (NK), first published in 1991 – initially bi-weekly, and then, from 1998, monthly. Particularly helpful was a history of Soyuz by S. Shamsutdinov, published in four parts in 2002.

1985 *Kosmonavtika Entsiklopediya*, Moscow, Sovetskaya Entsiklopediya (Russian space encyclopedia) edited by V. Glushko.

1995 *Strytyy kosmos, kniga pervaya (1960–1963)*, N. Kamanin (diaries, Vol.1), 400 pp.

1996 *Raketno-kosmicheskaya korporatsiya Energiya im. S.P. Korolyova, 1946–1996*, Moscow, RKK Energiya, Vol.1, 670 pp.

1997 *Strytyy kosmos, kniga vtorya (1964–1966)*, N. Kamanin (diaries, Vol.2), 444 pp.
Rakety i lyudi: goryachiye dni kholodnoy voyny, B. Chertok (memoirs), Moscow Mashinostroyeniye, 533 pp.

1999 *Strytyy kosmos, kniga tretya (1967–1968)*, N. Kamanin (diaries, Vol.3), 351 pp.
Rakety i lyudi: lunnaya gonka, B. Chertok (memoirs), Moscow Mashinostroyeniye, 569 pp.

2001 *Raketno-kosmicheskaya korporatsiya Energiya im. S.P. Korolyova, 1946–1996*, Moscow, RKK Energiya, Vol.2, 1,327 pp.
Strytyy kosmos, kniga chertvyortaya (1969–1978), N. Kamanin (diaries, Vol.4) 383 pp.
Sovetskiye i rossiyskiye kosmonavty, 1960–2000, I. Marinin, S. Shamsutdinov and A. Glushko, Moscow, published by *Novosti Kosmonavtiki*. A 'who's who' of Russian cosmonauts.

Press kits

The authors used the French (CNES) and ESA press kits from the missions to Salyut 7 and Mir space stations, the series of Interkosmos press releases from the missions to Salyut 6, and the NASA press kits and other documents relating to the Apollo–Soyuz Test Project (1975) and Shuttle–Mir (1995–1998) programmes. Also consulted were the BBC Monitoring Summaries: Space (Russia).

Index